Third International Conference on Information Technology and Digital Applications (ICITDA 2018)

IOP Conference Series: Materials Science and Engineering
Volume 482

Manila City, Philippines
8 – 9 November 2018

ISBN: 978-1-5108-8329-1
ISSN: 1757-8981

Printed from e-media with permission by:

Curran Associates, Inc.
57 Morehouse Lane
Red Hook, NY 12571

Some format issues inherent in the e-media version may also appear in this print version.

Printed by Curran Associates, Inc. (2019)

For permission requests, please contact the Institute of Physics at the address below.

Institute of Physics
Dirac House, Temple Back
Bristol BS1 6BE UK

Phone: 44 1 17 929 7481
Fax: 44 1 17 920 0979

techtracking@iop.org

Additional copies of this publication are available from:

Curran Associates, Inc.
57 Morehouse Lane
Red Hook, NY 12571 USA
Phone: 845-758-0400
Fax: 845-758-2633
Email: curran@proceedings.com
Web: www.proceedings.com

Table of contents

Volume 482

International Conference on Information Technology and Digital Applications (ICITDA 2018)

8–9 November 2018, Manila City, Philippines

Accepted papers received: 4 January 2019
Published online: 11 March 2019

Preface

Papers

Applications

Software Engineering

Hard Science and Engineering

IOP Conf. Series: Materials Science and Engineering **482** (2019) 011001 doi:10.1088/1757-899X/482/1/011001

PREFACE

It is indeed a pleasure for me to welcome you all to this Third International Conference on Information Technology and Digital Applications (ICITDA) in 2018 that is being held in this marvelous city of Manila.

As a chairman of this conference, I am very honored to greet you, who come from many parts of the world, to attend this conference. The ICITDA conference is an international forum for exchanging of ideas, knowledge, and experience on the latest development in the field of information technology and digital applications among researchers from academia, engineers, and practitioners from industries. This conference is also expected to enable opportunities for collaboration among the participants to advance the theory and practice in the field of information technology.

ICITDA has been a hub of many people from many countries, especially from Asia. Since its inception, ICITDA has been the work of many universities. It was started two years ago, in 2016, with a collaboration work of Informatics Department, Universitas Islam Indonesia, from Indonesia and Rajamangala University of Technology Thanyaburi, Thailand, and Youngsan University, South Korea. And this year, 2018, we are very proud to have our colleagues from University of the Cordilleras (Baguio, Philippines) to host this conference, extending our academic collaboration with more countries involved. I especially wish to thank to Ms Thelma D. Palaog, the organizing committee of this conference, from University of the Cordilleras, she has given an immense support to make our conference today happens. I also want to thank to Ms Nancy Flores, the Executive Vice President, University of the Cordilleras, whom without her support, all of us will not be here.

ICITDA 2018 received 103 papers from 6 countries, and 51 papers were accepted in various topics, and 51 delegates will present these papers. To make sure that we have a high quality paper, we also have 27 high quality reviewers from 7 countries: United Kingdom, Indonesia, Thailand, Australia, Taiwan, Netherlands, and Philippines.

This year's conference will feature two keynote speakers, Dr. Ma. Mercedes T. Rodrigo, Professor at Department of Information Systems and Computer Science Ateneo de Manila, Philippines and Dr. Teduh Dirgahayu, the Chair of Centre for Enterprise Information System Universitas Islam Indonesia.

In concluding the opening session, I wish you every success and make new collaboration in this conference and I wish you a very pleasant stay in Manila.

Mukhammad Andri Setiawan, Chair

November 2018

IOP Conf. Series: Materials Science and Engineering **482** (2019) 011001 doi:10.1088/1757-899X/482/1/011001

CHAIRS

Host

- University of the Cordilleras
- Department of Informatics, Faculty of Industrial Technology, Universitas Islam Indonesia

Editors

- Taufiq Hidayat
- Mukhammad Andri Setiawan
- Moh. Idris
- Fayruz Rahma

Organizing Committee

- Mukhammad Andri Setiawan (Chairman of The Committee)
- Thelma Palaoag
- Nancy M. Flores
- Hendrik
- Taufiq Hidayat
- Moh. Idris
- Fayruz Rahma
- Jefferey S. Ingosan

Program Committee

- Th.P. van der Weide (Radboud University Nijmegen, Netherlands)
- Reyer Zwiggelaar (Aberystwyth University, United Kingdom)
- Iping Supriana Suwardi (Institut Teknologi Bandung, Indonesia)
- Taufiq Asyhari (Cranfield University, United Kingdom)
- Wei-Chung Teng (National Taiwan University of Science and Technology, Taiwan)
- Jefferson Tan (IBM Research, Australia)
- Alvin Malicdem (Don Mariano Marcos Memorial State University, Philippines)
- Cecille Marcos (Adamson University, Philippines)
- Josephine Dela Cruz (University of the Cordilleras, Philippines)
- Felipe A. Pati (University of the Cordilleras, Philippines)
- Thelma D. Palaoag (University of the Cordilleras, Philippines)
- Agfianto Eko Putra (Universitas Gadjah Mada, Indonesia)
- Kridanto Suhendro (Institut Teknologi Bandung, Indonesia)

- Fitra Abdurrachman Bachtiar (Universitas Brawijaya, Indonesia)
- Samuel Mahatmaputra Tedjojuwono (Bina Nusantara University, Indonesia)
- Shofwatul 'Uyun (Universitas Islam Negeri Yogyakarta, Indonesia)
- Dhomas Hatta Fudholi (Universitas Islam Indonesia, Indonesia)
- Fathul Wahid (Universitas Islam Indonesia, Indonesia)
- Izzati Muhimmah (Universitas Islam Indonesia, Indonesia)
- Mukhammad Andri Setiawan (Universitas Islam Indonesia, Indonesia)
- Raden Teduh Dirgahayu (Universitas Islam Indonesia, Indonesia)
- Sri Kusumadewi (Universitas Islam Indonesia, Indonesia)

Keynote Speeches

USE OF EYE TRACKING TO CHARACTERIZE PROGRAMMER DEBUGGING BEHAVIORS

Dr. Ma. Mercedes T. Rodrigo

Professor at Department of Information Systems and Computer Science, Ateneo de Manila, Philippines

Professor at Department of Information Systems and Computer Science, Ateneo de Manila University. Head of the Ateneo Laboratory for the Learning Sciences. Writer of children's books including the recently published Women of Science series of Bookmark.

She is also a member of the Pittsburgh Science of Learning Center. Her areas of specialization are educational technology, intelligent tutoring systems and affective computing. In 2011, she established the Ateneo Laboratory for the Learning Sciences through a grant from the Department of Science and Technology's Engineering Research and Development for Technology program.

She is currently studying students' behavior and affect when using intelligent tutors, educational games, and other learning software.

INFORMATION IN LOCATION-BASED SERVICES: FROM PROVIDERS TO CONSUMERS

Dr. Teduh Dirgahayu

Chair of Centre for Enterprise Information System, Universitas Islam Indonesia

Dr. Teduh Dirgahayu is a senior lecturer and researcher at Department of Informatics, Universitas Islam Indonesia. He obtained his doctoral degree in Computer Science from University of Twente, The Netherlands.

He received best paper awards in the ACM Symposium of Applied Computing in 2010 and the IEEE International Conference on Cybernetics and Computational Intelligence in 2017. He is now the chair of Centre for Enterprise Information Systems.

His research interests are enterprise engineering, software engineering and service computing. His current research is on a context-aware systems.

IOP Conf. Series: Materials Science and Engineering **482** (2019) 011002 doi:10.1088/1757-899X/482/1/011002

Peer review statement

All papers published in this volume of *IOP Conference Series: Materials Science and Engineering* have been peer reviewed through processes administered by the proceedings Editors. Reviews were conducted by expert referees to the professional and scientific standards expected of a proceedings journal published by IOP Publishing.

IOP Conf. Series: Materials Science and Engineering **482** (2019) 012001 doi:10.1088/1757-899X/482/1/012001

A unique message encryption technique based on enhanced blowfish algorithm

Godfrey L Dulla[1], Bobby D Gerardo[2], and Ruji P Medina[3]

Department of Graduate School, Technological Institute of the Philippines, Quezon City, Philippines

gldulla@gmail.com[1], bgerardo@wvsu.edu.ph[2], and ruji_p_medina@gmail.com[3]

Abstract. Online transaction especially sending of messages has become easier and simpler to be manipulated. Nonetheless, security is the important aspect to secure its content and raise difficulty to unauthorized attempt. This study aims to secure both plaintext and file message content through encryption technique which is based on enhanced Blowfish algorithm. An enhanced Blowfish algorithm is developed to improve its performance by reducing the number of rounds and by increasing the block length with fixed length during encryption and decryption with added transformation method on selected rounds. The study used Visual Studio application running at x64 operating system.

1. Introduction

There are various techniques and methods designed for specific problem. Encryption of data over the network is one of the key method which translates information that only authorized user has the knowledge of the secret key. The purpose of encryption is to protect and secure confidential information transmitted over communication channel/medium. Not all encryption initiatives gained acceptance and popularity because of different reasons, Blowfish is not an exemption. Blowfish algorithm used 64-bit block cipher which is said to be vulnerable to attacks as opposed to newer algorithm that uses larger block sizes. Many cryptographers examined blowfish but few have published. Some have examined and detected weak keys on certain rounds and adjustments on variable sizes but others have successfully implemented in software and hardware with less memory consumption and improved execution time. Nonetheless, Blowfish potentially offers significant exploration consideration given that various researches suggest to further work on variable size, expansion of key length and iterations.

Thus, this study aims to explore the processes involved by reducing the number of rounds and increasing the block variable length and introducing transformation method on selected rounds as attempt to introduce new method in encrypting the message and compare the results with the existing algorithm. The enhancements to the cryptanalysis provide better security for the message and amplify potential advantages to Blowfish algorithm.

2. Review of Related Literature

There is a general need for enciphering all data sent between computers connected to a network. The communication lines between different terminals are wide open not only to tapping but also to deliberate or accidental alteration and corruption of traffic [1]. Computer systems must have a means to prevent unauthorized access to avoid any alteration or worse loss of data between devices connected to the network.

The Blowfish Encryption Algorithm contains 16 rounds; each round consists of XOR operation and a function (F). Blowfish is fast, compact, simple and variably secure. Significantly faster than DES and optimized for application where key does not change like communication link or file encryptor and not suitable for packet switching with frequent key changes or one-way hash function. A feistel network is consist of two parts – key expansion and data encryption. The key expansion converts a key of up to 448 bits into several subkey arrays totaling to 4,098 bytes. Data encryption consists of a simple function iterated 16 times. Each round consists of permutation and key. And, all operations are additions and XORs on 32bit words [2].

A combined hybrid cryptosystem along with splitting and merging mechanism enhanced the speed, time and security of the file using SRNN cipher algorithm. It includes file splitting and merging mechanism where each slice is encrypted by its corresponding key [3][4] developed a tool designed for file encryption. That is, the file was divided into number of pieces which depends on user's specification with application of encryption algorithm. They modified the F-function defined as F(X)=((S1+ S2 mod 232) XOR S3+ S4 mod 232 into F(X) = ((S1 XOR S2 mod 232) + (S3 XOR S4 mod 232)). As a result, they proved that the total execution time of the modified algorithm is 14% lesser than the original Blowfish algorithm. [5] presented a method for image encryption/decryption designed to increase the security and improve the performance using a variable key size of 448 bits employing 16 iterations. The data image served as the plaintext with two input of encryption process. The image data bit stream is divided into the block length and excludes the image header. The start of the bitmap pixel or array begins right after the header of the file. The byte elements of the array are stored in row order from left to right with each row representing one scan line of the image and the rows of the image are encrypted from top to bottom. They concluded that both colored and white image of any size saved in tagged image file format (TIF), bitmap (BMP), portable network graphics (PNG), joint photographic experts group (JPG), etc can be successfully encrypted and decrypted. The algorithm cannot be broken until an attacker attempted 28r+1 combination (r is the number of rounds). Furthermore, they concluded that the algorithm becomes stronger if the number of round are increased thereby considered as an excellent standard encryption algorithm. Ghorpade and Talwar [6] and Sowbarnika, Balasubramani, and Varatharajan [7] pointed out that text and pictures/video files have different prerequisite in the context of data encryption. Usually text is smaller than image and when both use the same algorithm to encrypt and decrypt, the data image will more likely perform slower than text. This is true with data on the internet where images need to be protected more than text. The Blowfish was utilized by the authors as they consider it as the best encryption algorithm with a good performance and only consume as low as 5K of memory. The authors have compared this algorithm with other symmetric encryption algorithms like AES and DES but to them Blowfish are capable of variable length key which makes it more difficult to decode the key from attackers. Enhanced-Blowfish is in par with Blowfish algorithm only that it is more secured due to block switching as an additional means of scrambling data especially information must be the topmost security concern in today's business and industry.

Agrawal and Mishra [8] enhanced the security level of Blowfish algorithm and decrease the time for encryption and decryption. They generate random number from 0 to 65535 and number into 16 bit binary form and find the positions that are holding 0 entries. If the flag is 1 then the F-function will not work, if the flag is 0 then the F function will work. In every round a new random number is generated and this gives different results in the application of F function. They analysed that the encryption and decryption time and concluded that it was reduced compared to the original Blowfish algorithm. Similar study was conducted [9], the generated random numbers come from an image. It reads the picture by pixel, selects the specific location randomly picked at any two colors then applied XOR between two selected colors then specify the length of the key. On the other hand, dividing the original image into a random number of blocks and then transformed the blocks into new locations give better transformation because aside from fewer pixel it kept their neighbors and difficult to predict neighbor pixels. The entropy was increased providing better security for the image [10].

Manku and Vasanth [11] developed a system using blowfish algorithm that reduces the rounds and made modification in each round. They proposed single blowfish round with two S-boxes connected with XOR same as the other 2 S-boxes. The 2 XOR was added then form the plaintext. The system was simulated and implemented in VHDL and Java jdk1.7 using Eclipse 4.2.0 integrated environment. In the study of [12] the implementation of the Function F was changed because the change in the total time taken for encryption and decryption cannot be seen on software implementation. The authors implemented VHDL application to show the differences in the delay. [13] They concluded that the improved modified algorithm has enhanced the performance over existing blowfish algorithm by reducing the number of clock cycles required for the execution of Blowfish function by 33% and hence reducing the overall execution time of the modified Blowfish algorithm by 14%. A sample waveforms were used because the modified Blowfish function F' executes both the summations in parallel, whereas the existing function executes the sums in sequential fashion. [14] in their comparative analysis that increase in processing rounds strengthens the security as single Fiestel round provides inadequate security.

Panda [15] studied to evaluate the amount of computing resources consumed namely, CPU time, memory, and battery power using AES, DES, Blowfish and RSA algorithms. The researcher used three different types of files (text, binary and image) using encryption time, decryption time and throughput as performance parameters. The AES provided the better performance in terms of both throughput and encryption-decryption.

Ramesh and Suluriandi [16] yielded a different result. The algorithms tested were DES, AES and Blowfish against certain performance metrics namely, execution time, required memory for implementation and throughput. The experiments resulted in concluding that Blowfish is the best performing algorithm among the tested algorithms. The paper further supports this by stating that Blowfish performs approximately 4 times faster than AES and 2 times faster that DES while consuming less memory in the process.

There have been studied conducted on comparison of the different algorithms such as DES, 3DES, AES, Blowfish, RC4, MD5, IDEA, TEA, CAST, RC2, RC5, RC6, Serpent, Twofish, Threefish, Mars, ECC, DHA and SHA. The blowfish and ECC provided the faster encryption speed. ECC is having some attacks on it but on Blowfish no attack was successful [17]. And, blowfish is much faster than DES but the speed increasing for Blowfish is slower compared to DES because it needs much more memory for sub-key and S-box initialization [18]. But, there is a minimal performance and low overhead cost achieved by Elliptic Curve Digital Signature Algorithm (ECDSA) while still providing protection for the outsourced data [19].

Priya and Arun [20] showed that Blowfish is most efficient and consumes less memory. This is a surprising and interesting result since AES supersedes Blowfish in speed but considering the efficiency of Blowfish algorithm where even AES could not count on. It cannot be taken for granted especially the need to conserve power consumption in mobile and ubiquitous computing is something Blowfish algorithm is worth considering as a valuable option. Enhanced-Blowfish wraps this up for an added security as the core means of enhancing the stability of Blowfish algorithm.

Ahmad, Manaf and Ismail [21] found that Blowfish has minimum power consumption (microJoule/Byte) and percent battery consumed in his evaluation of performance of the symmetric key algorithms: DES, 3DES, AES and Blowfish. Also, [22] found blowfish proved to be superior for secure wireless communication with high speed using low power source of energy efficiency. They concluded that the algorithm is best embedded in mobile devices and it was verified through a reprogrammable FPGA. The optimal performance is defined strictly by the least number of hardware requirements, highest throughput, and lowest power consumption. The results showed the algorithm has the smallest design core and highest throughput, with low power consumption. The same with the study made by Mandal, it was found that Blowfish is the most efficient one in terms of power consumption, security and throughput.

3. The Enhanced Encryption Technique

In this study, the concept follows the reverse, swapping and shifting of plaintext to its binary equivalent. A series of steps have been implemented to produce the encrypted values. The 32-bits representation in each block carries the plaintext in byte before converting it to its equivalent ASCII representation. The final result shall be transformed into its equivalent encrypted values.

3.1. Design Concept

In Figure 1, the uploaded file and inputted key supplied by the user served as baseline parameters for the entire process. Each phase has its own set of scheme allowing the result to be a strong potential ciphertext of that given process. Reducing the number of rounds and increasing the block length to four (4) blocks with 32-bit fixed length during encryption and decryption with added transformation method on selected rounds are main components to transform the message to generate the secured ciphertext.

3.2. Encryption/Decryption

In the encryption and decryption process, the series of steps follows the shifting, XOR and switching scheme. The block length will be increased to 128-bit and the number of rounds will be reduced from 16 to 14 rounds. The key will be provided by the user and generation of random characters and numbers that will form the plaintext. The integration of ShiftLeft and ShiftRight (according to the file size) and swapping method in each round posed potential adjustment on the speed, efficiency and generation of strong ciphertext.

In Figure 1, the plaintext will be divided into 4 blocks with 32-bit fixed length size. Each block has each own structure and operation. In the first 10 rounds, the operation will consist of XOR, Shifting and swapping scheme. The remaining rounds include the swapping of the center blocks with combination XOR and another swapping operation.

Many algorithms like DES, AES, XXTEA, MARS and many others make use of shift as part of the cycle (rounds). The shift method was used in the new version of blowfish called Twofish using 128-bits. The shift was introduced in K1 to K3 using <<<8, <<<1 and >>>1 respectively. In the proposed study, the shift method will be used in each block directed to the function with the S-box. Figures 4 and 5 shows the proposed enhancement to the algorithm. This gives a conditional process that will depend on the file input. There will be two sets of operation involved in the process. If the file size of the input resulted as even value the shifting will use the ShiftLeft method. This scheme will provide additional security as it will be difficult to predict the file size.

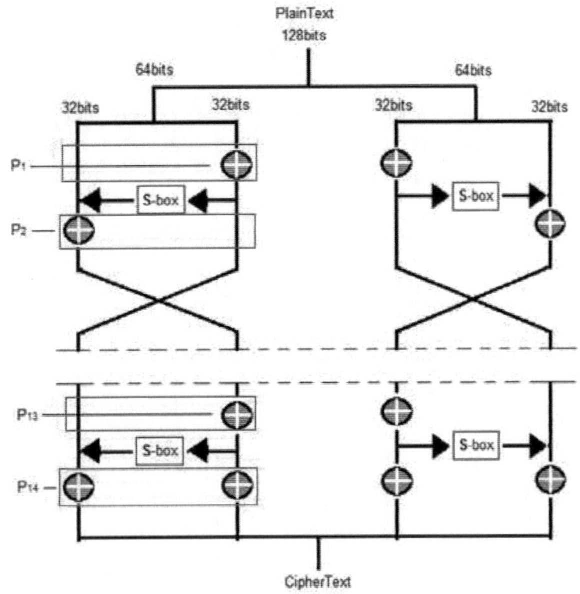

Figure 1. Enhancement Concept for Encryption and Decryption

The enhancement in the S-box perceived to be interesting that requires activity as compared with the iterations

$$F = ((S4 \text{ XOR } S3, \text{Shift2}(a) + S2), b \text{ XOR } \text{Shift4}(S1) \qquad (1)$$

The input will be divided into four (4) 32bits block and each block consists of 8bits. From the S-boxes the input will be transformed into 32bits following the lookup table that consists of 256 hexadecimal values (00 to FF) and the output will be 32bits. One thing that this proposed study is interesting (as mentioned earlier) is the double layered process (from 32bit to 8bit operation). The process embedded in the 32bits process and 8bits process presents the inner and outer process with parallel operation.

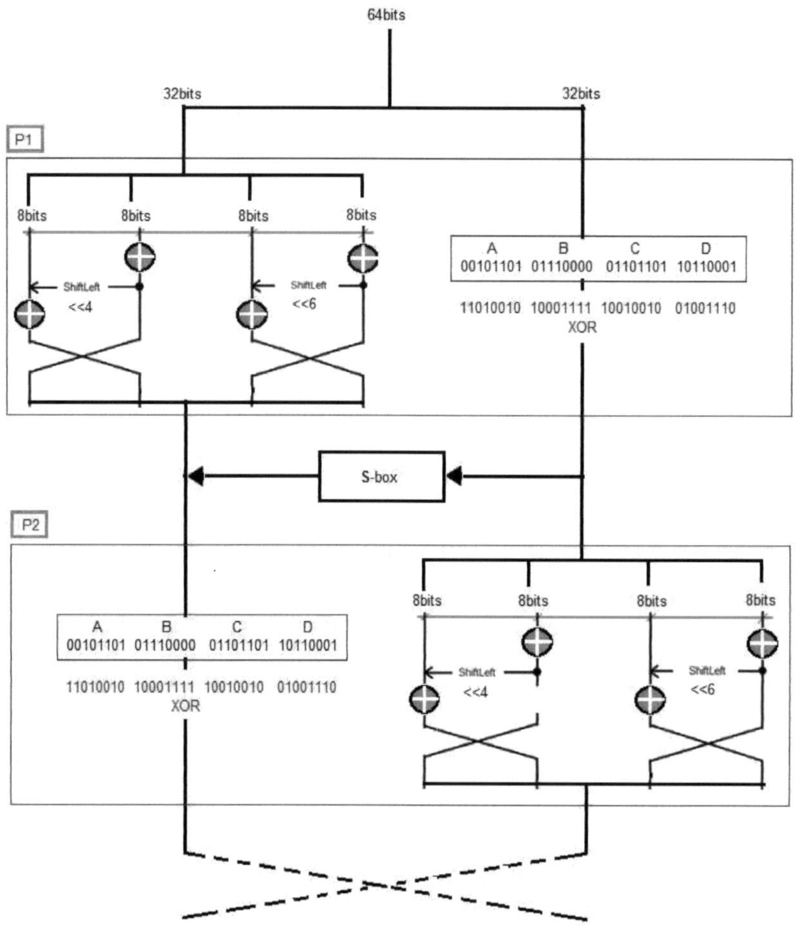

Figure 2. Detailed Proposed Enhancement Showing the Coverage of Each P-Array

Figure 2 illustrates the detailed proposed enhancement showing the coverage of each P-array. Each P-array consists of XOR and shifting of bits in the 8-bits phase which completes the cycle. After, the S-box will be performed, preparing each 32bits rounds for another P-array procedure.

4. Results and Discussions

The encryption/decryption process provides the file content to be transformed in such a way that only the receiver enables to reveal its content. The parameter filesize and generated key (from combination string of characters, numbers, and location) offer to enhance the algorithm leading to the generation of ciphertext. Figure 3 shows the main user interface for encryption and decryption process.

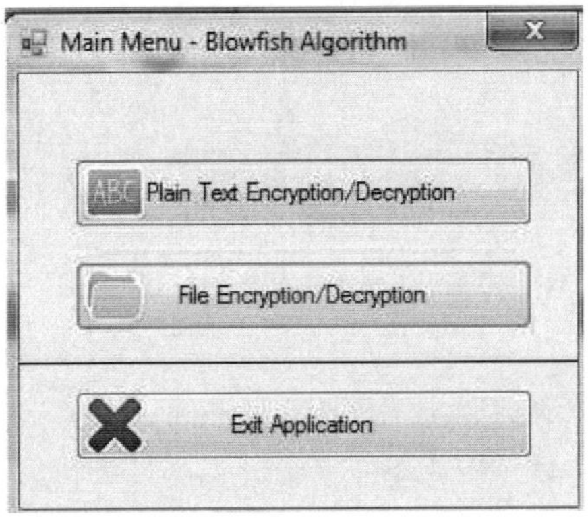

Figure 3. Main interface of the Program

The user submits the file for encryption and generates the ciphertext. The proposed algorithm was simulated in two options: plaintext and file content. In plaintext message, the user needs to input the text to be processed. The input contains any symbols or characters and holds closed to 33,000 characters (Figure 4).

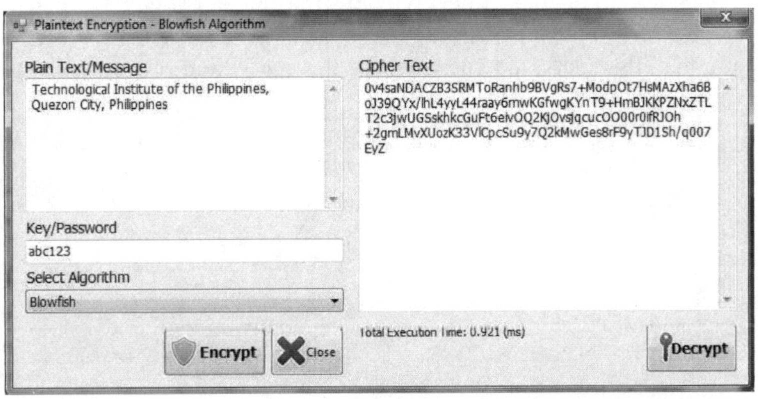

Figure 4. Encryption interface showing an encrypted message

The characters with its equivalent encrypted values formatted in hexadecimal format (00 to FF) follows the proposed enhancement algorithm as provided in the proposed section. A key or password provided by the user served as added information to the encryption process allowing the user to redisplay the original message. As an example, the user inputted the key as abc123 and the message translated into its corresponding equivalent. In the decryption process, in the event that user failed to submit the correct key or password, the message will not be reverted back to its original message but rather different ciphertext message will be generated (Figure 5).

IOP Conf. Series: Materials Science and Engineering **482** (2019) 012001 doi:10.1088/1757-899X/482/1/012001

Figure 5. File encryption interface showing the encrypted file content

The inputted key or password (reflected as abc1) which is incorrectly typed generated different results. In the case of file content encryption, the user needs to select and open the file for processing. In the example, the user opened the file named sample.txt together with the key or password ABCDEFGH and displayed in the file content window, the cipher text window generates the encrypted text of the file (Figure 6). The encrypted file can be saved in .txt file format. To decrypt the file content, the user needs to upload the encrypted file to allow the decryption process decrypt said file content (Figure 7).

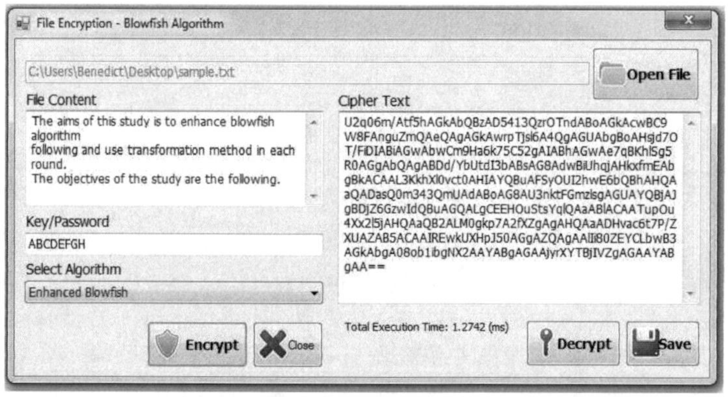

Figure 6. File encryption interface showing the encrypted file content

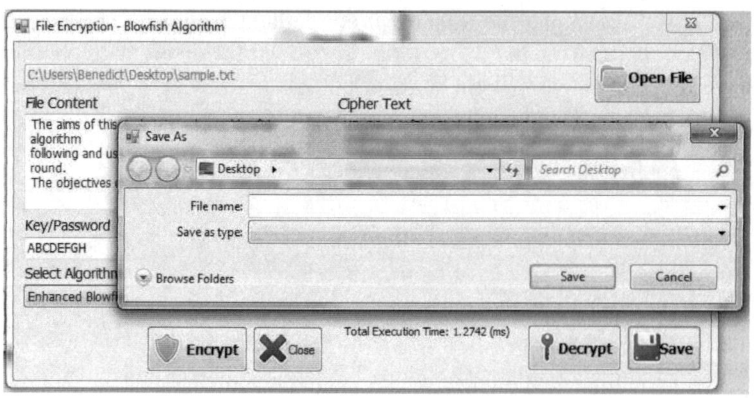

Figure 7. Saving ciphered text to assigned file name

The succeeding section presents the simulation results of the algorithm both for plain text and file content showing a significant change in the execution time following the modification in the iterations and block length expansion. Table 1 shows the comparative runtime performance of encryption and decryption process. The program used different filesize in text file format. The improved modified algorithm has enhanced the performance by reducing the number of rounds, expanded block size along with the different activities of bit-byte manipulation, shifting and block swapping by 11.37% or 2.69 msec advantage over the classic algorithm.

Table 1. Encryption comparative runtime performance (in milliseconds)

Key	filesize (kb)	Classic (ms)	Enhanced Blowfish (eBf)	Difference
file1	64682	32.64	5.34	16.37%
file2	17951	13.40	1.74	13.02%
file3	35902	22.99	2.06	8.95%
file4	29960	21.40	2.22	10.35%
file5	37840	28.08	2.11	7.51%
	AVG	**23.70**	**2.69**	**11.37%**

This shows that the modification made in the iterations affects the execution time without compromising the intricacy of the results (as previously presented). The file with smaller file size (file2) having 17951kb was executed in 13.40ms using the classic Blowfish algorithm whereas the enhanced algorithm has a result of 1.74ms or a difference of 13.02%. The file1 has the highest file size (64682kb), in classic algorithm it was executed in 32.64ms, and 5.34ms for the enhanced algorithm. The table presents that the enhancements made show significant change in execution time.

In decryption performance, the process was improved by 10% or 2.29msec of the average runtime performance over the classic algorithm. The algorithm depends on the filesize, that is, it delivers an improved performance in terms of smaller file size. For example, the file size with 172505 (being the higher file size) performed 9% over 34.16 (classic) and 3.12 (enhanced) decryption processing time. This provides the same performance result that was observed when a file has a file size of 100k-175k. This was also observed in file size 95k-80k which performed 10% over the existing algorithm.

Table 2. Decryption comparative runtime performance of classic and Enhanced Algorithm (in milliseconds)

Key	filesize after encryption (kb)	Classic (ms)	Enhanced Blowfish (eBf) (ms)	Difference
efile1	172505	34.16	3.12	9%
efile2	47873	13.87	1.72	12%
efile3	95745	21.60	2.27	10%
efile4	79897	21.78	2.08	10%
efile5	100909	25.56	2.28	9%
	AVG	**23.39**	**2.29**	**10%**

The results showed that the enhanced encryption to the algorithm has provided the fastest execution time compared with the classic encryption algorithm (Figure 8). The reduced number of iterations to the enhanced algorithm affects this performance.

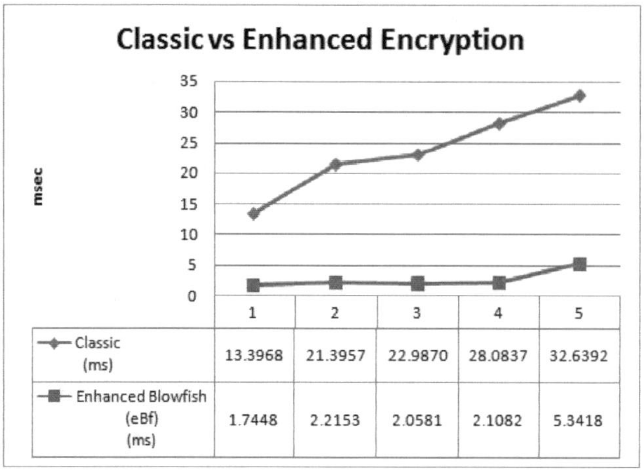

Figure 8. Simulation Results of Classic vs Enhanced Encryption

In decryption results, it showed that same with the encryption process, the enhanced algorithm provided the fastest execution time in dealing with decrypting file content. In the enhanced algorithm, the filesize plays significant factor in translating the file content. The higher the value of the decrypted file produced, the longer processing time. There was a small difference in processing the file with size 47k-79k which is 1.72ms and 2.08ms having a difference of only .36ms (Figure 9).

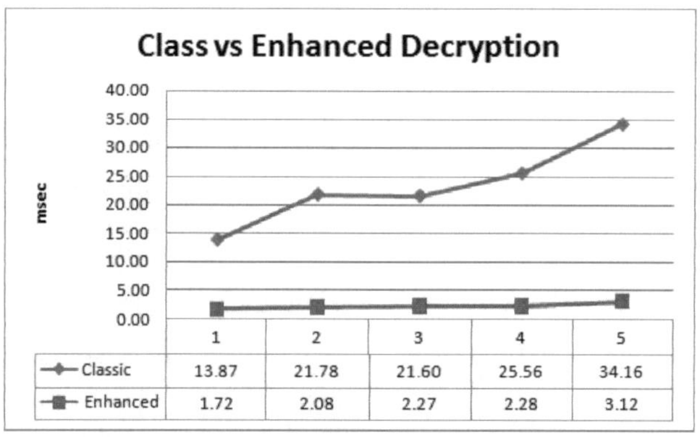

Figure 9. Simulation Results of Classic vs Enhanced Decryption

5. Conclusion

The enhancements made in the Blowfish Algorithm have provided a significant changed in the execution time without compromising the complexity of the encrypted file content. The reduce in the number of iteration gives the algorithm a faster execution time. And, the enhancements in the block size and other manipulation such as application of byte splitting, block size transposition, shift row and mix row methods provided complexity to the encrypted file content. Thus, it offers a considerable encryption performance of 2.69ms or 11.35% advantage over the existing Blowfish algorithm. And, a considerable decryption performance of 2.29ms or 9.8% difference.

The researchers concluded that this paper presents an enhanced Blowfish cryptanalysis that will improve its performance without compromising the complexity of the entropy which can be used in securing data transmitted over an unsecured communication channel.

Acknowledgment
The author/s would like to extend their sincere thank you first to Almighty God for all the blessings. Secondly to my family, friends, and relatives who extended their full support in the fulfillment of his Doctor in Information Technology and this paper. Lastly, to the three Institutions, St. Joseph's College of Rodriguez, Technological Institute of the Philippines and Commission on Higher Education Department (CHED) for all the push and scholarship grant that made this a reality.

References
[1] Feistel H 1973 Cryptography and computer privacy *Scientific American* vol **228** no **5** pp 15-23
[2] Schneier B 1996 *Applied Cryptography, Second Edition: Protocols, Algorithms, and Source Code* (John Wiley & Sons)
[3] Sharma M and Sharma V 2016 A hybrid cryptosystem approach for file security by using merging mechanism *2nd Int. Conf. on Applied and Theoretical Computing and Communication Technol. (ICATCCT)* (IEEE) 978-1-5090-2399-8 pp 713-7
[4] Manikandan G, Rajendran P, Chakarapani K, Krishnan G and Sundarganesh G 2012 A modified cryptoscheme for enhancing data security *J. of Theoretical and Applied Inform. Technol.* vol **35** no **2** pp 149-54
[5] Singh P and Singh K 2013 Image encryption and decryption using blowfish algorithm in Matlab *Int. J. of Sci. & Eng. Res.* vol **4** no **7** retrieved from website: https://www.ijser.org
[6] Ghorpade A and Talwar H 2016 *The Blowfish Algorithm Simplified* ISSN: 2278-8875
[7] Sowbarnika A, Balasubramani M and Varatharajan N 2017 Evaluating the effects of cryptographic algorithms on different sets of text data *Int. J. of Comp. Sci. and Eng.* (IEEE) pp 152-60
[8] Agrawal M and Mishra P 2012 A modified approach for symmetric key cryptography based on blowfish algorithm *Int. J. of Eng. and Adv. Technol.* vol **1** no **6** pp 79-83
[9] Tahseen I and Habeeb S 2012 Proposal new approach for blowfish algorithm by using random key generator *J. of Madent Alelem College* vol **4** no **1** pp 1-10 retrieved from website: https://www.iasj.net
[10] Devi A, Sharma A and Rangra A 2015 A review on DES, AES and Blowfish for image encryption and decryption *Int. J. of Comp. Sci. and Inform. Technol.* vol **6** no **3** pp 3034-6 ISSN:0975-9646
[11] Manku S and Vasanth K 2015 Blowfish encryption algorithm for information security *ARPN J. of Eng. and Applied Sci.* vol **10** no **10**
[12] Krishnamurthy G N, Ramaswamy D V and Leela M G 2007 Performance enhancement of blowfish algorithm by modifying its function *Innovative Algorithms and Techniques in Automation, Industrial Electronics and Telecommunications* pp 241-4
[13] Kondawar S and Gawali D 2016 Security algorithms for wireless medical data *Int. Conf. on Green Engineering and Technologies (IC-GET)* (IEEE)
[14] Zaran H, Druga G and Varazdin C 2016 Comparative analysis of cryptographic algorithms *Int. J. of Digital Techn. and Economy* vol **1** no **2** pp 127–34
[15] Panda M 2016 Performance analysis of encryption algorithms for security *Int. Conf. on Signal Processing, Communication, Power and Embedded System (SCOPES) 2016* pp 278-84
[16] Ramesh A and Suruliandi A 2013 Performance analysis of encryption algorithms for information security *Int. Conf. on Circuits, Power and Computing Technologies (ICCPCT-2013)* pp 840-4
[17] Bhanot R and Hans R 2015 A review and comparative analysis of various encryption algorithms *Int. J. of Security and Its Applications* vol **9** no **4** pp 289-306 http://dx.doi.org/10.14257/ijsia.2015.9.4.27
[18] Nie T and Zhang T 2009 A study of DES and blowfish encryption algorithm *A Project of Shandong Province Higher Educational Science and Technology Program* (No. J09LG10) (IEEE 978-1-4244-4547-9)

[19] Gajra N, Khan S and Rane P 2014 Private cloud security: Secured user authentication by using enhanced hybrid algorithm *2014 Int. Conf. on Advances in Communication and Computing Technologies*

[20] Priya S and Arun B 2017 Implementation of blow fish algorithm in crypto system for text and file data hiding *Int. J. of Sci. Res. Organization* vol **1** no **3** pp 29-35 e-ISSN: 2456 6942

[21] Mandal P C 2012 Superiority of blowfish algorithm *Int. J. of Adv. Res. in Comp. Sci. and Soft. Eng.* vol **2** no **9** pp 196-201

[22] Ahmad R, Manaf A A and Ismail W 2016 Implementation of a high-performance blowfish for secure wireless communication *J. of Telecom., Electronic and Comp. Eng.* vol **8** no **6** pp 147-51 ISSN: 2180–1843, e-ISSN: 2289-8131 retrieved from http://journal.utem.edu.my/index.php/jtec/article/view/1264/755

[23] Mandal P C 2012 Evaluation of performance of the symmetric key algorithms: DES, 3DES, AES and Blowfish *J. of Global Res. in Comp. Sci.* vol **3** no **8** pp 67-70

The sentiments of Philippines' underground artist and technology intervention

R J M Ventayen[1], S Kupsch[2] and T D Palaoag[3]

[1]Pangasinan State University, Philippines
[2]Don Mariano Marcos Memorial State University, Philippines
[3]University of the Cordilleras, Philippines

[1]dayjx@yahoo.com, [2]skupsch@dmmmsu.edu.ph, [3]tdpalaoag@gmail.com

Abstract. Listening to music is a significant and essential part of the daily life of many citizens, it is not just a hobby, but it reflects one's personality. In the new age of technology, several ways of playing and producing music became a trend. Thus cultural and underground artist in the Philippines is still left behind. The objective of this study is to determine the current situation of the unsigned artist in the Philippines by means of interviewing unsigned artist, bands, and independent music practitioners. This study uses the descriptive method of study that answered the questions regarding the current situation and sentiment of the unsigned artist. The study also discussed the technological trends could be considered to address problems of the Unsigned artist based on the current situation, and the layout of the proposed portal and lastly the overview of the architectural model that shall be adopted in designing the proposed portal. Based on the result of the study, the Unsigned Artist are in need of a platform for proper distribution of music. In order to satisfy the needs, this paper also discusses the overview of the proposed portal.

1. Introduction

Music is one that makes life colorful, every living person in the world has listened to music, and some tried singing despite a wrong tune. It is evident from the past that music is a hobby for everyone and has an effect on the emotions of the individual [1]. The philosophy of Music is the study of fundamental questions that are related to the nature and value of Music. Most people do enjoy music on a daily basis thus has a listening preference that is very strong. Music is said to be a collection of distinct sounds. Musicologists and Philosophers might think that well-defined as necessary, but that is not a sufficient basis for a set of sounds to be called Music. Over the years, music was associated with voice, but it is known more about the details of music apart from the musical traditions that were written or passed on. Music may likely be one of those numerous things which you know when you experience it, but you cannot define. Music enhances a garden variety of emotions such as fear and movies excitement. Music is indeed an essential part of societies around the whole world. There are recent survey that music was even used in medical treatment such as the effect of music therapy with relaxation imagery in the management of patients undergoing transplantation of bone marrow [2] that could provide relaxation to the patients. In this recent study, the value of music cannot be underestimated, it is not just a hobby for an individual, but music also heals our soul [3].

In the United States alone, as a form of art and cultural activity, listening to music is a significant and essential part of the daily life of many citizens. The needs for music has created a billion-dollar global music industry, which encompasses music production and distribution, major record companies that includes other, other music-related activities such as concerts and shows. In the same country, the music industry was estimated to generate about 17.2 billion U.S. dollars in the year 2016. Forecasts show a growth in the coming years by 2021; it is expected that the music industry revenue just in the single country will total over 22.6 billion U.S. dollars. Because of the new rise of technology, In the last few years, digital formats saw a rise in popularity in the U.S. While on the other hand, sales of other music formats – albums, digital albums, CD and digital track sales – have been continuously declining since 2012 due to availability of the digital formats [4].

In the Philippines, over the past 90 years, we saw the birth and end of different media that affects OPM (Original Pilipino Music). From the 45-rpm singles and 33-rpm long-playing vinyl albums from the year of the 1950s to the old type such as the cassette tapes of the 1980s, and the rise of compact discs of 1990s. There is a need to contend with another seismic shift by the local recording industry and artist where the expansion of the digital media will be only the one to survive [5].

For this generation of music players and downloaders, interesting issues are still unresolved, where despite the development in the recent technology, several artists are still left behind, they are known as Unsigned Artist, where those who have great compositions and music that can contribute to the music industry was not heard.

1.1. The objective of this study

The objective of this study is to determine the current situation of the unsigned artist in the Philippines by means of interviewing unsigned artist, bands, and independent music practitioners. This paper will also discuss a few opinions about the situation of the Philippines in the unsigned music industry. Discussion includes an examination of literature and contextual works relevant to the practice and process of the unsigned artist as a basis for the development of Online Music Distribution Portal (OMDP).

1.2. Statement of the Problem

This paper will also answer the following question: (1) What are the current situation and sentiment of the unsigned artist. (2) What could technological trends be considered to address the problems of the Unsigned artist based on the current situation (3) What is the layout of the proposed portal? (4) What is the conceptual framework that shall be adopted in designing the proposed portal?

1.3. The significance of the Study

The result of this study will be a basis for the development of the Online Music Distribution Portal in the Philippines. While this is just an initial study regarding the proposed system, this study investigates the current situation in the country and be an eye-opener in the music industry.

2. Music and Its Contribution

The philosophy of Music is the study of fundamental questions that are related to the nature and value of Music. Most people do enjoy music on a daily basis thus has a listening preference that is very strong. Music is said to be a collection of clear sounds [6]. Musicologists and Philosophers might think that well-defined as necessary, but that is not a sufficient basis for a set of sounds to be called Music. The Music industry is one of the cornerstones of the entertainment industry along with other entertainment industries such as movies, television, and radio [7]. The music industry includes many different companies and individuals whom all share the fact that they make money from music one way or the other. The list of people who are involved in the music industry is long and apart from the artists themselves includes managers, agents, publishers, producers, distributors, retailers and those who are involved in presenting live music to name a few [8]. Since the music industry has evolved over time with the progress in technology, new ways to participate in the music industry and generating revenues are continually being created. Over the years, music was associated with voice, such as rap music. Music may likely be one of those numerous things which you know when you experience it, but you cannot actually define. Music enhances a garden variety of emotions such as fear and movies excitement. We can conclude that Music is indeed an essential part of societies around the whole world.

2.1. Music advantage towards developing the economy

The existence of music is felt in our daily activities. People tend to listen to music while driving, relaxing, studying and sleeping. Music tends to move people in numerous unimaginable ways. Music is one of the leading industries in this 21st century. The International Federation of Phonography Industry discovered that the sales of the music industry were 5.8 billion dollars and the revenue for accurate performance is rapidly growing to 943 million dollars and up to 862 million dollars in 2011 [9]. In the present world, people are exploring different kinds of music that are best suitable for their taste. Apart from the rapid use of the internet, it is straightforward to spread and open a new movement in music. Asia music can get bigger by gaining more popularity all over the world. In addition to the record sales for Asian music, music can create other sources of revenue like concerts, tourism, sales of the band, etc. which would create more employment in the local market and Asia as a whole. Asides the financial effect of music on the national development, Music can also affect the mindset of people. Music is the channel for people to share their feelings with others. Asian Musicians encourages their citizens to keep trying harder in order to overcome their daily life challenges an achieve higher goals. Music also attracts attention to Asian countries and enlighten citizens in other counties that Asia is a progressing continent but not sad information for the world.

3. Methodology

The research design of this study is descriptive and experimental. It gathers information based on observation, interview, and documentation. The unsigned artist who was interviewed is those who won the battle of the band from a different province in the region. While gathering data was a challenge, the researcher utilized the internet to contact the artist to know the exact and accurate sentiments.

3.1. Sources of Data

The primary sources of the data are the unsigned artist who is playing in different places in the region and is known locally. Some of them are searched from social media for data gathering purposes, but the researcher personally interviews most of them in order to answer the constructed questions.

3.2. Instrumentation and Data Collection

Personal observation was used by the researcher while attending several band play. The researcher personally experiences the need to conduct this study due to the observed needs of the band. Another method is the Interview. This method is done by gathering information by asking a series of questions by means of oral communication. It was conducted by the researcher with several possible stakeholders to determine the insights to be able to identify the needs of the unsigned artist. A structured interview was constructed.

A following constructed interview Questions asked (1) May I know necessary information about you? (2) Where do you rehearse and how? (3) What music have you released so far? (4) What is your current label situation? (5) What is your current publishing situation? (6) What social media platform do you use in publishing your music? (7) How do you gain profit from music? (8) What musical technology do you use in recording music? (9) How do you sell music? (10) What are the challenges and problems encountered?

A system comparison was also made, where it is a technique that is used to gather requirements for the elicitation phase of the future project. It is the process of reviewing the existing documentation of related business processes or systems in order to import detailed pieces of information that are related and important to the current project, and therefore should be considered as project requirements for the proposed portal in the future study.

To understand the proposed portal, a conceptual framework which is IPO was adopted in order to explain the development of the system for future research.

4. Results and Discussion

4.1. Profile of the unsigned artist
A total of 18 unsigned artist which includes 12 bands, 3 solo singer, and 3 one-man bands. Most of the respondents are joining the battle of the bands conducted by an alcoholic beverage in the country. Majority of the band members and solo singers are single, while all one-man band is married.

4.1.1 Practices in Rehearsing music
Majority of the unsigned artist is rehearsing in the rented band room with half of the respondents. While 22 percent of the respondents shows that they rehearse at home. Also, 11 percent responded that they are rehearsing at Friend's Home, and 6 percent in rented spaces. This shows that not all unsigned artist has studio or equipment.

4.1.2 Music released situation
As shown in Figure 1, most of the respondents have not released music. Some of them released singles, but it does not mean that they are signed. Based on the interview, that artist or unsigned artist who has released, are an artist who created the original composition and plan for personal distribution.

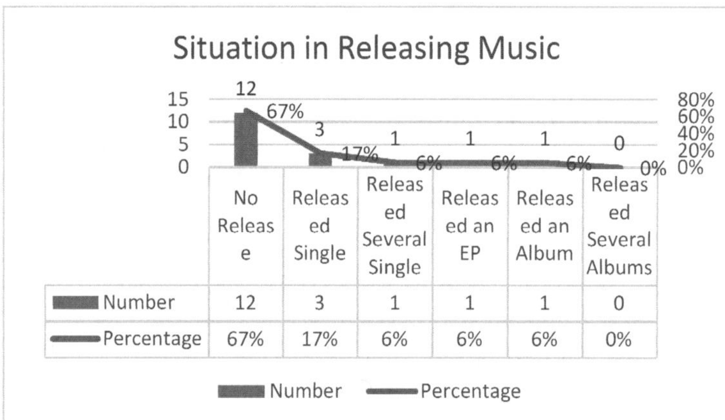

Figure 1 The situation in releasing music

It is expected that there will be no Signed artist in the unsigned community. Where only one is signed in an independent label, it does not mean that being signed in an independent label make you indeed signed. Independent distribution does not have the power to distribute music in large scale as aimed by the artist. Based on the respondents, 94% of the artist has no publication.

4.1.3 Social media in Promotion
Although the majority of the artist does not have released, it does not stop the artist to promote music such as cover songs. As shown in Figure 2, most of the Unsigned Artist post information on social media, all of the artists use Facebook as an online marketing tool for gigs and another announcement. 78 percent of the artist created a personalized video and posted it on video-sharing site such as Youtube. There are also 28% of the artist who publishes cover music and original composition in Soundcloud, and only one has their own band website.

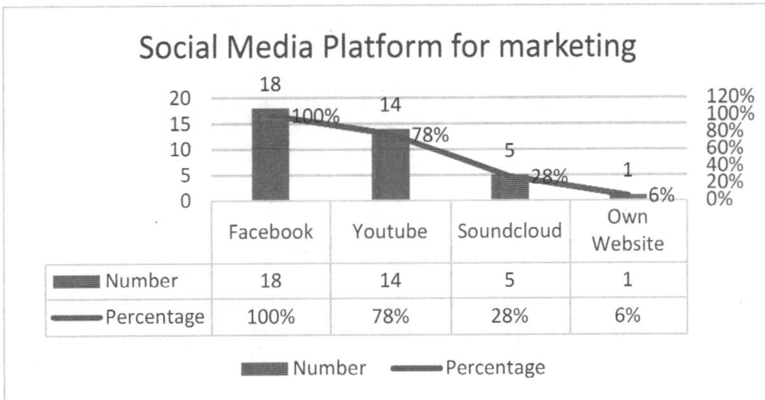

Figure 2 The platform used in marketing

4.1.4 Source of Income

Music is a business in the eye of the artist, and 94 percent of the respondents earn money thru Music Play and Gigs, while only 6 percent of the respondents earn thru CD sales and Merchandise Sales. As shown in the Figure 3, 94 % of the artist does not sell music because they do not have any idea or there is no professional recording. While only one of the respondents sells album thru personal selling.

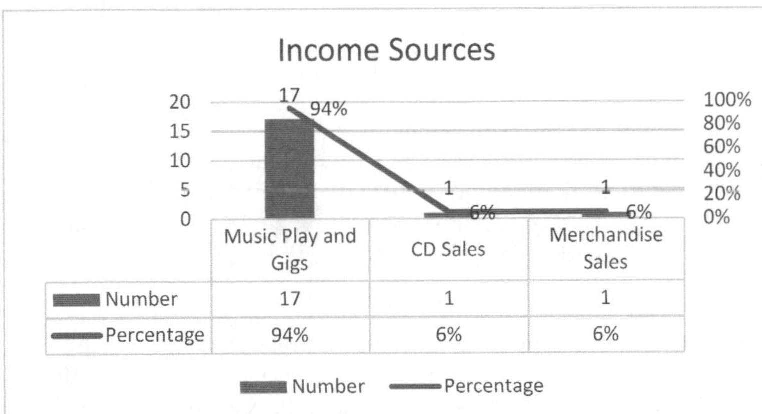

Figure 3 The sources of income

4.2 Technological Trends needed by the Unsigned Artist

Based on the result of this study, there is a technological advantage that technology may bridge the gap between unsigned artist and music distribution. The unsigned artist is in need of a distribution portal that may help them to distributed music. It is also visible in the result that most of the unsigned artist earn money from Musical Play/Gigs and not in the selling of music.

The proposed portal will help the unsigned artist not only to earn revenue from the music but also increase the popularity of the Philippine music industry and also promote the culture of the Philippines.

4.3 The need for Web in Philippine Unsigned Artist

The internet is a means of sharing ideas and connecting with other people. This birthed illegal sharing of files such as Digital music files. Piracy of copyrighted materials has economically affected the music industry; moreover, with the introduction of web development, the music industry should be able to trace its path back. Streaming music through mobile development would allow the lovers of the internet and music to enjoy listening to their favorite artists without harming the industry. Music Piracy will probably to exist [10] who is an advocate of streaming services opined that the websites have begun to reverse the harmful effects of piracy. Currently, the effect of streaming on the music industry is at a neutral level.

The internet era, like all periods in the past, promised a redesign of the revenues structure for the recorded music industry. Physical media like CDs continued to decline with the introduction of MP3 which became famous, although it seemed that the MP3 was beautiful to the average consumer as legal music, the late-2000s introduced a new concept of music consumption.

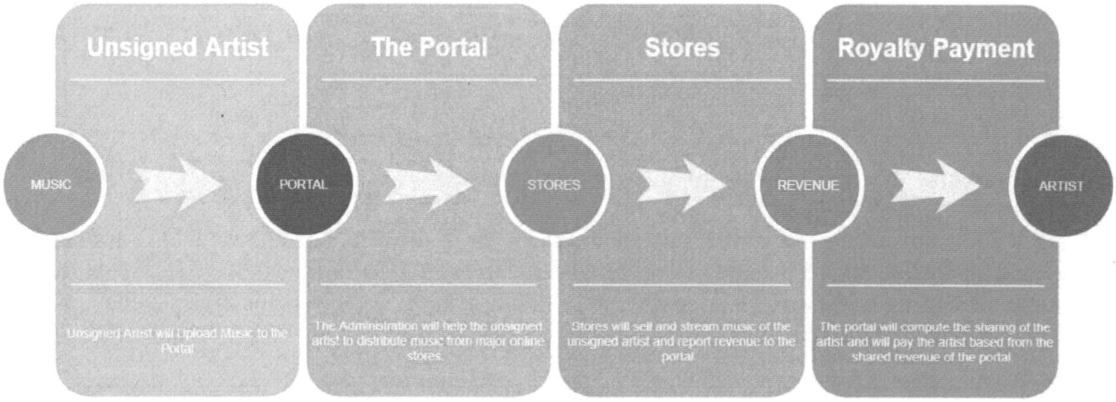

Figure 4 Process of the Proposed Portal

Based on the result of the study, the unsigned artist is looking for a possible way to promote music. A proposed Online Distribution Portal was conceptualized. Figure 4 shows the general process of the proposed portal. First, the unsigned artist will prepare the MP3 format of their recorded music. Next, the management will approve the music for distribution. After approval, the portal will distribute the music in different music stores. Lastly, a royalty payment will be given to the artist.

4.4 Overview of the Systems and Conceptual Framework

The conceptual framework is shown in Figure 5, where the proposed portal has three account level; this includes the Unsigned Artist account, the Web administrator, and the Manager or the Distribution label. The user or the artist will create an account and will provide information based on the requirement. The administrator approved the account of each of the unsigned artist and verified the authenticity of the account. Next, the unsigned artist will upload music and can monitor the revenue of the submitted music upon the approval of the manager. Lastly, the Administrator will generate payouts for the Unsigned Artist. The proposed portal solves the problems of the unsigned artist regarding music sales, promotion, and marketing with the intervention of technology.

To elaborate on the technological intervention of the proposed system. The projected system flow starts with account registration and verification. The Unsigned Artist can Upload music, check account balance and request for pay-out. The proposed system has two-way verification using CAPTCHA in order to avoid bots and attackers to unauthorized access to the portal.

It is also proposed that in the payout and account balance section, the Artist can add Paypal Account in order to be paid, funds from the PayPal account can also be withdrawn with any bank accounts locally. A minimal amount of 10% will be deducted from the royalty for the sustainability of the portal.

Lastly, in the upload section, the Unsigned Artist, can upload the music and verified by the site administrators. Site administrators will review the submission before submitting to the site partners, and the site partners will report royalty back to the site administrators for payout to the artist.

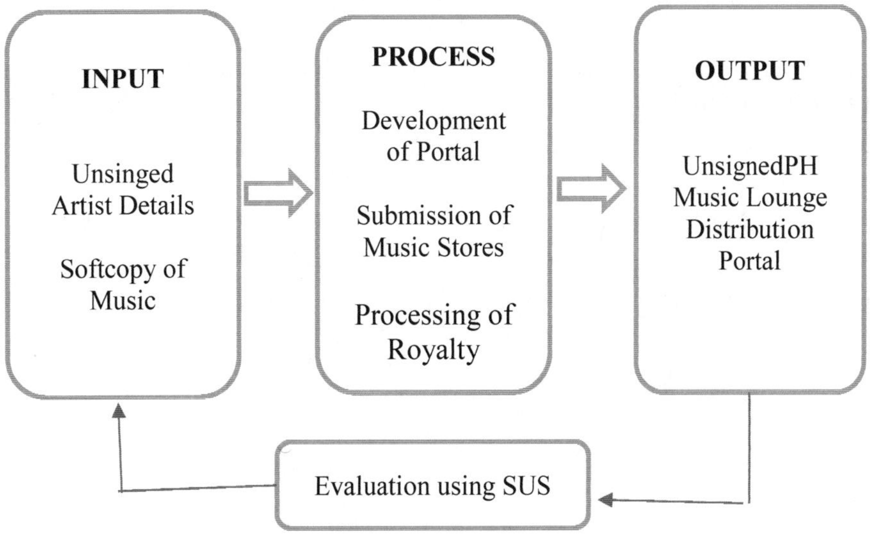

Figure 5. The figure shows the Conceptual Framework

5 Conclusions and Recommendation

With the result of this study include the brief overview of the system, it is concluded that there is a need to develop a portal for an unsigned artist. Based on the result of the study, it is visible that most of the unsigned artist is not aware of distributing music online. Filipino Artist A proposed portal is needed in order for them to distribute music. As the primary objective of this study is to determine the current situation of the unsigned artist in the Philippines, the unsigned artist in the country is struggling to gain popularity. As also shown in the result of the study where Artist earns more from performance and not from the revenue of music itself, where the traditional way of earning income is still present and technological use still not practices. It is recommended that the proposed portal should be developed since it will benefit the unsigned artist to disseminate music. It will also benefit the listeners to discover the music of unsigned. For the other band, this is a ground for collaboration to improve the music industry. This would also help the country to promote music worldwide from an unsigned artist. As for the business part of the proposed portal, this will help the proprietor and owner of the business to gain profit. It is also a promotion ground for the Original Pilipino Music to the world.

References

[1] Juslin, P N and Sloboda J A 2001 Music and emotion: Theory and research *Oxford University Press*

[2] Sahler O J Z, Hunter B C and Liesveld J L 2003 The effect of using music therapy with relaxation imagery in the management of patients undergoing bone marrow transplantation: a pilot feasibility study *Alternative Therapies in Health and Medicine* vol **9** no **6** pp 70-4

[3] de Bruin N, Doan J B, Turnbull G, Suchowersky O, Bonfield S, Hu B and Brown L A 2010 Walking with music is a safe and viable tool for gait training in Parkinson's disease: the effect of a 13-week feasibility study on single and dual task walking *Parkinson's Disease 2010*

[4] Statistica 2018 *Topic: Music Industry* www.statista.com retrieved 24 May 2018 from https://www.statista.com/topics/1639/music/

[5] Arceo-Dumlao T (n.d.) Filipino musicians, artists demand support from gov't, people retrieved from http://business.inquirer.net/166464/filipino-musicians-artists-demand-support-from-govt-people\

[6] Alperson P (Ed.) 2010 *What is Music?: an Introduction to the Philosophy of Music* (Penn State Press)

[7] Stanford Encyclopedia of Philosophy 2017 plato.stanford.edu retrieved 24 May 2018 from https://plato.stanford.edu/entries/music/

[8] Krasilovsky M W and Schemel S 2007 *This Business of Music: the Definitive Guide to the Business and Legal Issues of the Music Industry* (Billboard Books)

[9] IFPI Global Music Report 2011 ifpi.org retrieved 24 May 2018 from http://www.ifpi.org/news/IFPI-GLOBAL-MUSIC-REPORT-2011

[10] Faughnder 2015 *Federation of the Phonographic Industry* https://www.latimes.com/business/la-et-ct-state-of-stealing-music-20150620-story.html

Ensuring learning gain using webeditor: an experiential learning approach

J T Trinidad Jr.[1] and E P B Guidang[2]

Abra State Institute of Sciences and Technology, Lagangilang, Abra, Philippines

[1]deune3nidad@gmail.com, [2]rieann12@gmail.com

Abstract. Student-centered learning is a wide-ranging teaching approach that involves replacing lectures with dynamic learning, incorporating self-paced learning courses and/or cooperative group situations, ultimately holding the student responsible for his own developments in education. This study aimed to ensure learning gain using WebEditor in light of the experiential learning approach. Specifically, it achieved the following objectives: a) determined the prior knowledge of BSIT students in HTML5 using Notepad and a Web browser, b) determined the individual student gain score or normalized gain and total average gain score after using WebEditor, and 3) determined if there is a significant difference between HTML5 coding skills using Notepad and Web browser and WebEditor. The study is quantitative and descriptive in nature. The study found that the total average gain score of all the student-participants after using WebEditor is 0.32 or a 32% learning gain. 86% of the student-participants had a positive gain, 7% had a zero gain, and also 7% had a negative gain which is the common case. The study concludes that the entire population on average gained 32 percentage points out of a possible100 percentage points that they may have gained from pre to post assessment. Experiential learning helped the students fully learn new skills and knowledge using WebEditor. Moreover, all of them agreed that WebEditor is more efficient in displaying the output. Aside from their learning gain from using WebEditor, the performance of the participants in HTML5 coding indeed improved with the introduction of WebEditor in place of using a separate text editor (Notepad) and a Web browser.

1. Introduction

Student-centered learning is a wide-ranging teaching approach that involves replacing lectures with dynamic learning, incorporating self-paced learning courses and/or cooperative group situations, ultimately holding the student responsible for his own developments in education. The primary focus of the student-centered learning environment is need satisfaction while transmission of a body of knowledge is that of a subject-centered environment [1]. With the advent of technology, it is quite easier to overcome many barriers to implementing innovative alternatives [2].

Experiential learning is simply learning by doing but there is more to the process. In its simplest form, experiential learning means learning from experience or learning by doing. Experiential education first immerses learners in an experience and then encourages reflection about the experience to develop new skills, new attitudes, or new ways of thinking. [3].

An example of experiential learning is hands-on learning that might involve working with a trained carpenter to learn how to build a chair. In this case, the learner has gained some skills but has not taken part in experiential learning. The next step would involve taking time to reflect on the experience and compare chair-building to other building projects. Based on reflection, the learner would then develop new ideas about how best to go about building a chair, and return to chair building with new insights and ideas.

The same goes true with this study wherein hands-on learning involved HTML5 coding using Notepad and Web browser first. Experiential learning has not yet taken place at this point so the next step was to experience HTML5 coding using another tool. Therefore, when they were introduced to WebEditor, the learners then developed new ideas about how best to code HTML5 and return to HTML5 coding with new insights and ideas.

These scenarios prompted the researchers to experiment on the possibility that experiential learning be implemented at the Abra State Institute of Sciences and Technology. In this study, Notepad WebEditor was used to ensure learning gain through experiential learning among student-participants. This study aimed to ensure learning gain using WebEditor: An experiential learning approach. Specifically, it achieved the following objectives: 1) determined the prior knowledge of BSIT students in HTML5 using Notepad and a Web browser, 2) determined the individual student gain score or normalized gain and total average gain score after using WebEditor, and 3) determined if there is a significant difference between HTML5 coding skills using Notepad and Web browser and WebEditor.

1.1. Conceptual Framework
Figure 1 shows the conceptual framework of the study. Fourteen (14) third year BSIT students of the Abra State Institute of Sciences and Technology participated in the demonstration of their skill in HTML5 coding in the pre-assessment by performing what are being asked in the questionnaire through an individual hands-on assessment using Notepad and a Web browser and answering the additional question added to the pre-assessment questionnaire. It is divided into two parts: code snippet writing and constructing an HTML design. Questions aimed at measuring the HTML5 coding skills of the student-participants. To determine the individual student gain score or normalized gain and total average gain score, a post-assessment was administered using WebEditor.

WebEditor is an open-source portable application that runs through any Web browser which, as shown in Figure 2, allows input of (1) HTML (body) codes, (2) CSS codes, (3) JavaScript codes, and displays responsive (4) Output. The authors did not develop the application rather they made use of WebEditor since it is readily available and open source.

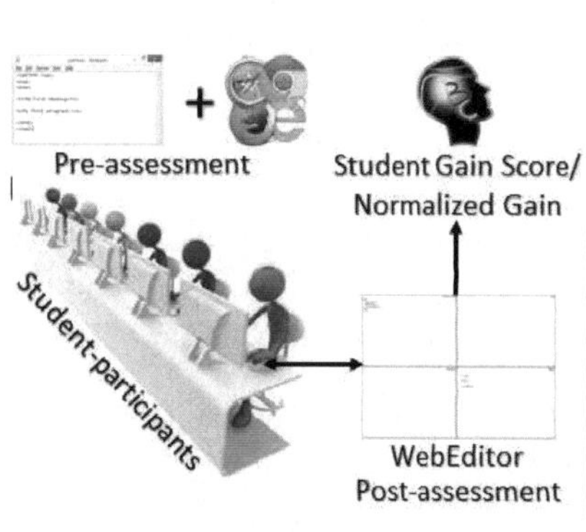

Figure 1. Research Paradigm

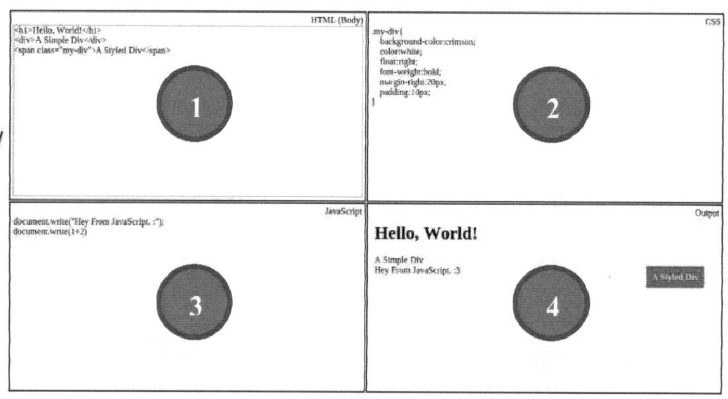

Figure 2. Web Editor User Interface

2. Methodology
2.1. Research Design
The study is quantitative and descriptive in nature. Since the study is concerned with the determination of the prior knowledge of BSIT students in HTML5 using Notepad and a Web browser and determination of the individual student gain score or normalized gain and total average gain score after exposition of the BSIT students to a WebEditor application that both accepts HTML5 input in the first

quadrant and displays responsive output in the fourth quadrant, the descriptive method is the most appropriate method to use. This type of research includes administration of a survey questionnaire.

Fourteen (14) third year BSIT students of the Abra State Institute of Sciences and Technology served as the respondents of the study. The sessions were divided into two three-hour duration which is the standard laboratory hours duration in a week. The post-assessment took place a week after the pre-assessment was administered in time with their succeeding scheduled laboratory session. The respondents were asked to demonstrate their skill in HTML5 coding by performing what are being asked in the questionnaire through an individual hands-on assessment and answering the additional question added to the pre-assessment questionnaire. To determine the prior knowledge of BSIT students in HTML5, a pre-assessment questionnaire was administered during the first session. It is divided into two parts: code snippet writing and constructing an HTML design using Notepad and Web browser. Questions aimed at measuring the HTML5 coding skills of the student-participants.

During the second session, participants then interacted with WebEditor, an application that both accepts HTML5 input in the first quadrant and displays responsive output in the fourth quadrant. The same questions in the pre-assessment were asked but this time, a different HTML coding tool, WebEditor, was used. To determine the individual student gain score or normalized gain and total average gain score, a post-assessment was administered. Moreover, a third part was added to the pre-assessment questionnaire to elicit their preferred tool(s) in HTML 5 coding because it was their first time to use WebEditor.

2.2. Data Analysis

To determine the prior knowledge of BSIT students in HTML5, frequency count and percentage were used. To evaluate the HTML 5 coding skills of the student-participants, Website Project rubric [4] developed by Portfolio Village was used as shown in Table 1. However, the design criterion was omitted because it is not evident in the skills to be demonstrated. To determine the individual student gain score and normalized gain of the student-participants, the following formula was used:

$$\text{Gain} = \frac{\text{Post-test} - \text{Pre-test}}{100\% - \text{Pre-test}}$$

Where: pre-test is the percentage of correct items on pre-unit test
post-test is the percentage of correct items on the post-unit test

The study used the normalized learning gain because using a pre-assessment/post-assessment model with calculation of normalized gain was found effective by [5] on a one-day introductory bronchoscopy curriculum.

Table 1. Website Project Rubric

Points	Concept	Content	Execution
1 – Not acceptable	The Website has no clear visual concept and does not relate to the task goals.	The Website has no appropriate content for the project goals and does not present sufficient quantity of output samples and related content.	The Website contained many spelling errors, many image errors, navigation errors, and had broken/missing content.
2 - Needs improvement	The Website almost has a clear visual concept but does not relate to the task goals.	The Website almost has appropriate content for the task goals but does not provide sufficient quantity of related content.	The Website contained a lot of spelling errors and a lot of image/text errors (wrong size or scale, typographical, blurry or inappropriate images).
3 - Good	The Website has a clear visual concept that	The Website provide has appropriate content for the	The Website contained only a few spelling

Points	Concept	Content	Execution
	relates to the task goals.	task goals and provides sufficient quantity of related content.	errors or image errors.
4 - Outstanding	The Website has a very solid and clear visual concept that relates to the task goals.	The Website provide has highly realistic content for the task goals and presents sufficient quantity of related content.	The Website contained no spelling errors, navigation errors, or broken/missing content.

3. Results and Discussion

3.1. Prior Knowledge of BSIT Students in HTML 5 Using Notepad and a Web Browser

Figure 2 shows that 11 (or 79%) of the student-participants demonstrated a passing score while three (or 21%) of them failed in the pre-assessment. This implies that majority of the student-participants have adequate skills in HTML 5 using Notepad and a Web browser.

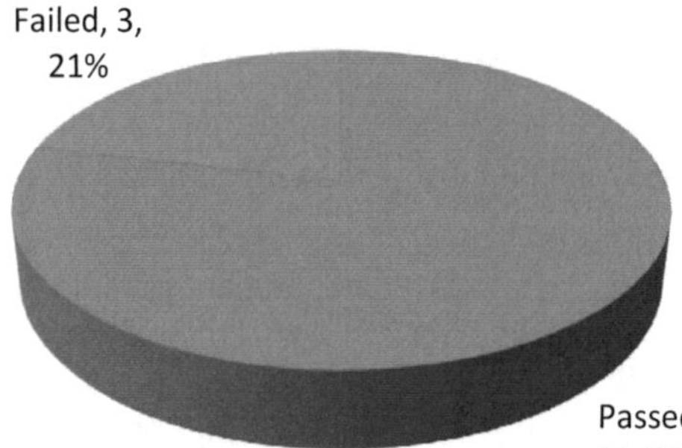

Figure 3. Distribution of Remarks in the Pre-Assessment

3.2. Individual Student Gain Score or Normalized Gain and Total Average Gain Score After Using WebEditor

Table 2 shows that the total average gain score of all the student-participants after using WebEditor is 0.32 or a 32% learning gain. This means that, in general, they gained 32 percentage points out of a possible 100 percentage points that they may have gained from pre to post assessment. This implies that experiential learning through WebEditor helped them fully learn new skills and knowledge because [6] states that it involves a few steps that allow students a hands-on, collaborative and insightful learning experience.

This findings is similar to that of [7] wherein the quantitative analysis revealed significant improvement of the learning outcomes when the technique is implemented using state-of-the-art teaching technologies and hands-on laboratory sessions.

Figure 4 shows that 14 (or 100%) of the student-participants passed the post-assessment and none of them failed. This means that experiential learning through WebEditor had a significant impact on their HTML 5 coding skills that from 79% passing rate in the pre-assessment, all of them already passed in the post-assessment.

As shown in Figure 5, it is notable that 12 (or 86%) of the student-participants had a positive gain, one (or 7%) had a zero gain, and also one (or 7%) had a negative gain which is the common case. This means that most of them have scored higher on their post-test than on their pre-test. Only one participant scored the same on his/her pre-assessment and post-assessment and also only one participant scored lower on his/her post-assessment than he/she did on his/her pre-assessment.

According to [8], a positive gain score indicates that the post-test score was greater than the pre-test score; a negative gain score indicates that the post-test score was less than the pre-test score. A zero gain score indicates that the post-test score is equal to the pre-test score. Student-participant 5 demonstrated the highest gain of 45 percentage points out of a possible 60 percentage points that he/she may have gained as shown in Figure 6. Therefore, he/she gained 0.75 (or 75%) of the potential percentage points he may have gained from pre to post test. On the other hand, student-participant 9 may have gained up to 15 percentage points, but rather lost 20 percentage points (or 133%) of what he/she may have gained.

Table 2. Individual Student Gain Score/Normalized Gain

Student-Participant	Pre-assessment			Post-assessment			Student Gain Score/ Normalized Gain	Classification of Gain
	Raw Score	Percen-tage	Remarks	Raw Score	Percen-tage	Remarks		
1	14	70	Passed	16	80	Passed	0.33	Positive
2	14	70	Passed	15	75	Passed	0.17	Positive
3	10	50	Passed	17	85	Passed	0.70	Positive
4	11	55	Passed	15	75	Passed	0.44	Positive
5	8	40	Failed	17	85	Passed	0.75	Positive
6	15	75	Passed	15	75	Passed	0.00	Zero
7	12	60	Passed	13	65	Passed	0.13	Positive
8	11	55	Passed	13	65	Passed	0.22	Positive
9	17	85	Passed	13	65	Passed	-1.33	Negative
10	10	50	Passed	17	85	Passed	0.70	Positive
11	11	55	Passed	17	85	Passed	0.67	Positive
12	8	40	Failed	15	75	Passed	0.58	Positive
13	11	55	Passed	17	85	Passed	0.67	Positive
14	8	40	Failed	14	70	Passed	0.50	Positive
Total Average Gain Score							0.32	Positive

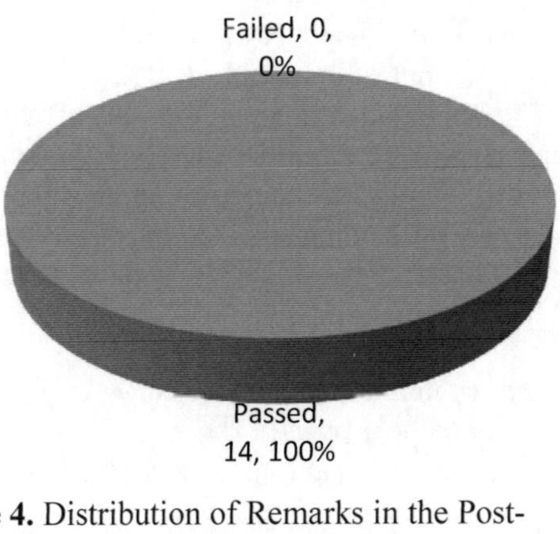

Figure 4. Distribution of Remarks in the Post-Assessment

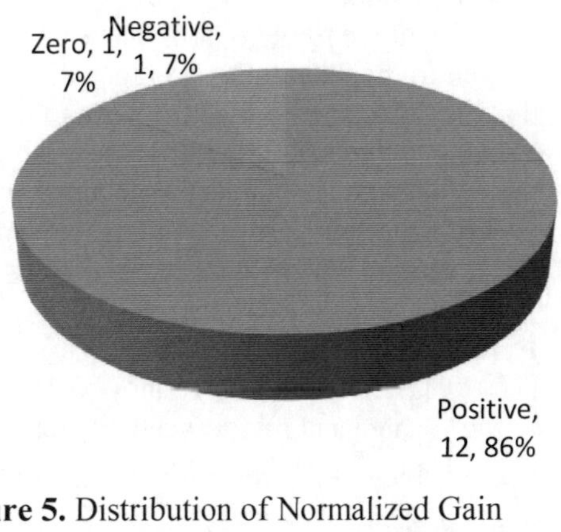

Figure 5. Distribution of Normalized Gain

Figure 7 shows that 14 (or 100%) of the student-participants prefer to use WebEditor in HTML 5 coding. All of them agreed that WebEditor is more efficient in displaying the output because there is no need for them to save the document as HTML and view the page in a Web browser. The process often times causes them not to view the page because Notepad appends the default .txt file extension name if the file type has not been changed to All Files and if .html extension name has not been appended.

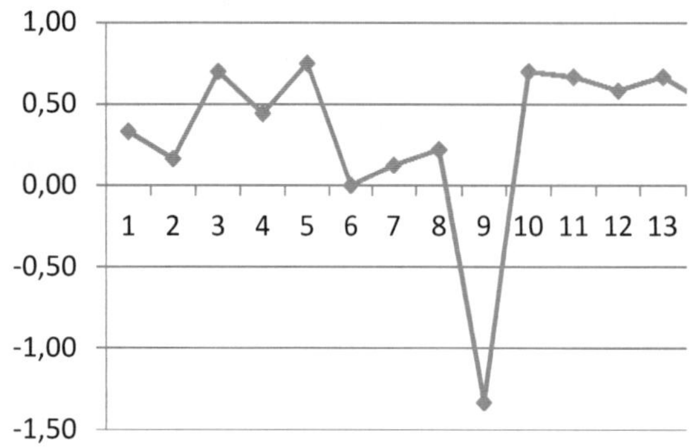

Figure 6. Student Gain Score/Normalized Gain

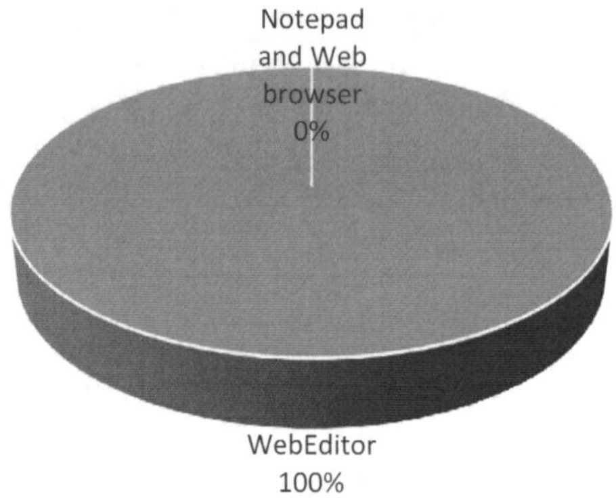

Figure 7. Preferred Tool in HTML Coding

3.3. Difference between HTML5 Coding Skills Using Notepad and Web browser and WebEditor

Table 3 shows that there is a significant difference between the HTML5 coding skills of the participants using Notepad and Web browser (pre-assessment) and WebEditor (post-assessment). It is indicated by the probability associated with t-value of 0.002 at 0.05 level of significance. This further affirms that aside from their learning gain from using WebEditor, the performance of the participants in HTML5 coding indeed improved with the introduction of WebEditor in place of using a separate text editor (Notepad) and a Web browser.

Table 3. Difference between the Pre-Assessment and Post-Assessment

	t-value	df	Probability associated with t-value
Pair 1			
Pre-assessment Post-assessment	-3.945	13	0.002

4. Conclusion

Majority of the student-participants have adequate skills in HTML 5 using Notepad and a Web browser. The entire population on average gained 32 percentage points out of a possible 100 percentage points that they may have gained from pre to post assessment. Experiential learning with the use of WebEditor helped the students fully learn new skills and knowledge using WebEditor. Moreover, all of them agreed that WebEditor is more efficient in displaying the output. Aside from their learning gain from using WebEditor, the performance of the participants in HTML5 coding indeed improved with the introduction of WebEditor in place of using a separate text editor (Notepad) and a Web browser.

References

[1] Clasen R E and Bowman W E 1974 Toward a student-centered learning focus inventory for junior high and middle school teachers *J. of Edu. Res.* vol **68** no **1** pp 9-11

[2] Hannafin M J and Land S M 1997 The foundations and assumptions of technology-enhanced student-centered learning environments *Instructional Sci.* vol **25** no **3** pp 167-202

[3] Lewis L H and Williams C J 1994 In Jackson L and Caffarella R S (Eds.) *Experiential Learning: A New Approach* pp 5-16 (San Francisco: Jossey-Bass)

[4] PortfolioVillage 2017 *Website Project Rubric* retrieved from http://learning.portfoliovillage.com/rubrics/4589978208

[5] Colt H G Davoudi M Murgu S and Rohani N Z 2011 Measuring learning gain during a one-day introductory bronchoscopy course *Surgical Endoscopy* vol **25** no **1** pp 207-16

[6] Haynes C 2017 *Experiential Learning: Learning by Doing* retrieved November 19, 2017

[7] Abdulwahed M and Nagy Z K 2009 Applying Kolb's experiential learning cycle for laboratory education *J. of Eng. Edu.* vol **98** no **3** pp 283-94

[8] Becker L A 2000 Analysis of pretest and posttest scores with gain scores and repeated measures *FontPage Workshop* retrieved from Lecture Notes Online Web site: http://www.ucs.edu/lbecker/gainscore.html

Exploring constructivism learning theory using mobile game

L M Padirayon[1], M V Pagudpud[2] and J S D Cruz[3]

[1] Cagayan State University, Sanchez Mira, Cagayan, Philippines
[2] Quirino State University, Cabarroguis, Quirino, Philippines
[3] University of the Cordilleras, Baguio City, Philippines

desmpadirayon@gmail.com

Abstract. Experience as our way of constructing our understanding of the world we live in is what constructivism theory advocates. In this study, the Decimal FunPro, a mobile game application which is designed to learn number systems was used to explore the occurrence of constructivist learning theory by employing clustering algorithm on the dataset by means of RapidMiner instrument. The performance of the students in using the mobile game shows that during the first tries from the different game categories, students spent more time finishing the activity as compared to the other trials. This means that the students learned from the first trials. A t-test was also conducted to define the significant transformation between the post-test and pre-test. The result showed that there is a significant difference between the posttest as well as the pre-test signifying that students acquired knowledge in number systems when they used the mobile game application. With the evidence stated, it can be concluded that using the Decimal FunPro mobile game application reveals that students can build new knowledge based on prior experience.

1. Introduction

Academic interest evolved and is being adopted today in different approaches. With the escalating demand for high quality education and the emergence of smart mobile phones technologies, educational institutions are taking other forms of learning. Laptops, personal digital assistants, and mobile devices are becoming learning instruments posting great help inside the classes [1].

On the other hand, a clear learning theory improves significance to lesson study, as the different theories of learning serve as foundation in managing values for the teachers when employing academic plans, lesson studies, and assessments [2]. Playing with a yo-yo in the years before the digital age is like observing a youngster with a smartphone in their hand nowadays [3]. Behaviorist, constructivist, positioned, combined, relaxed and enduring learning, and learning and teaching sustenance have momentarily recognized six main theories and ranges of knowledge related to learning with portable technologies [4][5]. Students are stimulated to be dynamic builders of information when students with difficulties are encouraged to be answered to some extent than bringing them straight information. Empowering the learners to be rooted in the good framework over mobile devices and offer them with quick assistant [6] can improve learning.

As mobile technologies are becoming more persistent, the greatest challenge is to determine how to use mobile technologies to transform learning into a continuous component of daily life to the point where it is not recognized as learning at all[7]. One good example of this is the use of a mobile learning application called Decimal Binary Fun Pro. This mobile application is very engaging and challenging for learners learning binary to decimal conversion and vice versa. The said mobile application activates the logical thinking of the learner and trains the brain in converting binary to decimal numbers. The application is used to learn binary coding and decoding. It will help computer science students to do faster binary conversions.

In learning, constructivist paradigm says that an individual construct his own understanding and knowledge of the world, through experiencing things and reflecting on those experiences [8]. The constructivism emphasizes on the situation and content in need in mobile learning such as queries for investigation and tricky answered and decision-making applications, as well as teamwork and active collaboration in mobile learning like communication via mobile phones or contact among learners. The Decimal Binary Fun Pro gamified mobile application was used in the classrooms and this study investigates that learners aggressively concept their knowledge and figure out problems related to binary and decimal number systems interactively. It sought to substantiate that learners build or construct new knowledge created in their prior learning. Specifically, it required explanations on the learner's accomplishments by clustering the learners and by comparing their pre-test and post-tests.

2. METHODOLOGY

The study is experimental. The approach employ in this study is defined in this section.

2.1. The Decimal Fun Pro Mobile Game

Decimal Binary Fun Pro gamified mobile application used in this study for one of the main beliefs of the constructivist is that student develops and builds understanding from their own personal experiences. Students bring their own past experiences in the binary system into their Intro to Computer Subject with binary system topic, use it to enhance their learning in a binary system by gaining more knowledge and build upon their previous knowledge. Constructivist learning theory enables the Decimal Binary Fun Pro gamified mobile application to prove on the student's ability in a binary system to be self-directed and draw their decisions [9]. Likewise, students finally learn more when they have to explore the binary system in Decimal Binary Fun Pro gamified mobile application rather than being told how to solve. When students can use what they are learning in real world situations, it sinks in and stays with them for a longer period of time

Decimal Binary fun pro game is an interesting game wherever players require placing one of the zeros in the grid to arrive a correct illustration of the decimal number assumed at the end of row and column. The game has three categories, the 4x4, 6x6 and the 8x8. The accomplishment is measured in seconds. It indicates how fast a player finishes the game. Figure 1 shows the screenshots of the game.

Figure 1. The Decimal Fun Pro Mobile Game

2.2. Data Gathering

The target group was students of Quirino State University and Cagayan State University who were enrolled in Computer 101 – Computer Fundamentals class because the course contains the topic regarding decimal and binary number systems. A 30 item pre-test was intended and given to the learners. The pre-test includes items which ask students to convert binary numbers to decimal numbers and vice versa. After checking, the result of the pre-test was recorded. After the pre-test, a series of lectures about number systems were given to the students. At the same time, the mobile game Fun-pro mobile game was downloaded and installed to the mobile devices of the students. Board works and exercises were given to the students after the lecture. This time, there are fewer board works and practices given compared to the classes in the previous years. A quiz was given after the lectures to gauge the knowledge acquired from the lectures. After that, students were taught about using the mobile game and the rules to follow. They were also instructed to record game information. Each of the students was given 3 hours as laboratory time in using the mobile game. During the laboratory time, the students played the different categories of the mobile game, the 4 x 4, the 6 x 6 and the 8 x 8 categories. Each of the categories was played five times. The researchers observed the students and recorded notable acts that can be used in the analysis of the experiences of the students. A post-test was given to each of the learners after the given laboratory exercises. The result was recorded.

2.3. Data Set

The dataset gathered 52 examples with 19 features. The succeeding is the features included:

- Pre-test (1) – scores from the first test given to the students before introducing the topic.
- Pre-test (2) – scores this is the quiz given to the students used the mobile application.
- 4 x 4 (1) – time spent during the first try in the 4 x 4 category.
- 4 x 4 (2) – time spent during the second try in the 4 x 4 category.
- 4 x 4 (3) – time spent during the third try in the 4 x 4 category.
- 4 x 4 (4) – time spent during the fourth try in the 4 x 4 category.
- 4 x 4 (5) – time spent during the fifth try in the 4 x 4 category.
- 6 x 6 (1) – time spent during the first try in the 6 x 6 category.
- 6 x 6 (2) – time spent during the second try in the 6 x 6 category.
- 6 x 6 (3) – time spent during the third try in the 6 x 6 category.
- 6 x 6 (4) – time spent during the fourth try in the 6 x 6 category.
- 6 x 6 (5) – time spent during the fifth try in the 6 x 6 category.
- 8 x 8 (1) – time spent during the first try in the 8 x 8 category.
- 8 x 8 (2) – time spent during the second try in the 8 x 8 category.
- 8 x 8 (3) – time spent during the third try in the 8 x 8 category.
- 8 x 8 (4) – time spent during the fourth try in the 8 x 8 category.
- 8 x 8 (5) – time spent during the fifth try in the 8 x 8 category.
- Post-test – score on the test given after using the mobile game.

Table 1. Sample Dataset

Student	Pre-test(1)	Pre-test(2)	4 x 4 (1)	4 x 4 (2)	4 x 4 (3)	4 x 4 (4)	4 x 4 (5)	6 x 6 (1)	6 x 6 (2)	6 x 6 (3)	6 x 6 (4)	6 x 6 (5)	8 x 8 (1)	8 x 8 (2)	8 x 8 (3)	8 x 8 (4)	8 x 8 (5)	Post test
1	53	70	130.2	94.2	103.5	88.9	60.1	153.7	128.9	122.5	134.2	142.5	183.8	196.7	143.2	127.3	123.2	82
2	55	67	115.4	110.3	90.4	85.3	69.5	110.4	116.9	100.3	98.7	104.3	105.8	99.6	93.2	95.7	99.2	92
3	50	80	96.3	91.1	88.3	64.5	67.6	99.8	95.6	102.3	100.6	99.4	93.2	90.5	100.2	92.0	88.3	93
. . .																		
52	52	75	91.9	112.2	122.5	134.2	142.5	183.8	94.2	103.5	88.9	142.5	183.8	177.5	100.3	97.3	92.1	80
Ave.	53.75	77.85	96.25	100.94	99.19	96.36	98.54	108.26	105.36	101.61	99.48	104.41	105.61	109.41	105.87	104.38	102.79	87.72

Actual values ranging from 1 to 100 are recorded the times recorded are floating values measured in seconds as shown in table 1.

2.4. Data Analysis
To prove that students can build new knowledge based on previous knowledge, the following data analyses were conducted:

2.4.1. Clustering Techniques
One of the primary methods frequently used in studying datasets is clustering where individual cluster formed is a group of data objects that are similar to another place within the same cluster but different objects in another cluster [10]. This study used a clustering analysis to segment students into groups based on their scores and fun-pro time achievement. There are numerous present clustering algorithms which possibly produce the entire dissimilar segments on the similar set of data. Various groups of data set based on the standard of getting the most out of the intra-class similarity and reducing the inter-class similarity where clustering algorithm partitioned employed. In this study, the clustering algorithms are assessed using internal validity measures to select which of the algorithms is utmost suitable for the fun-pro dataset. Three clustering algorithms were emphasized to the internal evaluation which comprises k-means, k-medoids, and expectation maximization (EM) clustering.

The procedure for clustering and assessing the clustering algorithms with RapidMiner is given in figure 2.

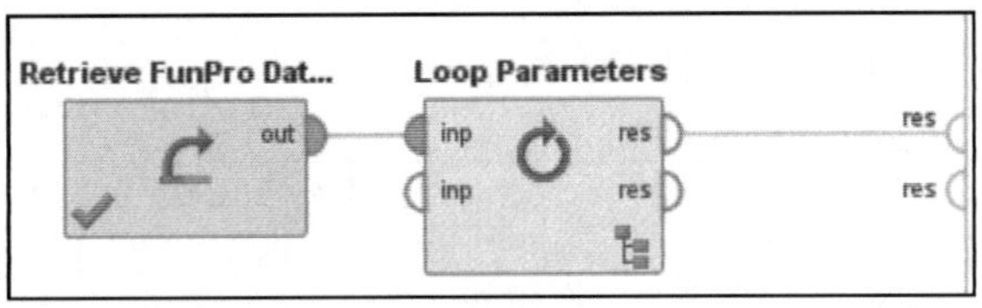

Figure 2. RapidMiner process for clustering and evaluating the clustering algorithm

The operator 'Retrieve FunPro Data Set' is an operator used to get the repository where the FunPro dataset was saved. This allowed the RapidMiner to access the data and automatically converted it into complete metadata which helped as the output of the operator. Also, the 'Loop Parameters' operator was defined to execute a loop process of evaluation of the different algorithm used. The 'Loop Parameters' operator comprises a 'Clustering (Subprocess)' and 'Silhouette Index' as inner operators as shown in figure 3.

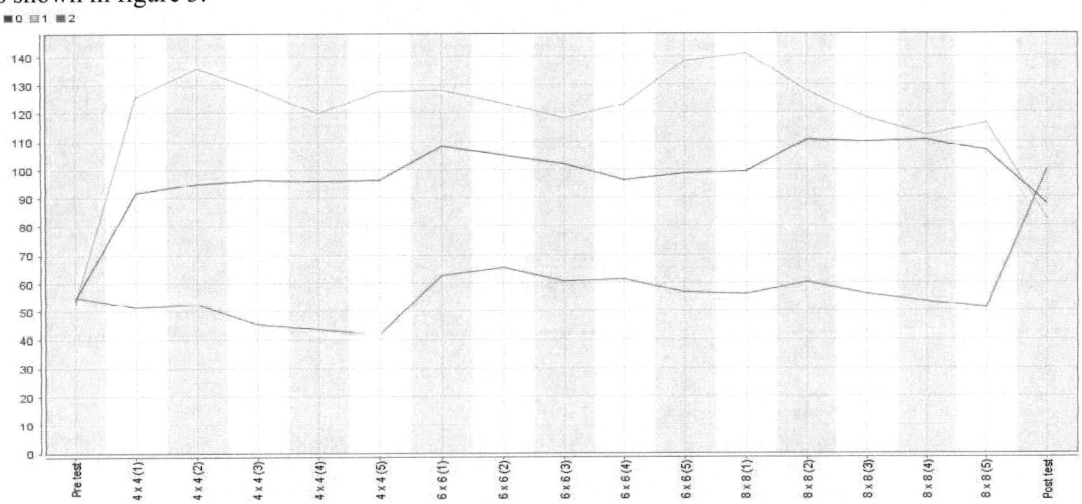

Figure 3. K-means Clustering Centroid Plot

Result shows that K-means provided the best model as related to the other clustering algorithms tested. Cluster 0, cluster 1 and cluster 2 formed by the k-means algorithm were composed of 32, 14 and 6 students correspondingly as the larger value of the silhouette index specifies a better clustering model. Likewise, the centroid plot formed through the k-means algorithm, where k = 3 is shown in figure 5. Centroid plot is used to review the clustering outputs. It can be concluded from the figure that the students who belong to Cluster 0 (blue line) with 32(62%) for number of instances are the AVERAGE LEARNERS, as performance compared to the other two clusters for they ranked second on both Pre-test(2) and Post-test, spent longer time finishing the activities using the mobile application compared to cluster 2.

Besides, cluster 1 (green line) are students as the TOP LEARNERS with 14(27%) number of instances compared to with other clusters. They spent lesser time finishing each category in the mobile app, and they scored best in their post-test. In contrast, Cluster 2 (red line) characterizes the students as POOR LEARNERS with 6(11%) number of instances. They spent more time finishing the activities given using the FunPro game with these students, and they scored lowest during the post-test. It can also be inferred from the figure that during the first tries from the different categories, students spent more time finishing the activity as compared to the other trials. Table 3 shows the output of the t-test.

Table 2. Type of Learners Based on the Clustering Result

Cluster Number	Number of Instances	Type of Learners	Behaviour
0	32(62%)	Average Learners	- They ranked second on both Pre-test(2) and Post-test - Spent longer time finishing the activities using the mobile application compared to cluster 2
1	14(27%)	Top Learners	- Highest score in the Pre-test(2) as compared to the other clusters - Highest score in the Post-test as compared to the other clusters - Lowest time spent in completing all the mobile application activities as compared to the other clusters
2	6(11%)	Poorest Learners.	- They spent more time finishing the activities given using the FunPro game with these students, and they scored lowest during the post-test. It can also be inferred from the figure that during the first tries from the different categories, students spent more time finishing the activity as compared to the other trials

Table 3. t-test result

Samples of Paired Statistics			
	Mean	**N**	**Standard Deviation**
Pair Pre-test (2)	77.852564	52	9.8149354
Post test	87.724359	52	8.3952834
Paired Samples Test			
Pair 1	**Mean**	**t**	**Sig. (2-tailed)**
Pre test (2) – Post test	-9.8717949	-13.130	0.000

Based on the result, the mean of the post-test is greater than the mean of the pre-test. A t-test was conducted to check for the significant difference between the post-test and the pre-test. Table 4 shows the result of the t-test where the p-value computed is .000 which is lesser than 0.05, and the computed t=13.13 is greater than t critical = 2.008. This means that there is a significant difference between the post-test and pre-test.

3. CONCLUSION

This study focuses on the exploration of the constructivist learning theory in the Decimal FunPro mobile game application by applying clustering algorithm on the dataset using RapidMiner tool to derive with the clusters of students. By using an internal validation measure which is the silhouette index measure, the output presented that the K-means algorithm, using k = 3, as long as the best clustering model to cluster students requiring a value of 0.253. Three groups were made having 32 students in the cluster 0, 14 students in cluster 1 and six students in cluster 2. The performance of the students in using the game shows that during the first tries from the different game categories, students spent more time finishing the activity as compared to the other trials. This means that the students learned from the first trials. A significant difference between the post-test and pre-test as the t-test result shows. The post-test is higher than the pre-test indicating that the students acquired knowledge in number systems when they used the mobile game application. Having the said evidence, it can be concluded that using the Decimal FunPro mobile game application reveals that students can build new knowledge based on previous knowledge.

Acknowledgment
The authors would like to express their deepest and sincerest thanks to the developer of the FUN PRO game to the individuals who extended their hands of kindness for the completion of the study.

Above all, the **Father Almighty**, for showering His everlasting grace and love that made the researchers face and accept all the obstacles and challenges of life.

References

[1] Sung Y T, Chang K E and Liu T C 2016 The effects of integrating mobile devices with teaching and learning about students' learning performance: A meta-analysis and research synthesis *Comp. & Edu.* vol **94** pp 252-75

[2] Pang M F and Ling L M 2011 Learning study: Helping teachers to use theory, develop professionally, and produce new knowledge to be shared *Instructional Sci.* vol **40** Issue **3** pp 589–606 DOI 10.1007/s11251-011-9191-4

[3] Nielsen Company 2017 Mobile kids: the parent, the child, and the smartphone *Nielsen's Fourth-Quarter 2016 Mobile Kids Report* (©The Nielsen Company)

[4] Futurelab N, Naismith L, Lonsdale P, Vavoula G and Sharples M 2004 *Literature Review in Mobile Technologies and Learning* ISBN: 0-9548594-1-3 (Futurelab © 2004)

[5] Bruner J S, et al. 1966 *Studies in Cognitive Growth* Harvard Book List (edited) 1971 #314 (PsycINFO Database Record (c) 2016 APA, all rights reserved)

[6] Kirriemuir J and Mcfarlane A 2004 *Literature Review in Games and Learning* (Futurelab Series)

[7] De Simone G C 2016 Mobile learning: extreme outcomes of everywhere, anytime *12th Int. Con. Mobile Learning* pp 139-43

[8] Olusegun B S 2015 Constructivism learning theory: a paradigm for teaching and learning *IOSR J. of Res. & Method in Edu. (IOSR-JRME)* e-ISSN: 2320–7388 p-ISSN: 2320–737X Vol **5** Issue **6** Ver. I (Nov-Dec 2015) pp 66-70 www.iosrjournals.org

[9] Karagiorgi Y and Symeou L 2005 Translating constructivism into instructional design: potential and limitations *J. of Edu. Technol. & Society* vol **8** no **1** (January 2005) pp 17-27

[10] Shuler C 2009 Informatique[cs]/environnements informatiques pour l'apprentissage humain *Sciences de l'Homme et Société / Education* HAL Id : hal-00696254, version 1

Improving tourism experience in open data environment with mobile augmented reality: needs and challenges

A J Ocampo [1] and T Palaoag [2]

[1]Lorma Colleges, Carlatan, San Fernando City, La Union 2500, Philippines
[2]University of the Cordilleras, Gov. Pack Road, Baguio City 2600, Philippines

[1]ajocampo@lorma.edu, [2]tpalaoag@gmail.com

Abstract. Tourism industry plays a significant role in the economic growth of every country. It is one of the fastest growing sectors that provides continuous growth and deepening diversification that leads to socio-economic progress. This study gives an overview of the needs and challenges of incorporating Mobile Augmented Reality in an open data environment to improve tourism experience. The use of open data in tourism and mobile augmented reality was presented to have an overview of its impact on tourism. Mobile Augmented Reality applications are one of the best suitable applications to boost tourism activities. However, there are still existing needs and challenges that need to be addressed, these are a. technical requirement of the application, b. acceptability of users to utilize MAR, c. availability of data specifically on the tourism site.

1. Introduction

Tourism industry plays a significant role in the economic growth of every country. It is one of the fastest growing sectors that provides continuous growth and deepening diversification that leads to socio-economic progress. The global state of tourism has produced economic and employment benefits in many related sectors. In 2015, the international arrival of tourists grew by 4.65 % which is roughly equivalent to 1,184 million and has generated 1.5 trillion US $ in export earnings and UNWTO is predicting that by 2030, arrivals of international tourist will reach 1.8 billion [1]. The importance of tourism activities is evident in the numerous numbers of tourism applications available for Android users such as Trip Advisor, Explore Macau and various map portals. Also, in its IOS counterpart which, tourism related applications are evident examples of which are MetroAr, Waalkz, Layar, and Wikitude. The tourism industry thrives on information [2] and a vast number of data repositories contributed by tourists thru reviews, ratings and social media posts and comments provides researchers and policymakers the data needed in decision making.

The principles of open data [3] provide the capability of free access to relevant data. Reusability [4] of these data promotes public engagement and efficient use of public data resources. [5] argues that it is possible for businesses in all industries to find and utilize open data to improve their products and services. In the tourism sector, the presence of open data is deemed necessary specifically in locating tourist destinations and their exact landmarks, providing information on transportation, accommodations, food, and business establishments. New technologies such as Geographic Information

Systems (GIS), Mapping, Virtual Reality (VR) and Augmented Reality (AR) are significant in the delivery of timely and improved tourist experience. Utilization of both open data and AR offers holistic tourism environment.

Concurrently, new consumer-level media platforms like virtual and augmented reality are evolving rapidly and show promise of mass adoption [6]. A visual AR System [7] enhances the visual environment of the user in par with the real-world environment. The integration of AR improves tourism experience [8], the presence of Mobile Augmented Reality (MAR) delivers an interesting way of accessing relevant information at the same time have an on-hand visual grasp on that specific tourist destination. The utilization of MAR is an avenue of reaching wider scope and different types of audience as it delivers an increased level of interest throughout the process, also, it enables citizen participation and contribution to touristic public data. These new technologies may inspire increased citizen engagement with public data and affairs and may even encourage citizens to empower themselves by developing their own media based on publicly available data.

Tourist, sightseers or travelers usually don't want to utilize multiple applications, compare result and interconnect various information from different domains. Majority of data could not be found in one specific domain [9], a minimal number of applications cater to just one domain. From navigation tools for transport and travel, booking an accommodation and locating a specific tourist destination exist in separate applications. The problem in accessing multiple applications is time-consuming and oftentimes produce incomplete or irrelevant data. Also, the heterogeneity of data [10] plays a significant problem in the integration of data from multiple sources.

This study determines the existing needs and challenges of incorporating open data and MAR in improving tourism experience. In the following sections, the use of open data in tourism and the use of mobile augmented reality for tourism was discussed to specifically determine the need for and challenges in tourism.

2. Open Data in Tourism

Open government data typically comes in three sources: authorized statistical data, sensor-based data, and user-generated content [11]. Authorized statistical data includes population, the existence of businesses, employment, crimes, and health, etc., and can also come in geo-mapping formats such as Shapefiles or Geo-json. Sensor-based data can include real-time data from sensors such as street lighting, humidity, temperature, air pressure, traffic etc. User-generated content can often include GPS-based data from mobile devices or user-interactions such as text input. Open government data can thus include static content as well as real-time feeds [8].

There is already a considerable number of actors providing open data in the field of tourism. Majority of the data is from governments and cities. the open tourism data has been seen as a data source for mobile applications and websites. Already there are many successful and well-documented examples of the efficient use of open data in various fields [12]. However, when for example the idea of Smart Tourism Cities develops and data is utilized as a part of the whole ecosystem, then the real benefits of open data can be utilized [13], the classification of open data are shown in Table 1. Business level data is not as openly available as government data. Tourism businesses should at least provide the information that is available on their websites in a machine-readable format. This content could include among others GPS-coordinates, opening hours and products descriptions. Opening such data could provide a possibility for innovative tourist services and mobile applications [14]. Especially destination management organizations are in a crucial role in collecting and opening data from their region for innovation purposes. If a destination wants to open its data suitable resources have to be allocated and openness should be included in the destination management strategy. Table 2 provides an example of the different applications and websites that utilize open data.

Classification of Open data in tourism presented in Table 1 provides an overview on the existing applications and websites that utilize the concept of open data, the majority of this applications are collaborative information from different individuals, organizations and government agencies. Table 2 shows some sample applications and websites that apply the open data concept that becomes relevant to the tourism industry.

Table 1. Classification of Open Data in Tourism

Open Data Type	Description	Where data has been used
Geographic Data	GPS-Location	Mobile App, websites
Event Data	Description of events, bands playing, timetables, event types	Mobile App, websites
Visitor statistics	Number of overnights	Mobile App, websites
Supply statistics	Number of businesses, types of business, number and information on attractions and museums	Mobile App, websites
Survey data	Data from survey studies	Mobile App, websites, academic, and business research
Supply information	Information about travel, destinations, attractions, restaurant, and happenings	Mobile App, websites, academic, and business research
Transit Data	timetables	Mobile App, websites
Governmental Data	Tax distribution and collection	Mobile App, websites, academic, and business research
All of the above		Smart Tourism City, Augmented Reality Applications, services that combine data from several sources

3. Mobile – Augmented Reality in Tourism

Mobile tourism applications [15] provides the capability of providing information to users while navigating at a certain place. Applications such as location-based services [16], social networking sites as mobile guides [17] and navigation maps contain information to help users navigate via thru data from different users, places and any point of interest (POI). This information is integrated into digital maps that are on par with the real-world environment which is suitable for travelers while they are on the move and this application is termed as Mobile Augmented Reality (MAR). The integration of MAR [18] in tourism offers an opportunity for tourism agencies, government entities and tourism sites to deliver relevant information in an interactive form rather than visiting a specific website. The existence of MAR helps boosts the marketing strategies of the different tourism destinations at the same time improving tourism experience of travelers without moving away from their current location and be able to have an overview on the place they intend to visit.

3.1. Mobile Augmented Reality Tourism Applications

Numerous MAR applications have been developed for tourism, applications such as Google Translate, Apple Maps: AR flyover, HolloMaps, and Hudway are just some of the AR apps that are useful in traveling. However, there are also MAR applications that are designed for the specific city, such application is Tuscany+ [6] an iOs based AR application that is used as a digital tourist guide in the Tuscany region. Another is the Basel AR [7] an AR tourist guide that is dedicated to the contents of the Basel city, it can run thru the Layar AR browser, it contains information on tourist sites, accommodation and food establishments in the city of Basel.

Table 2. Sample applications and websites that utilize Open Data in Tourism

Open Data Type	Sample Application	Mobile Platform	Sample Website	URL
Geographic Data	Google Trips	Android and IOS	Czech Tourism	www.czechtourism.com
Event Data	Tourist: New Guide Around Me App GoOut	Android Android and IOS Android and IOS	Geneva Tourism	https://www.geneve.com/
Visitor statistics	Sense-T	Android and IOS	gov.wales the visitor journey	https://gov.wales/statistics-and-research/great-britain-tourist-survey/?lang=en https://www.visitscotland.org/research-insights/about-our-visitors/visitor-journey
Supply statistics	Klook	Android	The Faroe islands World Tourism Organization	https://visitfaroeislands.com/ http://www2.unwto.org/
Survey data	-	-	Business Victoria	http://www.business.vic.gov.au/tourism-industry-resources/research/tourist-accommodation
Supply information	Klook, Tourist: Travel Lovers,	Android Android	Booking.com	https://www.booking.com/
Transit Data	Walks and Trails App Waze Rome2Rio Transit	Android and IOS Android and IOS Android and IOS Android and IOS	TRAVIC	https://tracker.geops.ch
Governmental Data			World Tourism Organization Open Data of Singapore	http://www2.unwto.org/ data.gov.sg
All of the above	Smart Tourism app Smart Tourist Guide	Android Android	Center Smart Tourism European Capital of Smart Tourism	www.centersmarttourism.com http://smarttourismcapital.eu/

These Mobile Augmented Reality applications provides an interactive form of navigating a certain place. It has the capability to display relevant information about the place and explore around the area in an interactive and dynamic manner.

3.2. Mobile Augmented Reality Applications in an Urban Open Data

Integration of open data in mobile augmented reality provides a visualization of useful information needed in exploring a specific destination. Accurate information on accommodation, tourist spots, restaurants, and such other information is very important in saving time, effort in money. Linked Open Data [19] integration was utilized to draw point of interest (POI) coming from several data sets and was displayed thru the MAR. Example of MAR application that utilizes linked open data is the LOD4AR [19] and MOBIAR [20].

Also, several applications that incorporates open data are as follows, a good example of this is the AR app Tunnel Vision [21] it is an AR application that helps commuters navigate the subway system of New York thru a physical map, it also provides statistical data on number of commuters in every substation thru the smart transit data feeds. Also, it allows commuters to explore the city via providing data on income levels, real – estate prices and etc. Another AR App is the Oculus Rift, it is an AR application that displays the virtual world of Canada. It allows tourists to have a view on the economic data of each province.

4. Needs and Challenges in Mobile Augmented Reality Tourism App

Mobile Augmented Reality applications are one of the best suitable applications to boost tourism activities. However, there are still existing needs and challenges that need to be addressed, these are a. technical requirement of the application, b. acceptability of users to utilize MAR, c. availability of data specifically on tourism sites.

4.1. Technical Requirement of the Application

The technical requirements in running MAR tourism applications vary in terms of the type of mobile platforms [22] the users currently have, the majority of AR applications are designed for iOS platforms and Apple mobile devices because of the capability of this devices to run AR applications without the need for additional hardware requirement. Another issue in terms of the technical requirement is the device capability and specifications since AR utilizes GPS and Internet Connectivity, one of the major drawbacks is the accuracy of GPS signal that is transmitted thru the device. Majority of GPS depends on the transmission of signal with the WIFI connectivity of the device. Based on experience conducted, the researcher utilized Pokemon Go as a testing application to test the accuracy of AR and GPS signal, it was found out that there are instances that the current location of the player and the location on the app differs, which concludes that the internet connection has a major impact on the GPS signal of the device thus it provides different display in the AR feature of the game. Thus, this technical requirement needs to be addressed in implementing the MAR for tourism to avoid wrong information in terms of location.

4.2. Acceptability of users to utilize MAR

Since the researcher currently conducts the study in the Province of La Union, based on an initial interview conducted, the citizens and some tourists utilize mobile applications related to tourism such as Google map, Waze and Pokemon Go. However, the majority of the respondents made mention that the utilization of the AR is challenging at some point because of problems in interconnectivity and mobile platform compatibility. Since the majority of mobile users interviewed utilize Android, the challenge in developing the MAR should be compatible with both Android and IOS devices. Though they intend to utilize the AR because of its interactivity features, the challenge on their device compatibility is one of the considerations they asked to be addressed.

4.3. Availability of Data

Majority of mobile-based tourism applications provides minimal data about a tourist destination; general information is only provided. One of the challenges that tourists experience is the accuracy of information provided by tourist applications. Information on transportation including fare is not updated, map navigation only displays main roads, accommodation, food establishments, and business establishments are non-existent. Also, tourist destinations are not updated as seen on the official website of the province of La Union as the target respondent of this study.

5. Conclusion and Future Direction of the Study

The concept of MAR provides an interactive medium for tourists to explore different destinations they want to visit because of its capability to provide a visual perspective. However, in terms of open data, it is clear through research on the use of open data in tourism is needed. Especially business and destination management perspectives are still in study and there is a considerable gap in research in studying the benefits of open data for tourism destinations and businesses. Also, there is a gap between benefits and adaptability of citizens to open data for tourism specifically in implementing AR also in consonance with the compatibility requirement of the different mobile platform.

Future research on the following needs and challenges should be conducted and addressed to better improve tourism experience. The integration of open data to MAR should be addressed so as with the technical requirements, user adaptability and availability of data.

References

[1] World Tourism Organization 2018 *Why Tourism?* http://www2.unwto.org/content/why-tourism

[2] Benckendorff P J, Sheldon P J and Fesenmaier D R 2014 *Tourism Information Technology* (Wallingford: Cab International)

[3] Ubaldi B 2013 Open government data: towards empirical analysis of open government data initiatives *OECD Working Papers on Public Governance* no **22** (OECD Publishing) (http://dx.doi.org/10.1787/5k46bj4f03s7-en)

[4] Davies T, Perini F and Alonso J M 2013 Researching the emerging impacts of open data - ODDC conceptual framework *ODDC Working Papers* **#1**

[5] Deloitte 2012 Open data - Driving growth, ingenuity and innovation *A Deloitte Analytics Paper* https://www2.deloitte.com/content/dam/Deloitte/uk/Documents/deloitte-analytics/open-data-driving-growth-ingenuity-and-innovation.pdf

[6] Deogirikar A 2014 *The Impact of Open Government on Innovation: Does Government Transparency Drive Innovation* Thesis (Georgetown University)

[7] Azuma R, Baillot Y, Behringer R, Feiner S, Julier S and MacIntyre B 2001 Recent advances in augmented reality *IEEE Computer Graphics & Applications* vol **21** no **6** pp 34-47

[8] Carnevale M 2017 *Toronto Augmented Reality Map: Enhancing Citizen Engagement with Open Government Data Using Contemporary Media Platforms* Thesis (OCAD University) Open Research Repository

[9] Cerba O, Jedlicka K, Charvát K, Ježek J, Janecka K and Mildorf T 2012 The overview of spatial data harmonisation approaches and tools *Proc. 4th Int. Conf. on Cartography and GIS* vol **1** ISSN: 1314-0604

[10] Cerba O 2013 Information around us: questions connected to information and data heterogeneities in planning activities *Proc. REAL CORP 2013: Planning Times, Acquario Romano* (Rome: House of Architecture) ISBN: 978-3-9503110-4-4 (CD-ROM); ISBN: 978-3-9503110-5-1 (Print)

[11] Hartmann S, Mainka A, Stock W G 2016 Opportunities and challenges for civic engagement *Int. J. of Knowledge Society Res.* vol **7** Issue **3**

[12] Wiggins A and Crowston K 2011 From conservation to crowdsourcing: a typology of citizen science *Proc. of the 44th Hawaii Int. Conf. on System Sciences*

[13] Pesonen J and Lampi M 2016 Utilizing open data in tourism *ENTER2016*
[14] Vert S and Vasiu R 2015 Integrating linked open data in mobile augmented reality applications- a case study *TEM J.* vol **4** no **1** pp 35- 43
[15] Gavalas D and Kenteris M 2011 A web-based pervasive recommendation system for mobile tourist guides *Pers. Ubiquit. Comput.* vol **15** issue **7** pp 759-70 doi:10.1007/s00779-011-0389-x
[16] Noguera J M, Barranco M J, Segura R J and Martínez L 2012 A mobile 3D-GIS hybrid recommender system for tourism *Information Sciences* vol **215** pp 37-52 doi:http://dx.doi.org/10.1016/j.ins.2012.05.010
[17] Viana W, Miron A, Moisuc B, Gensel J, Villanova-Oliver M and Martin H 2011 Towards the semantic and context-aware management of mobile multimedia *Multimed Tools Appl.* vol **53** no **2** pp 391-429 doi:10.1007/s11042-010-0502-6
[18] Kounavis C D, Kasimati A E, Zamani E D and Giaglis G 2012 Enhancing the tourism experience through mobile augmented reality: challenges and prospects *The Proc. of the 2nd Advances in Hospitality and Tourism Marketing & Management (AHTMM 2012) Conf.*
[19] Vert S, Dragulescu B and Vasiu R 2014 LOD4AR: Exploring linked open data with a mobile augmented reality web application *13th Int. Semantic Web Conf. (ISWC 2014)* POSDRU/159/1.5/S/137070
[20] Marimon D, Sarasua C, Carrasco P, Alvarez R, Montesa J, Adamek T, Romero I, Ortega M and Gasco P 2010 MobiAR: tourist experiences through mobile augmented reality *Telefonica Research, And Development*
[21] Tunnel Vision http://www.tunnelvisionapp.com/
[22] Langer A 2008 *Analysis and Design of Information Systems* 3rd ed (Berlin: Springer)

The empirical study on the impact of student-centered learning application to cognition and social learning

D I Cendaña[1], A B Ocay[2], N V Bustillo[1] and J. S. Dela Cruz[3]

[1]PHINMA-University of Pangasinan, Dagupan City, Philippines 2400
[2]Urdaneta City University, Urdaneta City, Philippines 2428
[3]University of the Cordilleras, Baguio City, Philippines 2600

dicendana@up.phinma.edu.ph, limmuelocay@gmail.com, nvbustillo@up.phinma.edu.ph, and delacruzpen@gmail.com

Abstract. Cognition is the umbrella term for learning skills-it is the ability to relate and comprehend what was learned. Whereas social learning is the ability of learner to grasp ideas shared by others and from a community of shared knowledge resources. The main objective of this study was to explore the impact of technology-mediated learning to the cognitive thinking of learners in terms of their ability to understand computer-assisted instructions, adopt new knowledge and apply mental activities and skills to perform reasoning and grasp what is being taught using the Student-Centered Learning Software (SCLS). Also, it aims to understand the impact of technology-mediated learning to social learning in terms of learner's ability to learn and perform activities with peers or in a group-related work. Experimental research was applied and a pre-tested questionnaire was used for data collection from 300 tertiary learners in 2 private universities in the Philippines. Using one-way ANOVA and regression analysis, study has revealed that with lower limit of 95% confidence level, the learner's frequency of use, time spent in taking activity and their level of understanding and comprehension have significant difference to their cognition where $p=0.000<0.05$. Whereas using T-test, learner's engagement to peer and group related activities has significant difference of pvalue<0.05 and variance=0.20 to their social learning. Thus, integration of SCLS may positively uphold active learning, increase cognitive thinking, stimulate enthusiasm and social learning. Hence, learner's interactivity to computer mediated technology may lead to a productive learning process.

1. Introduction

Information and communication technologies (ICT) have transformed lives and reshaped the nature of everyday activities and contemporary times as it plays important role in our lifestyle [1] and [2] collectively state that education is becoming an abundant service delivered anywhere and anytime over the world, the higher education institutions promotes the implementation of e-learning in traditional course delivery to prepare their learners and institution for the future participation in new educational evolution. This is why e-learning has become an important means of gaining knowledge with the help of electronic media, typically on the Internet [3].

E-Learning is a technology known as web-based learning, online learning, distributed learning, computer-assisted instruction or Internet-based learning. It is primarily a web-based system of education that makes information or knowledge available to users or learners [4] [5]. The strongest argument of E-learning for distance education is its potential to provide instruction to learners who, because of distance, time, or financial constraints, do not have access to traditional learning opportunities or specialized courses. Distance education, in which the teacher and learners are separated in place and/or time, may be able to provide access for them. It also mentioned that e-learning can help lecturers to save time and effort [6]. Many studies have established proofs about the viability of e-learning as a means for distance education. A research study results show that most of learners who were exposed to e-learning preferred to learn via e-learning because it provides them greater flexibility to learn at any time and any place [7]. Similarly, it was conceded in the Philippines that the design for a good formula to suit the Filipino preference would open the opportunity for growth of e-learning in the country that large organizations such as universities, big communities, large and medium-sized businesses. It could enhance the efficacy of knowledge and qualifications via ease of access to a huge amount of information and reduce their training costs [8]. Consequently, today's modern classroom, whether online or campus-based, uses e-learning tools and Learning Management Systems (LMS) which aimed to facilitate cognitive process and improve student's performance on classroom activities [9]. Thus, majority of higher educational institutions in the Philippines already embraces LMS. However, a successful e-learning system should consider different factors such as personal, social, cultural, technological, organizational and environmental because the success implementation of e-learning tool depends of student's adaptability and gratification [10].

1.1 The Role of e-learning in improving cognitive thinking

Studies associated to online learning have begun to turn its focus to better explore the unique needs associated to online learners including their ability to perform online learning, understand what is being taught using e-learning, understand distributed learning, and use computer-assisted instruction or internet-based learning in adopting new knowledge [11]. Among the significant role of e-learning according to recent studies is its positive impact to improving the cognitive thinking of individuals. Cognitive thinking is referred to as the process of using mental activities and skills to accomplish tasks such as learning, reasoning, understanding, remembering, and paying attention [12]. As to cognition, it refers to process of acquiring knowledge and understanding through thought, experience, and the senses [13]. Further, it was also verified that the actual use of e-learning has significant effect on student's academic performance while attitudes have influence on intention to use it. Similarly, [14] as per study with chemistry learners, e-learning is effective for knowledge and creativity acquisitions and the greater e-learning opportunities should be provided for a wider audience. While [15] has recently discovered that learner's employs variety of learning strategies in cognitive learning to be able for them to complete their coursework in a semester. Whilst, time management was recorded as the learner's primary cognitive learning strategy. Aforementioned studies only prove that finding out the impact of e-learning to the cognitive thinking of learners is a need. Therefore, following the initial studies on cognitive presence, as defined by [16], cognitive presence is the ability of participants in any particular pattern of Community of Inquiry (COI) to conceptualize a meaning through persistent communication. Additionally, [17] confirms that in the works and idea of John Dewey, the two sides of education concerns about psychological and social sides. Hence, collaborative reconstruction of experience has a deep essence in education because practical inquiry model operationalizes cognitive presence in the COI. Consequently, [18] also attest that social cognitive view of self-regulated learning has positive impact to learners who uses online learning environment wherein learners who embraces high level of motivation employs better learning strategies and react appropriately to environmental influences such as activity feedback. Since social learning was also discovered to have an association with cognitive learning, [19] [20] conforms that peer, teacher's collaboration and encouragement of families contributes a lot to the cognitive learning of learners.

1.2 Social Learning and student's engagement

Social interaction is one of the main focus and objective of e-learning. In addition, [21] mentioned that in the actual process of learning, the most active factor is the learner. When the E-learning application is well designed and specifically suited to a type of learners, enhancements in learning most specially using technology gives learners the opportunity to interact and become more engaged [22]. Further studies also confirmed that e-learning promotes a learner-centred instructional strategy that allows the learners to do independent explorations of a given topic [23]. Thus, e-learning developments should take high consideration of the learner's ability to acquire the required and needed knowledge. To achieve this, appropriate pedagogical and didactical approaches is the key aspects of all e-learning technology [24]. Moreover, [25] mentioned that inquiry-based learning is a pedagogical approach that enables the learners to develop critical thinking through social interaction. Likewise, [16] said that the role of interaction in an online learning environment does not only refer to student-student interaction but also interaction with the teachers and even with the content.

1.3 Research Objectives

The objective of the research was to uncover the impact of a developed e-learning tool to learner's cognition and social learning. Specifically, the study focuses on utilizing a student-centered learning environment that will be used to cross examine the student's profiles such as engagement in e-learning, frequency of use, time spent in doing learning activities, and the student's comprehension level. These factors shall then be used to determine whether the participant's profiles in using e-learning has a significant relationship to the impact of utilizing technology in cognition and social learning

2. Methodology

Experiment was conducted within three (3) weeks during June to July 2018 to get students assessment result in using the SCLS and a pre-tested questionnaire which was reviewed and approved by the Institutional Review Board (IRB) of University of the Cordilleras, Philippines with an adaptation of Cognitive Presence Scale [21] and Social Skills Rating System [22] from ASPECT [23] was used for data collection. The data was collected from freshmen learners in two (2) private universities in the Philippines, i.e PHINMA-University of Pangasinan (UPang) and PHINMA-UPang College Urdaneta. The non-probability purposive sampling technique was used and the learners were approached during their class break to fill-in online survey. A total of 300 questionnaires were distributed and 271 filled-in useable questionnaires were received back resulting in a response rate of 90.33%. One-way ANOVA, regression analysis and T-test were used to analyze the significant difference between the variables.

2.1. The e-learning tool

In the course of the study, the researchers utilized SCLS which was used by teachers and learners in Information Technology Education (ITE) subjects of PHINMA-University of Pangasinan. SCLS is an online application developed by IT faculty members to promote active learning [9] and is being used by 5 universities among PHINMA Education Network. SCLS has 3 user level: Admin, Teacher and Student. The admin account has the overall control over the system including module contents; Teacher's account is capable of adding student's list, view learners' scores, activate module and activities, and manually check essay-types questions; students can view activated modules, can answer activities, quizzes and view scoreboards and badges. Figure 1 shows the user's interface of student's account. Contemplating to the percentage on student's engagement to the use of e-learning application [26], it was discovered that the integration of e-learning application like SCLS in private universities was to promote active learning where student learn by applying what they have learned based from a given concept notes and guided activities. This is the reason why SCLS was designed to have 8 components: learning objectives, review lessons, concept notes, developmental activity, guided practice, independent activity, closure activity and enrichment activity. The first five (5) modules mentioned were to support student's comprehension to the topic matter. Independent and closure activities were to be graded by the system or the teacher; these modules pertain to the assessment of the student's learning. Lastly, enrichment activity is a supplementary module for the teacher to create strategies to improve student's learning through the SCLS. Figure 2 shows the user interface for learners as to taking up of quizzes and activities.

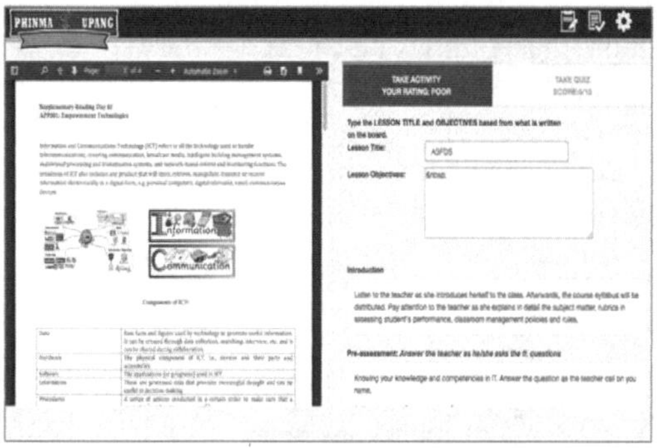

Figure 1. Student's account page to take activity

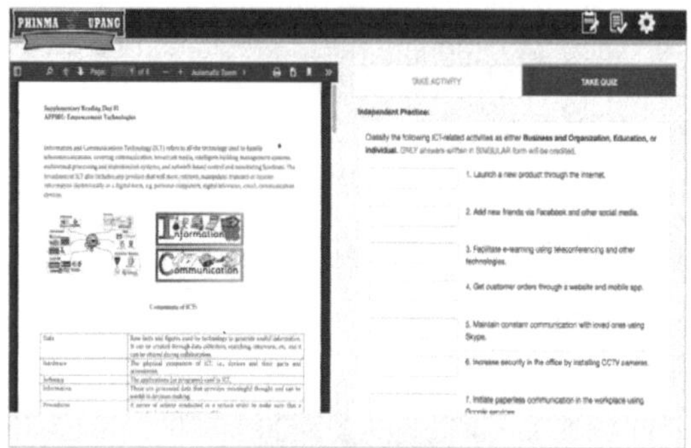

Figure 2. User interface for student to take quiz

3. Discussion

A response rate of 90.33% (N=271) was obtained from 300 participants. Among 5 PHINMA Education Network, 2 institutions participated in the study wherein 64.21% of them came from PHINMA-University of Pangasinan and 35.79% were from PHINMA-UPang College of Urdaneta. There was almost an equal distribution of participants to gender 49.82% male and 50.18% female. As to nationality, there were N=269 Filipinos, N=1 Filipino-Chinese, and N=1 Filipino-American.

3.1. Student's engagement to e-learning and their frequency of use

Student's engagement from their previous schooling was collected to understand their exposure to e-learning prior to the use of SCLS in 2 universities. Noticeably, it was found out that 67.16% learners had never experienced e-learning from their previous year of study because they were enrolled in public school where their learning materials and facilities were limited to pen/chalk and board and while 32.84% who went for schooling in private school had experienced e-learning tool where technological innovation is positively embraced to better improve their quality education. Surprisingly, learner's exposure to e-learning has evidently increased to 70.85% where 36.16% of them used e-learning applications in their major subjects in an ITE course and 34.69% used e-learning tools in minor subjects when enrolled in college. Also, student's engagement to SCLS, was further attested by getting the frequency of their exposure to the learning system. Wherein 39.85% of learners use the SCLS twice a week. This supports the teacher's implementation of active learning, which suffice that the teacher requires the learners to access the SCLS for the distribution of concept notes and supplementary reading. Whereas 22.14% among respondents uses SCLS once a week for the purpose of downloading the lecture notes, supplementary reading and submitting the given activities online. Amongst 21.40% learners occasionally access the SCLS only when prescribed by teachers. Thus, student's engagement and exposure to technology-mediated learning maybe influenced by teacher's initiative and teaching strategy. Further, learners were asked a Likert- type of question to understand their engagement to SCLS. Table 1 shows the factors that influence learners to engage in technology-mediated learning.

3.2. Student's understanding and comprehension vs. time spent in assessment

Analysis has revealed that 61.62% (N=167) of 271 respondents consumes >=1hour in taking the activity. It was found out that with 4 different activities namely (a) review lesson, (b) developmental practice, (c) guided practice and (d) independent activity, student spends more time in reading and understanding the concepts notes and instructions so as for them to independently answer the questions and perform hands-on activities in independent activity section. Whilst, 29.15% (N=79) spent 31-59 minutes in doing the same set of activities and 9.23% (N=25) spent <=30mins in doing the activity. It was also discovered that learners who consumes less than an hour in taking the activity have already experienced using e-learning tool even before coming to college.

Table 1. Factors influencing learners to be engaged in technology-mediated learning

	Statements	(TCS)	SCLS
a.	I was given the opportunity to review notes from time to time.	48.71%	51.29%
b.	I can finish the activity within a given period of time.	37.64%	62.36%
c.	Topics are better understood when supported by digital resources and examples.	27.68%	72.32%
d.	I don't need to repeatedly approach the teacher if I was not able to understand the instructions and the concept notes.	42.07%	57.93%
e.	I can easily absorb the idea and create new knowledge.	43.54%	56.46%
f.	It allows me to collaborate with my peers if I need to get additional knowledge.	39.11%	60.89%
g.	It allows me to perform creative imagination while reading the concept notes because I can clearly understand each sentence.	35.79%	64.21%
h.	It gives me ample time to think critically and logically.	39.48%	60.52%
i.	I can think and work independently.	36.90%	63.10%
j.	I am properly guided thus I was able to develop new knowledge.	38.38%	61.62%
k.	I immediately get feedback and result to my quiz/activity.	29.52%	70.48%

A Likert-type of question was asked to assess their understanding and comprehension. Analysis has revealed that learners who consumed more than an hour in taking the activity, with weighted mean (WM=3.43) learners were able to explain, (WM =3.24) discuss, and (WM =3.47) briefly summarize independently what they have learned in the class through the use of SCLS. This can be supported via their responses on the (WM =3.97) content of the module where N=147 of them agreed that the content is what they really expected to learn; N=117 (WM =3.58) agreed that the content of concept notes is good enough to represent what they have learned in the class in a form of graphs and diagram and with that, they were able to reorganize (N=132 and WM =3.66) and outline (N=112 and WM =3.51) all the topics.

As to student's level of constructing knowledge, learners who spends >=1 hour in taking the activity, N=145 with WM =3.94 felt that they were actively learning in the class because with WM =3.91 they were able to excellently do their assignment and (WM =3.9) apply in real world through hands-on activity what they have learned. With that, getting new perspective (N=140; WM =3.80) and in-depth understanding (N=117; WM =3.62) is what learners value when assessing their level of constructing their knowledge similar to the study of [27]. Using an existing tool named as Cognitive Presence Scale for measuring student's involvement [21], student's level of understanding was analyzed and the analysis revealed a WM =3.57 which means that learners positively agree on the appropriateness of the SCLS's content while with a WM =3.80, learner's response on their level of constructing knowledge using the SCLS is evidently positive. Figure 3 shows a graphical representation on student's responses for their understanding and comprehension.

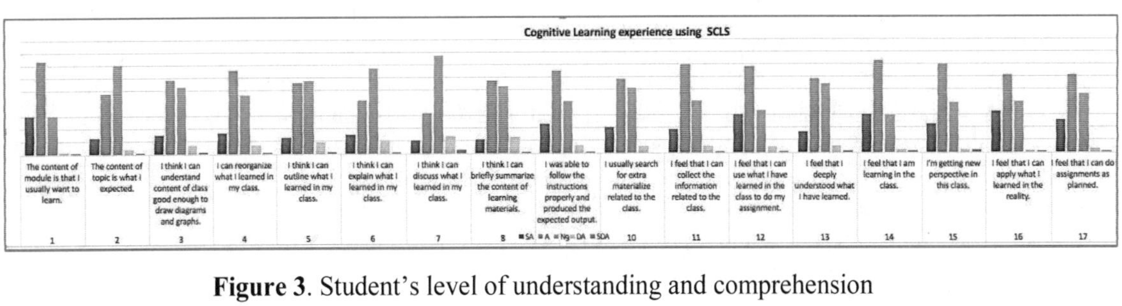

Figure 3. Student's level of understanding and comprehension

3.3. Analysis on the impact of SCLS to learner's cognition

To analyze the impact of SCLS to student's cognition, results on their activity was used as Dependent Variable (DV) while Independent Variable (IV) includes (a) their frequency of use to SCLS, (b) Time spent in activity, and (c) level of understanding and comprehension. The result has provided a scientific explanation to compare the hypothesis where H_0: The SCLS has no impact to the learner's cognition; H_1: SCLS has an impact to the learner's cognition. One-way ANOVA and regression analysis was used to produce scientific data. As to the frequency of use and activity result, analysis has resulted to coefficient=3.95 where a lower limit of the 95% confidence interval level was used and t-stat=24.01 which resulted to $p=0.002<0.05$, thus H_0 can be rejected because there is a significant difference between frequency of use of SCLS and the score obtained in their activity. As to the time spent in taking the activity with M=2.53 (>=1 hour) and result in their activity with M=12.96 where 24 is the highest score, there is a significant difference of 0.43 and 30.62 correspondingly which has a $p=0.000<0.05$ with $\Omega^2=0.0636$. Further, Figure 4 shows a positive relationship between time spent in activity and quiz where the value of y at x-axis =0.067 +1.906 and $R^2=0.124$. Also, the impact of SCLS to cognition using the presence scale where (a) level of understanding was analyzed, a M=3.56 with 0.24 variance to that of score in their activity with M=12.96 and 30.62 variance has resulted to a $p=0.000<0.05$. Correspondingly, SCLS has also an impact to the cognition of learners where learners score to activity has a variance of 0.30 with Mean=3.80 and their level of comprehension with 30.62 variance of Mean=12.62 and a $p=0.000<0.05$. With all the computed results, the SCLS has a constructive impact to the learner's cognition. With all the computed results using DV and identified IV's, the SCLS has a constructive impact to the learner's cognition. Noticeably among the learner's profile, analysis has revealed that student's engagement to SCLS has no impact to the learner's cognition where $p=0.686>0.05$ in contrast to the study to [28].

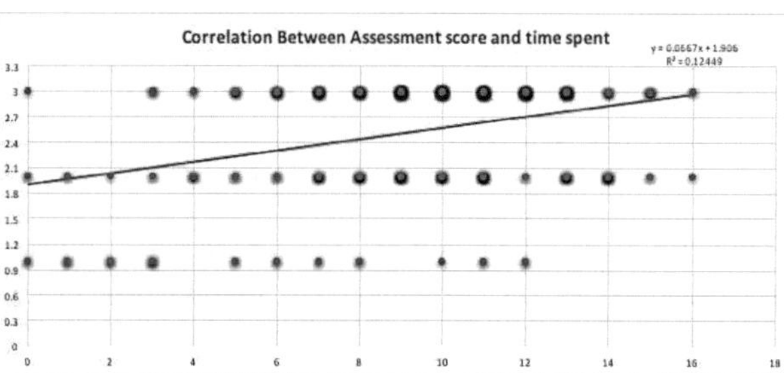

Figure 4. Correlation between time spent and activity scores

3.4. Analysis on the impact of SCLS to learner's social learning

Learners were asked on their social learning experiences as to the use of SCLS during the day 2, 3, 4 activities and a nominal question was used where 5 is the highest and 1 is the lowest. The collected Mean on student's social learning experience was used as the dependent variable while learners profile on their (a) engagement (b) frequency of use (c) time spent and (d) understanding and comprehension was used as independent variables.

With the overall weighted mean on student's responses using T-test assuming equal variances, analysis has revealed that student's engagement has significant difference to social learning where a pooled variance=0.07 to that of social learning Mean=3.72 and a variance of 0.20 was uncovered. The computed $p=0.000<0.05$ with critical value =1.96 further shows a significant difference. Similarly, using the analysis of variance (one-way) ANOVA, the frequency of use revealed a significant difference to social learning experience where Mean=3.16 and total variance of 1.37, resulting to $p=000<0.05$ with $\Omega^2=0.23$. Consequently, time spent to take the activity and their social experience was analyzed which has resulted to a Mean=1.86 and a variance=3.57 with $p=0.000<0.05$ and $\Omega^2=0.97$. The result further showed that time is an important factor to consider student's engagement and social learning when working in a group. Also, allowing learners to work in a group improves their ability to understand the content of concept notes where Mean=3.64 and variance=0.23 with computed p=0.000<0.05 and Ω^2=0.024. Correspondingly, analyzing student's level of constructing knowledge has a significant difference to the social learning experience of learners where a computed Mean=3.76 and variance=.0.25 with p=0.000<0.05 and Ω^2=0.004. Further, to analyze learners score after taking 3 activities where activity requires a team work, Figure 5 shows a significant effect of group activity score as compare to individual activities.

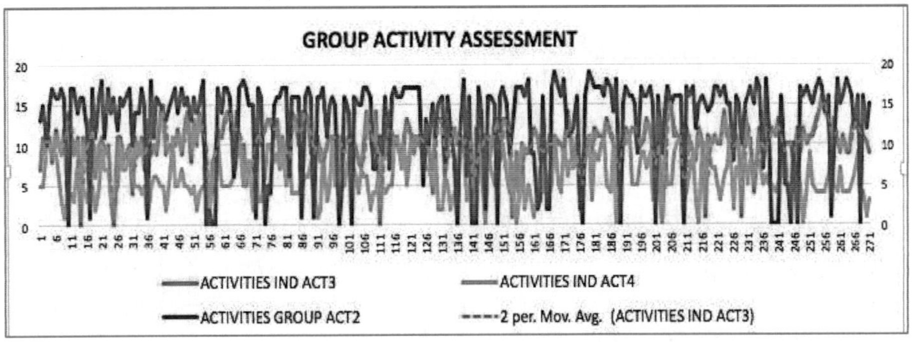

Figure 5. Comparison of assessment for individual and group activities

4. Conclusion

The research study has been strongly motivated to answer the query on knowing the potential qualities of a technology-mediated learning environment as an aid to increase learner's cognition and social learning capabilities. It is mainly guided to look at the significant difference of the several learners' profile variables which has resulted to p-value<0.05 as to their cognitive and social learning abilities in the presence of an e-learning application tool.

Moreover, the SCLS impact to student's cognition and social learning has shown significant difference to their level of engagement and compression. Specifically, student's profile such as engagement, frequency of use, time spent in an activity and level of comprehension have positively indicated that there is an impact of using the SCLS to student's cognition. The study also emphasizes the utilization of the SCLS which greatly impact the social learning of the learners when correlated with the same variable as mentioned in the cognition. This would simply mean as the learners are more engaged in the SCLS, the higher the outcome they get. Thus, SCLS is an effective tool in enhancing student's capabilities in cognition and social learning. The study however does not include factors influencing social-emotional learning or the affective domain of the learners. Therefore, further studies are highly encouraged. As a supplemental effort in the paradigm-shift of teaching and learning approaches in higher educational institutions, taking into consideration of the affective domain of the learners would essentially give the teachers newer perspective in making the learning environment truly conducive to all type of learners with the integration of technological tools. Further, since the learning environment is dominantly heterogeneous, knowing the varying emotions of the students could led to an effective means of learning process.

References

[1] Bonk C J 2009 The world is open: how web technology is revolutionizing education *Proc. of ED-MEDIA 2009--World Conf. on Educational Multimedia, Hypermedia & Telecommunications* pp 3371-80
[2] McCradie J 2003 Does IT matter to higher education? *Educause Review* vol **38** Issue **6** pp 15-22
[3] Makkar L, Alsadoon A, Prasad P W C and Elchouemi A 2016 Impact of e-Learning on students: A proposal and evaluation of enhanced e-learning model to increase the academic performance of university students *2016 Sixth Int. Conf. on Digital Information Processing and Communications (ICDIPC) 21-23 April 2016* DOI: 10.1109/ICDIPC.2016.7470797

[4] Franklin U E and Nahari A A 2018 The Impact of e-learning on academic performance: preliminary examination of King Khalid University *DEVELOPMENT* vol **7** Issue **1**
[5] Yacob A, Kadir A Z A, Zainudin O and Zurairah A 2012 Student awareness towards e-learning in education *Procedia-Social and Behavioral Sci.* vol **67** pp 93-101
[6] Islam N, Beer M and Slack F 2015 E-learning challenges faced by academics in higher education *J. of Educ. and Training Studies* vol **3** Issue **5** pp 102-12 ISSN 2324-805X E-ISSN 2324-8068
[7] Luaran J, Samsuri N N, Nadzri F A, Baharen K and Rom M 2014 A study on the student's perspective on the effectiveness of using e-learning *Procedia - Social and Behavioral Sci.* vol **123** pp 139-44
[8] Arkorful V and Abaidoo N 2014 The role of e-learning, the advantages and disadvantages of its adoption in Higher E-ducation *Int. J. of Educ. and Res.* vol **2** no **1**
[9] Magno C 2017 Facilitating student learning in schools through a learning management system: an action research *The Int. J. of Res. and Review* vol **12**
[10] Kattoua T, Al-Lozi M and Alrowwad A 2016 A review of literature on e-learning systems in higher education *Int. J. of Business Manag. & Econ. Res.* vol **7** Issue **5** pp 754-62
[11] Richardson J C and Newby T 2006 The role of students' cognitive engagement in online learning *American J. of Distance Educ.* vol **20** Issue **1** pp 23-37
[12] https://www.learningrx.com
[13] www.oxforddictionaries.com.
[14] Sarikhani R, Salari M and Mansouri V 2016 The impact of e-learning on university students'academic achievement and creativity *J. of Technical Educ. and Training* vol **8** Issue **1**
[15] Kiener M and Weaver C 2011 Examining how cognitive and affective learning strategies change as students complete coursework *Transformative Dialogues: Teaching & Learning J.* vol **5** Issue **1**
[16] Joksimovic S, Kovanovic V, Gasevic D and Hatala M 2014 Psychological characteristics in cognitive presence of communities of inquiry: A linguistic analysis of online discussions *The Internet and Higher Education* vol **22** pp 269-72 10.1016/j.iheduc.2014.03.001.
[17] Rhalmi M 2016 *John Dewey's Educational Philisophy* retrieved from https://www.myenglishpages.com/blog/john-deweys-educational-philosophy/
[18] Goradia T and Bugarcic A 2017 A social cognitive view of self-regulated learning within online environment *Adv. in Integrative Medicine* vol **4** Issue **1** pp 5–6 DOI: https://doi.org/10.1016/j.aimed.2017.05.001
[19] Genlott A A and Grönlund Å 2013 Improving literacy skills through learning reading by writing: The iWTR method presented and tested *Computers & Educ.* vol **67** pp 98-104 DOI= https://doi.org/10.1016/j.compedu.2013.03.007
[20] Ballera M, Lukandu I A and Radwan A 2014 Improving learning throughput in E-learning using interactive-cognitive based assessment *The Int. J. of E-Learning and Educational Technologies in the Digital Media (IJEETDM)* vol **1** pp 32-49
[21] Kang M, Park J U and Shin S 2007 Developing a cognitive presence scale for measuring students' involvement during e-learning process *EdMedia: World Conf. on Educational Media and Technol. (Association for the Advancement of Computing in Education (AACE))* pp 2823-8
[22] Costley K C 2014 *The Positive Effects of Technology on Teaching and Student Learning* (Arkansas Tech University) DOI = https://files.eric.ed.gov/fulltext/ED554557.pdf.
[23] Wiggins B, Eddy S, Fligner L, Freisem K, Grunspan D, Theobald E, Timbrook J and Crowe A 2017 ASPECT: a survey to assess student perspective of engagement in an active classroom *CBE – Life Sciences Education* vol **16** no **2**
[24] Gagne E D, Yekovich C W and Yekovich F R 1993 *The Cognitive Psychology Of School Learning* 2nd edition (NY: Harper Collins)

[25] Bunwirat N 2018 Inquiry-based learning: an effective pedagogical approach for empowering 21st century learners and education 4.0 *Thailand Political Sci. and Public Administration J.* vol **9** no **1**

[26] Raspopovic M, Cvetanovic S, Medan I and Ljubojevic D 2017 The effects of integrating social learning environment with online learning *The Int. Review of Res. in Open and Distributed Learning* vol **18** no **1**

[27] Roseli M H M and Umar I N 2015 Students' levels of knowledge construction and cognitive skills in an online forum learning environment *Procedia – Soc. and Behavioral Sci.* vol **197** pp 1983-9 DOI= https://doi.org/10.1016/j.sbspro.2015.07.574

[28] Olojo O J, Adewumi M G and Ajisola K T 2012 E-learning and its effects on teaching and learning in a global age *Int. J. of Academic Res. in Business and Soc. Sci.* vol **2** no **1** ISSN: 2222-6990

Environmental acoustic transformation and feature extraction for machine hearing

R Catanghal Jr[1], T Palaoag[2], and C Dayagdag[3]
[1]University of Antique, Sibalom, Antique, Philippines
[2]University of the Cordilleras, Baguio, Philippines
[3]Romblon State University, Odiongan, Romblon, Philippines

catanghal.ric@gmail.com, tpalaoag@gmail.com, carlwindayagdag@gmail.com

Abstract. This paper explores the transformation of environmental sound waveform and feature set into a parametric type representation to be used in analysis, recognition, and identification for auditory analysis of machine hearing systems. Generally, the focus of the research and study in sound recognition is concentrated on the music and speech domains, on the other hand, there are limited in non-speech environmental recognition. We analyzed and evaluated the different current feature algorithms and methods explored for the acoustic recognition of environmental sounds, the Mel Filterbank Energies (FBEs) and Gammatone spectral coefficients (GSTC) and for classifying the sound signal the Convolutional Neural Network (CNN) was used. The result shows that GSTC performs well as a feature compared to FBEs, but FBEs tend to perform better when combined with other feature. This shows that a combination of features set is promising in obtaining a higher accuracy compared to a single feature in environmental sound classification, that is helpful in the development of the machine hearing systems.

1. Introduction

A significant number of information is carried by sounds in connection with our daily surroundings and physical happenings that take place in it. In terms of efficiency, humans are very effective and accurate in identifying the typical general feature of the soundscape surrounding him, whether it is an office environment, a quiet park, bustling city lanes, beach fronts and deciphering individual sources of sound in the scenes like coughing, rooster crowing, crying baby, glass breaking, running cars, dog barks, or thunderstorm. The automatic recognition of sounds in the environment for example in the multimedia applications is an expanding and increasing problem of research. The comprehension of this sounds in the environment are an important factor in the perception of the content in multimedia. This environmental sounds are very much divergent and varied sound events occurrence from day to day and cannot just be confined between the music or speech. For this reason, the technological development of the classification of sounds in the environment is best for describing large numbers of applications in multimedia, such as classification of scenes in the audio, hearing aid systems, intelligent house monitoring, generation and highlighting of video content, surveillance system through audio, and a lot more [1].

Sound is frequently a valuable supplement to modalities, for example, a video conveying information not generally present, for instance, data from discourse and birdsong and it can likewise be more easier to gather or collect, such as on a cell phone or portable recording device. Collected information from a semantic acoustic examination can be valuable for additional transformation, for example, robotic navigational route, user alarms, or prediction and analysis in pattern occurrence of events. Apart from being a listening machine, similar advancements have applications in cataloging/searching sound files, whose digital compilations have developed massively in decades. Often sound or acoustic collection contains a rich assorted and variety of music, bio-acoustics, speeches, city and rural soundscapes, environmental sounds, ethnographic archives to name only some, in spite of the accessibility at present, it is behind compare to text [2].

Researchers have attempted to use information extracted from various sounds to enrich various applications, such as monitoring systems for older adults or infants, automatic surveillance systems, automatic lifelogging systems, and advanced multimedia retrieval systems. The techniques used in these applications include acoustic scene analysis that analyzes scenes regarding the places, situations, and user activities they depict and acoustic event analysis that analyzes various sounds, such as footsteps, the rumble of traffic, and scream [3].

This paper explores and presents the current feature extraction algorithm and methods for environmental sound recognition that would be utilized in machine hearing systems, further objective of this research paper is the comparative feature extraction and methods. A summary of the architecture of the machine hearing system is discussed. A thorough analysis of the most current developed feature was discussed, namely the gammatone filterbank and the Mel filterbank energies (FBEs). The most relevant result of its evaluation shall be shown and suggested implementation for the environmental sound recognition.

2. The Machine Hearing

2.1. The Machine Hearing Design

Even what particular type of problem is applied in the acoustic field, the fundamental structure of the system can be characterized using a general or universal pattern design as shown on Figure 1.

Figure 1. Architecture of Machine Hearing

The process of windowing is the first stage; this is accomplished by capturing a continuous stream of audio using a device usually a microphone and partitioning this into a block of the shorter signal. To accomplish the process of windowing, a hypothetically endless sample of streams from the signal input will be slided by a window function. Moreover, in order to facilitate the successive signal analysis, contingent on the window function duration it is presumed that a commonly non-static sound signal within the individual frame is quasi-stationary.

The problem of buffering can be overcome with the use of the window function. This window function works by fetching the audio chunks at a certain intervals of time, even not knowing with regards to their completeness. To further address the issue, the edge of the buffer is being polished using the window functions. Through this, the entire important periods in the audio signal is required to be completed. In this study the Hamming window was used and is characterized as the equation 1 below.

$$HammingWindow(s) = 0.54 - 0.46\cos\left(\frac{2\pi s}{BufferSize}\right) \quad (1)$$

2.2. Feature Set Extraction

Features extraction, for the most part, is carried out in three main stages and is the series of feature vector calculation that accomplishes for a given signal a condensed representation. The first stage is called acoustic front-end or analysis of the sound, in here where the raw feature are produced characterizing the container of the power spectrum of abrupt intervals of sounds, and in this stage where the sound signal spectra temporal analysis is carried. In the next stage is where the compilation of a continued feature vector which comprises of the two features: dynamic and static. Then lastly the transformation stage, where the continued feature vectors is converted further to a much compressed and concentrated vectors as a recognizer input.

2.3. The Mel Filterbank Energies (FBEs)

Research experiment noted that human ear works as a filterbank, that is a bank of subband filters. In this human perception experiment where the analysis of the Mel frequency was based. This Mel frequency focuses only on some particular or specific components of frequency. It is in the frequency-axis where the filters are space unevenly or overlapped, further in research of processing of sounds; it exhibited that the signal in the range of 10 – 30 ms duration is considered to be in static and in this instance a window that is smaller is considered. A proposed phase encoded filterbank energies (PEFBEs) were even proposed as an enhancement technique [4] as shown in Figure 2.

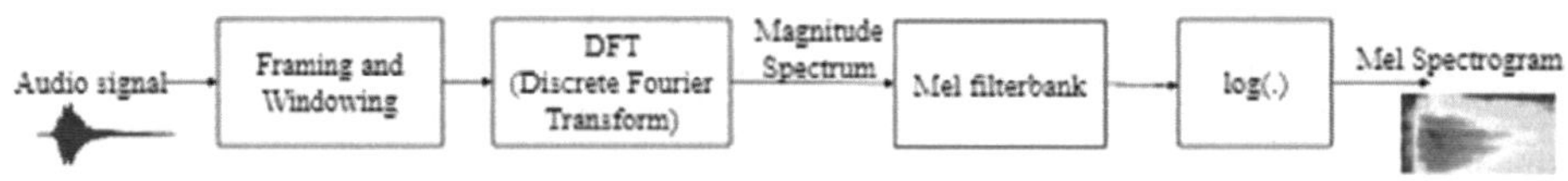

Figure 2. Mel spectrogram diagram of sound signal [4].

2.4. *The Gammatone Filterbank and Teager Energy Operator*

The Gammatone filter impulse response is a multiplication of the Gamma distribution function and a sinusoidal tone centered at a particular frequency [6]. Given by the equation 2:

$$g(f,t) = t^{a-1}e^{-2\pi bt}\cos(2\pi ft), t > 0, \tag{2}$$

where f is defined as the center frequency, b is rectangular bandwidth, and a is the filter order. The concept of the Gammatone filter was motivated by the biological studies it is where it stems. In the modeling of the filter response of the auditory of humans, this is where the Gammatone function is utilized [7]. There is a great similarity on the how the reox function (human auditory filter response in the cochlea) to the Gammatone filter for its magnitude response. In the human hearing system the Gammatone filter bandwidth resembles the basilar membrane placement of filters. The scale to where it is measured is called equivalent rectangular bandwidth (ERB) [5].

The temporal modulations, among the many perceptual and acoustic features of the sound, is one of the essential parametric representations of the sound signal and it describes the signal variations of sounds regarding the frequency modulation (FM) and amplitude modulation (AM). Using the cochlea as the filter bank model, where the feedback of the AM-FM can be drawn, and it was noted that AM-FM is an integral combined feature of the sound signal and occurs at the same time. The researcher [5] believed that adopting an operator like Teager Energy Operator (TEO) which will trace the energy coming from the two elements which is the AM and FM. This Teager Energy Operator represents the necessary energy for the generation of a signal or the system energy that produces the signal. The Teager Energy Operator not like the standard or typical energy, it is roughly equivalent to the result of squaring the product of frequency and amplitude. The signal was separated for refinement through the narrowband filter bank before applying the teager energy operator, the application filtering was done because teager energy operator is not applicable for direct application in the sound signal. The reason for this process is that Teager energy operator, for the most part, functions on the bandpass filtered signal or on the monocomponent. The directly short-term spectral feature used in the CNN is called as TEO-based Gammatone spectral coefficients (TEO-GTSC) [5] as shown in Figure 3.

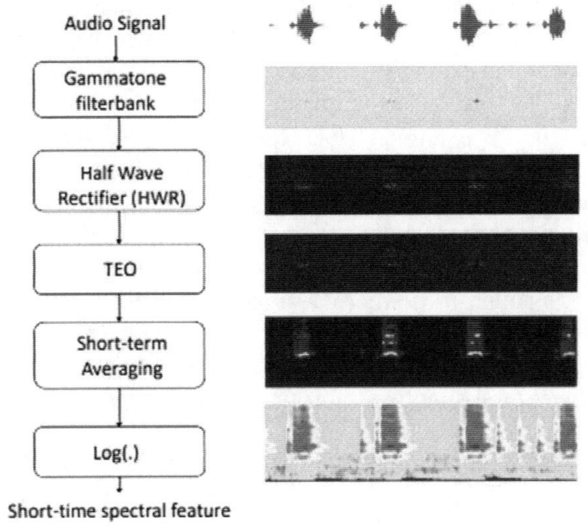

Figure 3. Block diagram of TEO-GTSC [5]

2.5. The Convolutional Neural Networks

The bottom line of the convolutional neural networks is that they are an elementary extended version of the model multilayer perceptron. Nevertheless, there are important and substantial pragmatic implication and results due to their design or architectural differences. The usual and conventional neural network in a deep architecture is composed of a few distinct layers stacked collectively. That is a layer for an input, an array of pooling and convolutional layers (there are several ways it can be mixed), completely linked layers that are hidden which are limited in numbers, and a layer for output (loss) as shown in Figure 4. In contrast with the multilayer perceptron, the actual distinction lies in addition of the pooling and convolution procedures [8].

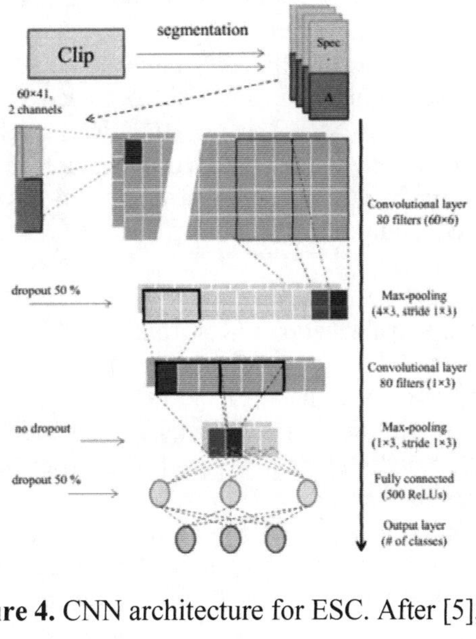

Figure 4. CNN architecture for ESC. After [5], [8]

3. Results and Discussions

A meta-analysis of the different experiments implemented [9][4][5][8] in the literature using the environmental sound classification dataset [1], with 2000 labeled environmentally recordings that are sampled at 44.1kHz and comprises of 50 equally balanced classes; each clip is about 5 seconds. The 50 classes was categorize into five (5) main divisions: natural soundscape, the animal sounds, different water sounds, urban/exterior sounds, domestic/interior noises and the non-speech by humans. An array of diversified audio sources from the dataset was used as a training, several are very recognizable such as the brushing of teeth, while for example airplane and helicopter noise in which the distinction or differentiation is somewhat slight, there are very familiar sounds like that of a barking dog, laughter or cat meowing are among the few examples.

An acoustic transformation was made before feature extraction was executed. To have an analogous comparison with the baseline, the entire sound files were resampled/down sampled to 22.05kHz [8]. The feature extraction of the sound files where accomplish by dividing the frames into windows by applying the 25 ms Hamming method and using a fifty percent (50%) window overlap. In removing the regions of silence, an algorithm called simple energy thresholding was used. The algorithm works by removing the frame if there exist silence greater than three frames, otherwise, frames are preserved.

Table 1. Classification accuracy(%) of different feature with CNN classifier

Feature Sets	Accuracy (%)
FBE ⊕ ConvRBM-BANK [9]	86.50
FBE ⊕ PEFBEs [4]	84.15
GTSC ⊕ ConvRBM-BANK [9]	83.00
GTSC ⊕ TEO-GTSC [5]	81.95
GTSC [5]	79.10
FBEs [5]	67.80
Baseline [8]	64.50

The performances of various feature sets were evaluated through 5-fold cross-validation. Table 1 shows that using Mel Filterbank Energies alone yield the lowest accuracy which is 67.80%, although higher compared to the reported baseline [8] of 64.50%. When compared as feature set without combination with other systems Gammatone spectral coefficients yields higher accuracy 79.10% compared to Mel Filterbank Energies which is 67.80%. It is noteworthy that even though Mel Filterbank Energies yields lower accuracy taken as the only feature when combined with other features it yields the highest accuracy which is 86.50%, higher compared to Gammatone spectral coefficients even when combined with other feature sets.

4. Conclusion and Recommendation

This paper aim is to explore the transformation of environmental sound waveform and feature set into a parametric type representation to be used in analysis, recognition, and identification for auditory analysis of machine hearing systems. We analyzed and evaluated the different most current developed feature algorithms and methods explored for the acoustic recognition of environmental sounds, the Mel Filterbank Energies (FBEs) and Gammatone spectral coefficients (GSTC) using Convolutional Neural Network (CNN) as a pattern classifier.

It seems that a combination of features set results to better performance as compared to single features. The Mel Filterbank Energies (FBEs) combined with Convolutional Restricted Boltzmann Machine filterbank (ConvRBM-BANK) shows a promising result as the highest accuracy in classifying the environmental sound. Studies show that a combination of features improves the performance in the classification of the environmental sounds.

Several probable questions that needs to be answered for future investigation and analysis is whether these features can be successfully applied in other machine learning classifiers and how will they perform in terms of accuracy. Given the complexity of the process and model, how will it perform regarding processing time when applied with the machine hearing systems.

References

[1] Piczak K J 2015 ESC: Dataset for environmental sound classification *Proc. of the 23rd ACM Int. Conf. on Multimedia* (ACM) pp 1015-8

[2] Stowell A D, Giannoulis D, Benetos E, Lagrange M and Plumbley M D 2015 Detection and classification of acoustic scenes and events *IEEE Trans. on Multimedia* vol **17** no **10** pp 1733-46

[3] Imoto K 2018 Introduction to acoustic event and scene analysis *Acoustical Sci. and Technol.* vol **39** no **3** pp 182-8

[4] Tak RN, Agrawal DM and Patil HA 2017 Novel phase encoded mel filterbank energies for environmental sound classification *Pattern Recognition and Machine Intelligence PReMI 2017* ed Shankar B et al (Cham: Springer)

[5] Agrawal D M, Sailor H B, Soni M H and Patil H A 2017 Novel teo based gammatone features for environmental sound classification *European Signal Processing Conf. 2017* pp 1809-13

[6] Valero X and Alias F 2012 Gammatone cepstral coefficients: biologically inspired features for non-speech audio classification *IEEE Trans. on Multimedia* vol **14** no **6** pp 1684–9

[7] Carney L H and Yin T 1988 Temporal coding of resonances by low-frequency auditory nerve fibers: single-fiber responses and a population model *J. of Neurophysiology* vol **60** no **5** pp 1653–77

[8] Piczak K J 2015 Environmental sound classification with convolutional neural networks *25th Int. Workshop on Machine Learning for Signal Processing (MLSP), Boston, MA, USA, 2015* pp 1–6

[9] Sailor H B, Agrawal D M and Patil H A 2017 Unsupervised filterbank learning using convolutional restricted Boltzmann machine for environmental sound classification *Proc. Interspeech 2017* pp 3107–11

Students' perception on e-learning: a basis for the development of e-learning framework in higher education institutions

MM Daniels[1], E Sarte[2] and J Dela Cruz[3]

[1]College of Computing Sciences, Ifugao State University, Lamut, Ifugao, Philippines mylemelodydaniels@gmail.com
[2]College of Information Technology and Engineering , Our Lady of the Pillar College Cauayan, Cauayan, Isabela, Philippines vanzsarte@yahoo.com
3College of Information Technology and Computer Science, University of the Cordilleras, Baguio City, Philippines delacruzpen@gmail.com

Abstract. As learnings styles evolve along with modernizing society, educational technology also expands. A current trend in education brought about by technological advances is the e-learning system where teachers and students can discuss lessons online and exchange learning resources. This study explored on the areas of e-learning and provided a review on current e-learning frameworks from different studies. A recent study of Debattista presented a comprehensive rubric for e-learning and it is adopted by this paper as basis for gathering student expectations, feedback, and problems encountered in e-learning. These rubrics were rated by students according to importance. Statistical findings show a significant difference between ratings of students from public and private institutions. Similarly, there is a significant difference between the ratings of male and female students. The difference might spring from the level of interest of students towards learning as factored in by type of institution and gender. Students' learning expectations in an e-learning environment were also gathered in this study as a basis for a proposed e-learning framework. All specific standards presented by Debattista were labelled very important by respondents and are therefore adopted into the proposed framework. Along with these rubrics are proposed additional standards that focus on the enrichment of student experience and enhancement of learning. It is still highly recommended that strict and proper implementation of such standards are supervised by concerned administrative departments.

1. Introduction

In this millennial age where learning style has shifted from traditional to a more student-centered learning, e-learning became widely used. Not only does it enhance classroom instruction [1], it has also helped address the problem with conflicting time and distant spaces among teachers and students. Collaboration became possible online that learning is no longer confined in the classroom. Specialized educational software that come in many forms like e-learning [2], 3D virtual lab [3], virtual environment, and mobile games [4] can supplement teaching aids to cater to the different learning styles of students. In its broadest sense, Abbad et al. [5] defined e-learning to mean any learning that is enabled electronically. It was also described by Jethro et al. [6] as a computer assisted learning, and as pedagogy for student-centered and collaborative learning. In some definitions e-learning encompasses more than just the offering of wholly online courses. For instance, Oblinger and Hawkins [7] indicated that e-learning has transformed from a fully-online course to using technology to deliver part or all of a course independent of permanent time and place. Also, the European Commission [8] describes e-learning as the use of new multimedia technologies and the Internet to increase learning quality by easing access to facilities and services as well as distant exchanges and collaboration. Brown and Voltz [9] presented six elements of e-learning, namely, activity, scenario, feedback, delivery, context and influence. These elements stem from a focus on student experience, while taking into account the broader networks that contribute to and are influenced by that experience.

E-learning provide facilities for sharing of learning resources and easy interaction among learners and teachers. Other e-learning software also provide immediate result on student performance [10] which makes it easier for students to track their progress. Parents could also be invited into the online platform to see their child's performance. Findings from empirical research by Riasati et al. [11] showed that technology integration in teaching, particularly in language, is advocated for reasons including 'engagement', 'improvement in academic ability', 'paradigm shift', 'assessment shift' and 'collaborative learning enhancement'. However there are some barriers hindering the use of technology. Riasati et al. [11] presented barriers from their literature reviews: 'lack of accesses', 'lack of time', 'lack of effective training', 'teachers' attitude', and 'students' attitude'. Aside from these, complexity of e-learning materials may also have effect on student learning especially so that they are presented in diverse designs and features. For example, complex design could distract learners from their focus on lessons. Another factor that could affect effectiveness of an e-learning materials is on how the teacher facilitate learning online. Implementation on the use of these useful materials may still be improved to optimize its use in education. This study focuses on gathering student's experiences on the use of e-learning. Their personal views on how implementation and use of e-learning materials could be a significant knowledge that will help teachers and also developers enhance its use.

Various frameworks from research studies were presented by authors to propose assessments rubrics for online courses. In an effort to provide a centralized point of reference for educators in the field of e-learning, a synthesized approach from four studies and an in-depth review of literature was presented by Debattista [12]. In his paper, he discussed four frameworks from which he based his proposed comprehensive rubric for instructional design in e-learning, these are, Graham's [13] Seven Principles for Good Practice in Undergraduate Education, the rubric provided by the Quality Online Course Initiative of the University of Illinois [14], California State University's Quality Online Learning and Teaching instrument to measure the effectiveness and quality of online courses [15], and University of Malta's [16] Minimum Standards for Study Units in the VLE . The synthesis of the four rubrics has produced the following comprehensive rubric that covers all the aspects mentioned by the four institutions. Debattista specifies that this comprehensive rubric is not a collation of the four rubrics but a synthesis of the separate approaches that – in the author's view – reflects the context of e-learning. He presented 10 main standards and specific standards under each. The standards are as follows: Instructional design, Course opening, Assessment of learning, Interaction and community, Instructional resources for teaching and learning, Learner support, Technology design, Course evaluation, Course closing, and Instructional design cycle. The specific standards can be found in Debattista's paper. This rubric has not been tested in laboratory setting or a real-life scenario. However, since the rubric was a synthesis of both traditional and online frameworks, it was adopted by this study as a reference in assessing e-learning areas. Additional important rubrics were also collected from respondents as basis for the development of an e–learning framework particularly for Higher Educational Institutions (HEI). An overview of the study is illustrated in Figure 1.

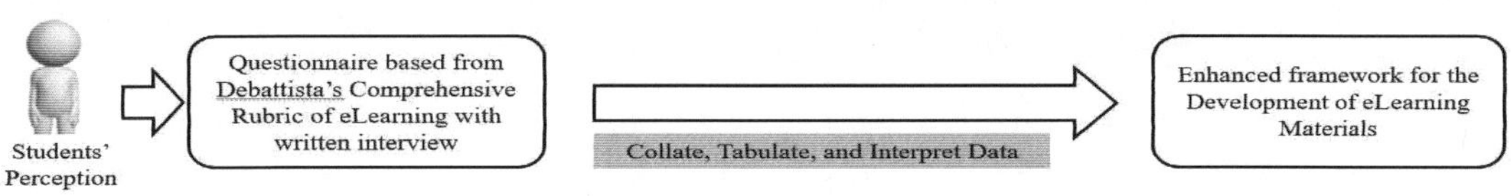

Figure 1. Conceptual Framework of the Study

2. Methodology

This study employed both qualitative and quantitative form of gathering and analysing student perceptions on the implementation and use of e-learning materials. For the respondents of the study, a sample size of 80 students from private and 80 students from public higher educational institutions were randomly selected. A semi-structured questionnaire was prepared based on Debattista's comprehensive rubric for instructional design in e-learning. In Debattista's rubric, he presented 10 main standards and each has its specific standards. For simpler terms, the main standards are referred to as 'e-learning areas' in this study, while specific standards are sometimes called 'items' or 'rubrics.' The questionnaire prepared in this study is intended to measure the importance of each of the specific standards of e-learning as perceived by students. The researchers provided descriptions for some standards that are not very much familiar with respondents, such as 'structure of learning' which pertains to how lessons are presented. Respondents rated each item using a Likert scale. Open-ended questions were also asked per area to gather as much detail on their perception on e-learning areas. Students were free to indicate other functions and features that they feel are useful to enhance their learning experience in e-learning environment. For easier collection of data, an online form of the survey questionnaire was also prepared and sent to respondents who opted to respond outside the school campus.

Ratings from each of the school categories were collated and presented in a table for easier comparison. Ratings from the Likert scale were summarized for each rubric. A statistical analysis was performed on the data to determine the level of significance of each. Further, t-test was performed to see if there is a significant difference between data from private and public respondents. Gender was also used in determining significant differences with regards to their preferences on the items stated in Debettista's rubric.

3. Results and Discussion

This study presents students' perception on each of the 10 main standards presented by Debettista. Rubrics under each area was rated by respondents as they see its importance in learning. Table 1 shows the respondents' ratings grouped according to type of school, and also according to gender. The overall ratings reflect that all items in Debattista's rubric is indeed relevant in the development and implementation of an e-learning environment as perceived by students. Mean rating of each is above 4, with opennes, resolution and interface having the highest mean of 4.48. This means that respondents believe that technical infrastructure used to deliver the teaching and learning has to be procured and implemented according to open standards and formats that maximise the value for money and the range of options to fulfil the learning outcomes and the academic needs of faculty and learners, which is opennes. They also believe it is very important that all pending issues between the instructor and the learners are resolved at the end of the course, referred to as resolution. Interface and navigation in the learning area should be simple enough to be conducive to teaching and learning without the need to possess advanced ICT skills and competences.

Table 1. Test of difference on the perception on e-learning of respondents when grouped according to type of school and gender.

Variable		Mean	Std. Dev.	df	t Computed Value	P value	Interpretation
Type of School	Public	4.46	0.45	158	2.15	0.03	Significant
	Private	4.29	0.54				
Gender	Male	4.26	0.58	158	2.76	0.01	Significant
	Female	4.48	0.40				

For the testing of hypothesis, t-test was used to analyze gathered data. Statistical analysis of these ratings show SIGNIFICANT DIFFERENCE among the ratings of students from public school and from private school with P value 0.03. A test based on ratings of each gender also shows significant differences with P value 0.01.

It can be noted from Table 1 that the mean ratings from public school respondents is higher than those from private school. This may imply that respondents from public school has higher expectations in e-learning than those from private school. Gender was also used as a factor in determining the level of difference among the responses with regards to Debattista's rubric. Further, Table 1 shows that female respondents has higher mean than male respondents. These ratings could be influenced by student's interest in studying and may reflect that female respondents are more serious in their studies even through online platform. Most male respondents admit that they are tempted to play online games when given access to internet before they even access e-learning sites.

3.1 Learning Expectations of Students in e-Learning

E-learning environment has apparently widened the borders of education. Respondents strongly agree that they are able to gain education anywhere at their most convenient time because of this technology, more so with the help of smartphones that made online resources available to users' fingertips. Collected feedback, expectations and problems encountered by students are discussed per area in this section.

On the area of *instructional design*, respondents appreciate current lesson presentation where layout is simple, files could be arranged according to lesson sequence and sorted according to type of file (text file, PDF, audio, images, etc.) for easier browsing. Learning objectives and outcomes are sought to be relevant for learning since it directs students' focus during the learning process.

Course opening covers the role of the teacher in the e-learning environment, the description of the course and behaviour on how it will be utilized. There is, however, lack of student orientation on how to operate the e-learning environment and how it is to be utilized in the duration of the course. For the e-learning system functions and features, students are usually oriented only concerning central modules like learning resources module and assessment module, and the rest is for them to discover. Students expect a run-through of the whole e-learning environment so they can fully utilize it. It then proposed to add *orientation* as one specific standard under Course Opening. As to the role of the instructor during the learning process, students expect a more descriptive instruction. For example, if they knew information like schedule of giving announcement or posting of new lessons, or even teacher's schedule for online consultation, they would know when to go online.

With *assessment of learning* available in e-learning, paperless assessment has become possible. While it is undeniable how e-learning benefits its users in this area, students has mentioned several observations in the implementation process that need improvement. One scenario expressed is that teachers announce requirements or even assessments on the e-learning forum without informing the class during the previous meeting. In this setting, students are not set to accomplish tasks. In other words, assessments should be given at agreed schedules for the students to be prepared whenever assessments are made. Another is with regards to narrative assessments where some teachers do not provide their rubric for grading answers. Transparency in this case will prevent conflicts that may arise and will allow the students to accomplish the assessment strategically. *Rubrics for grading* should then be included under this standard. Moreover, scores for narrative assessments are seldom disseminated unlike multiple choice assessments. Along with assessment results, teachers should find time to discuss answers to the assessment given so that students will learn from their mistakes.

Interaction and community provided by e-learning environment allows communication between instructor and learners. Respondents recommend a *real-time* interaction like a chat-box between student and teacher for a faster consultation. And when teacher is offline, teacher should be notified via sms.

The exchange of *instructional resources* for teaching and learning is a vital function of e-learning. Students feel relieved from the burden of carrying too much books and photocopying documents. There was a minimal concern on this area. With the numerous available online resources that can be linked as learning resources, students expect that relevant sections in these resources are identified by the teacher so they would know where to focus when reading. Too much resources could distract learners from focus, and they lose interest even before getting to the main text. Along with this, respondents find it crucial that learning resources are up-to-date and relevant to the course, to ensure *validity* of learning contents. Dynamic content also caters to different types of learners. A combination of text, images, audio and videos can meet the varying learning preferences of students.

To help learners achieve their full potential, *learner support* must be available. New users expect to find frequently-asked-questions page and forums where they could read seek answers on their own. For further assistance, respondents suggested a separate help desk or an *offline support* that could be reached via sms or call. This is for technical problems where user is unable to login into the systems. For students who live in the areas with poor internet connection, offline communication is a good alternative in getting their concerns addressed.

Technology design should support the achievement of the aims and objectives of the instructor and the learning outcomes for learners. The current e-learning design is generally acceptable to respondents. The separate module for learning resources are helpful in organizing their files into folders. However, few respondents say they are annoyed by the long tail of button that pop out after the other. While this is a design intended to categorize contents, too much specifications can be strenuous on users. Respondents suggest a search box for keywords that would lead them to the specific page they seek instead of browsing through buttons. Additionally, the interface design should be user-friendly across all the available platform, especially for smartphone that has limited view. A mobile app version of the e-learning materials that could be accessed *offline* could foster easy-access and portability that come handy for students who live off-campus. Since e-learning is deployed online, administration should support the provision of wifi connectivity in student lounges. Student often encounter administrative excuses such as lack of funding when requesting for such.

At the end of the course, students expect to provide a *course evaluation* where they give feedback on the whole learning experience. Some e-learning materials do not provide this feature. For those that do have this, the course evaluation entail too much questions that become exhausting to answer. Respondents suggest a more concise form of evaluation and easier scheme to answer like simply clicking on options or rating scale for satisfaction rating. This would make the course evaluation easy to accomplish and would encourage more accurate answers. Student who give their honest feedback and constructive comments also expect their evaluation to be forwarded to the concerns administrative departments for proper action. Suggestions that only require a few days to integrate are expected to be *implemented* in time for the succeeding semester.

Course closing includes assessment, resolution, and archiving. Student expect to see their grades at the end of the course as they are highly concerned with their performance. This will provide transparency on their class performance like assessment scores, missed requirements, and other details. An alternative option is for their performance report to be emailed directly to them.

Instructional design cycle is more of an administrative responsibility. Although the students do not see this part of the e-learning platform, they rated the specific standards of this area as very important. They appreciate the relevance of this practice where the quality of the e-learning system is maintained and improved. Respondents suggest a regular review of the e-learning system, at least every semester, when feedback on the course are still fresh (*frequency*). Administrative duty of supervising the utilization of e-learning makes sure that learning resources are regularly updated by teachers.

3.2 Proposed e-Learning Framework based on Learning Expectations

A summary of proposed additional specific standards to Debattista's comprehensive rubric based on student learning expectations is shown in Figure 2. All specific standards presented by Debattista is proven to be relevant in e-learning as perceived by students. They are therefore fully accepted by this paper and are adopted into the proposed e-learning framework.

There are 8 additional specific standards proposed in this framework. First, under Course Opening, there is a need for student *orientation* since students claim they lack knowledge on how to operate the e-learning environment. This basic knowledge could help them fully benefit from the features of the system. Second, under Assessment of Learning, students need to know the *rubrics* for grading their answers. This will guide them on their approach in answering especially for narrative assessments. Third, under Interaction and Community, students assert that a *real-time* interaction facility like a chat-box could get their concerns addressed immediately. Fourth, under Instructional Resources, instructional materials' *validity* needs to be ensured for an updated learning content. Fifth, under Instructional Design Cycle, *frequency* refers to the number of times that a review is conducted on the e-learning design and content. Students believe there should be a regular session set by concerned administrative departments to review the course content. Sixth, on Course Evaluation, proper *implementation* is expected by students, whereby their feedback on the e-learning course experience could actually be considered by the institution in improving their e-learning system. Seventh, an *offline version* of the e-learning system could cater to students who have little access to the internet. A mobile app that could be accessed even offline could later be synchronized to the online version for necessary updates. Finally, *offline support* like sms and phone call are viewed by students as a very convenient means to reach for help.

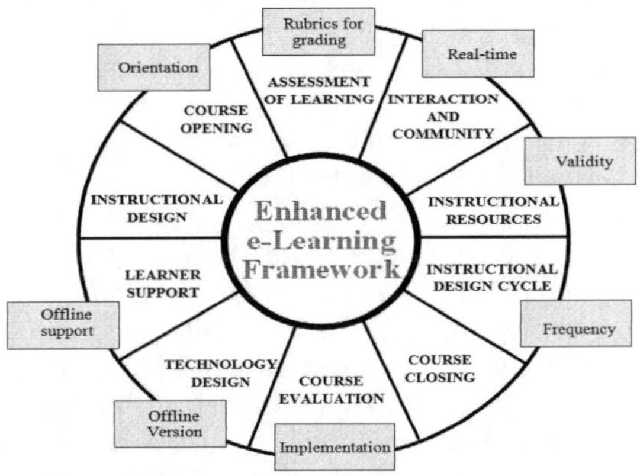

Figure 2. Proposed e-Learning Framework

4. Conclusion and Recommendation

Areas of e-learning environment are relevant for teaching and learning. Debattista's specific standards are highly accepted by students as relevant factors to be considered in the development and implementation of e-learning environment. They are therefore fully accepted by this paper and are adopted into the proposed e-learning framework. Students' learning expectations in an e-learning environment were also gathered in this study. Out from these expectations, several standards are proposed. These standards focus on the enrichment of student experience and enhancement of learning. It is still highly recommended that strict and proper implementation of such standards are supervised by concerned administrative departments.

Finally, statistical analysis on respondent ratings shows a significant difference between ratings from public and private institutions. Students of public institutions show higher expectations on e-learning than students from private institutions. Moreover, there is a significant difference between ratings of male and female students which may be due to respondents' attitude toward learning when devices are involved.

References

[1] Arkorful V and Abaidoo N 2014 The role of e-learning, the advantages and disadvantages of its adoption *Higher Educ. Int. J. of Educ. and Res.* vol **2** no **12**

[2] Doculan J 2016 E-learning readiness assessment tool for Philippine higher education institutions *Int. J. on Integrating Technol. in Educ.* vol **5** no **2**

[3] Amorima T, Tapparob L, Marranghellob N, Silvac A and Pereirab A 2014 A multiple intelligences theory-based 3D virtual lab environment for digital systems teaching *Procedia Comp. Sci.* vol **29** pp 1413-22

[4] Chuang T and Su S 2012 Using mobile console games for multiple intelligences and education *Int. J. of Mobile Learning and Organisation* vol **6** pp 204-17

[5] Abbad M, Morris D and De Nahlik C 2009 Looking under the bonnet: factors affecting student adoption of e-learning systems in Jordan *The Int. Review of Res. in Open and Distance Learning* vol **10** no **2**
[6] Jethro O, Grace M and Thomas A 2012 E-learning and its effects on teaching and learning in a global age *Int. J. of Acad. Res. in Business and Soc. Sci.* vol **2** no **1**
[7] Oblinger D and Hawkins B 2005 *The Myth about E-learning* (Educause review)
[8] European Commission 2001 *The eLearning Action Plan: Designing Tomorrow's Education* (http://www.elearningeuropa.info)
[9] Brown A and Voltz B 2005 Elements of effective e-learning design *The Int. Review of Res. in Open and Distributed Learning* vol **6** no **1**
[10] Mastilak C 2012 First-day strategies for millennial students in introductory accounting courses: It's all fun and games until something gets learned *J. of Educ. for Business* vol **87** pp 48-51
[11] Riasati M, Allahyar N and Tan K 2012 Technology in language education: benefits and barriers *J. of Educ. and Practice* vol **3** no **5**
[12] Debattista M 2018 A comprehensive rubric for instructional design in e-learning *Int. J. of Information and Learning Technol.* vol **35** Issue **2** pp 93-104 https://doi.org/10.1108/IJILT-09-2017-0092
[13] Graham CR, Cagiltay KC, Lim B, Craner J and Duffy T 2001 *Seven Principles of Effective Teaching: A Practical Lens for Evaluating Online Courses* (www.technologysource.org/article/274/?utm_content=buffere64be&utm_source=buffer&utm_medium=twitter&utm_campaign=Buffer) accessed 20 October 2016
[14] University of Illinois 2015 *Quality Online Course Initiative* (www.ion.uillinois.edu/initiatives/qoci/rubric.asp, accessed 7 January 2017)
[15] Christie B 2014 *CSU Quality Online Learning and Teaching (QOLT) Evaluation Instrument Sections and Objectives* (California State University) (http://qolt.sfsu.edu/sites/default/files/QOLT2-pgnumberslogoCore24CClicense111715.pdf) accessed 20 November 2016
[16] University of Malta 2015 *Minimum Standards for Study-Units in The VLE* (www.um.edu.mt/vle/staff/minimumstandards) accessed 10 January 2017

Assessing the impact of e-learning system of higher education institution's instructors and students

M A P Burac[1], J M Fernandez[2] , M M A Cruz[3], J Dela Cruz[4]

[1]College of Information Technology and Engineering International School of Asia and the Pacific Penablanca, Cagayan, iamgelotron21@gmail.com
[2]College of Computer Studies Panpacific University Urdaneta City, Pangasinan, Janefernandez2122@gmail.com
[3]Institute of Engineering and Applied Technology Bulacan Agricultural State College Bulacan, Philippines, melanieablazacruz@gmail.com
[4]College of Information Technology and Computer Science, University of the Cordilleras, Baguio City, delacruzpen@gmail.com

Abstract. Through technological and globalization innovation, the traditional mode of instruction in higher education institutions has been transformed through e-learning. It is a new educational trend which created a new medium in the delivery of instruction. The researcher attests by determining the e-learning systems influence on students and instructors is vital to the advance implementation of effective and suitable in e-learning systems. In order to know the perceptions of the students and instructors from higher education, the researcher conducted a survey regarding their current experiences. The main purpose of the study was to investigate the impact of e-learning in higher education. This study shall document the e-learning systems impacts which had on Instructors and Student participant's performance about the level of user satisfaction and productivity, utilization and problems encountered. The study found out that the use of e-learning systems shows a positive influence on student learning. Most instructors utilized e-learning system as presentation and preparation tool in teaching and learning. Evidently, most instructors positively confirm that e-learning supports teaching and learning effectively implemented.

1. Introduction

One of the trending educational tools in the market today is the E-learning system which is used as an alternative to the traditional learning. E-learning connects two areas like learning and with the use of technology. Nowadays online teaching and learning are used in other prestigious universities. In achieving knowledge, a combination of learning and use of educational technology is a component of the learning process. E-learning systems have faster delivery cycle time than traditional classroom-based instruction because system combined with tools such as writing technologies, communication technologies, visualization and data storage. [1]. As of now the perception of conducting classes is face to face interaction inside the classroom with teacher and group of students has been the common practice in some schools, however with the emergence of computer technology and internet the traditional approach is already starting evolving into e-learning. [2]. With the new approach and environment for learning the used of electronics networked allowed learners to interact between instructors, teachers and peers will receive individualized support that is suitable for the students [3]. E-learning will also support the scheduling and time management for the learning process of the learners.

This research study assessed the impact of the e-learning systems used in three universities in the Philippines namely University of the Cordilleras, Panpacific University and Bulacan State University. The main purpose of this study is to evaluate the learner's perception of the frequency and nature of use of e-learning, Instructors satisfaction with access to e-learning resources and Instructors views on the impact of e-learning on learners' experiences of higher education.

The traditional process of teaching in other developing country has been started to evolve with the new technologies and other learning tools [4]. A study showed that the learning process in Ajman University with the used of Learning Management System there is a positive response and increase collaboration of learners and peers [5]. With the use of information and communication technology e-learning has been started to develop because it helps a lot especially on distance learning where e-learning is one of the solutions. Colleges and Universities use e-learning to improve the classroom experience and to deliver over the Web. E-learning is a learning process that is increasingly being used in an institution, however, e-learning is still at the primary stage and the adoption of e-learning continues to rise [6].

Nowadays, the use of electronic learning is not restricted to the students. Educators and institutions can also realize the great benefits from the adoption of e-learning programs like students can perform better when online learning is incorporated because the student can study the lesson in advance and will increase interaction and collaboration of students, peers, and instructors compared to traditional learning. The use of e-learning can lower the operating cost while also lowering the costs associated with the production of materials for the activities, hand-outs and laboratory manuals. Today's students and instructors engaged to work on their mobile devices and other digital tools. By incorporating technology into our daily activities, classroom instruction, educational materials it can give more improvements in student learning [7]. The use of e-learning has a great help rather than the traditional techniques [8].

The challenges of educators and learners is the lack of knowledge and skills of how to use the e-learning tools, however proper dissemination, focused, more time and efforts in using the e-learning and with the help of the top management for proper implementation with enough physical infrastructure will create a better opportunity for the user [9].

E-learning offers many opportunities for students in their learning process, but it has challenges for Instructors who forced to create new teaching methodologies [10]. Information technology tools for e-learning are available on the Web, however, engagement and more practice in using the e-learning will have a better opportunity to become an effective learner. [11].

Thus, this paper presents the results of the study and their indications for the impact of e-learning system on students and Instructors for higher education institution in University of the Cordilleras, Panpacific University and Bulacan State College.

2. Methodology

To attain the objectives of this study, the researchers used the descriptive survey method. It is applicable to this study primarily because it describes the influence of the E-learning system to higher education institution instructors and students using quantitative data. The respondents of the study came from the 17 instructors and 80 students of different schools namely Panpacific University, Urdaneta City, Bulacan Agricultural State College, Bulacan and University of the Cordilleras Baguio City. A questionnaire was used to gather data for this study. The tool used for this research has been lifted from the study "Impact of e-learning in Further Education: Survey of Scale and Breadth" [1]. The data gathering tool consists of the demographic profile of the respondents, questions on how e-learning system is being utilized, the number of occurrences in utilizing the e-learning system, the teachers' insight on the use of e-learning and the level of satisfaction. The researchers will then give the questionnaire among the respondents, give them ample time to answer it and retrieve the data gathering tool upon completion. The researchers then tabulated the data; interpreted the data. Response to the questionnaires by the respondents was statistically analyzed; descriptive statistics which is the frequency count and percentage distribution were used to statistically quantify the satisfaction and impact of using e-learning for teachers and students.

3. Results and Discussion

As gleaned in table 1, most students utilize e-learning system to submit their assignments or work all the time which represents 47% of the respondents as supported in the study entitled 'Student Awareness Towards E-Learning in Education' that the application of e-learning helped the students in doing their homework's efficiently and effectively [12]. It also noticeable that students perceived to be that they are well organized in their work as they responded frequently (52%). Students occasionally present written work/data research topic to the e-learning system they used most probably because they want to personally interact with the instructor. It is also worth noting that students collaboratively work with peers in the classroom.

As gathered in table 2, the instructors mostly utilized e-learning to prepare and present their lessons. It is noticeable that teachers perceived to be that e-learning are used to make course materials available to students and present it to the class through these students understanding their lessons are being developed. Additionally, giving assistance or academic counseling is also being performed using the system. One feature of the e-learning is the forum in which it gives the students an avenue to discuss and solve their academic concerns. Thus, learning is implemented in this very effective medium [13].

Table 1. Learners' insight into the number of occurrences based on how e-learning is used

use of e-learning to:	Constantly %	Frequently %	Occasionally %	Never %	No response %
Present written work/data Research topics	21	42	37	0	0
Submission of work assignment on time	47	29	23	0	2
Participation in classroom activities	29	45	23	0	3
Organize activities	31	52	13	0	5
Consultation with teachers	29	34	31	3	3
Collaboration for class activities	21	47	32	0	0

Table 2. The frequency of use of e-learning by Instructors

use e-learning to:	Constantly %	Frequently %	Occasionally %	Never %	No response %
make course materials available to learners	75	19	6	0	0
discuss information in the class	75	19	6	0	0
develop learners' understanding of the subject	69	25	0	6	0
communicate with learners outside of the classroom	56	38	6	0	0
test learners' understanding	50	31	13	0	6
give one-to-one consultation to students in the classroom	31	50	13	6	0

Table 3, as shown below, instructors perceived to be that most learners utilized e-learning system for them to engage in the class. Students also used it in presenting their output such as researches and other written works. It is also being noticed that students used the e-learning system to catch up with their lessons. However, working collaboratively of students to their peers is occasionally practiced. E-learning provided a new form of learning where it uses internet technologies. It does not require the attendance of the teacher and the students in the educational activities at the same time [13].

Table 3. Instructors' insight into the number of occurrences based on how e-learning is used

Students use e-learning to:	Constantly %	Frequently %	Occasionally %	Never %	Don't know %	Not applicable %
Discuss written work/data Research topics	13	63	25	0	0	0
Create visual presentations	38	50	13	0	0	0
Submit homework's' on time	44	50	6	0	0	0
Engage with the subject in the classroom	63	31	6	0	0	0
Work collaboratively with peers in the classroom	38	38	25	0	0	0
Catch up on missed lectures	44	31	25	0	0	0

Table 4 shows the satisfaction if instructors in terms of the access to e-learning resources. It can be observed that instructors had enough access to the e-learning system for them to design their research and make their lessons. However, they have less fulfillment with the use of e-learning to communicate and provide support for their learners. It is evident that the learning experiences of the students were efficiently and effectively enhanced with the utilization of blended learning [14].

Table 4. Instructors' approval with access to e-learning resources

Enough resources to	I have enough access %	I have less access %	I have no access %	Unsure %
Design research and make lessons	69	31	0	0
Share course materials with colleagues	56	38	6	0
Share course materials with learners	50	50	0	0
Communicate and provide support for learners	38	63	0	0
Monitor and assess learners' progress	44	56	0	0
Ensure that learners can use e-learning whenever required in the classroom	63	38	0	0

Table 5 shows, that, generally, instructors are contented with the sustenance available in terms of delivery of technical support and access to enough and relevant e-learning training. Additionally, there is also satisfaction in the availability in incorporating learning into the process. Satisfaction in learning using blended learning environment greatly affects students learning styles and performance [15].

It can be observed that instructors have a positive response when asked about their present involvement and training in using e-learning. As can be gleaned in table 6, 63% percent of the instructors strongly agreed that they have enthusiasm in using e-learning in teaching and learning, 50% of them agreed, to some extent, that they can recognize opportunities in the use of e-learning. On the other hand, there is a certain 13% percentage of the instructors that felt e-learning consumes so much time in teaching and risks and issues has been created. This means that instructors must have much time as they must prepare and upload all materials to be used by the students in the e-learning.

Table 5. Contentment with the availability of assistance in the use of e-learning

Contentment with support	Very Contented %	Contented %	Uncertain %	Dis-Contented %	Very Dis-Discontented %
Provision of technical assistance	25	56	19	0	0
Enough access and relevant e-learning training	31	56	13	0	0
Provide time for attending e-learning training opportunities	19	50	25	6	0
Provide time to incorporate e-learning into teaching and learning	25	56	19	0	0
Reliability of equipment	31	50	13	6	0

Table 6. Instructors' views on the knowledge of using e-learning

Instructors' opinions on e-learning	Strongly Agree %	Agree %	Uncertain %	Disagree	Strongly Disagree
I show enthusiasm about the use of e-learning in teaching and learning	63	38	0	0	0
I can willingly recognize opportunities in my subject for the use of e-learning	50	44	6	0	0
I show determination to use e-learning to its full potential	31	56	6	0	0
E-learning is too overriding to use in teaching and learning	25	31	25	13	6
E-learning has several risks	19	38	25	13	6

It can be seen in table 7, that instructors have the most similar views on the impact of e-learning in teaching and learning activities. Most of the instructors responded that e-learning has helped them in making the course materials in their subject available to students which they find it more effectively. Additionally, Instructors perceived to be that communicating with students outside the classroom and sharing course materials to their colleagues the impact of e-learning is more effective. On the other hand, there is a small percentage of instructors who perceived on the impact of e-learning less effectively in managing individual student activities and outputs and giving them assistance one-to-one.

Table 7. Instructors' views on the effect of e-learning on their teaching

Teaching activity	Very effective%	No Change %	Less effective%	unsure %	Not applicable %
Make course materials available to learners	94	6	0	0	0
Course materials sharing with colleagues	69	19	19	0	0
Lesson plan preparation	75	19	6	0	0
A medium of communication to the students	69	25	0	0	6
Management of individual activities of the students	63	19	13	0	6
give one-to-one consultation to students in the classroom	50	31	13	0	6

It can be gleaned in table 8, that instructors view on the impact of e-learning on learners' experiences shows that, it developed students understanding of the subject more effectively which represents 75%

of the total response. It also impacts the students conduct of research, helped reinforcement in their knowledge, create visual presentations more effectively. Exercises and activities provided in the system made the students self-reliant, involved in working collaboratively and have good time management [16]. E-learning became an effective tool in training teachers. It promotes improved teaching and learning process [17].

Table 8. Instructors' views on the effect of e-learning on learners' experiences

Learning activity	Very effective%	No change %	Less effective %	Don't know %	Not applicable %
Doing research	69	6	0	0	6
Reinforce their knowledge	63	13	0	6	0
Develop their understanding of the subject	75	13	0	0	0
Work independently Contact me with queries	63	13	6	0	0
Engage with the subject in the classroom	63	0	13	6	0

Table 9. Instructors' perceptions of the possible advantage of e-learning

E-learning has the possible advantage to:	Strongly Agree %	Agree %	Uncertain %	Disagree %	Strongly Disagree %
Provide flexibility in learning the process	94	6	0	0	0
Provide students with a better understanding of the lessons	81	13	6	6	0
Provide students better employability	56	38	6	6	0
improve communication of teachers and students.	81	13	6	0	0
Identify learning to individual learners' needs	75	25	0	0	0
save lecturers' time by using online resources	81	19	0	0	0

Most respondents strongly agreed that e-learning has a higher possibility to enhance the teaching and learning experience in all the aspects listed in table 9. Ninety-four (94%) of the respondents strongly agreed that e-learning provides flexibility in the learning process. Also, 81% have a strong belief that e-learning help students better understand their lessons, improve how staff communicate with learners and saved lecturers time by using online resources. Importantly, E-learning improves the educational system as it provides better teaching and learning experiences. It bridges a way to alleviate barriers to education as it delivers new and creative method to implement educational practices [16].

4. Conclusion

The focus of this research is to assess the impact of e-learning on teachers and students in higher education. Based on the result on the conducted survey, different Instructors e-learning had aided them to be more effective in their teaching strategy. Many Instructors point out that e-learning had helped them to prepare their teaching and learning scheme. Also, they used this application to prepare their presentational tool to be more effective in giving lessons and learning materials to their students. E-learning tools also aid the instructor in strengthening the interface between learners and lectures. Based on the result of the survey, noticeably fewer indicated that e-learning tools helped them as a medium for assisting and handling learning more effectively. However, numerous instructors stated that this particular tool had helped instructors in preparing learning materials and presentation to be more convenient of work in strengthening and developing knowledge. Based on the result of the survey, it shows that fewer instructors point out that this tool helped both instructors and students in undertaking collaborative activities.

To further maximize the potential use of e-learning, enhancement of faculty multimedia skills must be undertaken by providing them training about how to use multimedia resources for them to create their own media for additional teaching materials which will be made available to the learners. Greater use of e-learning tools by instructors for sharing educational materials and developing understanding requires good access. The benefits of e-learning tools, such as improved performance, engagement in school activities, increased motivation, and development of other skills like interpersonal communication, let particular indirect ways in which this certain tool may impact on achievement, contribution, and retention [18].

5. References

[1] Aparicio M, Bacao F and Oliveira T 2016 An e-learning theoretical framework *Educational Technol. & Society* **19(1)** pp 292–307

[2] Alkhalaf S, Alhussain T and Drew S 2012 Assessing the impact of e-learning systems on learners: a survey study in the KSA *Procedia - Social and Behavioral Sciences* vol **47** pp 98-104

[3] Tao Y H, Yeh C R and Sun S I 2006 Improving training needs assessment processes via the Internet: system design and qualitative study *Internet Res.* **16(4)** pp 427–49

[4] Wang Y S, Wang Y M, Lin H H and Tang T I 2003 Determinants of user acceptance of internet banking: an empirical study *Int. J. of Service Industry Management* **14** pp 501–19

[5] Shishakly R 2016 Significance of Learning management systems (Moodle) on the students of Ajman University, United Arab Emirates *IAMURE Int. J. of Educ.* **18(1)** retrieved from http://ejournals.ph/form/cite.php?id=11654

[6] Holmström T and Pitkänen J 2012 *E-Learning in Higher Education a Qualitative Field Study Examining Bolivian Teachers' Beliefs about E-Learning in Higher Education* Bachelor Thesis (Department of Education, Umeå University)

[7] Qiyun Wang, Zhiting Zhu, Li Chen and Hanbing Yan 2009 E-learning in China *Campus-Wide Information Sys.* vol **26** Issue **2** pp.77-81 https://doi.org/10.1108/10650740910946783

[8] Macgregor G and Turner J 2009 Revisiting e-learning effectiveness: proposing a conceptual model *Interactive Technol. and Smart Educ.* vol **6** Issue **3** pp 156-72 https://doi.org/10.1108/17415650911005375.

[9] Sawang S, Newton C and Jamieson K 2013 Increasing learners' satisfaction/intention to adopt more e-learning *Educ. + Training* vol **55** Issue **1** pp 83-105 https://doi.org/10.1108/00400911311295031

[10] Mainka C and Benzies A 2006 E-learning: vision to reality *Interactive Technol. and Smart Educ.* vol **3** Issue **2** pp 101-11 https://doi.org/10.1108/17415650680000056

[11] Mohammad M 2012 The impact of e-learning and e-teaching *Int. Sci. Index, Educational and Pedagogical Sci.* vol **6** no **2** waset.org/Publication/194

[12] Yacob A, Kadir A Z A, Zainudin O and Zurairah A 2012 Student awareness towards e-learning in education *Procedia – Soc. and Behavioral Sci.* **67** pp 93–101

[13] Yucel A S 2006 E-learning approach in teacher training *Turkish Online J. of Distance Educ. (TOJDE)* vol **7** no **4** Article 11 ISSN 1302-6488

[14] Garrison R and Kanuka H 2004 Blended learning: uncovering its transformative potential in higher education *The Internet and Higher Educ.* Vol **7** Issue **2** 2nd Quarter 2004 pp 95-105

[15] Wu J H,Tennyson R D and Hsia T L 2010 A study of student satisfaction in a blended e-learning system environment *Comp. & Educ.* vol **55** Issue **1** pp 155-64

[16] Tegenge K 2014 The Influence of E-Learning on the academic performance of mathematics students in fundamental concepts of algebra course: the case in Jimma University *Ethiop. J. Educ. & Sc.* vol **9** no **2**

[17] Ghavifekr S and Rosdy W A W 2015 Teaching and learning with technology: effectiveness of ICT integration in schools *Int. J. of Res. in Educ. and Sci. (IJRES)* vol **1(2)** pp 175-91
[18] Harris R, Hall J, Muirhead A, McAteer E, Schmoller S and Thorpe G 2004 *Impact of E-Learning on Learner Participation, Attainment, Retention, and Progression in Further Education: Report of A Scoping Study* (University of Strathclyde Glasgow)

Categorizing mobile user interface patterns using EFA

Almed Hamzah[1]

[1]Department of Informatics, Universitas Islam Indonesia
Jalan Kaliurang KM 14,5 Sleman Yogyakarta Indonesia

Email: almed.hamzah@uii.ac.id

Abstract. A mobile user interface is a specialized form of a user interface as it runs on mobile devices which has some limitations. It must be designed in a simple and engaging appearance. There is a guideline available that consists of eleven specific patterns of mobile user interface design to achieve this objective. This paper aims to regroup these patterns into smaller groups so that there is a simpler guideline. To do so, the Exploratory Factor Analysis (EFA) approach is being used. The result shows that the eleven patterns could be grouped into four categories. These categories depict the spectrum of mobile user interface design from mandatory to optional. A prototype is built based on this categorization.

1. Introduction

The mobile application has several characteristics that differentiate it from a conventional PC-based application. These kind of application are popular in several areas such as traveling [1], healthcare [2], [3], and e-commerce [4], [5]. The interaction between user and application takes places in a fast and limited environment. This is mainly because the mobile hardware is smaller as a trade-off to mobile ability, especially the screen size of the device [6], [7] and the battery size [8]. Moreover, it becomes a means to support people doing their task every day [9]. However, there are some advantages of a mobile application. According to [10], mobile applications have main characteristics such as pervasive, useful, affordable, personal, and context-aware. It also becomes a means to deliver contents and functions of mobile applications [4].

Mobile User Interface (MUI) is an interface of an application that runs on mobile devices. The application itself will operate the mobile devices in terms of its hardware [9]. Designing mobile application, including the interface, requires different pattern than the other type of application. Based on previous research [11], Mobile User Interface (MUI) should be simple, consistent and requires less user cognitive ability. In addition, any content that displayed on the screen should be the most important one to the user [7]. This will allow the user to finish their task quickly. It should be developed so that the user can interact easily, naturally, and engaged in the application [12]. Moreover, a well-designed user interface is directly correlated with the application's usability [4].

In [7] the interface of a mobile application is characterized by eleven patterns which are Vertical Stack, Filmstrip, Touch Tools, Bottom Navigation, Thumbnail-and-Text List, Infinite List, Generous Borders, Text Clear Button, Loading Indicators, Richly Connected Apps, Streamlined Branding. The aim of this paper is to categorize these patterns further so that there is a simpler understanding of the patterns which will, in turn, assist the developer in focusing the work of mobile user interface development. Exploratory Factor Analysis (EFA) is used to categorize the already known mobile user interface pattern. It is a common tool to reveal the unknown characteristics or factors from the known one [13].

This paper consists of five sections. After the introduction part, section two covers literature review about the development of a mobile user interface based on previous publications. Section three describes the methods employed in this research. Section four explains the research results followed by section five that will end this paper with concluding remarks and future works.

2. Related Works

Mobile devices are becoming a popular tool that has been used to finish various daily task [14]. A mobile application needs to appear beautifully in a mobile device [15] even though there is some limitation regarding to screen size, battery life, processing power, and memory capability. A good application design could increase the user's engagement in terms of social interaction and experience [1]. In addition, the websites or applications that run on the mobile devices should be able to provide various information in a various form regardless of the limitation. The mobile user interface is a part of a mobile application. Its function is crucial in terms of the interaction of the application with the user. The design of a mobile user interface becomes more complicated as the variety of contents and functions are increasing [4].

In [6] the pattern of a mobile user interface is divided into three main dimensions, i.e. utilizing screen space, interaction mechanism, design at large. Meanwhile, other works are categorizing mobile devices based on CPU, communication resources and LCD usage [8].

From the literature [7], it has been known that there are eleven patterns that specifically characterized mobile user interface, i.e. Vertical Stack, Filmstrip, Touch Tools, Bottom Navigation, Thumbnail-and-Text List, Infinite List, Generous Borders, Text Clear Button, Loading Indicators, Richly Connected Apps, Streamlined Branding.

Vertical Stack means that all the elements in the interface should be arranged in a vertical order to anticipate the various screen width of mobile devices. Filmstrip means that all elements containing the same kind of information should be arranged horizontally in multiple pages. Touch tools mean that the interface should only expose the touch interaction elements in order to help the user focus on the elements.

Bottom navigation means that the main application navigation menu should be arranged at the bottom of the screen. This is quite different with the PC-based application or website that placed the main navigation on the top of the page. Thumbnail-and-text List means that the element chosen by the user should have an image as a thumbnail to represent the text related to the element.

Infinite list means that all the elements that categorized in one group should be arranged infinitely vertical. Generous borders mean that every element should have enough space in its surrounding to separate it with other elements so that user can easily identify and interact with those elements. Text clear button means that the application should provide a button to delete all the contents in the textbox.

Loading indicators means that the application should show a loading sign when a particular element is being downloaded. This will help the user to know that application is still working to perform a task. Richly connected application means that the application should be able to interact with other application by providing an initial and basic data during the interaction process. Streamlined branding means that every identity element should in a small size so that it can be loaded and appear quickly in order to help the user easily recognize the application.

3. Methods

Ten groups of observer consist of 4-5 people each is taking a role as examiner of the mobile appearance of popular sites. There are 32 sites that have been visited. Most of the sites are social networking sites and news portal.

These websites are then being analyzed by observers to identify whether they have characteristics of a mobile user interface or not. The evaluation is based on mobile user interface pattern published in [7].

After examine and interact with the mobile version of the application, the observers than decide the existence of patterns. If the pattern exists, they will score the websites with 1 value, otherwise 0.

After that, we held an exploratory factor analysis (EFA) to categorize the eleven factors into several groups. There are five steps in EFA as adopted from [16] and depicts in Figure 1.

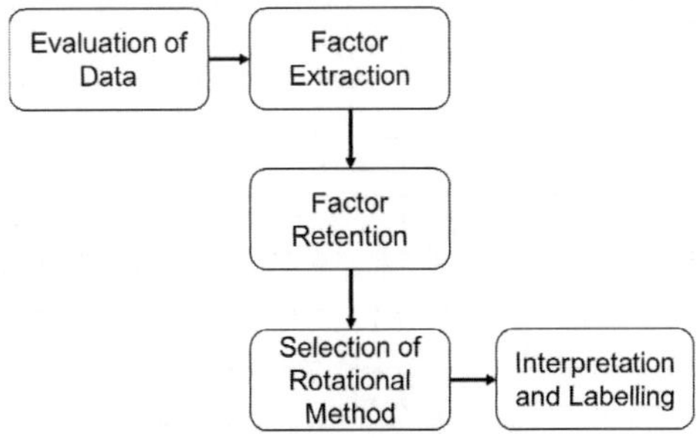

Figure 1. Steps of EFA

According to [16], the first step of EFA is the evaluation of the data. This is mainly a process to determine whether the data is appropriate for EFA analysis. The data will be examined using KMO (Kaiser-Meyer-Olkin) and Bartlett's Test. KMO evaluates the correlation between an item with another item in EFA. The score of KMO is between 0 and 1. The bigger score indicates the stronger correlation between the particular item with another item. The KMO test is supported by Bartlett's Test that evaluates the relationship's strength among variables. The significance score of Bartlett's Test should be less than 0.05 that shows that there is a strong relationship.

The second step is factor extraction. This is the main process of EFA. The result from this phase is a new model of categorization. In this research, the principal component analysis is employed as it is the most used method. The third step is factor retention. The objectives of this phase is to determine which factors are maintained for the next step with the cumulative percentage of variance test.

The fourth step is a selection of the rotational method. This step aims to identify the correlation between a variable with more than one construct. The most used method for this step is the Varimax rotation. Finally, the last step is the interpretation which labeling the group of constructs from the EFA process. All the process is done by SPSS.

4. Result and Discussion

This research used 32 data as a source for EFA analysis. Based on KMO and Bartlett's Test, this source is adequate for EFA. As depicted in Table 1, the score of KMO reaches 0.579. The minimum KMO value allowed to run EFA is 0.5 [16]. The score is relatively low since it is close to the minimal value. However, it is still enough to run EFA. The Bartlett's Test shows a significance score that less than 0.05. Thus, there is a strong relationship between variables under evaluation of EFA.

Table 1. EFA Results

Kaiser-Meyer-Olkin Measure of Sampling Adequacy.		.579
Bartlett's Test of Sphericity	Approx. Chi-Square	101.046
	df	55
	Sig.	.000

After six iterations of EFA using SPSS 16, there are four groups of mobile user interface patterns coming up as depicted in Table 2 that will labeled and interpreted in the following section. Each pattern is categorized based on the highest loading factor within a group.

Table 2. EFA Results

No	Patterns	Categories			
		1	2	3	4
1	Vertical Stack	**0.682**			
2	Filmstrip		**0.730**		
3	Touch Tools		**0.775**		
4	Bottom Navigation	**0.768**			
5	Thumbnail-and-Text List		**0.528**		
6	Infinite List		**0.779**		
7	Generous Borders	**0.805**			
8	Text Clear Button			**0.716**	
9	Loading Indicators			**0.863**	
10	Richly Connected Apps				**0.875**
11	Streamlined Branding	**0.693**			

The EFA result is returning four groups of user interface pattern. Group 1 consists of four patterns, i.e. vertical stack, bottom navigation, generous borders, and streamlined branding. Group 2 consists of four patterns, i.e. filmstrip, touch tools, thumbnail-and-text list, infinite list. Group 3 consists of two patterns, i.e. text clear button, loading indicators. Group 4 consists of one pattern, i.e. richly connected applications.

Based on the result above, these four groups could be seen as the design spectrum of user interface design.

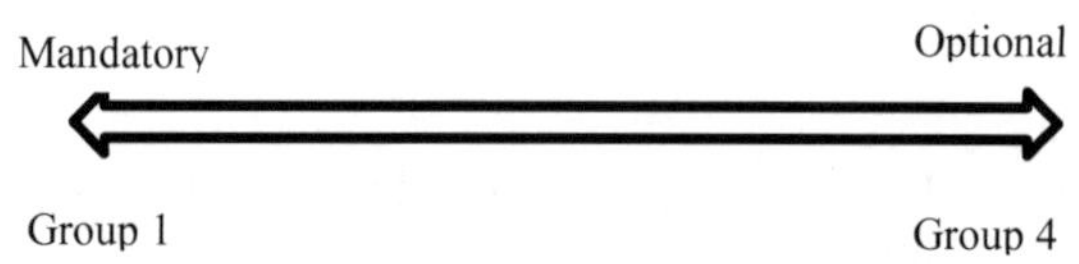

Figure 2. The spectrum of MUI design

As depicts in Figure 2, the spectrum spans from mandatory in Group 1 to optional patterns in Group 4. Mandatory means that the patterns should be adopted in the application interface while the optional means that the pattern is adopted when it needed only.

5. Case Study

A case study is conducted to implement the result from the previous section. A prototype of a mobile application was built based on four new categories. The development of the prototype is focusing of the design of mobile user interface and to put the required elements in an appropriate way. The application that was built here is mobile learning. It has several main features which include downloading course material, monitor the announcements, doing online quizzes and also monitor their score.

The development of the prototype is divided into four phase. Phase 1 consists of four activities, i.e. designing all elements in a vertical stack, put all the link of main navigation in a bottom of the screen which is in a fixed position. Then, make sure to spare enough space for the finger to click on the required elements. Finally, put the application logo on small-sized elements. The result of phase 1 is depicted in Figure 3.

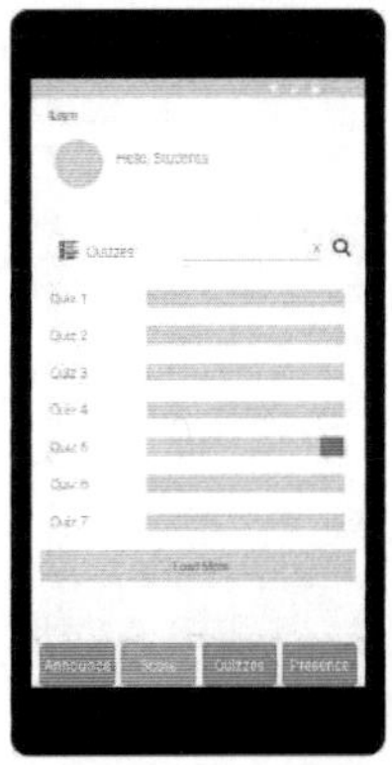

Figure 3. The user interface from step 1 (Overall look)

Phase 2 consists of the design of similar screen that must be displayed parallel in filmstrip format, provide a prioritizing interface on touch elements, put a thumbnail on the side of each text element to help the user better understand the element, and finally create the infinite list of contents. The result from this phase is depicted in Figure 4.

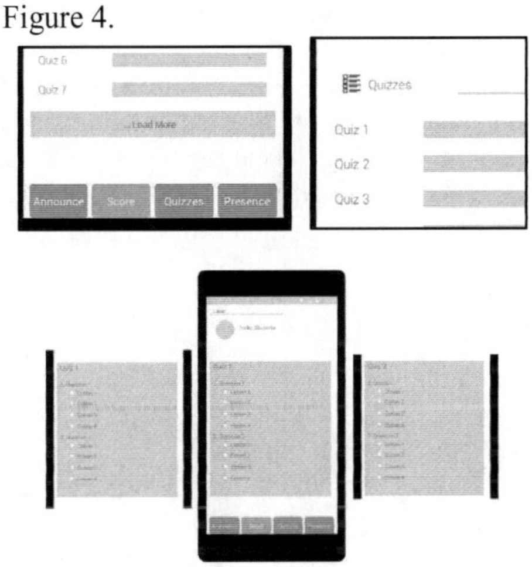

Figure 4. Phase 2

In phase 3, text clear button and loading indicators are added to the design. It is located in the component that needs a search feature, such as multiple page quizzes as depicted in Figure 5.

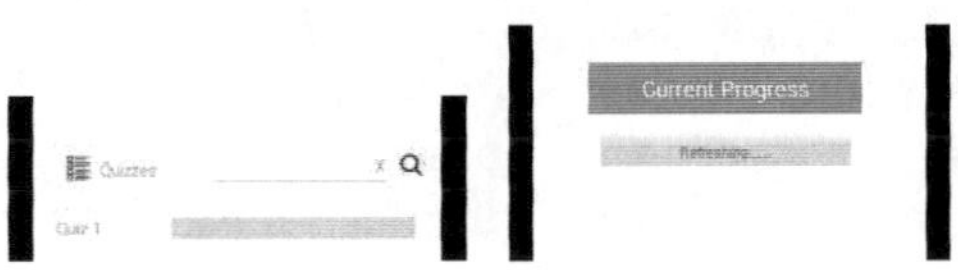

Figure 5. Phase 3

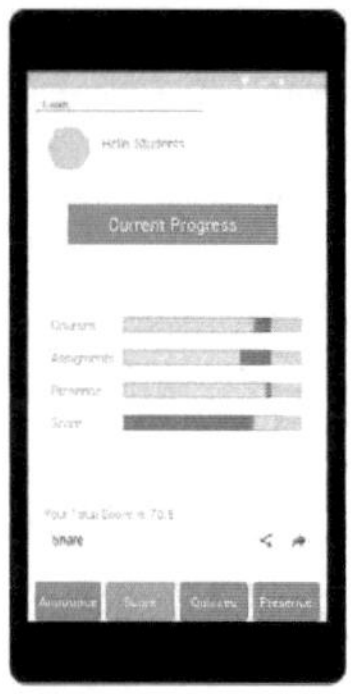

Figure 6. Phase 4

The last phase is to connect the application with other application by capturing and exchanging basic application data, such as user profile. This could be done by utilizing embedded phone feature as a hub between applications. The design from this phase could be seen in Figure 6.

6. Conclusion

Mobile user interface design needs a different approach than conventional PC-based application. The main constraint of the design is the devices' screen size. Moreover, it needs to be simple so that the content to appear can be loaded quickly. According to EFA process conducted in this research, there is a spectrum of mobile user interface design from the most mandatory features to the optional features.

Future research will be done on expanding the application under evaluation. There are various categories of application that will enhance the analysis result. Moreover, the study could be done to make an inquiry regarding the effect of a well-designed mobile user interface on the usability of the application as a whole.

References

[1] Fang J, Zhao Z, Wen C and Wang R 2017 Design and performance attributes driving mobile travel application engagement *Int. J. Inf. Manage.* **37**(4) pp 269–83 available from: http://dx.doi.org/10.1016/j.ijinfomgt.2017.03.003

[2] Kao HY, Wei CW, Yu MC, Liang TY, Wu WH and Wu YJ 2018 Integrating a mobile health applications for self-management to enhance Telecare system *Telemat. Informatics* **35**(4) pp 815–25 available from: https://doi.org/10.1016/j.tele.2017.12.011

[3] Sarwar CMS et al 2018 Mobile health applications in cardiovascular research *Int. J. Cardiol.* available from: https://doi.org/10.1016/j.ijcard.2018.06.039

[4] Jung W 2017 The Effect of representational UI design quality of mobile shopping applications on users' intention to shop *Procedia Comput. Sci.* **121** pp 166–9 available from: https://doi.org/10.1016/j.procs.2017.11.023

[5] McLean G, Al-Nabhani K and Wilson A 2018 Developing a mobile applications customer experience model (MACE)- implications for retailers *J. Bus. Res.* **85** (February) pp 325–36 available from: https://doi.org/10.1016/j.jbusres.2018.01.018

[6] Nilsson EG 2009 Design patterns for user interface for mobile applications *Adv. Eng. Softw.* **40(12)** pp 1318–28

[7] Tidwell J 2010 *Designing Interfaces: Patterns for Effective Interaction Design* (O'Reilly Media, Inc.)

[8] Mehrotra D, Srivastava R, Nagpal R and Nagpal D 2018 Multiclass classification of mobile applications as per energy consumption *J. King Saud Univ. – Comput. Inf. Sci.* available from: https://doi.org/10.1016/j.jksuci.2018.05.007

[9] Xu K, Zhang W and Yan Z 2018 A privacy-preserving mobile application recommender system based on trust evaluation *J. Comput. Sci.* **26** pp 87–107 available from:

http://linkinghub.elsevier.com/retrieve/pii/S187775031731428X

[10] Aluko R 2017 Applying UNESCO guidelines on mobile learning in the South African context: Creating an enabling environment through policy *Int. Rev. Res. Open Distance Learn.* **18(7)** pp 24–44

[11] Alnanih R and Ormandjieva O 2016 Mapping HCI Principles to design quality of mobile user interfaces in healthcare applications *Procedia Comput. Sci.* **94** pp 75–82 available from: http://dx.doi.org/10.1016/j.procs.2016.08.014

[12] Kamaruzaman MF, Rani NM, Nor HM and Azahari MHH 2016 Developing user interface design application for children with autism *Procedia – Soc. Behav. Sci.* **217** pp 887–94 available from: http://linkinghub.elsevier.com/retrieve/pii/S1877042816000471

[13] Tarute A, Nikou S and Gatautis R 2017 Jul Mobile application driven consumer engagement *Telemat. Informatics* **34(4)** pp 145–56 available from: https://www.sciencedirect.com/science/article/pii/S0736585316307006

[14] Kalimullah K and Sushmitha D 2017 Influence of design elements in mobile applications on user experience of elderly people *Procedia Comput. Sci.* **113** pp 352–9 available from: http://dx.doi.org/10.1016/j.procs.2017.08.344

[15] Charland A and Leroux B 2011 Mobile application development : web vs native *Commun. ACM* **54** pp 0–5 available from: http://search.ebscohost.com/login.aspx?direct=true&db=bth&AN=60863975&site=ehost-live

[16] Taherdoost H, Sahibuddin S and Jalaliyoon N 2014 Exploratory factor analysis: concepts and theory *2nd Int. Conf. Math. Comput. Stat. Sci.* pp 375–82

Investigating students' engagement in a hybrid learning environment

A Eliveria[1], L Serami[2], LP Famorca[3], JS Dela Cruz[4]

[1] Eastern Samar State University, Guilian, Eastern Samar, Philippines
[2] King's College of the Philippines, Bambang, Nueva Vizcaya, Philippines
[3] Saint Louis University, Baguio City, Philippines
[4] University of the Cordilleras, Baguio City, Philippines

[1]aidaf.ae95@gmail.com, [2]leodser@gmail.com, [3]lpfamorca@slu.edu.ph, [4]delacruzpen@gmail.com

Abstract. Higher learning institutions (HEIs) are using different learning modalities to encourage better performance from students in accomplishing course requirements. A hybrid learning environment introduces opportunities for educational leaders, teachers and learners in finding alternative approaches to enhance traditional brick and mortar setting. A hybrid learning environment was introduced to a group of information technology (IT) and computer sciences (CS) students through a learning management system. The mixed methods study sought to investigate students' participation and enthusiasm in a hybrid learning environment. A survey was presented to the students at the end of the semester. The resulting analysis revealed that students prefer online activities but acknowledged that in-class activities help students learn and understand course work. Results further revealed students' insight of how a successful hybrid learning environment should be handled. The results from this research can provide understanding to help guide future attempts of a hybrid learning environment.

1. Introduction

The persistent improvement of the internet and the continuous development of new knowledge to be shared and delivered has afforded higher education institutions (HEIs) another means to improve and expand the delivery of curricula to learners. The conduct of education in the 21st century applies technology in the learning process. New learning environment and teaching technologies use information and communication technologies such as e-learning, open and distance learning, web-based learning, blended learning and hybrid learning have been introduced in many courses. It allows learners to learn anywhere, usually at anytime, as long as they have a computer and e-learning application [1]. [2] claimed that systematic and strategic combination of ICT tools into academic courses introduces a new way to approach instructional goals.

Hybrid learning, also known as blended learning, is a new approach in delivering programs to learners. As discussed by [3], hybrid learning offers an increased flexibility for students with the online component while maintaining the personal connection with teachers and students in the classroom. However, [4] contests that hybrid learning is not realized by a simple transfer of teaching materials online, but rather involves redesign of delivery and pedagogy. Figure 1 shows the model that the study used in delivering a hybrid learning environment. Activities were divided as in-class activities and online activities. A class regularly meets 3 times a week for face-to-face meetings where each meeting is equivalent to 1 hour. In-class activities included lectures, attendance, laboratory exercises, computer-assisted examinations, seatwork, recitation and submission of assignment, case studies, research work and a project. Online activities were posted for the week but activities were controlled to a 1-hour duration. The online component included online quiz, posting of assignment, case study, research work and project specification, repository of lecture materials, posting of related course video materials, homework, links to related websites.

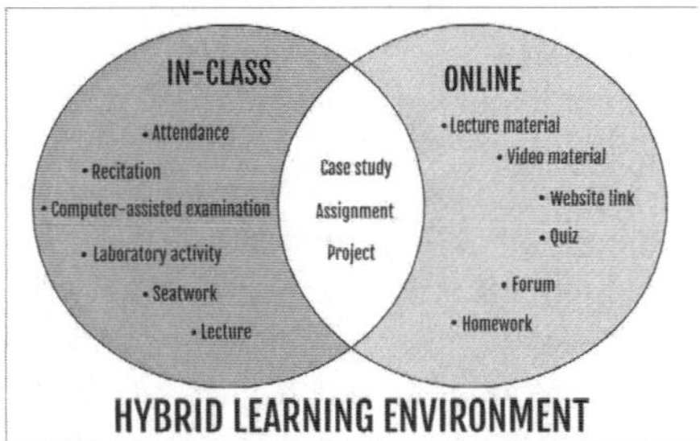

Figure 1. Hybrid learning environment model

Several HEIs in the Philippines have started to implement web-based learning environments capable of delivering online education in a blended learning academic setting [5]. With the current thrust and direction of HEIs towards personalized learning, hybrid learning is gaining popularity [6] and HEIs are using hybrid instruction to improve pedagogy, enhance students' learning outcomes and achievements, and improve cost-effectiveness [7][8]. A hybrid learning environment enables institutions to conduct a learner-centered approach to teaching where learners are given space and flexibility to indulge with effective learning activities [9] [10]. Several local universities are acknowledging the supposed benefits of hybrid learning and educational leaders are consenting educators to investigate further.

This study addresses the general research question: to identify the students' engagement of hybrid learning. Specifically, the researchers would like to look into the significant difference between the level of engagement of students in brick and mortar and learning management system, significant factors that can be drawn from students' engagement in hybrid learning and the relationship between predictor factors and the level of engagement of students.

The results of the study can provide educators the necessary requirements needed to engage students towards a successful delivery of a hybrid learning environment. Educational leaders could use the information to establish policies and guidelines in the delivery of hybrid learning.

2. Related Literature

Hybrid learning became renowned in the early 2000's and many researchers conducted studies to measure its effectiveness. [11] defines hybrid or blended learning as the integration of face-to-face classroom instruction learning with distance/e-learning. [12] added that hybrid learning is a blended or mix of conventional face-to-face instruction and web-based distance learning. Hybrid learning refers to the addition of online learning as a supplement to the brick and mortar face-to-face instructional method [13] [14]. According to [15] effective practice has demonstrated, that the best hybrid instruction allows the students to interact with content and engage in learning activities before, during, and after the face-to-face class. It could be further described as a method of instruction that combines online with face-to-face learning activities that are integrated in a "planned, pedagogically valuable" way and where some of the face-to-face time is replaced by online activities [16]. In a hybrid learning setting, part of the learning activities and assignments are transferred from the face-to-face classroom to the distance learning environment [17].

The prevailing attitude by HEIs has been to introduce and adopt e-learning merely as an addition to face-to-face teaching rather than as an integral component [18]. Apparently this approach introduces many of the problems related with technology use, and with the failure to realize full potential of technologies when combined to educational challenges [18] [19].

A learning management system (LMS) is an option for institutions to implement a hybrid learning environment. Recognizing and identifying face-to-face activities and to be complemented with the right online activities becomes a valuable measurement of hybrid instructional design [20]. Successful integration of face-to-face learning supplemented by online delivery of learning creates environments that are "highly conducive to student learning" [5].

3. Methodology/Procedure

A mixture of qualitative and quantitative research is used in this study. Through a learning management system (LMS), a hybrid learning environment was created and administered to 180 undergraduate students enrolled in information technology and computer science. There were 105 (58%) students in Information Systems Fundamentals class, 24 (13%) in Data Communications and Computer Networks class and 51 (28%) in Network Technology with Administration and Maintenance class for an entire semester which lasted for 16 weeks. The population consisted of 105 first year students, 5 second year students, 66 third year students and 4 fourth year students, of which 106 (59%) students are first time users of the LMS.

Students are required to attended regular face-to-face meetings. During in-class sessions, students were always reminded of the corresponding online component of the in-class activity while the LMS sent corresponding emails to remind students of activities for both online and in-class. The LMS is designed with notification areas where students can verify accomplishments of activities as well as progress indicators within the course. Additionally, students were provided with an option to verify activity completion status. These features were made available to students through the LMS in order to avoid students missing online activities which eventually resulted to missing the corresponding in-class activities. Figure 2 shows how the activities and resources are presented in the LMS. Summative assessment of coursework was performed throughout the semester and grades were computed based on completion rates and academic performance.

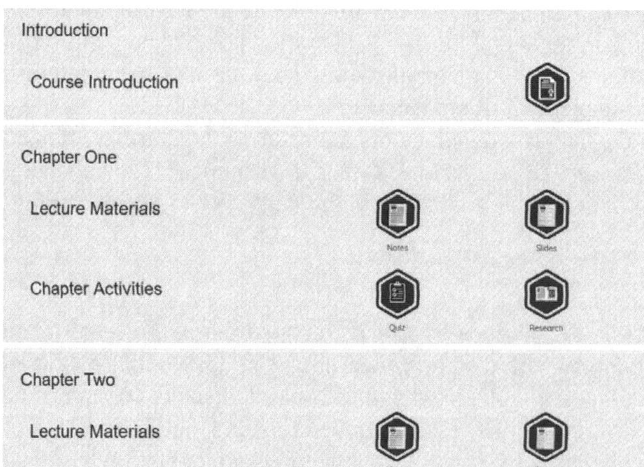

Figure 2. The interface of the LMS showing the different activities and resources

4. Measure

At the end of the semester, an online survey was administered to the students. The instrument was a combination of different e-learning surveys such as Web-Based Learning Environment Instrument (WEBLEI), Constructivist On-Line Learning Environment Survey (COLLES) and The Distance and Open Learning Environment Survey (DOLES). Several items were modified and combined to fit the requirements of the study. The instrument consisted of two parts which are the end of semester survey which comprised of 3 categories and 17 items and the post course survey which comprised of 3 categories and 13 items. To establish strict compliance, the survey was administered as a pre-requisite for students to accomplish in order to gain access to the reviewers and other materials needed for the final exams. A reliability analysis was carried out on the post course 13 items. Cronbach's alpha showed the questionnaire to reach acceptable reliability, α = 0.80. The high reliability of the scales for the survey instrument is most likely due to utilization of recognized surveys.

The results from the end of semester survey and the post course survey were subjected to descriptive analysis obtaining mean and standard deviation and results from the post course survey were subjected to one sample *t*-test. Additionally, results of the survey were introduced to IBM Watson Analytics for further analysis and interpretation of the relationship of the items. To obtain a more personal observation, students were asked to comment on the course using the hybrid learning environment. Results from students' comments were subjected to cleaning and text analysis. RapidMiner Studio trial version was used to perform the text analysis.

5. Results and discussion

The methodology used was a combination of qualitative and quantitative techniques. Quantitative analysis was performed through one sample t-test and hypothesis testing. Performing a one-sample *t*-test on the post course survey revealed that there is a significant difference on student's level of engagement between traditional brick and mortar and learning management system. Results revealed student satisfaction on on-line activities over the in-class activities, however results revealed that students acknowledge that in-class activities helped them learn and understand the topics more than on-line activities. Table 1 shows the values from the *t*-test.

Table 1. One-sample *t*-test results

	N	Mean	SD	SE
I liked the online activities	180	4.52	.728	.054
I liked the classroom activities	180	4.15	.801	.060
The online activities helped me learn and understand the topics	180	4.32	.758	.057
The classroom activities helped me learn and understand the topics	180	4.34	.750	.056

When queried why students enjoy learning in a hybrid learning environment, Watson analytics revealed the following discoveries (1) the combination of "I felt a sense of satisfaction and achievement about hybrid learning" and "I like the online activities", (2) the combination of "I understand why this course mixed online and classroom activities" and "I like the online activities" and (3) the combination of "the online and classroom activities complemented each other" and "I like the online activities". The discoveries revealed that students were more engaged with online activities posted in the learning management system. Figure 3 shows the resulting graph for these discoveries. Spearman's correlation revealed further that "I like the online activities" (ρ=0.389, α=.000) is significantly correlated to students' engagement with hybrid learning environment ("I enjoy learning in a hybrid learning environment").

Students were solicited a general comment on their experience, use and feedback of the hybrid learning environment. Results revealed that adequate technology skills and confidence is a prerequisite for students to achieve an acceptable learning experience (t = 3.659, ρ = .000). LMS activities and features (structure and organization) as well as coordination and timing between classroom activities and online activities are influences for student engagement (t = 2.836, ρ = .005). Technology quality especially internet connection is a lingering dispute for students since students are accessing the LMS from different locations away from the campus (t = -1.31, ρ = .896). Text analysis revealed a rather biased result as students' comments focused on the online influence of the hybrid learning environment and these results cannot provide a conclusive finding. A separate comment for online experience and in-class experience could have been in place. Figure 4 shows the text analysis results.

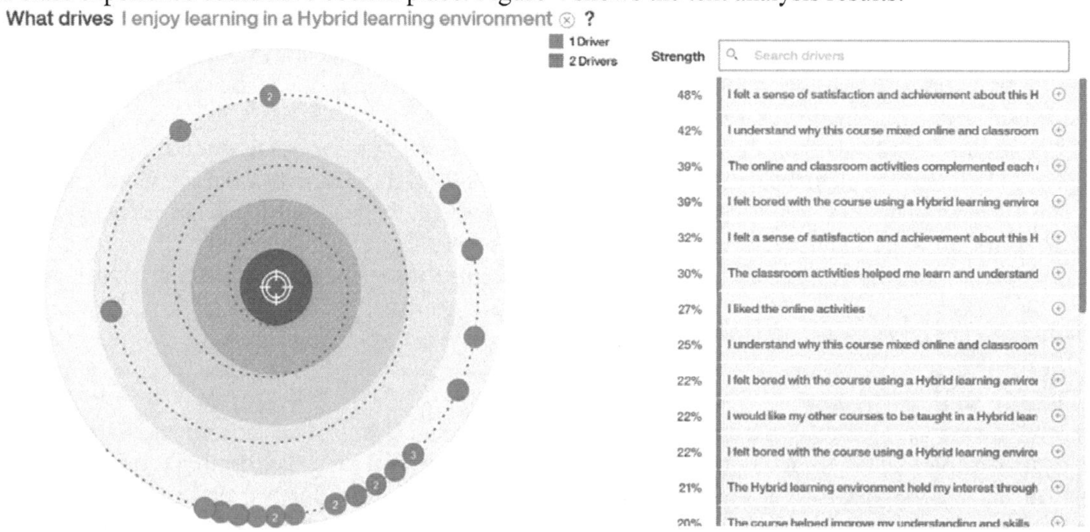

Figure 3. Discoveries in I enjoyed learning in an hybrid learning environment

Figure 4. Results of the text analysis

The study revealed that students' engagement in a hybrid learning environment was motivated by online influence ($\rho = 0.389$, $\alpha = .000$) and the motivation was derived from their level of knowledge in using technology in dealing with coursework ($t = 3.659$, $\rho = .000$). Figure 5 shows the predictors for a successful engagement derived from Watson's analytics.

6. Conclusion

A hybrid learning environment is truly a paradigm that challenges the education perspective of students and effectively engages students in accomplishing course requirements and by doing so, increases the level of knowledge creation and perception of students. The identification of level of engagement and factors can support the design and delivery of hybrid learning in the future.

The study hoped to investigate student engagement in a hybrid learning environment. However, findings showed that students focused on the online influence of the environment. This could have been a result in an improper design of the survey instrument or in the interpretation of results. The brick and mortar influence was not properly involved in the results and interpretation. For the hybrid learning environment to be successful, it should exhibit the strengths of both online and classroom engagement.

With the results revealing partial findings, further study should be undertaken to resolve several issues that were encountered such as: (1) a strong distinction between the online influence and classroom influence of a hybrid learning environment, (2) the impact of a hybrid learning environment to students' study habits, (2) the financial impact of a hybrid learning environment to both teacher and student, and (3) the influence of teachers in handling and delivering hybrid learning coursework.

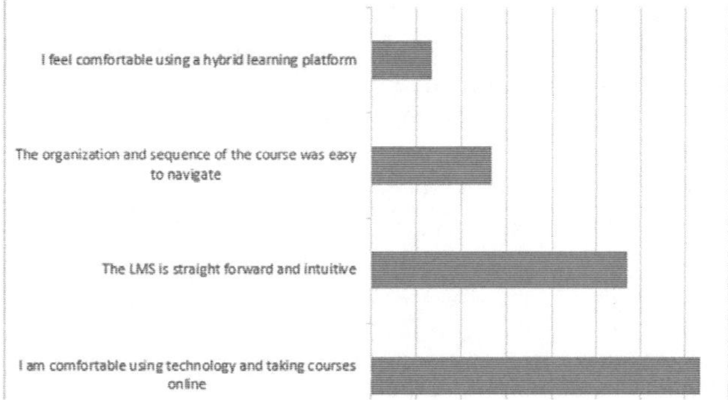

Figure 5. Predictors for student engagement in a hybrid learning environment

References

[1] Ziden AA, Rosli M, Gunasegaran T and Azizan SN 2017 Perceptions and experience in mobile learning via sms. a case ctudy of distance education etudents in a malaysian public university *Int. J. of Interactive Mobile Technologies* vol **11** no **1** pp 116-32

[2] Delialioglu O and Yildirim Z 2007 Students' perceptions on effective dimensions og interactive learning in a blended learning environment *Educ. Technol. & Society* **10(2)** pp 133-46

[3] Owston R, York D and Murtha S 2013 Student perceptions and achievement in a university blended learning strategic initiative *Internet & Higher Educ.* **18** pp 34-46

[4] Vaughan N 2007 Perspective on blended learning in higher education *Int. J. on E-learning* vol **6** no **1** pp 81-94

[5] Estacio RR and Raga Jr RC 2017 Analyzing student online learning behavior in blended course using Moodle *Asian Assoc. of Open Universities J.* vol **12** no **1** pp 52-68

[6] Mirriahi N, Alonzo D and Fox B 2015 A blended learning framework for curriculum design and professional development *Res. in Learning Technologies* vol **23**

[7] Graham C 2006 Blended learning systems: definition, current trends, and future directions *Handbook of Blended Learning: Global Perspective, Local Designs* ed Bonk CJ et al (San Francisco, CA: John Wiley & Sons, Inc)

[8] Reynard R 2007 *Hybrid Learning: Maximizing Student Engagement* (Campus Technology)

[9] Hughes G 2007 Using blended learning to increase learner support and improve retentions *Teaching in Higher Educ.* vol **12** no **3** pp 349-63

[10] Roby T, Ashe S, Singh N and Clark C 2013 Shaping the online experience : how administrators can influence student & instructor perceptions through policy and practice *The Internet of Higher Educ.* vol **17** pp 29-37

[11] Elearnspace 2005 *Blended. http://www.elearnspace.org/doing/blended.htm*

[12] Koohang A and Durante A 2003 Learners' perceptions toward the web-based distance learning activities/assignments portion of an undergraduate hybrid instructional model *J. Information Technol. Educ.* **2** pp 105-13

[13] Harding A, Kaczynski D and Wood L 2012 Evaluation of blended learning: analysis of qualitative data *Proc. of The Australian Conf. on Science and Mathematics Education* vol **11**

[14] Garrison D and Vaughan N 2008 *Blended Learning in Higher Education-Framework, Principles, and Guidelines* (John Wiley & Sons)

[15] Penn State University 2014 *Hybrid Learning at Penn State*

[16] Kochang A, Britz J and Seymour T 2006 Panel discussion hybrid/blended learning: advantages, challenges, design and future directions *Proc. of the 2006 Informing Science and IT Education Joint Conf.* pp 155-7

[17] Garrison D and Kanuka H 2004 Blended learning: uncovering its transformative potential in higher

education *The Internet and Higher Educ.* vol 7 no **2** pp 95-105
[18] Picciano A 2006 Blended learning: implications for growth access *J. of Asynchronous Learning Networks* vol **10** no **3** pp 95-102
[19] Laurillard D 2010 *Re-thinking University Teaching: A Framework for The Effective Use of Education Technology* (London: Routledge)
[20] Kaleta R, Skibba K and Joosten T 2007 Discovering, designing, and delivering hybrid courses *Blended Learning: Research Perspectives* pp 111-43

E-learning roadmap for open distance learning in Cordillera Administrative Region

D H Galeon[1], P G Jr. Garcia[2], J dela Cruz[3]

[1] College of Computing Sciences, Ifugao State University, Lamut Ifugao, Philippines
[2] Information and Communications Technology Office, Benguet State University, La Trinidad, Benguet, Philippines
[3] College of Information Technology and Computer Science, University of the Cordilleras, Baguio City, Philippines

[1]galeon.dh@gmail.com, [2]p.garciajr@bsu.edu.ph, [3]delacruzpen@gmail.com

Abstract. The eLearning technology and its integration in higher education's open and distance learning is generally claimed as matured, however actual information gathered on the ground states otherwise. This descriptive research investigated its implementation status in the Cordillera Administrative Region. It explored into the competency of faculty in course content development skills, their engagement and involvement in professional development activities, their integration of ICT in the curriculum and their eLearning culture particularly on their access of eLearning resources. The management support particularly on leadership and planning, support to ICT infrastructure, professional development, eLearning culture, and ICT in the Curriculum were also part of the investigation. Data analysis revealed that generally, faculty are ready to implement eLearning however, management should look into retooling its leadership to support the program in terms of leadership and planning, eLearning culture, and ICT infrastructure support. A roadmap was developed to serve as an implementation plan in addressing the identified issues in eLearning adoption to open and distance education. This roadmap may also service as guide for academic institutions, researchers, administrators and leaderships in implementing similar programs.

1. Introduction

1.1 Definition, Background and Application

E-learning is a teaching-learning engagement using electronic devices such computers, mobile devices, content management systems, internet and other information and communications technology (ICT) based technologies. ICT has always been beneficial to all aspects of human engagement. Academics is as one of the major beneficiaries with ICT being integrated ICT especially in open and distance education (ODE) to bridge the time and place barriers of learners and teachers. E-learning flourished in the beginning of the 20th century with the introduction of online and virtual learning systems [1] integrating various types of media including images, user interfaces, text, videos, hypertext, hypermedia, animations. Today, social media, video conferencing, collaboration tools, discussion tools, mobile applications and other related technologies are now important component of e-learning.

In the Philippines, e-learning started to popularized in early 2000 in parallel with the growing popularity of ICT in government and education. The implementation then has been a challenge with the absence of considerable ICT infrastructure to support the service was the greatest challenge to its early adoption. Today, the improvement of ICT infrastructure and the increased penetration of internet access in the country including remote areas is seen as an advantage to support and strengthen e-learning in open and distance learning (ODL). A January 2018 report [2] revealed that there are already 68 million internet/social media users out of the 105.7 million population, which is equivalent to 63% penetration.

1.2 eLearning Adoption and Open and Distance Learning
ODL existed in the Cordillera Administrative Region (CAR) since 1997 in fulfilment of article XIV (Education) of the 1987 Philippine Constitution and republic act number 7722[3]. Republic act 10650[4] was just enacted in 2014 to support public and private higher educational institutions to offer traditional and nonconventional open distance learning. This aims to address the educational needs of individuals who are deprived of time and distance to attend regular room-based lectures such as employed or working individuals, parents and the likes. With almost 2 decades of e-learning implementation, the integration of ICT in ODL seems not yet well embraced despite initial support provided by the government [5]. Directed to a quality e-learning, the national government through the Department of Information and Communications Technology (DICT) have initiated a free e-learning platform program called Gabay Aral for government agencies; it is accessible at http://i.gov.ph/gabay-aral/. While the government have these initiatives in place, the school e-learning administration should make a directive mandating the use of the e-learning platform and should not be used as an option [6] to support blended learning.

1.3 Related Studies
While proceeding onward to "hybrid" classes employing teleconferencing in the near future [7], the distance education industry needs to consider the most essential hindrances that affect quality and subsequently learning outcomes of ODE especially technological infrastructure as the main reason behind the slow progress of e-learning adoption in most third world countries[8], [9], [10], [11] and issues on lack of leadership[12], [13] as well. While some course specialists in the Philippines are fully confident in handling Open University System classes[14], there is still a need to stimulate innovative practice among faculties who stay unengaged in ODL[13], increase teacher's ICT competency[15], [16], support and sustain innovative practice among the innovators[13], and stimulate the support of the business area in creating competency measures and curricula[17]. While Higher Education (HE) are confronting real difficulties including faculty acceptance in an international perspective[18], faculty members are involved in numerous instructional design exercises without formal training in learning theories and the science of instruction[19], and studies shows that several countries[9], [10], [12] do not have the necessary conditions for the development of quality educational contents, an IT teaching-learning program for faculty development preferred by the faculty[20] that integrates course design[1], [21] and preferably conducted online[22], [23] should be in place as the focus of an e-learning program[24] followed by a performance assessment for quality assurance and continuous improvement[25]. Among other solutions for these barriers are: standardization; strategies; funding; integration of e-learning into the curriculum; blended teaching; user friendly packages; access to technology; skills training; support; employers paying e-learning costs; partnership with other sector, international organizations, and multinational companies[9] and prioritizing skills training and management support[26].

2. Methodology
This study used a descriptive research and unstructured survey method to understand the limited success to e-learning in the CAR. It focuses and investigated on the factors of lack of human resources with sufficient competence and motivation, technological infrastructure, and lack of leadership in providing adequate and appropriate support as inputs to coming up with an e-learning roadmap (see Figure 1). Based on a comprehensive faculty development framework for ODL[1], the competency level in e-learning course development of faculty deliberating their e-learning course development skills, engagement and involvement in professional development activities, their integration of ICT in the

curriculum and their e-learning culture particularly on access of e-learning resources in the ODE was assessed. The level of support from the e-learning administration was also assessed using an e-learning framework by NCTE [27] with e-learning support categories of leadership and planning, ICT infrastructure, professional development, e-learning culture, and ICT in the curriculum. Fifty-three participants composed of the ODE teachers, the ODE's academic management and the ICT units' staff of state-run universities in the CAR particularly Benguet State University and Ifugao State University; and who have an access to the Internet (100%) for 2-5 hours a day (68%) from their homes (87.45%) and offices (87.5%) primarily for social networking (90%) and research (80%) activities were selected using convenience sampling and were able to and provide data for analysis. These teachers are from various private/public agencies in Baguio/Benguet thus representing the region-wide academics. The data was analyzed with the use of office productivity tools and Statistical Package for the Social Sciences.

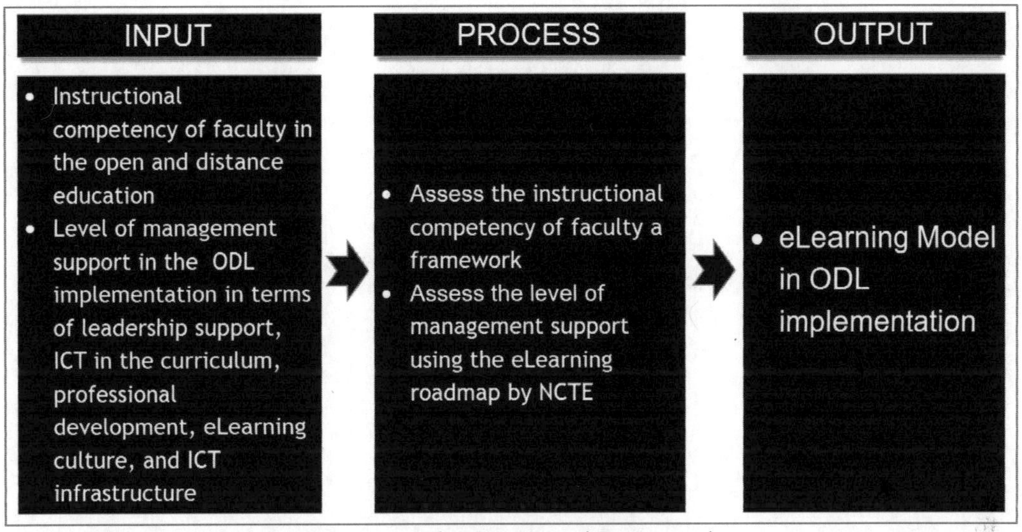

Figure 1. The Conceptual Framework

3. Findings and Discussion

3.1 Level of instructional development competency of faculty

Figure 2 shows the consolidated view of responses for the level of competency of faculty in e-learning course development. 2 out of 4 (50%) of the competency areas is reported competent level by the participants. It is evident from this study that almost majority of the faculty members are capable of completing tasks in developing e-learning courses without assistance.

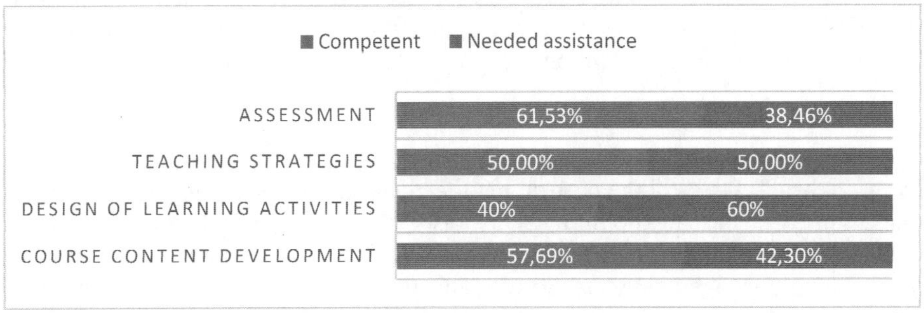

Figure 2. Questionnaire Responses for Level of Instructional Development Competency of Faculty

3.2 Level of management support in the implementation of ODE

This section presents a consolidated view of all findings for the level of management support in ODE. Figure 3 shows the percentage of management support reported initial stage by more than 50% of respondents. A baseline value of 50% is taken to portray the majority opinion against each management support. Any management support reported confident and mature by more than 50% of the respondents is not accounted for here. Research findings show the following overall management support results:

- Leadership and planning: 4 out of 5 (80%) supports was reported initial stage.
- ICT infrastructure: 2 out of 6 (33%) supports was reported initial stage.
- Professional development: 1 out of 6 (17%) supports was reported initial stage.
- E-learning culture: 2 out of 5 (40%) supports was reported initial stage.
- ICT in the Curriculum: 1 out of 4 (25%) supports was reported initial stage.

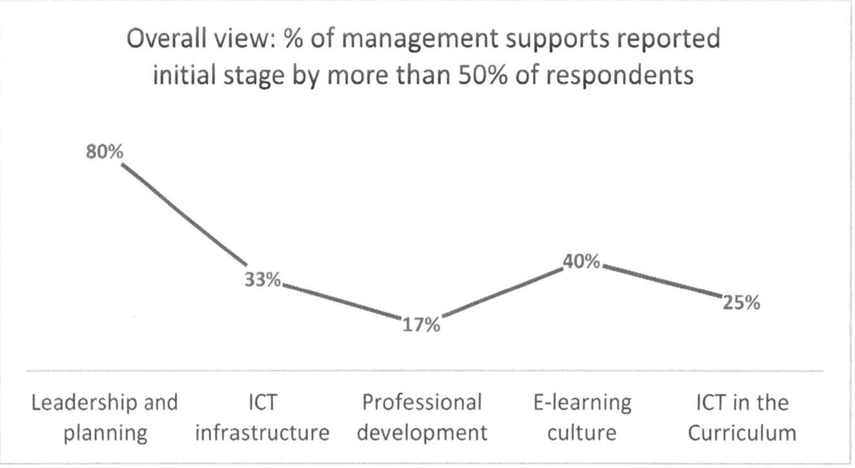

Figure 3. Overall View of Questionnaire Responses

From this study, it became evident that leadership and planning has majority of its management supports (80%) are in initial stage. Detail shows that the following management support in various categories are reported in initial stage: development of an e-learning vision, an e-learning plan and an Acceptable Use Policy (AUP) for Internet and ICT use and the support to Special Educational Needs (SEN) under leadership and planning; online presence and engagement to e-learning projects under the e-learning culture; school network and software and digital content under ICT infrastructure; integration of ICT to all aspects of SEN teaching and learning under ICT in the curriculum; and confidence and acquisition of skills by the faculty to use a wide range of technologies to facilitate the inclusion of students with SEN under professional development.

The fact that most management supports are at enabled stage, the administration needs to provide more support in advancing their e-learning engagement to confident and mature stages. The findings suggest the priorities in developing solutions for the advancement and for a more quality e-learning strategy.

3.3 An eLearning Roadmap for ODE

To go well with "hybrid" classes in the near future, the e-learning strategy must be well established the soonest possible time. Considering the findings discussed above in priority to the areas that needs improvement, an e-learning roadmap (see Figure 4) is developed to communicate direction and progress to stakeholders engaged in the e-learning. This tool which integrates the limitations of the faculty in instructional development to professional development matches the four-year long-term e-learning goals with specific objectives as milestones to help meet these goals. The goals in the roadmap are categorized as leadership and planning, e-learning culture, ICT infrastructure, professional development and ICT in the curriculum as adopted from the Digital Learning Framework of the NCTE which is a planning tool in effectively embedding digital technologies into teaching, learning and assessment and in lined with the objectives of the Ireland's Digital Strategy for Schools (2015-2020) [27]. While majority of the management supports are in enabled stage and after a deliberation by the e-learning administration heads from the participants on their e-learning engagement capability, achieving the confident stage in e-learning maturity takes the university a two years' time and two years more to complete the mature stage.

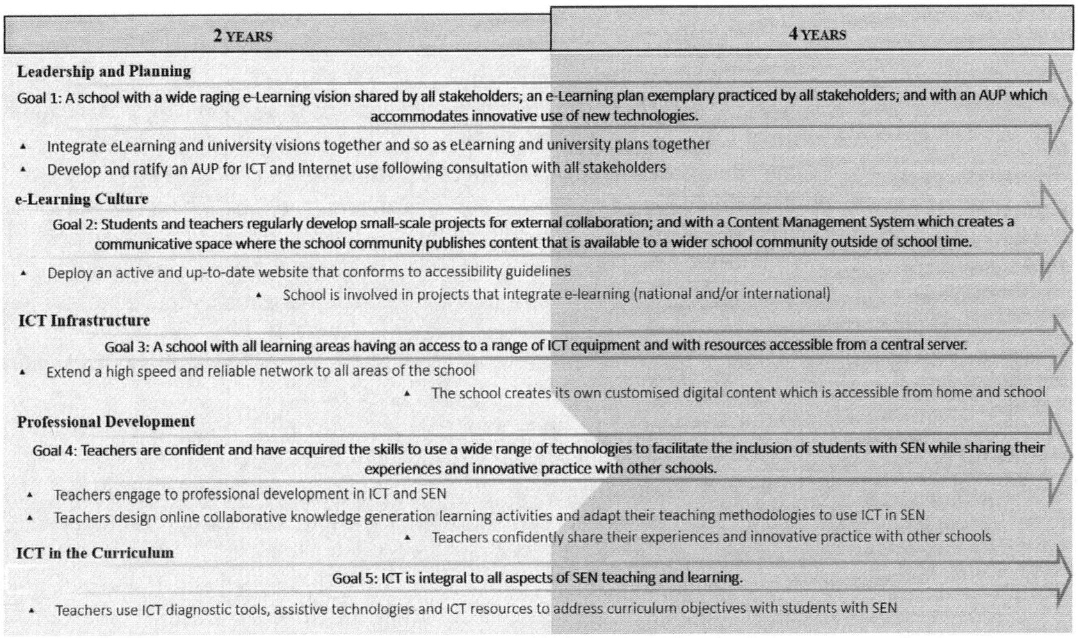

Figure 4. A roadmap to a matured e-learning in four years

4. Conclusion

This research investigated the adoption of e-learning in ODE in the CAR and confirmed its low actual adoption despite of the concerted efforts of various stakeholders such as the national level government, the academic institutions, the faculty and also the learners. Generally, the ICT infrastructure in CAR has at least improved however, various elements should be given attention to. The faculty should invest in their e-learning course development skills, their engagement and involvement in professional development activities, their integration of ICT in the curriculum and e-learning culture particularly on their access of e-learning resources. The management has to institutionalize the integration of e-learning by retooling its leadership to support the program in terms of leadership and planning, e-learning culture, and ICT infrastructure support. The implementation plan was presented in the e-learning roadmap, which is being recommended for implementation or adoption by the concerned universities and similar organizations involved in academics or similar programs.

5. References

[1] Arinto PB 2013 A framework for developing competencies in open and distance e-learning *The Int. Review of Res. in Open and Distributed Learning* vol **14** no **1** pp 167-85

[2] Digital in 2018 2018 *World's Internet Users Pass The 4 Billion Mark - We Are Social* retrieved June 10, 2018 from https://wearesocial.com/blog/2018/01/global-digital-report-2018

[3] Philippines 1994 *Republic Act no. 7722: An Act Creating The Commission on Higher Education, Appropriating Funds Therefor and for Other Purposes* (Metro Manila, Philippines: Congress of the Philippines, Republic of the Philippines)

[4] Philippines 2014 *Republic Act no. 10650: Open Distance Act* (Metro Manila, Philippines: Congress

of the Philippines, Republic of the Philippines)
[5] Gogos R 2013 *A Brief History of Elearning (Infographic)* retrieved June 10, 2018 from https://www.efrontlearning.com/blog/2013/08/a-brief-history-of-elearning-infographic.html
[6] Back DA, Behringer F, Harms T, Plener J, Sostmann K and Peters H 2014 Survey of e-learning implementation and faculty support strategies in a cluster of mid-European medical schools *BMC Medical Education* vol **15** no **1** p 145
[7] Kabanda G 2015 *Dimensioning Academic Strategies and Priorities with E-Learning Trends Technologies at The University of The Free State*
[8] Xanthidis D and Nikolaidis P 2014 A pilot study of the challenges associated with eLearning developments in Saudi Universities *Int. J. of Technol. Diffusion (IJTD)* vol **5** no **4** pp 63-79
[9] Al-alak BA and Alnawas IA 2011 Measuring the acceptance and adoption of e-learning by academic staff *Knowledge Management & E-Learning: An Int. J.* vol **3** no **2** pp 201-21
[10] Tarus JK, Gichoya D and Muumbo A 2015 Challenges of implementing e-learning in Kenya: A case of Kenyan public universities *The Int. Review of Res. in Open and Distributed Learning* vol **16** no **1**
[11] Weerasekera KS 2010 Looking for a quality assessment mechanism' for teaching engineering technology through open and distance education *Asian Assoc. of Open Universities J.* vol **5** no **2** pp 49-55
[12] Ahmed A and Phillips JA 2014 Elearning roadmap and initiatives in Malaysian higher education *ICT* p 65
[13] Arinto PB 2016 Issues and challenges in open and distance e-learning: Perspectives from the Philippines *The Int. Review of Res. in Open and Distributed Learning* vol **17** no **2**
[14] Sumande CT, Castolo CL and Comendador BE 2016 The ICT level of confidence of course specialists in distance education: The Polytechnic University of the Philippines experience *Turkish Online J. of Distance Educ.* vol **17** no **4** pp 175-89
[15] Red ER, Borlongan HG, Briagas TT and Mendoza MJ 2013 An assessment of the elearning readiness state of faculty members and students at Malayan Colleges Laguna *Int. J. of the Comp., the Internet and Management* vol **21** no **3** pp 20-6
[16] Marcial DE and Rendal JB 2014 Are teachers technophobic? Measuring information and communication technology competency among teacher educators in Central Visayas, Philippines *Malaysian J. of Distance Educ.* vol **16** no **2** pp 1-7
[17] Sabio RA and Sabio CJ 2013 Current challenges and prospects facing distance education in the Philippines *Asian J. of Distance Educ.* vol **11** no **1** pp 59-77
[18] Qureshi MF, Shah SM and Zaka B 2014 Towards providing better distance education: a roadmap for universities *J. of Strategy and Performance Management* vol **2** no **2** p 67
[19] Khalil MK and Elkhider IA 2016 Applying learning theories and instructional design models for effective instruction *Advances in Physiology Educ.* vol **40** no **2** pp 147-56
[20] Herman JH 2012 Faculty Development Programs: The frequency and variety of professional development programs available to online instructors *J. of Asynchronous Learning Networks* vol **16** no **5** pp 87-106
[21] Williams SW 2002 *Instructional Design Factors and the Effectiveness of Web-Based Training/Instruction*
[22] Barefield AC and Meyer JD 2013 Leadership's role in support of online academic programs: implementing an administrative support matrix *American Health Information Management Assoc.*
[23] Holmes B 2013 School teachers' continuous professional development in an online learning community: lessons from a case study of an eTwinning learning event *European J. of Educ.* vol **48** no **1** pp 97-112
[24] Marcial DE 2012 Teaching and learning with technology in higher education institutions in the Philippines *PeLS Online J.* vol **3** no **1** pp 50-66
[25] Belawati T and Zuhairi A 2007 The Practice of a quality assurance system in open and distance

learning: a case study at Universitas Terbuka Indonesia (The Indonesia Open University) *The Int. Review of Res. in Open and Distributed Learning* vol **8** no **1**
[26] Homod SA and Shafi MM 2013 Success factors of e-learning projects: a technical perspective *Turkish Online J. of Educational Technol.* vol **12** no **2** pp 247-53
[27] Anonymous 2018 *National Center for Technology Education* retrieved June 12, 2018 from http://www.ncte.ie/Planning/e-LearningRoadmap/

Dashboarding the maternal and child health profiles for health supporting system

Lizda Iswari[1], Dhomas Hatta Fudholi[2], Silfa Kurnia Aditya[3]

Department of Informatics
Yogyakarta, Indonesia
Universitas Islam Indonesia

[1]lizda.iswari@uii.ac.id, [2]hatta.fudholi@uii.ac.id, [3]silfa.kurnia.aditya@gmail.com

Abstract. Various types of health data are routinely collected and published as national health profiles. However, these vast amounts of data are not always used effectively since they are presented in raw tabular data sets, fragmented, and are not well-equipped with the statistical data inferences and the trend of changes which are essential to explore some unknown previous information. In this paper, we propose how to manage the national open health data sets as a dashboard application to support the decision making for maternal and child health-care services in Indonesia. The research was conducted in four steps. First, we described what maternal and child health parameters are involved in relation to the availability of open and accessible data set. Second, we designed the dashboard application based on Kimball's four step dimensional modelling. Third, we implemented the schema modelling as data warehouse by executing ETL (extract, transform and load) and OLAP (On Line Analytical Processing) cubes. Fourth, we analyzed the results with the business intelligence approach. We explored the open data sets in four consecutive years (2014-2017) and determined seven types of business process, i.e. analysis of health profile on neonates, infants, and children, analysis of maternal health profile before, during and after pregnancy period, and analysis of health supporting factors. By applying data warehouse and business intelligence, our proposed system has the ability to integrate and extract a number of tabular and fragmented data sets into an easy and quickly inferred information about the current situation of maternal and children health care services.

1. Introduction

Maternal and child health are still a major problem in Indonesia [1][2]. This is indicated by the level of Maternal Mortality Rates (MMR) and Infant Mortality Rates (IMR), which are still far below the target of Sustainable Development Goals (SDG)[3]. SDG is an international agreement among countries aiming for joint development by 2030. In 2015, it is recorded that MMR in Indonesia was 305 per hundred thousand live births and the IMR was 23 per thousand live births [1][3]. This rate still highly exceeds the maximum set by SDGs of 70 per hundred thousand live births, with the neonatal mortality rate of 12 per thousand live births, and the under-five mortality rate of 25 per thousand live births [4].

On this account, it is necessary to explore and analyze maternal and child health profiles in accordance with the current trends in maternal and child health issues in Indonesia [2][3][5]. Several studies have shown the importance of exploring the maternal and child health data sets, such as to identify the impact of measles vaccination on the prevention of pneumonia [6] and to find the relationship of maternal health services to neonatal mortality rates [7][8]. Profiling these data sets may also improve the health care environment and aid the government in budgeting and allocating finance and planning a better health system [9]. At the global level, exploration of these data sets may reveal the different quality of basic service between the public and private health facilities [10][11] and also show the health service gaps among districts in one country [12].

Annually, the Indonesian Ministry of Health publishes the national health profile which is open and accessible for anyone. This profile presents a comprehensive collection of health situation and condition from all 34 provinces in Indonesia, including a variety of data of maternal and children health care. A vast amount of health care data continued to be collected but these are not always used effectively due to its presentation technique. Data are published as raw tabular data sets, fragmented, and not yet equipped with the statistical data inferences and the changing trend. There are also some inconsistencies in how data are organized which were triggered by the incomplete and the poor quality of routinely collected data [7]. As a result, information that is needed by the decision makers to determine the direction of public health service policy has not been presented in an interpretable form. Hence, it is necessary to create a system capable of converging various types of health care data into meaningful information and presenting them in an appropriate and understandable mode. To achieve this, we propose a dashboard system for exploring the children and maternal health profiles. A dashboard aims to provide intense information to assist policy makers in recognizing and addressing maternal and children health problems and also to identify which parameters that require quick action to avoid adverse outcomes [13].

2. Literature Review

In this section, we elaborate a literature review in some existing dashboard systems, including dashboard system in maternal domain, in order to highlight the advantages of dashboard system as a decision support solution. Moreover, by doing so, it clarifies the contribution of our research in the area.

Dashboard has been widely used to give high supports in organizations' management. A dashboard allows simple, comprehensive, and interactive performance management. It also promotes continuous improvement in productive areas [15]. In [16], the developed dashboard provides real-time information on Emergency Department (ED) crowding, where the information given enables ED staff to judge whether corrective actions are required in an effort to avoid the adverse effects. The representation of the status of the entire ED and internal patients, allows providers to make more appropriate decisions. This will improved the awareness of the situation and allows rapid intervention when pre-determined care targets are not met and when a specific patient's status changes [17]. A novel concept on medical dashboard system called BESTBoard has been introduced [18]. The system is developed to cope with the user requirements at hospital ward, ICU, and ER. The system has been proven to be feasible and useful in delivering health information to healthcare professionals.

Getting more specific into the context of maternity dashboard, the Royal College of Obstetricians and Gynecologist (RCOG) in UK endorses an adoption of maternity dashboard in order to provide a local quality assurance system [19]. The dashboard helps to monitor outcomes on a monthly basis by comparing the observed outcomes with the pre-determined gold standards (i.e. goals) in three groups of parameters: activity; workload; and quality [13]. Moreover, it has been evaluated in term of its effectiveness, where it could improve clinical care, recognize good clinical care, disseminate information, and engages multidisciplinary team [13].

Our proposed system is developed with the spirit of getting the advantages of dashboard system in supporting organizational decision making. It is empowered by business intelligent engine. Our dashboard system is not only served as maternity/maternal dashboard, but also include children health profiles. Both information becomes a set of health monitoring system standard in Indonesia. Moreover, the parameter in the dashboard is specified based on the Indonesian's Ministry of Health guidance.

3. Methodology

The development of maternal and children health care dashboard is conducted on four activities as seen in figure 1. These activities include screening data sets, designing dimensional model, implementing data warehouse, and implementing business intelligence technology.

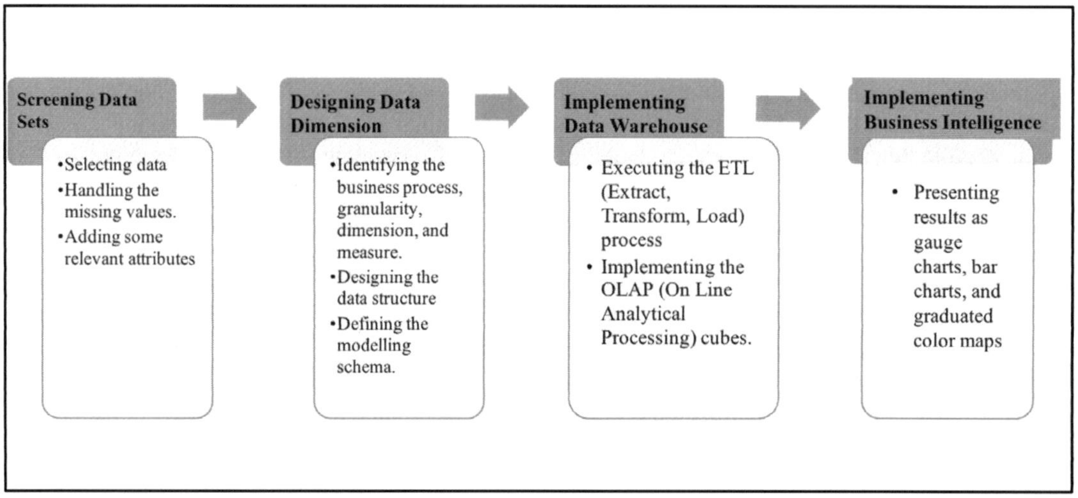

Figure 1. Four step of research methodology

3.1. Screening Data Sets

Data used in this paper was derived from the official website of the Ministry of Health's Data and Information Center, http://depkes.go.id. Data were displayed in tabular format and distributed as an open document, namely as Indonesia Health Profiles. An initial observation of health profiles in the last four consecutive years, 2014 – 2017, revealed some inconsistencies and incomplete of data reports. By applying a manual data screening, we found eight types of data in the section of family health that are consistently presented and in accordance with the Minimum Service Standard indicators set by the Health Minister's regulation in 2008 as shown by Table 1.

Table 1. Family Health Data Sets.

Nr	Main Theme	Supporting Data Set
1	Neonates visits	Two types of neonates visits, namely the first and the complete visits.
2	Vaccination for infants	Five types of compulsory vaccination on infants, namely BCG, Measles, Polio, DPT/HB1, and DPT/HB3.
3	Infectious diseases on children	Four types of infectious diseases on children, namely pneumonia, lepers, measles, and diphtheria.
4	Vaccination for fertile women	Five times of Tetanus Toxoid (TT) vaccination for childbearing women and two times of TT vaccination on pregnant women.
5	Maternity health service	Childbirth be assisted by trained health personnel and performed at at health-care facilities
6	Antenatal care	Two types of antenatal care, namely the first and the complete visits during pregnancy period.
7	Postpartum service	Midwifery complications and postpartum visits.
8	Family planning	Family planning participant and contraceptives users.

Data screening has also been done for data transformation by changing the initial data type that is a portable data format (pdf) to comma separated values (csv). The csv formats facilitate the data importing process and is compatible to be accessed by server machines. Another data pre-processing has also been performed by adding a relevant attribute of the island for each data and handling some missing values.

3.2. Designing Dimensional Model

Data management with integrity and consistency design is the key to generate reliable information. To get an optimum database structure, we mapped these data sets by following Kimball's four step dimensional design which consist of [14]: (a) business process as a specific target of operational activities, (b) granularity as low level of data that determine the depth of query detail that will be executed in a data warehouse, (c) dimension tables as collection of attributes that share similar entities and are used to filter and group the facts, and (d) fact table that store the measurement of business process event and is presented as numeric data types.

The business process is determined based on the main data categorization. We identified that data can be categorized into seven themes, i.e. health care on neonates, health care on infants, health care on children, health care on prenatal, health care on pregnant women, health care on post-pregnancy period, and supporting factor such as medical personnel and health facilities. Each business process is having a granularity in provincial level as the lowest level of data presentation. Furthermore, one business process may have a number of dimension tables. On the contrary, one business process may only have one fact table which store several measurable values. This research involves 31 types of dimension tables and 34 types of measurement values. The summary of dimensional model based on Kimball's design is presented in Table 2.

3.3. Implementing Data Warehouse

The development of data warehouse involves two activities, namely ETL (Extract, Transform, Load) and OLAP (On Line Analytical Processing). ETL works as a back-end room to integrate various data sets, while OLAP cubes works as a front-end room that enable users to perform queries [14]. The ETL process consists of three activities, i.e. retrieving data from the data source (extract), converting the obtained data set into a consistent data format (transform), and sending these formatted data set into a centralized data storage (load). The ETL process was done using Microsoft SQL Server Integration Services (SSIS). The OLAP cubes consists of four activities, i.e. defining the cubes dimensions, defining measure attributes, defining dimension hierarchy, and defining the relationship between dimensions and cubes. OLAP cubes were implemented using Microsoft SQL Server Analysis Services (SSAS).

3.4. Implementing Business Intelligence

Business Intelligence (BI) is an extraction process on a set of operational data and collected it in a data warehouse. During the extraction process, data can be transformed with various techniques in accordance with the interest of business analysis. This research applies three types of data analysis, i.e. (a) descriptive as a method to describes the tabular data in accordance with the characteristics of data, (b) comparative as a method to describe data by comparing data between regions or with specific targets, (c) tendency as a method to describe data by comparing data over time in a fairly long period. Eventually, the BI implementation was done by using Microsoft Power Business Intelligence, a proprietary software for data analysis and reporting. This tool has also been integrated with SQL Server for data processing.

4. Results and Discussion

Dashboard has been implemented to provide information in effective and efficient mode. The result of data analysis is presented into multiple data visualizations, such as graphics, charts, and maps. These analyses are arranged and put in a single screen that enables the delivery of important information in a quick and easy access. It consists of three main information, namely maternal health profiles, children health profiles, and health supporting factor profiles.

Table 2. Dimensional Model Design.

Nr	Business Process	Granularity	Dimension	Measure
1	Analysis of health care on neonates	Coverage of neonates visit for each province	Neonates visit; Baby born alive.	First neonates visit. Complete neonates visit.
2	Analysis of health care on infants	Coverage of infant vaccination for each province	Basic vaccinationn; Complete vaccination; Universal Coverage Vaccination (UCI) villages; Baby born alive; Number of villages.	Percentage of five types basic vaccination, i.e. BCG, Measles, Polio, DPT/HB1, and DPT/HB3. Percentage of complete vaccination. Percentage of UCI villages.
3	Analysis of health care on children	Coverage of infectious diseases on children for each province	Pneumonia realization; Pneumonia target; New lepers; Lepers in children; Measles sufferers in children; Vaccinated measles on children; Diphtheria sufferers in children; Vaccinated diphtheria in children.	Percentage of pneumonia in infants. Proportion of leprosy defects level 1 and level 2 in children. Percentage of measles vaccination on group age of less than 1 year, 1-4 year, 5-9 year, and 10-14 year old. Percentage of diphtheria vaccination on group age of less than 1 year, 1-4 year, 5-9 year, and 10-14 year old.
4	Analysis of health care on prenatal	Coverage of vaccination for child-bearing women for each province	Tetanus Toxoid (TT) vaccination; Childbearing women.	Percentage of five types TT vaccination on child-bearing women, i.e. TT1, TT2, TT3, TT4, and TT5.
5	Analysis of health care during pregnancy period	Coverage of pregnant and postpartum services for each province	Tetanus Toxoid (TT) vaccination; Antenatal care; Child-birth; Postpartum visit; Obstetric complication; Pregnant women; Postpartum mother; Delivery mother.	Percentage of TT1 and TT2 on pregnant women. Percentage of first antenatal care. Percentage of complete antenatal care. Percentage of delivery at health-care facilities. Percentage of obstetric complications. Percentage of complete postpartum visit.
6	Analysis of health care on post-pregnancy period	Coverage of family planning program for each province	Family planning participants; Active contraceptives users; New contraceptives user; Couples of childbearing age.	Percentage of new participant of family planning program. Percentage of active participant of family planning program. Percentage contraceptives users of condom, injection, pill, implant, MOW, MOP, and IUD.
7	Analysis of health supporting factors	Coverage of health facilities and health personnel for each province	Health facilities; Health personnel; General populations.	Ratio of hospital and community health centers per 30,000 populations. Ratio of doctors, nurses and midwives per 100,000 populations.

4.1. Dashboard of Maternal Health Profiles

The maternal health profiles are summarized as Key Performance Indicator (KPI). There are 11 parameters are involved, i.e. coverage of first antenatal care, coverage of complete antenatal care, coverage of TT Level 1 vaccination on pregnant women, coverage of TT level 2 vaccination for pregnant women, coverage of delivery at health facilities, coverage of postpartum visits, coverage of obstetric complications, coverage of new family planning participant, coverage of active family planning participant, coverage of TT level 1 vaccination on childbearing women, and coverage of TT level 5 vaccination on childbearing women.

To provide brief information, KPI is equipped with the traffic light colors associated with the level of parameter's achievement. Some parameters may have the national target, while others may not. However, we have set the upper and lower threshold for those without the national target. In this regard, green color indicates that goals have been met, yellow color indicates that the goals are not met but still within the upper threshold, and red color indicates that the lower threshold is violated. If the parameter is alerted as red, it means that there is a problem requiring a quick action. Meanwhile, the yellow color is a signal that some appropriate actions are needed as prevention prior to entering the red area.

As seen on figure 2, of these 11 parameters, only one parameter has performance above the national target that is coverage of active family planning participants. Meanwhile, there are five parameters are marked in red, i.e. coverage of TT vaccination level 1 and level 2 for the pregnant women, coverage of TT vaccination level 1 and level 5 for the child-bearing women, and coverage of new family planning participants.

One of the causes of maternal death and infant death is tetanus infection as a result of an unsafe childbirth process or from injuries obtained by pregnant women before delivery. To prevent tetanus infection, women of childbearing age are required to get five types of Tetanus Toxoid (TT) vaccination, i.e. TT level 1 to TT level 5, before she gets into the pregnancy period. TT vaccination is also given to pregnant women for two times, i.e. in the first and the second trimesters (TT1 and TT2). In this research, we found that coverage of TT vaccination level 1 for the pregnant women should receive more attention since the national average achievement is far below the national target, i.e. 19.25% of 85.35%.

A further analysis, i.e. comparative and tendency, of each parameter can be found on the coverage of active family planning participant based on the contraceptive method. As seen on figure 3, there is a comparison of contraceptive uses. Both active and new participant choose injection as the contraceptive method. Meanwhile, the tendency analysis can be found on the new family planning participant in the four years which present a similar trend of participant number from 2014 to 2015, which ranged from 75%. However, the number dropped in 2017 to 55%.

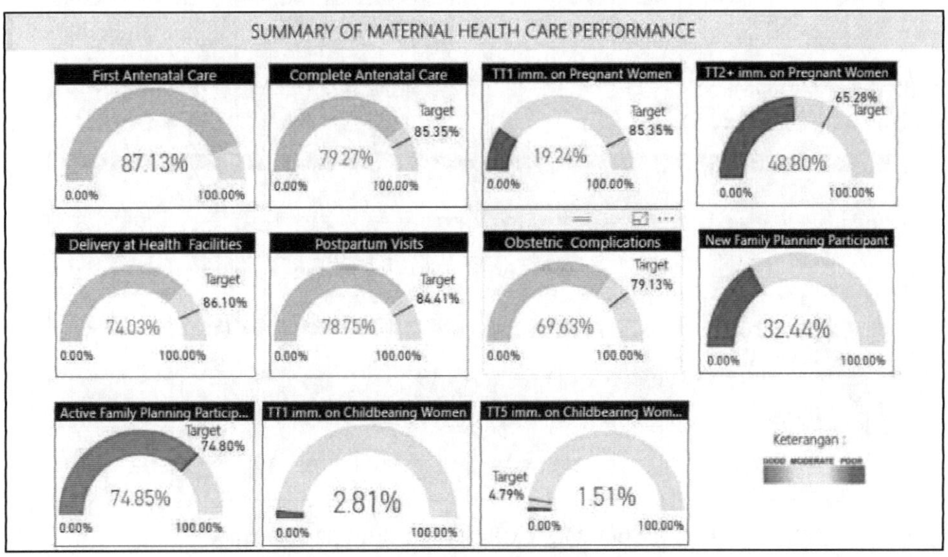

Figure 2. Summary of maternal health performance

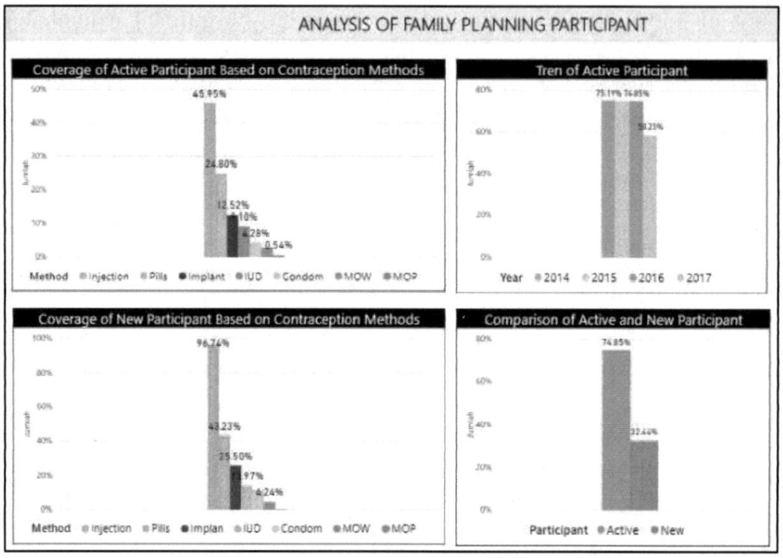

Figure 3. Comparative and analysis on family planning program

4.2. Dashboard of Children Health Profiles

The children health profiles are distributed into 12 parameters. It comprises coverage of five types of basic compulsory vaccination for infants, i.e. BCG, measles, DPT/HB1, DPT/HB3, and polio, coverage complete compulsory vaccination on infants, coverage of leprosy defect, the Universal Coverage Vaccination (UCI) for villages, coverage of vaccinated measles, coverage of vaccinated diphtheria, coverage of pneumonia findings, and coverage of first neonates visit.

As seen in figure 4, the summary of these 12 parameters shows that only two parameters that have performance above the national target, i.e. coverage of leprosy defects and coverage of UCI villages. Meanwhile, there are two parameters are marked as red, i.e. the proportion of measles vaccinated against measles findings and coverage of pneumonia findings.

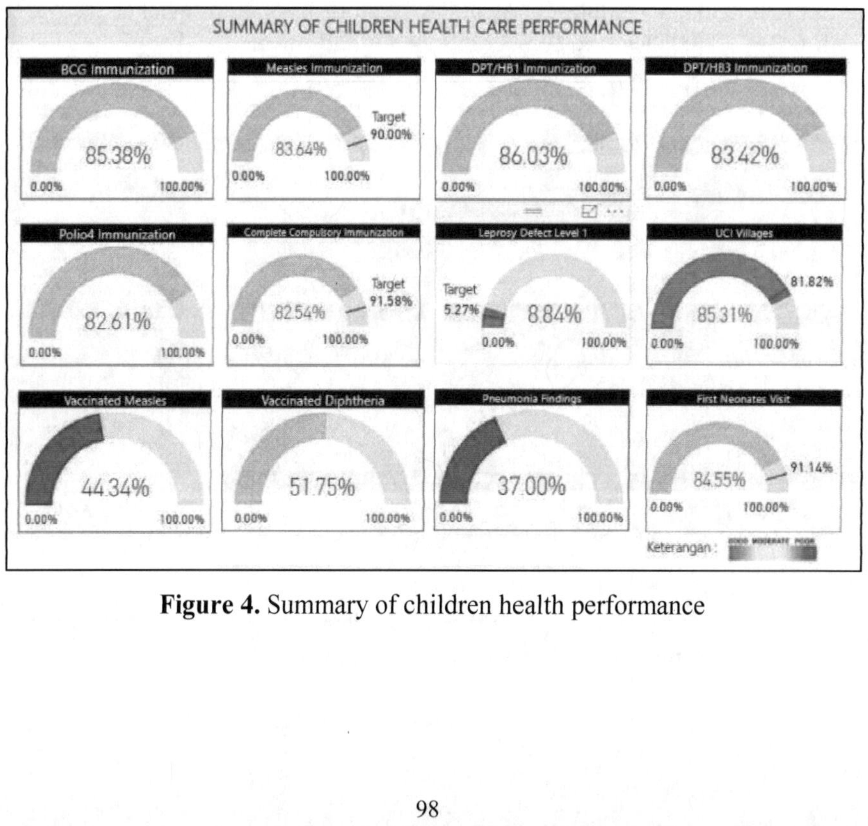

Figure 4. Summary of children health performance

The proportion of measles vaccination against measles findings can be explored further using the comparative analysis. It is based on the island and provincial entities as shown by figure 5. It is illustrated that among children aged 0 to 14 years, the highest proportion is found in Sumatra Island with the rate of 57.86% and the lowest proportion is found in Papua Maluku Island with the rate of 17.29%. However, this information only covers 24 of 34 provinces in Indonesia due to incomplete raw data set. For example, there is no data of measles findings for 3 of 4 provinces which belong to Papua Maluku Island. Therefore, comparison of the measles proportion rate based on island becomes invalid and unreliable information. On the other side, there is inferred information that can be drawn based on the comparative rates among the provinces. As seen on figure 5, there are 9 of 24 provinces (37.5%) are identified for having the proportion rate of more than 50%.

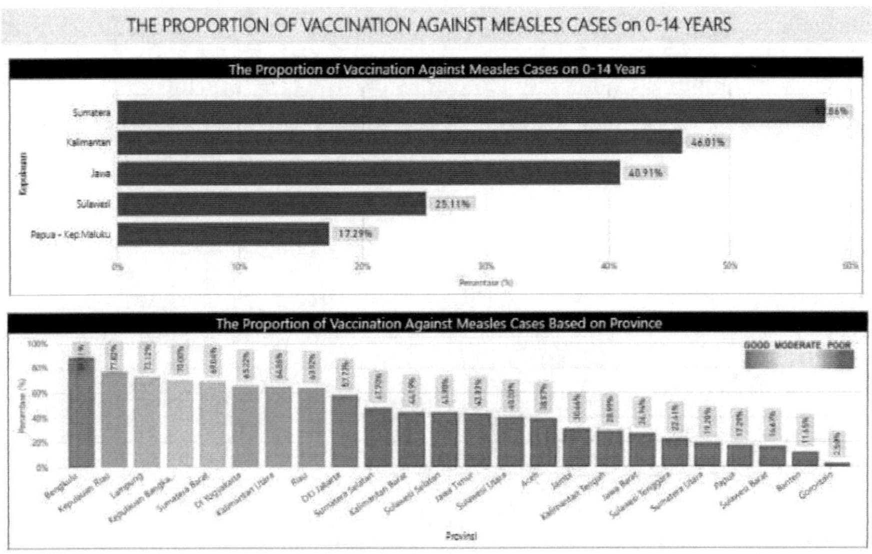

Figure 5. The comparative analysis of measles cases

4.3. Dashboard of Supporting Health Factors

The health support factor consists of five parameters, namely ratio of doctors per 100,000 population, nurse ratio per 100,000 population, midwife ratio per 100,000 population, hospital ratio per 30,000 population, and ratio of community health center per 30,000 population. Similar to the previous summary information, these parameters were also delivered as KPI to describe the current achievement. As can be seen on figure. 6, none of these supporting factors are able to achieve the national target. Nationally, the lowest performance belonged to the doctor's ratio of 19.86 from a target of 45 doctors per 100,000 population. It is also necessary to consider the ratio of midwife since its performance is still far from the target, i.e. 79.84 from 120 midwifes per 100,000 populations.

To illustrate the ratio distribution, we use a map as another visualization technique. This map consists of graded colors to highlight the health parameter performance. For example, we took the distribution of midwifes ratio for all 34 provinces. As seen on figure 7, most of Indonesia's territory is dominated by red color which indicates the low ratio of midwives to the number of population.

In general, we can state that this maternal and child health dashboard has a number of positive impacts, such as to aid the governments, both at central and regional levels, in identifying the health parameters with poor performance so as to determine the placement of health personnel and the development of health facilities can be determined more precisely. In addition, this application can also be an input for decision makers to define the priorities and the direction of health care policies for the community. In the end, government helped to identify what programs needs to better improved so that the quality of maternal and child health services can be better.

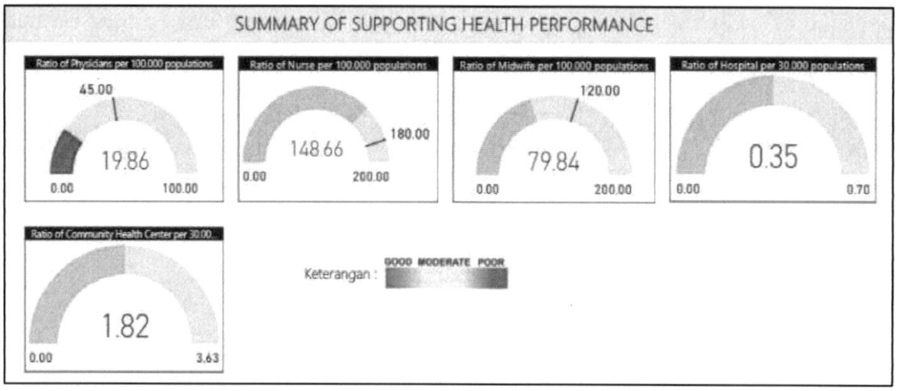

Figure 6. Summary of health supporting performance

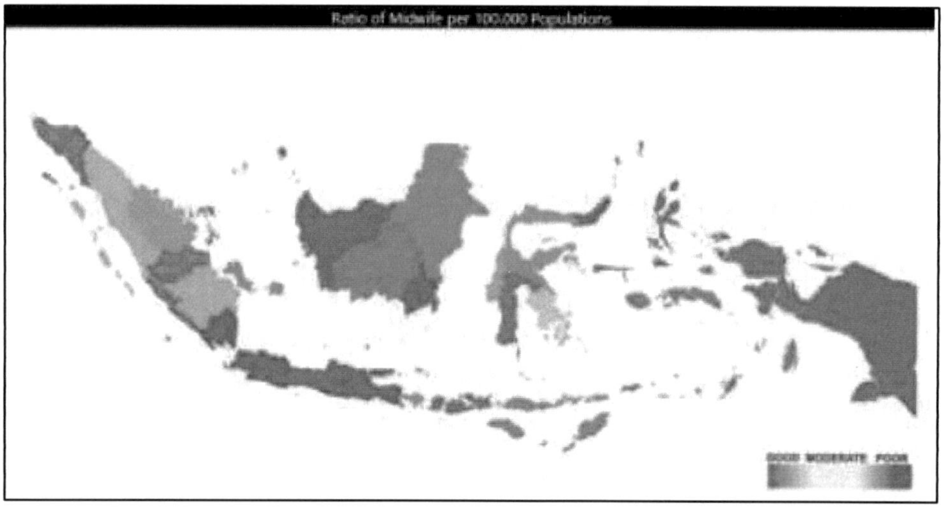

Figure 7. Distribution of midwives ratio as graded map

5. Conclusions

Based on the implementation of maternal and children health profile dashboard, there are several conclusions that can be drawn as follows:

- The research shows that the implementation of data warehouse and business intelligence as a dashboard application able to integrate and extract a number of fragmented data sets into an easy and quickly inferred information about the current situation of mother and child health care service.
- By applying an online analysis technology and several visualization techniques, our system has been able to recognize parameters and provincial entities that need performance improvement to conform the national standards of maternal and child health care services. We identify that child health services in Indonesia have seen a better performance than the maternal ones. Particularly in the provincial analysis level, we found that most provinces in eastern part of Indonesia have health service performances under the national standard.
- This research has applied a common data analysis method which have some limitations to extract new insight in the data set. Hence for the future studies, we recommend an expansion and deeper analysis of maternal and children health data set by applying some data mining techniques, such as associations to identify the rules for combining a number of health

parameters, clustering algorithm to understand the health condition in the past, and classification method to predict the future health events.

References

[1] Ministry of Health National *Health Profiles of Indonesia*

[2] Fitri D, Artha Y and Susila B 2013 Maternal Health Care and Neonatal Mortality *National Public Health J.* vol **7** no **8** pp 373–7

[3] Ministry of Health 2017 *Performance Report of the Directorate General of Public Health 2016 in Indonesia*

[4] UNDP 2016 *Sustainable Development Goals in Indonesia* http://www.id.undp.org/content/indonesia/en/home/sustainable-development-goals.html

[5] Iswarno M, Hasanbasri and Lazuardi L 2013 Analysis of policy implementation: stakeholder perspective in maternal and child health policy at Kepahiang Regency *J. Indones. Heal. Policy* vol **02** no **02** pp 77–85

[6] Widoyo R 2016 Measles vaccination and vitamin a for prevention of pneumonia in Indonesia *Natl. Public Heal. J.* vol **11** no **1** pp 46–50

[7] USAID 2015 *Technical Report: Strengthening Measurement of Pregnant and Newborn Babies Based on Facilities*

[8] Patwardhan M, Eckert LO, Spiegel H, Pourmalek F, Cutland C, Kochhar S, Gonik B and Brighton Collaboration Maternal Death Working Group 2016 Maternal death : Case definition and guidelines for data collection, analysis, and presentation of vaccination safety data *Vaccine* vol **34** no **49** pp 6077–83

[9] Aljunid SM, Srithamrongsawat S, Chen W, Bae SJ, Pwu RF, Ikeda S and Xu L 2012 Health-care data collecting, sharing, and using in Thailand, China mainland, South Korea, Taiwan, Japan, and Malaysia *Value in Health* vol **15** no **1** pp 132-8

[10] Kruk ME, Leslie HH, Verguet S, Mbaruku GM, Adanu RM and Langer A 2016 Quality of basic maternal care functions in health facilities of five African countries: an analysis of national health system surveys *The Lancet Global Health* vol **4** no **11** pp 845–55

[11] Kassebaum N 2016 Global, regional, and national levels of maternal mortality, 1990 – 2015: a systematic analysis for the global burden of disease study 2015 *The Lancet Global Health* pp 1990–2015

[12] Rai RK 2013 District level coverage gap in maternal, newborn and child health care services in India *J. Epidemiol. Glob. Health* vol **2** no **4** pp 221–4

[13] Muhammad S 2016 The maternity dashboard: how effective is it in improving maternity care? *Obstet. Gynaecol. Reprod. Med* vol **26** no **9** pp 276–9

[14] Kimball R and Ross M 2013 *The Data Warehouse Toolkit: The Complete Guide to Dimensional Modeling* (John Wiley & Sons)

[15] Vilarinhoa S, Lopesa I and Sousaa S 2017 Design procedure to develop dashboards aimed at improving the performance of productive equipment and processes *Procedia Manufacturing* **11** pp 1634-41

[16] Martin N, Bergs J, Eerdekens D, Depaire B and Verelst S 2018 Developing an emergency department crowding dashboard: A design science approach *Int. Emergency Nursing* **39** pp 68–76

[17] Franklin A et al 2017 Dashboard visualizations: Supporting real-time throughput decision-making *J. of Biomedical Informatics* **71** pp 211–21

[18] Lee K, Jung SY, Hwang H, Yoo S, Baek HY, Baek RM and Kim S 2017 A novel concept for integrating and delivering health information using a comprehensive digital dashboard: An analysis of healthcare professionals' intention to adopt a new system and the trend of its real usage *Int. J. of Medical Informatics* **97** pp 98–108

[19] Simms RA, Ping H, Yelland A, Beringer AJ, Fox R and Draycott TJ 2013 Development of maternity dashboards across a UK health region; current practice, continuing problems

European J. of Obstetrics & Gynecology and Reproductive Biology **170** pp 119–24

Surviving the disconnections: The use of information systems in Indonesian public health facilities

Fathul Wahid[1], R. Teduh Dirgahayu[2], Almed Hamzah[3], Hari Setiaji[4]

Department of Informatics, Universitas Islam Indonesia

{[1]fathul.wahid, [2]teduh.dirgahayu, [3]almed.hamzah, [4]hari.setiaji}@uii.ac.id

Abstract. The adoption of information systems in public health management in Indonesia aims to manage the data properly in order to improve the effectiveness of the provision of health services. However, based on previous works, it does not live up to expectation. This paper reports an insight coming from the field about the use of various and disconnected information systems in one place without sufficient coordination mechanism among those systems. The systems were developed by the superordinate body and it is compulsory for the public health facilities to adopt and implement the systems. This raises several problems related to the quality of the system itself, human resources capability, ambiguous procedures, various reports with the relatively same data, and unclear incentive mechanism for staffs. We conceptualize the problems of disconnections among three elements: actors, context, and information systems. These lead to different types of disconnections: between government agencies at national level, between information systems, between information systems and the context, between government agencies at national and local level, and between information systems and available staff capacity. This study also points out practical implications, which includes the use of more holistic approach to design and develop information systems, to develop an integrated information infrastructure for public health facilities.

Keywords: system integration, silo thinking, health information system, information infrastructure, Indonesia, developing countries.

1. Introduction

Previous works report that public health management in developing countries face several challenges, including lack of facilities and healthcare systems [1]. The use of information systems in the health sector has been advocated as this can help to improve the quality of healthcare service and policy making [1]–[3], which is referred as evidence-based public health policy [4]. This idea has also strong connection with public health surveillance systems [5]. Both surveillance and policy formulation are dependent on quality data [5] that can be provided with the help of well-designed and integrated information systems [6]. In this case, the data are presented in an electronic format so that they are relatively inexpensive to obtain and use [7]. From the extant literature, we understand that the improperly managed data will lead to the poor policy formulation in the health sector [3].

In this context, the availability of integrated health information systems is necessary. The question is, what happens if the information systems are in place, but the integration among them is absent. This study pays attention to the context where various information systems are already in place. The intriguing question here is: does the existence of many information systems really contribute to the better quality of public health management?

Previous study [8] pointed out that evidence-based public health policy modeling has to cope with several challenges in synthesizing and interpreting evidence, taking evidence into decisions, and using evidence at different levels. Our previous study [9] unveils that these challenges are found in the Indonesian context, where segregated health information systems are in place. However, there is no readily available information that explains this phenomenon. Hence, in this study we seek to answer the following question: *how to explain segregated health information systems in Indonesian public health facilities* (*Pusat Kesehatan Masyarakat* [Puskesmas])? Answers to this question will provide insights to the former question. In answering the question, we conducted an interpretive case study involving several public health facilities.

The remaining of the paper is structured as follows. Section 2 describes a literature review on the area of concern, i.e. the use of information systems in managing public health. In Section 3, we explain the research context and findings, which are then discussed in the subsequent section. Conclusion and recommendations in Section 4 bring this paper to end.

2. Conceptual Basis: Evidence-Based Public Health Policy

According to Brownson et al. [10], evidence-based public health policy has three interconnected domains: process, content, and outcome. Process domain is intended to understand approaches to improve the likelihood of policy adoption. Content domain is related to the identification of specific policy elements that are deemed to be effective. Outcome domain deals with the potential impact of implemented policy. Several actions need to be taken in this regard to prepare and communicate data, to use analytical tools in more effective way, to manage policy surveillance, and to monitor outcome of various types of evidence [10], [11].

Several factors influence evidence-based decision making [12]: research evidence, resources, and population characteristics (see Figure 1). These three factors altogether with environmental and organizational context inform the decision-making. However, we need to note that evidence may manifest in different forms, ranging from subjective (e.g. personal experience) to objective data (e.g. scientific literature). Recorded data by the public health facilities lies between these two extremes (see Figure 2). Administrative data from public health facilities are also beneficial as reference for planning and surveillance [7].

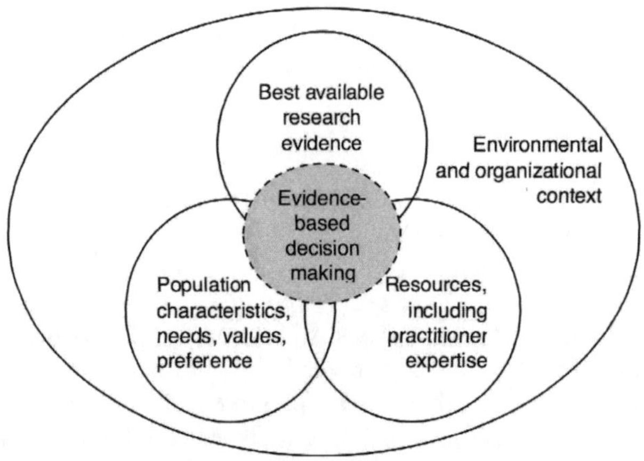

Figure 1. Determinants of evidence-based decision making.
Source: Satterfield et al. [12].

Further, public health surveillance systems may be evaluated by examining their properties, such as simplicity, flexibility, data quality, acceptability, sensitivity, predictive value positive, representativeness, timeliness, and stability [5].

In setting the roadmap for planning and improving evidence-based policy, Spencer [13] proposes a conceptual framework (see Figure 3) consisting of a continuum of quality of policy (as an example of practices; from emerging, promising, leading, to best) which is determined by two interconnected elements: (a) public health impact (effectiveness, reach, feasibility, sustainability, and transferability) and (b) quality of evidence (weak, moderate, strong, and rigorous). However, further conceptualization is needed, for example, to measure the quality of evidence.

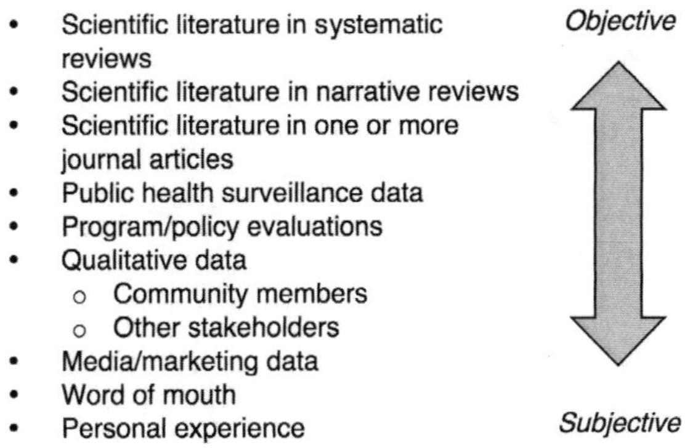

Figure 2. Different forms of evidence.
Source: Brownson et al. [10]

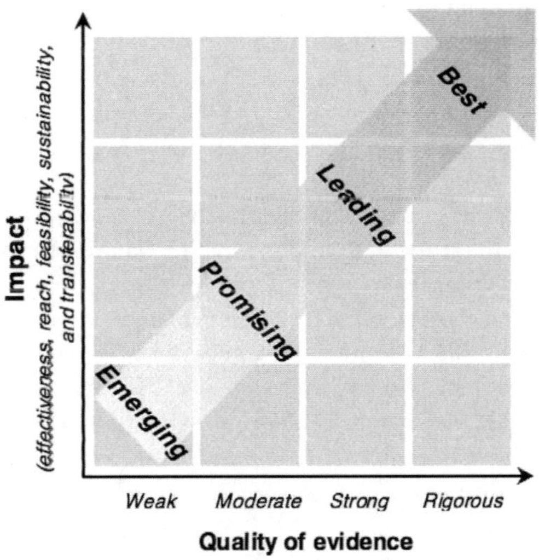

Figure 3. A conceptual framework for planning and improving evidence-based practices.
Source: Spencer [13]

Implementing evidence-based policy is not without barriers. Previous studies [10] has documented these barriers, such as lack of value placed on prevention, insufficient evidence base, mismatched time horizons, power of vested interest, researchers isolated from the policy process, and complex policy making process. In China, implementation of evidence-based policy faces some barriers which are related to (i) mentality and capability of individual and organisational actors and (ii) the absence of mechanism to translate scientific findings from research into policy [14]. In doing so, they develop several yardsticks, such as improvement of knowledge, attitudes, and practice; development of capability and intermediation; and establishment of a translation system [14].

This recent study can be considered as a preliminary stage focusing on mapping the use of information systems in public health facilities, which is needed in policy construction. In this case, an integrated system is necessary [15]. Otherwise, fragmented and overlapping systems will be in place and this often happens in some contexts [16]. This situation will lead to more severe problems, such as questionable data quality and unresponsive policy health policies [3].

3. Findings and Discussion

In the previous work [17], we found eight important insights, i.e. various information systems have been used by public health facilities, data entry redundancy, top-down decision model in developing information systems, lack of human resources capability, ambiguous job descriptions, overwhelming different reporting procedures, the absence of standard procedures to verify reports, and lack of incentive scheme.

These insights trigger some important issues that can be discussed further. It is obvious that the main problems here are 'disconnections'.

a. Types of disconnection

First, 'sectoral ego' or silo thinking is commonplace. This public secret in the public sector management in Indonesia[1] still extensively influences the design and the development of information systems. From inter-organizational perspective, this is problem of **disconnection between government agencies at national level**. As its consequences, a series of disconnected information systems was produced. From technical perspective, this creates **disconnection between information systems**.

To ensure the optimal benefits of the information systems, in the future, a better orchestration in designing and developing the systems is a must. One of our informant asserts and tells her expectation, "*... the systems should be crafted into a single system to provide more integrated data. But, this needs good infrastructure and qualified human resources*." Silo thinking has been found to a great burden in many context [18]. Various solutions have been advocated, include sticking on process thinking, defining standards and interoperability, developing ERP-like solution, and building enterprise architecture (EA) [18].

Second, it seems that the developers of new information systems always start from scratch and neglect the installed base, i.e. the information systems that have been in use to be developed further. This neglect has closed the opportunity to take the plethora of information systems into the next level as information infrastructure e.g. [19], [20]. This is a problem of **disconnection between information systems and the context**. The context here is the existence of installed-base, which is not taken into consideration when developing new information systems. This lack of orchestrated systems to some extent will create derived problems, such as energy consuming reporting systems. One informant unhappily criticized, "*if we use the systems, we should able to send the report immediately*." Another informant adds, "*... the report format should be unified*."

Third, from the findings, we understand that the public health facilities have been rarely involved in designing and developing information systems. Changes on the systems may take a long time to be accommodated as the bureaucratic communication between actors involved. This indicates another problem of **disconnection between government agencies at national and local level**.

Fourth, each information system needs to be operated and it needs specific attention form the staffs of public health facilities. What we found on the ground is that the staffs often have no sufficient skills to operate the information systems. This is the problem of personnel capacity, which leads to another **disconnection between information systems and available staff capacity**.

b. Conceptualizing the disconnections

---

[1] Be aware that 'sectoral ego' or 'ego sektoral' is an Indonesia-originated term to refer the similar phenomenon as silo thinking does.

Reflecting further on the types of disconnections that emerge from the findings, we may demystify the relationship between three elements: **actors**, **context**, and **information systems** (see Figure 4). We found four types of disconnections (X1, X2, X3, and X4). X1 is the most complex disconnections as it involves government agencies at the national and local level, including the public health facilities. In this point, we found two disconnections between actors, i.e. (a) **between government agencies at the national level** (X1a), such as between the Ministry of Health and the Ministry of Religious Affairs, in the case of provision of reports related to pilgrimage (hajj), and (b) **between government agencies at the national and local level** (X1b), Example of this disconnection is the absence of the involvement of public health facilities in designing information systems, which are elitist as it involves actors from upper level only.

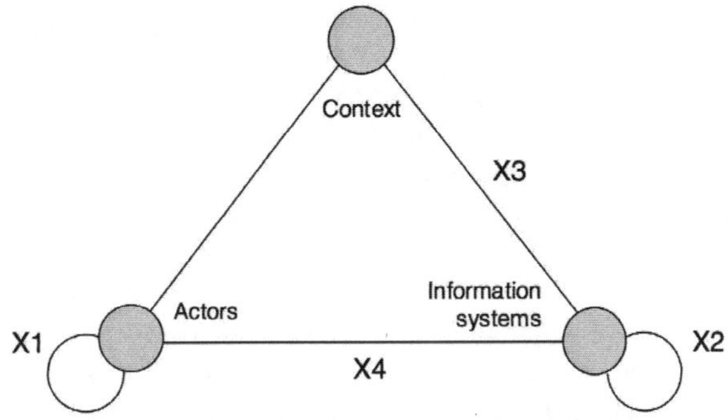

Figure 4. Elements of the disconnection.

As a result, fragmented information systems are developed and in place. This creates another problem, i.e. **disconnection between information systems** (X2). This disconnection has several derived consequences, including the validity of data, laborious data entry, and discouraged working morale.

Another unfavourable consequence of disconnection between actors, especially between national and local actors, is that the information systems do not fit with the context. Here, another problem arises, i.e. **disconnection between information systems and the context** (X3).

Each information system demands for specific skills to operate it. Our findings indicate that workarounds are taken to assign persons that are responsible for each information system. The problem is that they do not always have the needed skills. This creates **disconnection between information systems and available staff capacity** (X4).

4. Concluding Remarks

We have found many problems without clear and well-designed solutions regarding to the use of information systems in Indonesian public health facilities. These problems relate to lack of integration of various information systems used by health facilities, data entry redundancy, top-down decision model in developing information systems, lack of human resources capability, ambiguous job descriptions, overwhelming different reporting procedures, the absence of standard procedures to verify the reports, and lack of incentive scheme. We identified four factors that cause these problems, i.e. silo thinking, neglecting the legacy systems, uninvolved health facilities in the development of information systems, and lack of staff capability.

These problems create various disconnections among actors, information systems, and context. We conceptualize the problems into different type of disconnections: between government agencies at national level, between information systems, between information systems and the context, between government agencies at national and local level, and between information systems and available staff capacity.

The findings lead to several research avenues. First, our data does not reveal the motivation between these segregated information systems. Better information about the motivation may provide a more complete picture of the phenomenon under investigation. Second, further research may also pay attention to strategies to solve the problems on the ground, by for example, redesigning all the information systems into a more integrated one, or providing "a hub" that enables all the information systems to communicate to each other. Another systematic data collection is needed to provide contextualized design or solution.

References

[1] Cline GB and Luiz JM 2013 Information technology systems in public sector health facilities in developing countries: the case of South Africa *BMC Medical Informatics and Decision Making* vol **13** no **1** p 13

[2] Blaya JA, Fraser HS and Holt B 2010 E-health technologies show promise in developing countries *Health Affairs* vol **29** no **2** pp 244–51

[3] Pappaioanou M, Malison M, Wilkins K, Otto B, Goodman RA, Churchill RE, White M and Thacker SB 2003 Strengthening capacity in developing countries for evidence-based public health:the data for decision-making project *Soc. Sci. & Medicine* vol **57** no **10** pp 1925-37

[4] Rosen R 2000 Applying research to health care policy and practice: medical and managerial views on effectiveness and the role of research *J. of Health Services Res. & Policy* vol **5** no **2** pp 103–8

[5] German RR, Lee LM, Horan J, Milstein R, Pertowski C and Waller M 2001 Updated guidelines for evaluating public health surveillance systems *MMWR Recomm Rep* **50** pp 1-35

[6] Jacucci E, Shaw V and Braa J 2006 Standardization of health information systems in South Africa: The challenge of local sustainability *Inform. Technol. for Development* vol **12** no **3** pp 225–39

[7] Virnig BA and McBean M 2001 Administrative data for public health surveillance and planning *Annu. Rev. Public Health* vol **22** no **1** pp 213–30

[8] Law C 2010 Policy and evidence-based public health *Evidence-based Public Health: Effectiveness and Efficiency* ed A Killoran and M P Kelly (Oxford: Oxford University Press) pp 16–26

[9] Wahid F, Dirgahayu RT, Hamzah A and Setiaji H 2017 When many systems do not make you happy: The case of disconnected information systems in Indonesian public health facilities *Proc. of The Int. Conf. on Information Technology and Digital Applications (ICITDA) 2017*

[10] Brownson RC, Baker EA, Deshpande AD and Gillespie KN 2017 *Evidence-based Public Health* (Oxford: Oxford University Press)

[11] Brownson RC, Chriqui JF and Stamatakis KA 2009 Understanding evidence-based public health policy *Am. J. Public Health* vol **99** no **9** pp 1576–83

[12] Satterfield JM, Spring B, Brownson RC, Mullen EJ, Newhouse RP, Walker BB and Whitlock EP 2009 Toward a transdisciplinary model of evidence-based practice *The Milbank Quarterly* vol **87** no **2** pp 368-90

[13] Spencer LM, Schooley MW, Anderson LA, Kochtitzky CS, DeGroff AS, Devlin HM and Mercer SL 2013 Seeking best practices: a conceptual framework for planning and improving evidence-based practices *Preventing Chronic Disease* vol **10** no **E207**

[14] Jiang F, Zhang J and Shen X Towards evidence-based public health policy in China *Lancet* vol **381** no **9882** pp 1962–4

[15] De Kadt E 1989 Making health policy management intersectoral: issues of information analysis and use in less developed countries *Social Sci. & Medicine* vol **29** no **4** pp 503–14

[16] Okuonzi SA and Macrae J 1995 Whose policy is it anyway? International and national influences on health policy development in Uganda *Health Policy Plan* vol **10** no **2** pp 122–32

[17] Wahid F, Dirgahayu RT, Hamzah A, and Setiaji H 2018 When more is less: the case of disconnected information systems in Indonesian public health facilities *IOP Conf. Series: Materials Science and Engineering.*

[18] Bygstad B, Hanseth O and Le DT 2015 From IT silos to integrated solutions: a study in e-health complexity *Proc. of the European Conf. on Information Systems (ECIS) 2015*
[19] Star SL and Ruhleder K 1996 Steps toward an ecology of infrastructure: Design and access for large information spaces *Inf. Syst. Res.* vol **7** no **1** pp 111–34
[20] Wahid F 2013 A triple-helix model of sustainable government information infrastructure: Case study of the eProcurement system in the Indonesian public sector *Building Sustainable Information Systems* ed H Linger et al (Berlin: Springer) pp 555–67

IOP Conf. Series: Materials Science and Engineering **482** (2019) 012015 doi:10.1088/1757-899X/482/1/012015

Implementation and performance assessment of the enhanced RC5 (ERC5) algorithm based on addition-then-append key expansion technique

Excel B. Villanueva[1], Bobby G. Gerardo[2], Ruji P. Medina[1]

[1]Technological Institute of the Philippines, [2]West Visayas State University
[1]excel.villanueva.0816@gmail.com, [2]bgerardo@wvsu.edu.ph, [3]ruji.medina@tip.edu.ph

Abstract. In a business, it is important to note that owners should always make sure that they give a hassle free service to their customers to attract their attention and will eventually, gain their trust and continous patronage of their service and RFID offers a solution for this concern. Traditional way to collect payment is still in use up to this day. This paper intends to implement the use of the ERC5 algorithm in an RFID – based payment scheme for the DMMMSU-NLUC Fasfood center as an encryption algorithm for securing sensitive information. Moreover, it also intends to assess the performance of the said algorithm in terms of speed and work efficiency compared to the original algorithm. Result shows that the authors were successful in implementing ERC5 as an encryption algorithm. Result also shows that the ERC5 outperforms the classic RC5 algorithm in both speed and work efficiency category.

1. Introduction

Radio Frequency Identification, or famously known as RFID, has been part of our daily lives, may be it on our transportation, shopping or even parking. As a matter of fact, this technology has been explored for as early as in the year 1948 [1] but actually rose to fame when one of the biggest retailing company called Wal-Mart opened a new door for inventory tracking and supply chain management based on RFID when they announced the use of RFID-enable shipments in the year 2005 [2]. Since then, this technology has been an essential part of business enterprise.

One useful way to implement RFID technology is in payment systems. As a matter of fact, both government and business ventures have been eyeing on the implementation of RFID for further innovation that they could offer to their clienteles, for example cashless payments [3]. This breakthrough is evident with the use of RFID in different areas such as transportation, healthcare and other business related services.

In a business, it is important to note that owners should always make sure that they give a hassle-free service to their customers to attract their attention and will eventually gain their trust and continous patronage of their service [4]. RFID offers this convenience to business because aside from its benefits, it also offers low cost in which can make the business owners invest in a worthwile technology [5].

The Don Mariano Marcos Memorial State University (DMMMSU) is one of the prestigious state universities and colleges in the Philippines with Atty. Benjamin P. Sapitula as the current president. It has been founded by the late President Ferdinand Marcos on January 15, 1981 under Presidential Decree 1778 that merges 5 different schools in the entire province of La Union during that time. Since day one of the current president of the university, Atty. Sapitula, it has been his mantra to "Embrace World-class Standards," and it is evident with the different completed and ongoing projects in the entire university. Sadly, the university still operates in semi-automated processes. For example in DMMMSU-NLUC, one of its campus, the traditional way to collect payment for meals at the fastfood center is still in use. This means that the estabishment uses pen and paper in recording their sales for the day and this may result to inaccurate data records because it is prone to human error, lost paper, etc. On the other hand, it will also cause long queue in paying for the ordered meals.

This paper intends to implement the use of the ERC5 algorithm in an RFID – based payment scheme for the DMMMSU-NLUC Fasfood center as an encryption algorithm for securing sensitive information. They would like to propose an RFID-based payment scheme for DMMMSU-NLUC Fastfood center, in which they will call DMMMSU-NLUC E-canteen System. This system is proposed to help the university embrace world-class standards through the use of current systems the world has to offer. Moreover, it also intends to assess the performance of the said algorithm in terms of speed and work efficiency compared to the original algorithm.

2. Review of Related Literature

In their previous paper, the authors [6] suggested a modified technique to solve one of the weaknesses of the classic RC5 algorithm, the slow encryption speed [7]. The suggested enhancement is called addition-then-append key expansion technique, a unique solution to solve the above mentioned weakness of RC5 algorithm. This enhancement is very simple as it only requires common computer operation such as adding and appending of values to produce the needed key material. This enhancement also uses 2 more blocks and an additional combination of bitwise operations.

In business, it is always wise to keep sensitive information from anyone as secret as possible, which is why implementing a security protocol is important. In RFID, securing the tag unique number is the wisest way to secure information of the users [8].

To secure the unique tag numbers in an RFID-based transactions, various researchers have explored different ways to implement security protocols in their system and even suggested different methods. To name a few, references [9] [10] both suggested the use of lighweight algorithm for authentication purposes. For example, the use of a 128-bit AES algorithm was suggested. Their approach offers a lighweight implementation of the said algorithm to RFDI technology for authentication purposes. And results shows that they were able to successfully implement the said algorithm in authenticating RFID tags.

To determine the performance of an algorithm, assessment or evaluation has been a practice of various authors. For example, [11] [12] [13] [14] performed an evaluation in their own versions or enhancements in the existing algorithm and their results show that in determining the performance of an algorithm can help the author establish a concrete rebuttal or justification in their enhanced algorithm. Some of these researches also proved that their enhanced algorithm performed better than the existing algorithm. Moreover, evaluating the performance can pave a way for further improvement for a specific area in the algorithm thus opening a continuation of the research for the author themselves or other researchers.

3. Proposed Method

3.1. DMMMSU-NLUC E-canteen System

Figure 1 shows the diagram of the proposed method. As perceived in the diagram, each client or customer must scan his/her RFID-enabled card to the scanner to pay for his/her order. But it is important to note that the card should have a sufficient balance in order to avail the service, otherwise, he/she must pay in cash.

The manager can print report such as monthly and annual income of the canteen from the system. Aside from that, the clients can also request for their personal copy of their monthly or annual meal records. This feature will help both the client and manager keep tract of their meal spending and income respectively. Aside from these reports, additional feature like a notification text message will be sent to the client's personal phone number for each successful transaction and also if the remaining balance is below the allowable amount. This will increase the security of the system because, if the transaction is not legit or the card's owner did not make the transaction, he/she will be notified as soon as possible and can report it to the employee or management immediately. Reloading of the card can be done in different canteen's loading outlets.

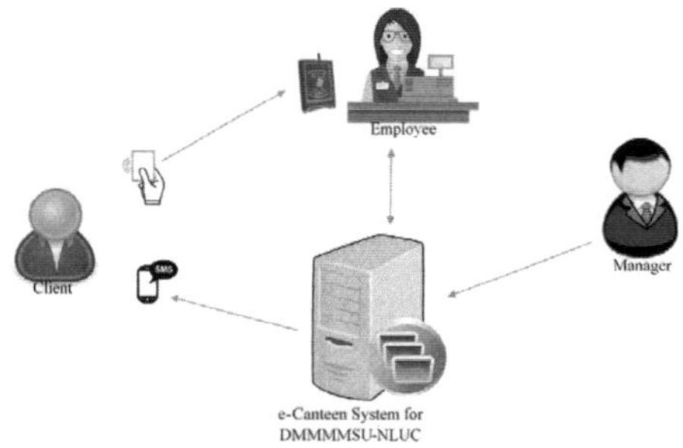

Figure 1. Proposed DMMMSU-NLUC *e*-Canteen System Diagram

3.2. ERC5 Algorithm based on Addition-then-Append Key Expansion Technique
The authors suggested an enhancement to solve the limitation of RC5 in their previous paper [6], which is slow in encryption speed. The enhancement is called addition-then-append key expansion technique, a unique way to expand the generated source key. Moreover, it also includes 2 more blocks and bitwise operations; the 1's complement and left & right bit rotate combination, this bitwise operation is responsible for producing a cipher text with the same size as the plaintext.

3.2.1 Encryption Phase Figure 2 shows the encryption process of the ERC5 algorithm. The algorithm comprises of several steps, which includes: (1) addition-then-append key expansion phase, (2) pre-processing of plaintext, (3) division of plaintext, and (4) block XOR-swapping. All these process will produce a cipher text with the same size as the plaintext.

- *Addition-then-append Key Expansion Phase*

Figure 3 shows the diagram of Addition-then-append Key Expansion technique. This technique involves 3 easy steps namely: (1) random number generation, (2) addition and (3) appending of values. The goal of this technique is to remove the use of word size, array, magic constant and golden ratio in which the original algorithm uses to generate its key.

The key expansion starts when the system generates a random number source key based from system date and time. The source key will be expanded through the 3 steps mentioned above. As perceived in figure 3, the source key will be added to any generated random number (which will be repeated twice). The algorithm will not generate the same random numbers for each two rounds, thus increasing the confusion of this phase. To produce the key material, all values from the previous steps will be appended with the original length of the source key. This key material will be needed in the next process, which is the pre-processing of the plaintext.

- *Plaintext Pre-processing*

This process accepts any input from the user, be it a password or series of random numbers, and it will be converted into its hexadecimal value. At the same time, the key material, which is the result of the previous phase, the key expansion, will also be converted in its hexadecimal value. Both values will be XOR-ed to produce the plaintext. This becomes an added confusion for the enhanced algorithm because without the knowledge to the value of the key material, the input from the user cannot be easily revealed. This phase is shown in the upper portion of figure 2.

- *Plaintext Division*

The plaintext, which is the result of the previous process, is divided into two major divisions, called left and right division. Furthermore, each major division will be divided into two more blocks, so each major blocks has its own pair of minor blocks and series of bitwise operations, which means that the both division does not possess the same arrangement of bitwise operations. The 2 major blocks will operate simultaneously for 12 rounds. This is shown in the middle portion of figure 2.

- *Left Division.* In this division, it will perform the combination of XOR and left & right bit rotate. This is almost the same as the original version of RC5 algorithm except that it only uses left bit rotate, wherein in the ERC5 algorithm the left & right bit rotation was used.
- *Right Division.* This division performs an additional bitwise operation along with XOR and left & right bit rotate, the 1's complement, or NOT bitwise operation.

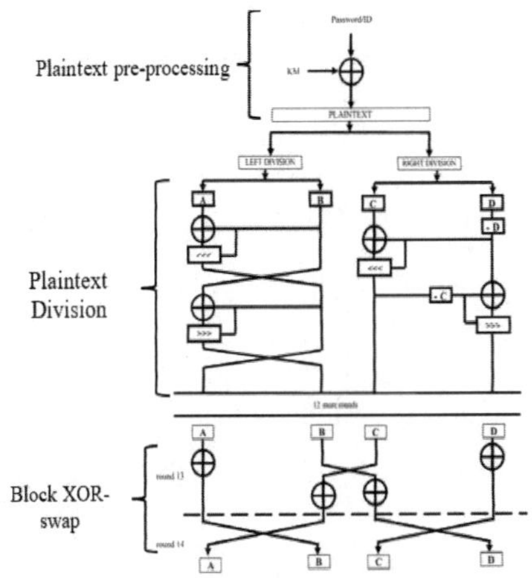

Figure 2. Encryption Phase of ERC5 Algorithm

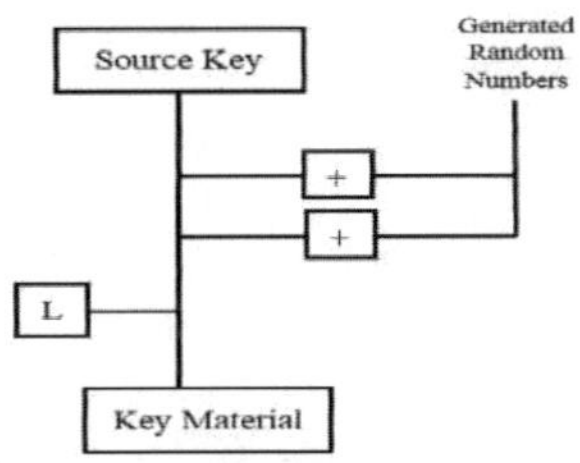

Figure 3. Addition-then-Append Key Expansion Technique

- *Block XOR-swap.*

When the previous steps are finished, additional 2 more iterations will be performed. In this last two rounds, the algorithm will perform another bitwise operation which is the XOR swap. This operation is responsible for juggling the values of the 4 minor block. This blocks wisll then be appended to produce the cipher text. This is shown in the lower portion of figure 2.

3.2.2 Decryption Phase This process is the inverse version of the encryption process. This means that the XOR swap will be performed first to place the minor blocks into its respective group. After the blocks are properly placed, the algorithm will now perform the corresponding inverse operation of each major blocks to produce the plaintext.

4. Results and Discussions

4.1 Implementation of ERC5 Algorithm in DMMMSU-NLUC E-canteen System

In developing the prototype, the authors made use of the latest version of PHP programming language, coded in a laptop with an Intel ® Core ™ i3-3217U CPU processor @ 1.80GHz (4 CPUs), ~1.8GHz with a memory size of 4096MB RAM. They also made use of a Mifare RFID reader and RFID cards for simulation purposes.

The authors were successful in implementing the ERC5 algorithm in encrypting the RFID card number in the database, as shown in Figure 4. In the prototype, once the user scans its RFID card in the reader, the system will encrypt its unique id using the ERC5 algorithm, that value will be stored in the database. The algorithm is used to protect sensitive information, such as the unique card number of the clienteles in order to prevent information thief.

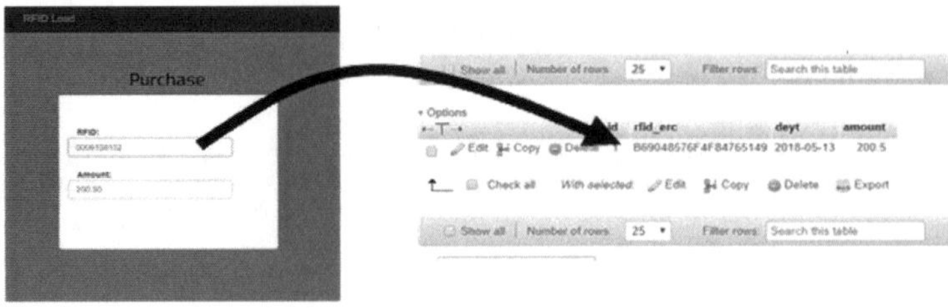

Figure 3. Implementation of ERC5 in DMMMSU-NLUC *e*-Canteen System

4.2 Performance Evaluation of ERC5 Algorithm

For algorithm speed comparison and evaluation, two sets of class/program were used, first is for the RC5 algorithm and second for the enhanced version of the algorithm. To evaluate the encryption and decryption speed of both algorithms, the authors made use of an embedded microtime () function in PHP**.** They ran the program simultaneously for 10 attempts for a common user input, say a 16-bit length password. The formula used to test the difference of the speed performance between the RC5 and ERC5 is shown in equation 1, where *r* stands for average time of the classic algorithm and *p* stands for average time of the ERC5. The formula used for ETPE is shown in equation (1). This evalution is used to determine the percentage of improvement on task execution [13]. The execution time average mean of the 2 algorithms was used as input data for the evaluation as shown in equation (2), wherein Latency (T_{old}) is the average mean of the classic RC5 algorithm and the Latency (T_{new}) is the average mean of the ERC5 algorithm.

$$\text{Performance Comparison} = \left[\frac{(R-P)}{R}\right] \times 100 \quad (1)$$

$$\text{ETPE }(\%) = \left[\frac{\text{Latency }(T_{old})}{\text{Latency }(T_{new})} - 1\right] \times 100 \quad (2)$$

4.2.1 Speed Performance The encryption speed simulation results with respect to the input size and required times of the algorithm compared with the existing algorithm is shown in table 1. As shown in the table, the ERC5 algorithm beats the original algorithm by 76% with an average speed of 56.38 milliseconds (ms) while the later has an average speed of 235.05 ms. On the other hand, the same table shows the comparison between the 2 algorithms in terms of decryption speed. As perceived in the table, the ERC5 algorithm outruns the original algorithm by 42% with an average speed of 147.31 ms while the later has an average speed of 260.80 ms.

4.2.2 Execution Time Performance Evaluation(ETPE) Table 2 shows the execution time performance efficiency results. It is notable that the ERC5 is more efficient in both encryption and decryption than the classic RC5 algorithm with a percentage of 316.90 and 77.04 respectively. This means that there is a big improvement when it comes to work efficiency in ERC5 algorithm as compared to the classic RC5 algorithm.

Table 1. Speed Performance Comparison

Speed	*RC5*	*ERC5*	*Difference*
Encryption	235.05 ms	56.38 ms	75.73%
Decryption	260.80 ms	147.31 ms	42.05%

Table 2. Execution Time Performance Evaluation

	Mean		**ETPE (%)**
	Latency (Told)	**Latency (Tnew)**	
Encryption	235.05 ms	56.38 ms	316.90
Decryption	260.80 ms	147.31 ms	77.04

5. Conclusion and Future Works

The authors were successful in implementing the ERC5 algorithm in an RFID-based payment scheme for DMMMSU-NLUC Fastfood center in which they called DMMMSU-NLUC *e*-Canteen System. Additionally, they were able to provide solution to replace the manual procedure of the said establishment. Moreover, the authors evaluated the performance of both the classic rc5 algorithm and the ERC5 in terms of its speed and work efficiency. Experimental shows that in both category, the ERC5 algorithm positively outruns the existing algorithm.

For future work, the authors intend to evaluate the security level of the ERC5 algorithm with the use of avalanche effect. Furthermore, the prototype used in this study will undergo usability testing to finalize its design and other system features according to the preference of the end-user.

References

[1] Landt J 2005 The history of RFID *IEEE Potentials* vol **24** no **4** pp 8-11
[2] Weinstein R 2005 RFID: A Technical overview and its application to the enterprise *IEEE Computer Society* vol **7** no **3** pp 27-38
[3] Lokshina I and Thompson W 2013 Analysis of design requirements for electronic tags from business viewpoint *IEEE 16th Int. Symp. on Wireless Personal Multimedia Communication* pp 1-6
[4] Caya MV, Losaria NF, Manzano MJ, Tan HK and Pellegrino RV 2017 Cashless transaction for resort club amenities using RFID technology *IEEE 9th Int. Conf. on Humanoid, Nanotechnology Information Technology, Communication and Control, Environment and Management* pp 1-5
[5] Weis SA 2007 RFID (Radio Frequency Indentification): principles and applications *System* vol **2** no **3** pp 1-23
[6] Villanueva EB, Medina RP and Gerardo BD 2018 An enhanced RC5 (ERC5) algorithm based in simple random number key expansion technique *IEEE Symp. on Computer Applications and Industrial Electronics (ISCAIE)* pp 134-8
[7] Bhanot R and Hans R 2015 A review and comparative analysis of various encryption algorithms *Int. J. of Security and Its Application* vol **9** no **4** pp 289-306
[8] Juels A 2006 RFID Security and privacy: a research survey *IEEE J. on Selected Areas in Communications* vol **24** no **2** pp 381-94
[9] Feldhofer M, Dominikus S and Wolkerstofer J 2004 Strong authentication for RFID systems using the AES algorithm *Springer* vol **3156** pp 357-70
[10] Juels A and Weis S 2005 Authenticating pervasive devices with human protocols *Springer* vol **3621** pp 293-308
[11] Hong L, Wan Y and Jain A 1998 Fingerprint image enhancement: algorithm and performance evaluation *IEEE Transactions on Pattern Analysis and Machine Intelligence* vol **20** no **8**
[12] Jamel T 2016 Performance evaluation of the fast euclidean direction search algorithm for adaptive beamforming applications *Int. Conf. on Multidisciplinary in IT and Communication Science and Applications* pp 1-6
[13] Wenceslao F 2015 Performance efficiency of modified AES algorithm using multiple S-boxes *Int. J. of New Computer Architectures and Their Applications* vol **5** no **1** pp 1-9
[14] Zheng X 2016 Research on performance evaluation method based on neural network algorithm *IEEE Int. Conf. on Smart City and Systems Engineering* pp 524-6

A study of public key 'e' in RSA algorithm

C Intila[1,2], B Gerardo[1,3], R Medina[1]

[1] Technological Institute of the Philippines Quezon City, Quezon City, Philippines 1000
[2] Poblacion Southeast Zone, Lemery, Iloilo Philippines 5043
[3] West Visayas State University Iloilo City, Philippines 5000

[1]carloski_0822@yahoo.com, [2]bobby.gerardo@gmail.com, [3]ruji.medina@tip.edu.ph

Abstract. RSA Algorithm is public-key cryptography which exposed from factorization attack based on the public key 'e' and modulus 'n'. The study aimed to modify an RSA Algorithm based on generation of different value of public key 'e' and show different result based on original equation of RSA Algorithm. A modified RSA Algorithm was developed to generate different value of public key 'e' and compared the result based on the output value and speed of execution. The results show that the modification of RSA based on public key 'e' provides additional security.

1. Introduction

The development of internet application has grown so fast but needs secured cryptosystem for data communication. To obtain the user's satisfaction, high speed algorithms with enhanced security have to be developed [1].

Public key cryptosystem is suitable and a better answer for more secured data communication and it comprises of encryption, decryption, authentication that is design for better data security and privacy [2] [3]. There two categories to select for encryption, the symmetric and asymmetric. Symmetric encryption algorithms are AES [2], DES [4], RC5 [5] , RC6 [6], Blowfish [7]. Asymmetric encryption algorithms are RSA [8], ECC [9], Diffie Helman [10].

The RSA algorithm is best solution in asymmetric cryptography not only for its mostly used but for its large scale of applications[11] and provides better security providing large prime numbers for public key, private key 'd' and modulus 'n' generation [12]. The Rivest Shamir Adleman (RSA) is an algorithm based on public-key cryptography and is considered one of the great developments in IT security due to its resistance to multiple attacks [13].

RSA Algorithm is the mostly used public key cryptography and the demand for a better security is one of the challenges faced to continue the reliance of the users [14]. Nevertheless, RSA suffers from slow speed, a demand for key deposit, and is generally inappropriate for use in several systems [3]. With the speed and security issues, the researchers are encourage to propose new algorithms with multiple parameters towards better performance of secured network communication [1].

In the RSA Algorithm shows that the public key 'e' can be used to factor *n* respectively [15]. Different studies and researches[16][17][18][19] in modification of RSA Algorithm have been conducted with the modification of modulus 'n' but disregard the modification of public key 'e'. With this, the study aims to modify the RSA Algorithm based on public key 'e' and show the importance of its security in the encryption and decryption process.

2. Related Literature

Several studies conducted to RSA Algorithm with different factors to modify. The following are the studies: A better version of RSA algorithm with new element 'f' as replacement for the exclusion of modulos 'n'. It increases the security against factorization but slightly increase the time consumed [17]; another paper presented a one sender to many receiver asymmetric cryptosystem that uses RSA as a base. The proposed approach used Chinese Remainder Theorem (CRT) to address the limitation of multi-asymmetric RSA Scheme [8]; a modification comparing to Original RSA Algorithm in terms of speed and security. Their Modified RSA Algorithm compare to Original RSA Algorithm is better in terms of speed and security [20]; a modification of RSA Algorithm that utilized four prime numbers in generating modulo 'n' to improve the security against factorization attack[21][18]; a study that proposed a RSA Algorithm that uses two public key which increase the security in decrypting message but decrease its speed [22]; a research conducted where the proposed Modified RSA algorithm improves the security against the brute force attack [3]; a study developed a Modified RSA Encryption Algorithm (MREA) which improved security against factorization attack and brute force attack[16]; proposed a Binary RSA Encryption Algorithm. They developed a modified RSA Algorithm that transform original message to binary code that gives better security in decryption process but gives moderate time and space difficulty [23].

3. Methods

3.1 Scope and Limitations

The study limits its content in the modification of RSA Algorithm based on public key 'e'. The comparison will limit only between the Original RSA Algorithm and the Modified RSA Algorithm. The objective of the study is to give importance of the public key 'e' in modification of RSA Algorithm.

3.2 The Modification

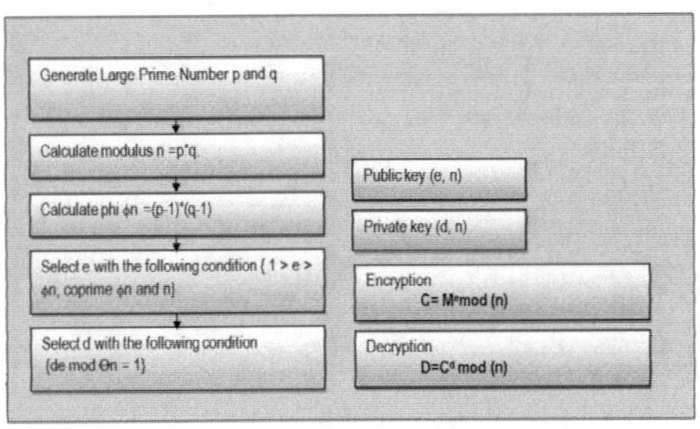

Figure 1.The Original RSA Algorithm Flow

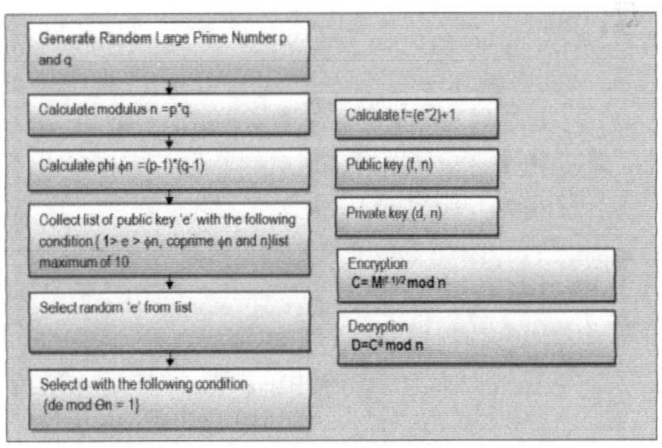

Figure 2. The Modified RSA Algorithm

Figure 1 shows the flow involved in the Original RSA Algorithm processes. Figure 2 shows the Modified RSA Algorithm of the study. Initial modification starts with the public key and followed by encryption formula.

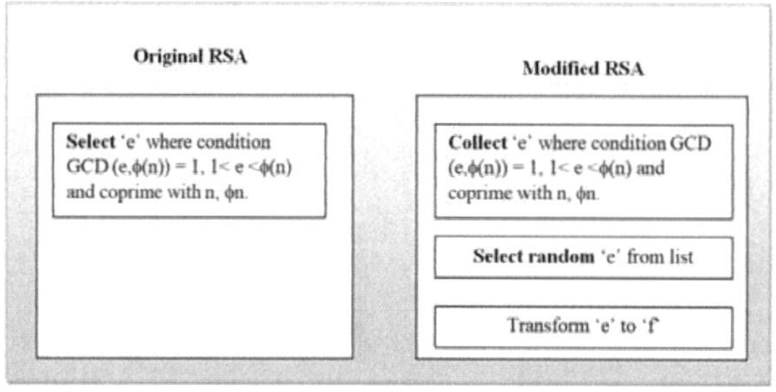

Figure 3. The Modification on Public Key 'e'

Figure 3 shows the modification of public key 'e' from original public key 'e' of RSA Algorithm. RSA includes three processes[24].

A. Key Generation

1. Select large prime numbers p and q.
2. Compute n = p * q.
3. Compute the $\phi(n) = (p-1) \times (q-1)$.
4. Select public key e.
   GCD $(e,\phi(n)) = 1$, $1< e <\phi(n)$ and coprime with n, ϕn.
5. Compute where condition "de(mod(phi n))=1"

B. Encryption

C=Me mod n

C. Decryption

$D=C^d \bmod (n)$

The steps below will show the modification process.

Modified RSA Algorithm Based on Public Key 'e'

The proposed study was to modify the Public Key 'e' value for more secure RSA Algorithm.

A. Key Generation

1. Select large prime numbers p and q.
2. Compute n =p*q.
3. Compute $\phi n =(p-1)*(q-1)$
4. **Collect e with the following condition { p > e >ϕn, coprime ϕn and n}(maximum of ten values)**
5. **Select random 'e' from list**
6. **Calculate f=(e*2)+1.**
7. Select d with the following condition {de mod n = 1}
8. Send Public key (f, n)
   The 'f' serve as a new public key which will hide the original e value.
9. Send Private key (d, n)

B. Encryption

$\mathbf{C = M^{((f-1)/2)} \bmod (n)}$

C. *Decryption*

$D = C^d \bmod (n)$

Sample

A. Key Generation

1. Select large prime numbers p and q. p=7 q=11
2. Compute modulus n =p*q. n
   =7*11

IOP Conf. Series: Materials Science and Engineering **482** (2019) 012016 doi:10.1088/1757-899X/482/1/012016

=77

3. Compute ϕn=(p-1)*(q-1)
   =(7-1)*(11-1)
   =6*10
   =60
4. **Collect list of public key 'e' with the following condition { p > e >ϕn, coprime ϕn and n}**
   e={13,17,23,27}(maximum of ten values)
5. **Select Random 'e' from the list.**
   e=23
6. **Calculate f=(e*2)+1.**
   = (23*2)-1
   = 46+1
   =47
7. Select d with the following condition {de mod n = 1}
   d=47
8. Send Public key (f, g) Public Key = (47,77)
9. Send Private key (d, g) Private Key = (47,77)

B. Encryption M=4

C = $M^{(f-1)/2}$ mod (n)
=$4^{(47-1)/2}$ mod 77
=4^{23} mod 77
= 70,368,744,177,664 mod 77
=9

D. Decryption

D =C^{d} mod (n)
= 9^{47} mod 77
= 7.0696504901e+44 mod 77
=4

The Original RSA Algorithm and Modified RSA Algorithm Based on Public Key Generation were developed using Java Programming Language. A modified key generation of RSA Algorithm will be implemented to improve its security. The method to modify the equation of public key will be utilized in concern with the process of encryption and decryption involved. The method to test the two algorithms is to differentiate the result of RSA in generating the public key 'e' between Original RSA Algorithm and Modified RSA Algorithm. Also, the two algorithms will be tested in the same hardware and software specifications. Figure 4 shows the method of the study.

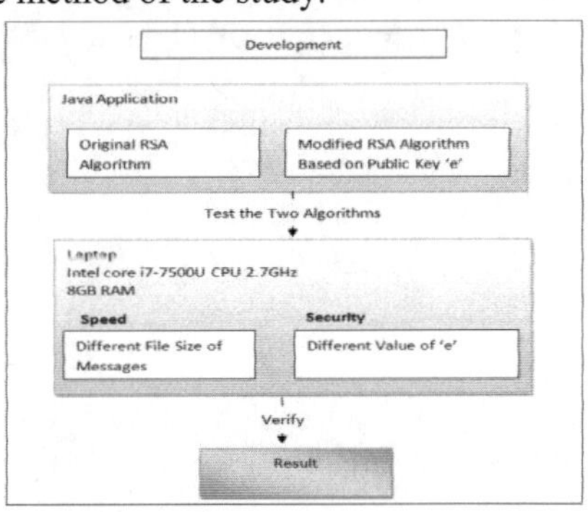

Figure 4. The Method of the Study

4. Result

The Original and Modified RSA Algorithm were implemented using Java Programming Language. The two algorithms were assessed by different file sizes of messages and recorded the efficiency in terms of encryption and decryption process. Both algorithms conducted a test to output cipher text based on formula of different related studies.

4.1 Result for Public Key 'e'

The first test conducted was generating different value of private key 'd' and cipher based on the different value of public key 'e' and message 'm'. The result of cipher text of the Original RSA and Modified RSA Algorithm on public key 'e' based on the formula "C=Me mod n" [16] [3][19].

The next three tables show the different result of cipher text with different value of public key 'e' in the Original RSA Algorithm.

Table 1. p=7 q=11

e	d	cipher		
		m=4	m=5	m=6
13	37	53	26	62
17	53	16	3	41
23	47	9	59	62

Table 2. p=5 q=11

e	d	cipher		
		m=4	m=5	m=6
13	37	9	15	51
17	33	49	25	41
23	7	9	15	51
27	3	49	25	41
29	29	14	25	29

Table 3. p=11 q=17

e	d	cipher		
		m=4	m=5	m=6
3	107	64	125	29
7	23	115	146	184
13	37	174	37	95
19	59	47	108	46
29	149	157	146	104

The Table 4 and Table 5 show the different result of ciphertext with different value of public key 'e' between Original RSA and Modified RSA Algorithm.

Table 4. Result for p=11 q=17

e(Original RSA)	e(Modified RSA)	d	Cipher	
			Original RSA	Modified RSA
3	7	107	64	115
7	15	23	115	166
13	27	37	174	115
19	39	59	47	0
29	59	149	157	0

Table 5. Result for p=5 q=11

e(Original RSA)	e(Modified RSA)	d	cipher		
			Original RSA	Modified RSA	A
13	27	37	9	49	49
17	35	33	49	0	0
23	47	7	9	0	0
27	55	3	49	0	0
29	59	29	14	0	0

4.2 Result for Performance

The following tables and graphs show the result of performance of the original and the modified RSA Algorithm. The two algorithms are developed using Java Programming Language. Table 6, Table 7, Graph 1, and Graph 2 show the encryption and decryption time using a Laptop with specification of Intel Core i7 2.7GHz, 8GB RAM, and with a 64 bit Windows 10 Operating System.

Table 6. Encryption Time

Decryption		
	Time in Second/s	
Size(KB)	Original	Modified
10KB	0	0
1000KB	2	2
5000KB	7	7
10000KB	16	16
20000KB	25	26
50,000KB	62	65
100,000KB	152	159

Table 7. Decryption Time

Encryption		
	Time in Second/s	
Size(KB)	Original	Modified
10KB	0	1
1000KB	2	3
5000KB	8	10
10,000KB	16	18
20,000KB	28	31
50,000KB	71	75
100,000KB	146	156

Figure 1. Encryption Time

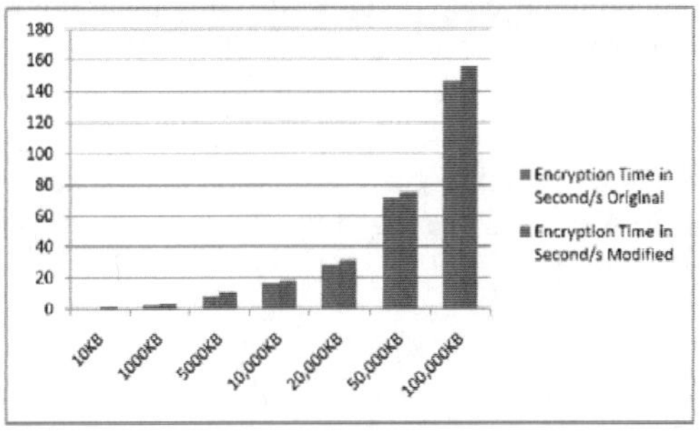

Figure 2. Decryption Time

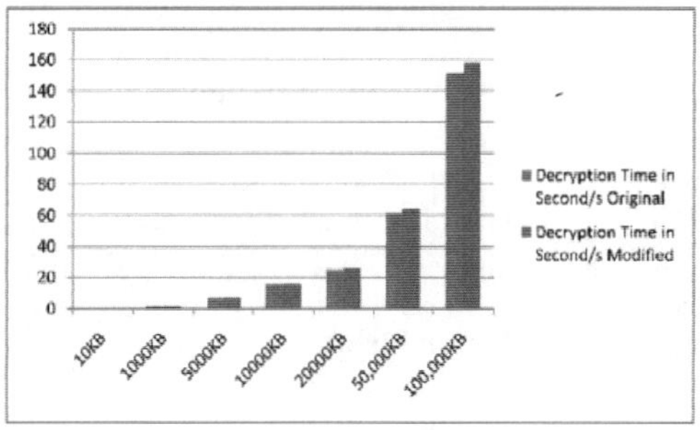

The table 6 and 7 show the execution time of different file sizes of messages of Original RSA Algorithm and Modified RSA Algorithm. The Graph 1 and Graph 2 display the comparison of execution time between Original RSA Algorithm and Modified Algorithm.

The table below shows the difference in time execution of encryption and decryption process between Original RSA and Modified RSA Algorithm.

Table 8. Result of Time Execution based on 'p' and 'q'

No	p	q	n	Time(second)			
				Encryption		Decryption	
				Original	Modified	Original	Modified
1	97	367	35,599	2	3	2	2
2	127	383	48,641	2	3	2	2
3	233	409	95,297	2	3	2	2
4	353	431	152,143	2	4	2	2
5	373	487	181,651	2	4	2	2
6	401	523	209,723	2	5	2	2
7	431	587	252,997	2	6	2	2
8	457	619	282,883	2	6	2	2
9	547	677	370,319	3	6	2	2
10	601	733	440,533	3	6	2	2

5. Findings and Analysis

5.1 Findings

The following are the results and findings of this study:

1. In the Table 1, Table 2 and Table 3, the result shows that with different value of public key 'e' designate a different value of private key and different value of public key 'e' will also produce different value of cipher text.
2. In the Table 4 and Table 5, the result shows that the Modified RSA Algorithm produce different value of cipher text from the original RSA Algorithm as the result for encryption process.
3. In the result of performance, Table 6, Table 7, Graph 1 and Graph 2, the Modified RSA Algorithm slightly increased its time complexity in encryption. In the decryption process, the Modified RSA Algorithm has almost the same time execution with the Original RSA Algorithm.
4. In the Table 8, the result shows that the Modified RSA Algorithm has slightly increased in time consumed in encryption but not in decryption process.

5.2 Analysis

Based on the findings of the study, in terms with the different valid value of public key 'e' designates a corresponding different value of private key 'd' and also produce a different value of cipher. In this regard, the level of security in encryption and decryption process increases. In terms of transforming the selected value of public key 'e' produced another different value of cipher which increases another level of security in the encryption and decryption process.

In terms of performance, the modification of public key 'e' by collecting the list of public key and transforming to new value will slightly increase the time complexity in the key generation and encryption process of RSA Algorithm but not in the decryption process.

6. Summary and Conclusion

6.1 Summary

Table 9. Summary of the Study

		Original RSA Algorithm	Modified RSA Algorithm
Performance	encryption	√	
	decryption	√	√
Security	Public key		√
	Encrypted Message		√

The Table 9 shows the summary result between the original RSA Algorithm and the modified RSA Algorithm. The table shows that the modified RSA Algorithm have increase the level of security in terms of public key 'e' and encrypted message while increase slight complexity in terms of performance.

6.2 Conclusion

This study presented different values of public key 'e' which were evaluated in the encryption and decryption process. Conclusion is established based on the two categories mainly on the security and performance and compared the Modified RSA Algorithm to Original RSA Algorithm.

In the result based on the performance, the modified RSA algorithm has slightly increased in time complexity in the encryption process. In the decryption process, the Modified RSA Algorithm is almost the same of time consumed compare to the original RSA Algorithm.

In the result based on the security, the Modified RSA Algorithm based on Public Key 'e' produced more complex result in terms of encryption process than the Modified RSA Algorithm. Different value of public key 'e' produced different value of private key and cipher and converting it to new values produced more complex values of cipher which makes the Modified RSA Algorithm based on Public Key 'e' more secured encrypted message which also increased security in the decryption process.

Finally, this study suggested that in the modification of RSA Algorithm, the Public Key 'e' is one of the factors that enhanced the security in the encryption and decryption process. Modification of public key 'e' will increase the security level of the RSA Algorithm

7. Future Enhancement

The implementation of this study uses the small list of public key 'e'. The next step of this work is to employ this to the bigger list and implement the modification of RSA algorithm based on public key 'e' with modification of modulus 'n' which are both exposed to public to increase the level of security of RSA Algorithm against factorization attack, and with the encryption and decryption process. Also, the Modified RSA Algorithm based on public key 'e' and modulus 'n' will be compared to other Modified RSA Algorithm in terms of security and performance.

References

[1] D. I. G. Amalarethinam and J. S. Geetha, "A Survey on Secured Communication with High Speed using Public Key Cryptography," vol. 7, no. 4, 2016.

[2] N. Khanezaei and Z. M. Hanapi, "A framework based on RSA and AES encryption algorithms for cloud computing services," *Proc. - 2014 IEEE Conf. Syst. Process Control. ICSPC 2014*, no. December, pp. 58–62, 2014.

[3] I. Jahan, M. Asif, and L. Jude Rozario, "Improved RSA cryptosystem based on the study of number theory and public key cryptosystems," *Am. J. Eng. Res.*, 2015.

[4] A. S. Alkalbani, T. Mantoro, and A. O. M. Tap, "Comparison between RSA hardware and software implementation for WSNs security schemes," *Proceeding 3rd Int. Conf. Inf. Commun. Technol. Moslem World ICT Connect. Cult. ICT4M 2010*, 2010.

[5] A. Biryukov and E. Kushilevitz, "Improved cryptanalysis of rc5," in *Lecture Notes in Computer Science (including subseries Lecture Notes in Artificial Intelligence and Lecture Notes in Bioinformatics)*, 1998.

[6] R. Rivest, M. J. B. Robshaw, R. Sidney, and Y. L. Yin, "The RC6 Block Cipher," *First Adv. Encryption ...*, 1998.

[7] B. Schneier, "The {Blowfish} encryption algorithm," *Dr. Dobb's J. Softw. Tools*, 1994.

[8] A. Mansour, A. Davis, M. Wagner, and R. Bassous, "Multi - Asymmetric Cryptographic RSA Scheme," pp. 1–8, 2017.

[9] P. Patel, R. Patel, and N. Patel, "Integrated ECC and Blowfish for Smartphone Security," in *Physics Procedia*, 2016.

[10] W. Diffie, W. Diffie, and M. E. Hellman, "New Directions in Cryptography," *IEEE Trans. Inf. Theory*, 1976.

[11] F. Lombardi and R. Di Pietro, "Transparent security for cloud," in *Proceedings of the 2010 ACM Symposium on Applied Computing - SAC '10*, 2010.

[12] R. Tripathi and S. Agrawal, "Comparative Study of Symmetric and Asymmetric Cryptography," *Int. J. Adv. Found. Res. Comput.*, vol. 1, no. 6, pp. 2348–4853, 2014.

[13] R. L. Rivest, A. Shamir, and L. Adleman, "A method for obtaining digital signatures and public-key cryptosystems," *Commun. ACM*, 1978.

[14] P. K. Arya, M. S. Aswal, and V. Kumar, "Comparative Study of Asymmetric Key Cryptographic Algorithms," *Int. J. Comput. Sci. Commun. Networks,* vol. 5, no. 1, pp. 17–21, 2012.

[15] D. Aggarwal and U. Maurer, "Breaking RSA Generically Is Equivalent to Factoring," vol. 62, no. 11, pp. 6251–6259, 2016.

[16] R. S. Dhakar, "Modified RSA Encryption Algorithm (MREA)," pp. 2–5, 2012.

[17] J. Sahu, V. Singh, V. Sahu, and A. Chopra, "An Enhanced Version of RSA to Increase the Security," vol. 7, no. 4, pp. 1–4, 2017.

[18] B. P. U. Ivy, P. Mandiwa, and M. Kumar, "A modified RSA cryptosystem based on ' n ' prime numbers," *Int. J. Eng. Comput. Sci.*, 2012.

[19] N. Kumar, "Implementation of Modified RSA Cryptosystem for Data Encryption and Decryption based on n Prime number and Bit Stuffing," 2016.

[20] S. A. Jaju and S. S. Chowhan, "A Modified RSA algorithm to enhance security for digital signature," *2015 Int. Conf. Work. Comput. Commun. IEMCON 2015*, 2015.

[21] A. Shamir and L. Adleman, "Introduction," *www.ijecs.in Int. J. Eng. Comput. Sci.*, vol. ISSN, no. 2, pp. 2319–7242, 2012.

[22] A. A. Ayele, "A Modified RSA Encryption Technique Based on Multiple public keys," vol. 1, no. 4, pp. 859–864, 2013.

[23] S. A. Nagar and S. Alshamma, "High Speed Implementation of RSA Algorithm with Modified Keys Exchange," pp. 639–642, 2012.

[24] "Design and Implementation of RSA Algorithm using FPGA Council for Innovative Research," no. September, 2015.

IOP Conf. Series: Materials Science and Engineering **482** (2019) 012017 doi:10.1088/1757-899X/482/1/012017

Analysis of the impact of social networking sites using web content mining and induction method

Renz Carlyle S. Bernados[1], Joshua O. Ty[2], and Angie M. Ceniza[3]

Department of Computer and Information Sciences, School of Arts and Sciences, University of San Carlos, Cebu City, Philippines

[1] bernadosrc@gmail.com, [2] joshuaty98@gmail.com, [3] amceniza@usc.edu.ph

Abstract. With the prevailing presence of social media, we cannot deny the fact that it influences the life of people. Usually teenagers tend to use this technology either to socialize or entertain themselves. In this paper, we have presented a framework to perform a qualitative analysis of social media networking sites using their social media reviews. The researchers gathered 503 links from Facebook, Twitter, Instagram and other related articles using web content mining. And, induction method was used to restrict and classify the harvested data. Researchers makes use of Word2Vec model that would find the similarity of one word to another. The experimental results show 81.39% precision, 67% accuracy, 70.93% recall and 75.58% F-Measure.

1. Introduction

Social media has gained popularity among young children, teens, and adults. Numerous social networking sites have emerged and has attracted many people into using these sites. The option to create accounts under a pseudonym or obscure username that separates users from their true identity may facilitate a sense of anonymity that allows for the disclosure of sensitive information on these sites [1]. Despite early concerns that adolescents might use the Internet to meet strangers, they use social media to interact with existing friends [2]. These social networking sites are useful as they can increase one's communication with one another, provide viable information and entertainment. These resources can take the form of useful information, personal relationships, or the capacity to organize groups [3].

The researchers can see and read public posts of people post online on their profiles. The post can be photographs, videos or texts. Some of these posts are considered inappropriate that may cause harm to those who sees these posts and put them at risk to mental illness or may possibly have a good influence to those who had read them. As there are benefits we can gain from using these sites, there are consequences as well. The goal of this research is to determine what are the possible kind of impact these social networking sites have on the people using them. These posts are important to this research since this is the only way of knowing what their activities are or the social interactions of a certain individual to others. We can gather the data we can get from these posts or different sites that have stated comments towards the subject through web content mining and induction method and then we can classify these posts on what are the possible impact of the social networking sites. The only assumptions we can classify them is that they are either positive or negative influences based on the result of the study.

Multiple solutions are presented to solve the problem devised to give classifications of sentiments in text. Most modern machine learning relies on good feature engineering or other high level of domain knowledge to produce good result [4], but in deep learning, there is feature hierarchy which algorithms have to learn automatically, hence the researchers will use Word2Vec since it implements algorithm for that hierarchy and is also computationally efficient.

This research aims to classify the gathered data and determine which social networking sites have a positive or negative impact on the users. By applying the concept of induction mining, there are rules made to extract data easily from the web without having to much "noise" or unnecessary data. This would also help raise awareness to the users to be more careful using social networking sites. To give preferences on what the users would want to use and avoid the less preferred choice of network.

2. Related Works

2.1 Social Networking Sites

In 2016, Stankovska [5] described that majority of children and adolescents in the world have access to the internet at an early age. Most of them use the Social Network Sites for academic purposes and connecting with other people. But the excessive internet usage can lead to negative outcomes such as depression, loneliness, and poor performance in school.

Another study was conducted to relate adults and adolescents having low self-esteem [6][7]. Adolescent who frequently use an SNS have more friends on the site and also more reactions on their profile. Having likes, comments or reactions on one's profile leads to higher self-esteem. Leading to anxiety and depression, thus having a negative impact on one psychological well-being. Furthermore, adolescents tend to socialize more on the internet rather than in social groups leading to depend on social networking friends that there is also a huge percentage where some users are "troll" leading to miscommunication and also depression.

2.2 Web Mining

An existing research focuses on making an algorithm that would extract and find relevant information hidden across a platform is the Web Mining Model and Its Applications for Information Gathering [8]. The unique feature of this research is that instead of using standard web mining techniques, it creates a new algorithm to approximately extract hidden contents in user profiles in certain Web or platform hence an abstract Web mining model. With this feature, the researchers have developed an abstract Web mining model on the semantic Web to apply the abstract Web mining model for information gathering on the Web.

2.3 Induction Method

According to Ceniza and Maderazo [9], since collection of data on a pre-built system considers the domain specific information on a certain individual, the gathered data would then be altered for the user's behalf. Also, when gathering data, some of the data is in different format depending on the user's preference, meaning data might be not organized and it would be more difficult to extract and less time efficiency. Wrapper induction provide the technique to extract effectively the information and it is also easy to learn.

3. Methodology

This section discusses the methodology used in the research as well as the different frameworks used in order to achieve the desired objective of the study.

3.1 Gathering of Data

As shown in Figure 1, the research chose 3 different Social Networking Sites, mainly Facebook, Twitter and Instagram since these social networking sites are currently popular towards the current generations and may be for future generations based on the statistic survey of Statista [10]. Although some of the social network sites are much higher than what the research have stated, the researchers had considered that some of the much higher social networking sites are not available or less used in other countries.

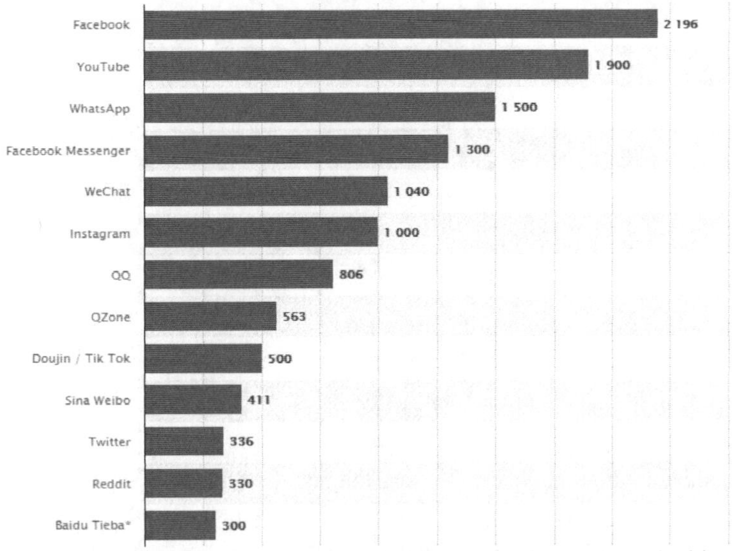

Figure 1. Most famous social network sites worldwide as of July 2018, ranked by number of active users (in millions)

The researchers make use of Google [9] as their platform for collecting data since Google Search Results provide a list of different reviews or topics relating to the subject to be researched. The data are then gathered using a scrapper tool, ScrapeStorm which extracts data or fields, after which those data would be then pre-processed based on the requirements needs of the research.

3.2 Pre-processing of Data

The data is pre-processed by removing unnecessary "stop-words" [11] such as "the", "is", "at", "which", and so on. A total of 709 reviews were used from Google Search Results, with 223(Facebook Reviews), 227(Twitter Reviews), and 259(Instagram Reviews). A word2vec related model was created and trained to identify word embeddings, since most data in the real world are unlabelled which aids in understanding the meaning of those words and also what other words have a relation to that word.

3.3 Continuous Bag-of-Words Model

The research make use of the extracted information based on the result of the pre-processing of the data and with the rules generated when gathering data. It generates "continuous bag-of-word", where in a bag of word model the occurrence of each word will be grouped and be classified as a positive or negative and with continuous bag of word model, there would be clusters of group having word analogies to be reproduced as vector math for prediction and translation. The cluster of words can also be classified as having the same word analogy depending on what the model generated based on the training set.

3.4 Forest Prediction

The research had used scikit-learn, a python library that has the classifier Random Forest in which it will use many tree-based classifiers to make a prediction based collected data. The clustering of the model for each words would be the basis for the many decision tree. Gathered data, pre-processing, sentiment and reactions are needed which aids in giving the prediction to data's whom are unlabelled as shown in Figure 2 below.

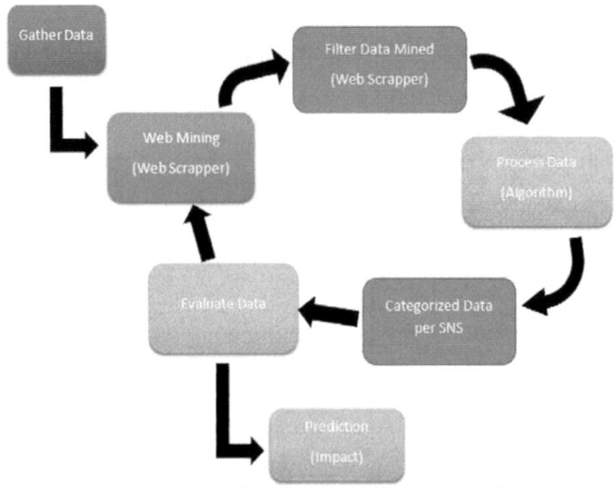

Figure 2. Conceptual framework of data gathering, training and prediction

The figure above shows the process of achieving the result of the study which is giving an analysis on the stated social networking medias. As shown as the title header of each sub-section on Section 3. Where the figure includes the gathering of data, then pre-processing or removing noise, then the classification and the evaluation of the said subjects.

Table 1. Word2Vec Results

Word	**Most Similar Word**	**Similarity**
Facebook	Twitter	0.878421
	Access	0.772450
	Media	0.724056
	Users	0.702237
Twitter	Facebook	0.878421
	instagram	0.776290
	Posted	0.689952
	Users	0.696966
Instagram	Twitter	0.776290
	Facebook	0.696815
	followers	0.654215
	Account	0.643138
Related Articles	Social	0.734260
	Public	0.723617
	propaganda	0.680053
	Society	0.705812

4. Findings and Discussions

Based on the data collected, mostly news media or different blogging sites have been collected during the process of data collection. However, not all these sites are giving bias to the said social media based on the different services the platform gives. Though Facebook, Twitter, and Instagram are mainly social networking services, these platforms also give news to different users all over the world. And with that news, the research could give an evaluation of the subject. Furthermore, within the training of the model, it also generated some various word analogy using the Word2Vec approach. Different words were evaluated but we chose those scored ranging from 0.60 and above as shown in the Table 1.

Based on the result on Table 1, we can see that there are words that has the same similarity or word analogy towards a subject. And we can get a general idea of what the analysis of each of the data gathered after training the model of the vocabulary. Table 2. Shows some of the review of each data gathered on each of the social media subjects. Considering the training and the model generated, if there are more negative than positive words from the content of each post, then that posts would then be considered negative and vice versa. Each word from each posts would be clustered to the model and shall be used for the Random Forest classifier.

From the sample result of Table 2, where each post is given sentiment. Table 3 shows the total analysis of the subject after the training had been completed showing the total percentage of each positive and negative words present on each of the social networking subjects corresponding to the final sentiment result of each subject.

Table 2. Harvested data with their corresponding review

Social Media Site	Content	Review
Facebook	Facebook executives have long called the company's top ranks a "family," but after a series of high-profile exits, the family is about to get a lot of...	1
	Section 230 of the Communications Decency Act (CDA), which provides platforms like Facebook with a broad shield from liability when a…	1
	Facebook says it has identified a covert Myanmar military propaganda campaign hosted on its platform, the first evidence that the country's...	0
	Research at Facebook just made it easier to translate between languages without many translation examples. For example, from Urdu to...	0
	Family of dead student criticise Facebook over 'sadist' troll	0
Twitter	A new Twitter update that allows people to see when other users are online has been criticised for leaving people vulnerable to cyber stalking.	1
	WASHINGTON: As Twitter CEO Jack Dorsey gets ready to testify before the US Congress on September 5, reports have surfaced that he is...	1
	This summer was an absolute mess, but at least there was one redeeming beacon of light on the internet: a Twitter thread of kittens growing up	0
	People across the globe took to Twitter to find out if the issue had affected ... the social media platforms, thousands took to Twitter to complain.	1
	Twitter couldn't help but complain about the amount of adverts they have to endure when they are intently focused on watching the drama...	1
Instagram	Another wrote: "Hang on, Facebook Instagram and WhatsApp are all down." Another said: "#Facebookdown It's down. Instagram appears to be...	1
	My Instagram motto is 'parenting the s--- out of life': what could be more befitting than rear-wiping for comedic effect? It was my most-watched...	1
	A man named Nathan, from Cardiff, Wales, tweeted the picture of Scarlett's post, using it to argue that "Instagram is a ridiculous lie factory made...	0
	Being the guilty Instagram addict that I am, I was intrigued when the New York Public Library (NYPL) announced its latest project, Insta Novels.	0
	AN INSTAGRAM model found dead on a billionaire's luxury superyacht died just hours before a reunion with her family. Sinead McNamara, 20...	0

Table 3. Sentiment Result of Each Social Media Sites

Social Media Site	Positive Words	Negative Words	Total # of Data Gathered	Sentiment
Facebook	0.52470	0.47530	223.0	1.0
Twitter	0.75770	0.24230	227.0	1.0
Instagram	0.73360	0.26640	259.0	1.0

The figure in Figure 3.0 shows the graphical representation of each of the Social media's sentiment result of the gathered posts relating to them. In Facebook, 48% of the posts were evaluated to have a negative review, 52% were positive review and no neutral review was evaluated. Meanwhile in Twitter, 24% evaluated to have a negative review, 76% were positive and 0% were neutral. Lastly Instagram, 27% were evaluated as negative, 73% were positive and 0% were neutral.

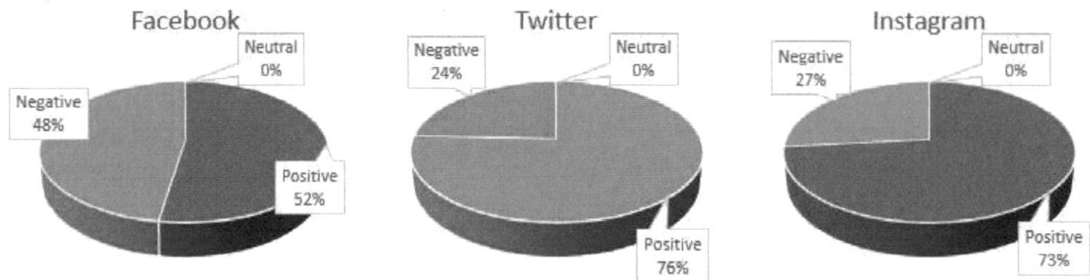

Figure 3. Sentiment result of the 3 social media sites

4.1 Result and Analysis

The measurement of performance evaluation is done by using precision and recall of the sentiment prediction.

$$\text{Recall} = \frac{\text{Number of Relevant Documents Received}}{\text{Total Number of Relevant Retrieved}} \quad (1)$$

$$\text{Precision} = \frac{\text{Number of Relevant Documents Received}}{\text{Total Number of Documents Retrieved}} \quad (2)$$

$$Accuracy = \frac{\text{Number of Relevant Documents+Numb of Irrelevant Documents}}{\text{Number of Relevant Documents+To Number of Documents Retrieved}} \quad (3)$$

$$\text{F Measure} = \frac{2* \text{ Number of Relevant Documents Recieved}}{2*\text{Total Number of Relevant Retrieved+Tot Number of Documents Retrieved}} \quad (4)$$

The researchers gathered 709 links that have contents towards the subject social media. Each link was then evaluated of the occurrence of positive and negative words based on the model. We have found out that 503(170 – Facebook, 150 – Twitter, 183 - Instagram) links were relevant to the gathered data since some of those remaining links are spam links and are not related towards the subject matter. These data are pre-processed and are being clustered to be evaluated upon the research model. The experimental results show that 81.39% precision,70.93% recall 67% accuracy and 75.58% score

5. Conclusion

In this paper, we have presented a framework to perform qualitative analysis of social media reviews. Using web content mining and induction method to generate rules and harvest data to remove unnecessary links gathered and also reducing the noise for it to be computationally efficient. Word2Vec models shows the similarity or relatedness of one word to another. The clustering of the words to a continuous bag of word model based on the relatedness of each word, is then evaluated to the Random Forest to give result. The result presented were positive results about the public perception of each of the said social networking sites from its current users.

6. Acknowledgments

The researchers would like to extend our deepest gratitude towards the Department of Computer and Information Science and University of San Carlos for supporting this research endeavour.

References

[1] Marwick AE and Boyd D 2011 I tweet honestly, I tweet passionately: Twitter users, context collapse, and the imagined audience *New Media & Society* vol **13** no **1** pp 114–33.

[2] Reich SM, Subrahmanyam K and Espinoza G 2012 Friending, IMing, and Hanging Out Face-to-Face *Developmental Psychology* vol **48** no **2** pp 356–68.
[3] Paxton P 1999 Is Social Capital Declining in the United States? A Multiple Indicator Assessment. *American Journal of sociology* vol **105** no **1** pp 88-127.
[4] Deng L and Yu D 2013 Deep Learning for Signal and Information Processing *Microsoft Research Monograph*
[5] Stankovska G, Angelkovska S and Grncarovska SP 2016 Social Networks Use, Loneliness and Academic Performance among University Students *Bulgarian Comparative Education Society*
[6] Ahn J 2011 The effect of social network sites on adolescents' social and academic development: Current theories and controversie *Journal of the Association for Information Science and Technology* vol **62** pp 1435-45
[7] Cavazos-Rehg P A, Krauss M J, Sowles S J, Connolly S, Rosas C Bharadwaj M, Grucza R and Bierut L 2016 An Analysis of Depression, Self-Harm, and Suicidal Ideation Content on Tumblr *The Journal of Crisis Intervention and Suicide Prevention*
[8] Li Y, Zhong N 2004 Web mining model and its applications for information gathering *Knowledge-Based Systems* vol **17** pp 207-17
[9] Ceniza A and Maderazo C 2016 An Analysis of Public Perceptions for K12 Implementation in the Philippines using Web Content Mining Techniques and Wrapper Induction Algorithm *National Conference on IT Education* pp 152- 57
[10] Statista 2018 Most famous social network sites worldwide as of July 2018, ranked by number of active users (in millions) *https://www.statista.com/statistics/272014/global-social-networks-ranked-by-number-of-users/*
[11] El-Khair IA 2006 Effects of Stop Words Elimination for Arabic Information Retrieval: A Comparative Study *International Journal of Computing & Information Sciences* vol **4** no **3** pp. 119-33

IOP Conf. Series: Materials Science and Engineering **482** (2019) 012018 doi:10.1088/1757-899X/482/1/012018

Computer simulation model for traffic enforcement using unity engine

Kasey G. Cuyos[1], Daniel R. Ubanan[2], Angie M. Ceniza[3]

University of San Carlos, School of Arts and Sciences, Department of Computer and Information Sciences

[1]cake.kgc@gmail.com
[2]du.ubanan@gmail.com
[3]amceniza@usc.edu.ph

Abstract. In this paper, we have presented a graphical representation of road traffic. The intersection rules can be manipulated to the configurations set by the user. In the real world, traffic congestion is inevitable. In fact, the Philippines is considered to have the 3rd worst traffic in Southeast Asia. This research aims to provide a system to help lessen the rate at which traffic congestion occurs. With the help of this simulation model, authorities may simulate situations that may aid in their decisions in order to improve traffic flow.

1. Introduction

The study was intended to have direct relevance to the development of a tool for preventative strategies in relation to traffic management. The fundamental reason for this was the major challenge of road traffic becoming a bigger and a more serious problem as time passes by. As the economy grows, so does the complications of road traffic [1].

Philippines is considered to have the 3rd worst traffic problem in Southeast Asia. This is evident as Filipinos waste 28,000 hours in traffic. That sums to wasting up to more than three years of a person's life. Considering that a person's average productive life is 30 years, this means that an individual has lost more than 10% of their working years due to traffic. Aside from that, according to Senator Ralph Recto, the Philippines suffers from a ₱2.6 billion economic loss each day. These issues are thanks to the massive problem of traffic congestion. One often hears suggestions that congestion may be solved with one or more big ideas but none of these can deliver a complete solution aside from provide only temporary relief. With that being said, building more roads will not necessarily be a solution for traffic congestion. Considering aforementioned issues regarding previous solutions, one would question as to what could be a better and a more effective remedy for traffic. The answer is to improve how management is done.

It was decided among the researchers to develop a traffic simulation because it is an effective solution to the current issue. Building more roads, particularly bigger ones, is not necessarily the best solution. A better approach of this would be to improve management. A simulation provides this vital improvement for the management of traffic. A few key factors as to why this is an effective solution are costs, efficiency, and scalability. This solution is cost effective as it is cheaper to develop simulations than to build roads. This is evident as building, repairing, or improving roads not only cost a hefty amount of funds but also costs a lot of time, manpower, as well as having to sacrifice functioning roads by closing them resulting in more traffic congestion during the operation. Using a simulation to diagnose traffic issues is efficient because it is more accessible for data gathering and manipulation compared to the vague real-life evaluation where unintended effects and possible hidden causes to problems are harder to identify [2]. Scalability could possibly the most impressive factor as to why this is a liable solution to traffic congestion. A computer simulated system can scale up to the demand quite quickly [3]. Given enough research and development, a computer simulated system will surely become a powerful tool to help solve, create, or operate something. This is not exclusively applicable to solving traffic issues, it is applicable to anything. In regards with solving traffic problems, the system's scalability will help by being able to keep up with the expanding complexity of the infrastructure that the network of roads is built upon [4].

This study aimed to develop a computer visual simulation model to simulate traffic situations so that better and more efficient actions may be made. The system will greatly improve road management as well as provide a sandbox for authorities to test their plans on road routes and the configuration of traffic regulations without the risk of negatively affecting the lives of the people. This research will also provide a model for future researchers as basis for their studies as well as provide a platform for testing their hypothesis and ideas. The simulation may also be used for testing by evaluating and comparing different configurations of traffic regulations to find which configuration is better and more applicable at certain conditions, to visually inspect and aid the detection of potential traffic congestion situations in which case the system acts as a tool for predicting outcomes of, and to better understand the flow of traffic in order to make better and more educated decisions to help with the problem regarding traffic [5].

Some of the few possible contributions or applications for this study would include research data, traffic prediction systems, and navigation systems (i.e. Google Maps, Waze). The most important contribution would be the data gathered from utilizing the system. The data gathered from using the system would help develop better navigation systems. It could also be possible to use the system to make predictions of upcoming traffic conditions such as simulating what would happen if a major highway were to be closed down. If the system were accurate enough, the data it will provide will prove useful and reliable.

2. Review of Related Literature

These are some of the review of related literature being reviewed in this research.

2.1. Adaptive Road Traffic Management

According to this research, the adaptive vehicle route guidance system integrated in the intelligent vehicle agent was developed based on a hybrid method [6]. This grants or allows for simultaneous and intelligent fine-tuning of the road traffic in the structure given by the real-time adjustments. They also proposed an algorithm where the road structure can uphold more vehicles without cutting down the average speed of traveling vehicles while taking into account the simultaneous road traffic information.

2.2. Traffic: Why It's Getting Worse, What Government Can Do

Traffic congestion was not mainly a problem, but rather the answer or solution to our essential mobility problem, which was that too large of a number of commuters want or need to travel at the same times each day [1]. Why is it like this? This is because adequate operations of both the economy and academic systems require people attend work and go to school during approximately the same hours. This results in the ever so hectic rush-hours commuters must endure to get to their destinations. With each car added into the equation, traffic problems increase exponentially, so much so that even something as simple as running errands prove difficult. People have no choice but to commute at approximately the same time that they end up interacting with each other. That essential requirement is impossible to alter without damaging our society and economy. The same issue persists in every major urban area in the world [7].

IOP Conf. Series: Materials Science and Engineering **482** (2019) 012018 doi:10.1088/1757-899X/482/1/012018

2.3. Managing Traffic Congestion
It is challenging to anticipate what instruments or tools to use that will aid or be helpful in clearing up the traffic issue or at least to just temporarily lessen it. Prices of road fees and parking payments may not be an economic problem once people are wealthier. It is unlikely that we will ever know when to say whether issues regarding traffic congestion are solved permanently or just solved for the time being. countless subsidies are required to further improve and enhance public transport facilities. Asking how to integrate or how to incorporate new and distinctive types of transport is another essential question, since there are distinct interest groups among different transit administrations, departments, and agencies. It proves challenging to influence or to persuade people to choose public transport rather than to use their own private cars [8]. So, when prices of road fees and parking payments are affordable for most of the public, a relevant number of them may still perhaps choose private cars.

2.4. Analysis of Human Driver Behaviour in Highway Cut-in Scenarios
The action, behaviour, and attitude of human motorists in response to cut-in vehicles moving at comparable speeds is what this research paper analyses. Regardless if automated or human-driven, both types of vehicles will inevitably encounter this situation in day-to-day highway commute [9]. Data have been collected from a diverse pool of human subjects using a driving simulator with pre-programmed scenarios[10]. To understand each driver's behaviour in response to cut-in vehicles, a novel means of visualization and analysis based on relative positions is proposed in this paper.

2.5. Application of Traffic Simulation Model
According to this paper, traffic simulation is generally adopted approach in terms of traffic arrangement, advancement, and imitation of existing systems. Its ambition is to carefully recreate traffic regulations and flow examined on the road model. It plays a significant part on traffic management and development as it can be practiced to create a strategy on administering the traffic flow [11]. The objective of this simulation model is to display the transport condition on changing design.

3. Methodology

This section discusses the methodology used in the research as well as the framework used in order to achieve the desired objective of the study.

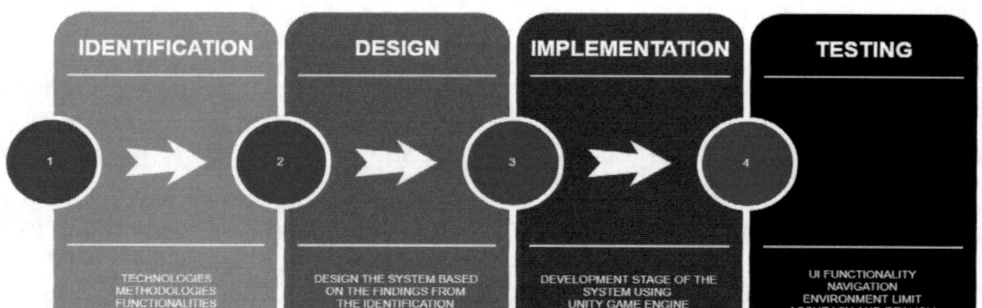

Figure 1. Conceptual Framework

The conceptual framework as seen in Figure 1 was originally designed by the researchers. This illustrates the basic steps necessary to developing the system. The system was conceptualized based on the related literatures. The papers the researchers found influenced the system by being the basis of its design. The related literatures also guided the researchers on deciding which elements to focus on such as which traffic regulations to be made modifiable.

3.1. Identification

Before proceeding with design and implementation, it was necessary to identify all functionalities of the system. Functionalities such as the necessary UI elements needed to allow the user to control the system. It was also important to identify all the algorithms, processes, and tools required to create the system. In order for us to identify all of the components needed for the system, the researchers must first understand how traditional traffic management is done as well as considering how existing adaptive traffic management is used in helping with coming up with solutions for traffic problems. They did this by researching traffic regulations most likely modified to change the flow of traffic. During their research, the researchers found that three regulations are modified to best suit the current situation of a particular traffic issue at hand. These regulations are namely No Left turn, No Right turn, and No Entry. The researchers concluded that the system will provide control of these three regulations making them modifiable. The researcher also concluded to use the built-in navigation functionality provided by Unity Game Engine. The available navigation agent provided by Unity Game Engine is called NavMesh(Navigation Mesh). The NavMesh agent utilizes the A* graph algorithm to enable movement and intelligent navigation of objects within the simulation. Once that was done, designing the system came next.

3.2. Design

Designing the system was much easier since the methods, functionalities, components, and other elements are broken down and identified. Each method was plotted out as to how it functions and how it interacts harmoniously with other functionalities. For this phase, the researchers decided upon which tool to use. As they were doing a 3D model, the researchers found Unity Game Engine to the best tool to use as it had available libraries and assets readily available to ease development of the system. Their design required an interactive but simple UI and the simulation model. Unity Game Engine provides all of these.

3.3. Implementation

As seen in Figure 1, implementation comes next after design. This is the phase of which the system is built with code. For this phase, instead of coding everything, readily available functions from the Unity Game Engine library were used. A Unity Game Engine asset called Playmaker was also heavily utilized as it provided an FSM interface which made the development of the functionalities much easier and more efficient. On top of Playmaker, Modular Traffic, another Unity Game Engine asset complementing Playmaker, provided the necessary models and object behaviour for the simulation. These assets and tools helped in the development of the system adequately. The previously mentioned NavMesh was utilized to implement intelligent navigation into the system. Although not heavily used in the navigation of the AI vehicles, NavMesh plays an important role in the navigation of the pedestrians. Instead of heavily relying on the NavMesh agent for the AI vehicle's movements, those objects use additional raycasting to guide navigation. This is because when left alone to rely on the NavMesh, the AI vehicles will behave undesirably. By using raycasting, the system can direct where the AI vehicles can and should go, giving the system further control of where the objects could go.

3.4. Testing

Testing came last. The system underwent unit testing of all the functionalities multiple times to identify bugs and other issues so that it will be fixed as soon as any were found. Unit testing was done by repetitively running the simulation given the circumstances that showed issues and then fixing those issues to move on to the next.

System testing was predominantly done visually since the system is a visual simulation. Object behaviours were observed and validated. The system would also be tested based from the expected functionalities. The Artificial Intelligence instances of each virtual car must behave as expected considering given parameters. Road regulation modifications must be implemented simultaneously with the AI cars responding by following the set rule. The following were tested to ensure the system's performance.

1. UI Functionality - It was expected that the user is able to control the module using the given user interface. Each button must perform the expected operation.
2. Navigation Functionalities - The roads must be used and followed by the virtual cars. It was expected that there shouldn't be any virtual cars going past their supposed limits within their routes.
3. Environment Limit - The elements of the simulation must behave accordingly. The actions performed by the virtual cars must be feasible in the real world and must be as realistic as possible.
4. Accuracy and realism of traffic rules - Physics elements that are crucial to this project are inertia and gravity. The simulation must be as realistic as possible. Traffic regulations must also be practical and realistic.

Unit testing for the AI cars behavior was done by visually inspecting if they reacted accordingly to inputs. The researchers examined the actions taken by the AI cars if the traffic regulations changed. Interaction between AI cars and AI pedestrians was also tested. In this case, the AI cars must stop if ever pedestrians were crossing the pedestrian lane. The AI cars must also never go out of bounds. This means that the AI cars should only traverse through the road provided to them. The AI cars was also tested on their reaction upon collision with other cars, in this case, they must stop.

To test whether the system was useful in real life applications, the researchers created a model of the road infrastructure of one of the busiest roads in the city of Cebu, Cebu Business Park. The control model was made to be, as much as possible, a copy of the real life road infrastructure with its respective configurations of road regulations. As a test subject, the researchers created an independent model with a road infrastructure based from the control model of which the researchers could then modify the traffic regulations and then perform comparisons between both models.

After numerous modifications, the researchers finally found a configuration of the road regulations that would improve the traffic flow in such a manner that would benefit more commuters compared to the control model. This comparison was done visually, and took several attempts. The unit of which the researchers compared the two models is the average time for all cars of a given road configuration to travel a certain distance as well as visually inspecting the number of cars moving versus the number of cars stopped at a given time.

After the tests, the researchers found that the simulation will display useful visual information such as where congestions start, traffic congestion density, bottlenecks, road infrastructure flaws and many more. Utilized properly by trained professionals, such as engineers and the authorities, the simulation can be used to remodel the existing road infrastructure and design to have the best possible traffic conditions beneficial to the majority of the people.

4. Result and Analysis

As shown on figures 2 and 3, the intersections' rules can be changed according to the configurations set by the user. This will greatly benefit the traffic enforcers; their role of enforcing traffic rules will be much easier and will be done more efficiently as they have basis for their decisions. It will show them the possible outcomes when the configuration is executed. The cars will follow the rules set by the user and will interact with each other as realistically as possible, using signal lights to indicate their current and intended actions. Pedestrians may also be generated to simulate an even more realistic environment but with the cost of graphical performance as it will demand more from the computer to simulate.

As shown on figure 4 and 5, the user is also able to view a 3rd person perspective of a random car or pedestrian. In this view, the user is able to move the view around using the mouse, however the camera will only stay focused on the car or pedestrian. The system also provides a Free Camera view which allows the user to freely move around and observe the simulation using the WASD keys. To edit an intersection's road regulations, it is recommended that the user select the Overhead Camera view to have a better perspective on the environment. To edit road regulations, the user must click a section of the road of which is to be edited. Green arrows will appear indicating that these are the paths the cars may take. If the user wishes to set a "no left turn" rule, then the user can simply click on the arrow pointing to the left. The arrow will turn red and the rule has been set. Given this type of interaction, the system is very straight forward to use.

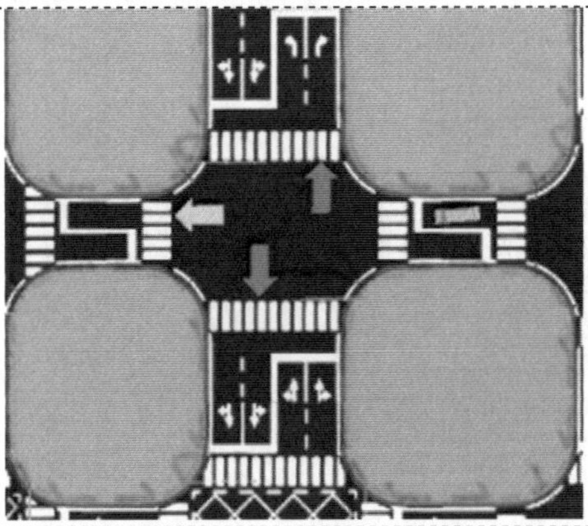

Figure 2. Intersection Configuration

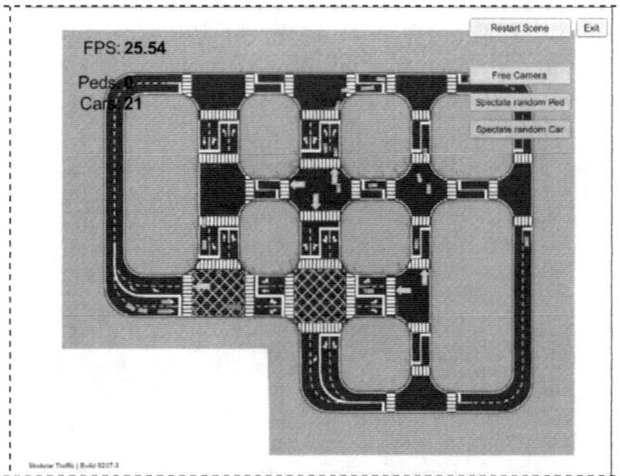

Figure 3. Road model and its configuration

Figure 4. Car Artificial Intelligence

Figure 5. Car Artificial Intelligence

As seen on Figure 6 and 7, important information is displayed such as the current number of pedestrians and cars. Information about the number of cars moving and stopped is also shown. More important information like the wait time of the traffic light is shown and is editable. The wait time for the traffic lights greatly influence the flow of traffic in the simulation. Lastly, the average distance travelled per minute of the cars is also displayed. The higher the average distance travelled per minute, the more desirable.

The system is simple and straight forward to use. Necessary information such as the traffic light wait time and average distance travelled per minute helps the user identify whether the traffic regulation configuration is more efficient or not. With the visual element of the system, the user will also be able to pinpoint problematic areas of the environment and address them accordingly without the risk of having to compromise the activities of real-life people.

5. Conclusion

In this paper, the researchers have presented a graphical representation of road traffic whose intersection rules can be manipulated according to the configurations set by the user. In the real world, traffic problems such as congestions are unavoidable, however, there are numerous manners of ways of decreasing the rate at which it intensifies as well as a better way to decide upon what action to take to relieve or lessen the congestion. With the help of this system, authorities may simulate situations that may aid in their actions to improve traffic flow. Researchers will also benefit from this as this will provide them with a model to aid with their studies. The traffic regulations available for change in this system is limited, however it is very possible to improve this by enabling more regulations to be changed. Potential improvements for the system would be allowing more control over more regulations, making the AI object behaviours of pedestrians and vehicles more realistic, and optimizing the system to run more efficiently to allow a simulation of a bigger scale.

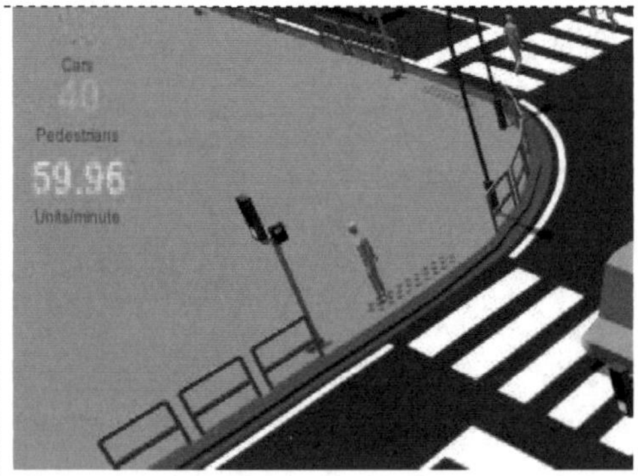

Figure 6. Important Information

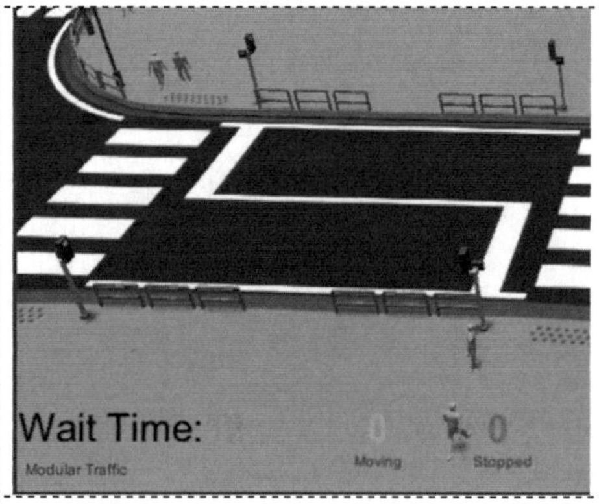

Figure 7. Important Information

Acknowledgements
We would like to express our greatest gratitude to our adviser as well as a special thanks to the members of the faculty who gave us the opportunity to perform the interesting study regarding the topic "Computer Simulation Model for Traffic Enforcement Using Unity Engine", which also helped in giving us the opportunity to do a lot of research on the topics regarding traffic congestion and explore the solutions on how to help solve them. As a result, not only have we made a simulation to help future researchers and professionals, but we also came to know about so many new things. We are really grateful to them.

Secondly, we would also like to extend our deepest gratitude towards the Department of Computer and Information Science and University of San Carlos for supporting this research endeavour.

References

[1] Downs A 2004 Traffic: *Why It's Getting Worse, What Government Can Do (No. Policy Brief# 128)* (Brookings Institution) https://www.brookings.edu/research/traffic-why-its-getting-worse-what-government-can-do/

[2] Azlan N N N and Rohani M M 2018 Overview of application of traffic simulation model *MATEC Web of Conferences* (EDP Sciences) vol **150** p 03006

[3] Abadi A, Rajabioun T and Ioannou P A 2015 Traffic flow prediction for road transportation networks with limited traffic data *IEEE Trans. Intelligent Transportation Syst.* **16(2)** pp 653-62

[4] Djahel S, Salehie M, Tal I and Jamshidi P 2013 Adaptive traffic management for secure and efficient emergency services in smart cities *IEEE Int. Conf. on Pervasive Computing and Communications Workshops (PERCOM Workshops)* pp 340-3

[5] Yue X D, Cao L B, Miao D Q, Chen Y F and Xu B 2015 Multi-view attribute reduction model for traffic bottleneck analysis *Knowledge-Based Systems* **86** pp 1-10

[6] Kammoun H M, Kallel I, Casillas J, Abraham A, and Alimi A M 2014 Adapt-Traf: An adaptive multiagent road traffic management system based on hybrid ant-hierarchical fuzzy model *Transportation Res. Part C: Emerging Technol.* **42** 147-67

[7] Wang J, Mao Y, Li J, Xiong Z and Wang W X 2015 Predictability of road traffic and congestion in urban areas *PloS one* **10(4)** p e0121825
[8] Zhang W 2011 *Managing Traffic Congestion-Case study of Hangzhou* Master Thesis (Karlskrona: Blekinge Institute of Technology)
[9] Kim S, Wang J, Guenther D, Heydinger G, Every J, Salaani M K and Barickman F 2017 *Analysis of Human Driver Behavior in Highway Cut-in Scenarios (No. 2017-01-1402)* SAE Technical Paper
[10] Reddy P S, Ramakrishna A and Ramji K 2015 Study of the dynamic behaviour of a human driver coupled with a vehicle *Proc. of the Institution of Mechanical Engineers Part D: J. of Automobile Eng.* **229(2)** 226-34
[11] Li H 2013 *Impacts of Traffic Interventions on Road Safety: An Application of Causal Models* PhD Thesis (London: Imperial College London)

Automatic music mood recognition using Russell's two-dimensional valence-arousal space from audio and lyrical data as classified using SVM and Naïve Bayes

K R Tan[1], M L Villarino[2], C Maderazo[3]

Department of Computer and Information Sciences - University of San Carlos – Talamban Campus, Cebu City, Philippines

[1]ringoreigns@yahoo.com.ph
[2]mlvillarino@yahoo.com
[3]cvmaderazo@usc.edu.ph

Abstract. Automatic music mood recognition is still a new field of research that is gaining attention in the last decade. This study created a system that predicts which of the four quadrants of the valence-arousal space the song belongs to. The system used support-vector machine (SVM) for audio features while Naïve Bayes was used for lyrical features. audio classification achieved a high accuracy for arousal while lyrics classification achieved a high accuracy for valence.

1. Introduction

The music industry has been facing a problem on annotating music tags to songs since the creation of large digital music collections used in radio broadcasts or for streaming. Having many songs available, it becomes difficult and time consuming to have experts listen to a song and then decide on what tags to put on it. In 2002, Tzanetakis and Cook [1] tried to address this issue by creating an automatic genre classifier. Their method of using features extracted from audio files has since served as a standard procedure for future research on the field. By 2007, the new field had added automatic emotion and mood classification.

Feature extraction is a process which reduces raw data to a smaller set of features such as tone, timbre, and tempo. This enabled researchers to use the data without having to sort through the whole song. The process of extracting these features require tools [2] [3] and finding the optimal features to use for a study requires further research on these features [4]. Feature extraction was first focused solely on audio data but in the recent years, lyrics [5] [6] [7] also had been used to detect the mood of a song by using dictionaries containing valence and arousal values [8] [9]. Other studies that used both audio and lyric features [10] [11] had also shown that lyrics showed higher accuracy when compared with audio features and the combination of both features yielded greater results [10] [11].

Research on this new field saw the use of machine algorithms to automatically detect mood using decision trees [13], neural network [14], SVM [15] [16], and Naïve Bayes classifiers [5] [17]. Other methods such as the use of weighted graphs [12] and linear regression on ratings of words from word embedding [16] had also been used.

The model that was used as the basis for classification was Russell's circumspect model [18]. It is a two-dimensional plane with valence and arousal serving as the axes which divides the plane into four quadrants. Figure 1 shows the emotional plane. Derivations of Russell's model have been created such as Thayer's model used in Bhat et al.'s study [14] [19] and Plutchick's model [7].

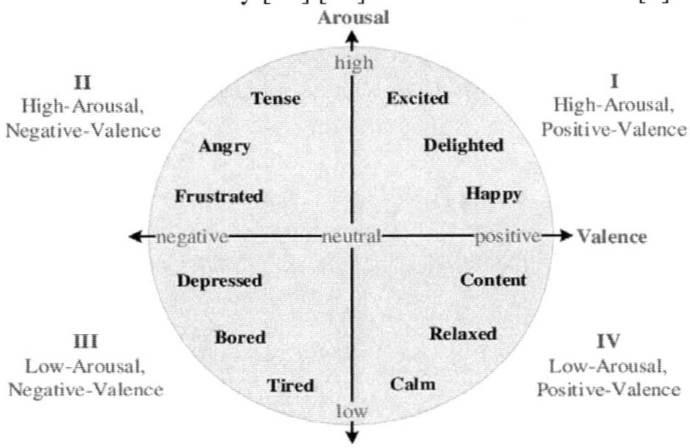

Figure 1. Two-dimensional valence-arousal space. [12]

For this study, the goal was to create a system that would be able to detect the mood of a song based on the four quadrants in Russell's model found at figure 1. To achieve this, the researchers used two classification algorithms SVM and Naïve Bayes to train separate classifier models for valence and arousal using selected audio features for SVM and lyrical features for Naïve Bayes. This process returns four trained models of valence and arousal for each algorithm. Valence is the positive or negative (pleasantness or unpleasantness) while arousal is the definition of how exciting or calm a person is towards a situation.

2. Design and Methodology

This section discusses the data processing and data handling.

2.1. Dataset

The dataset that was used for this study is the dataset used by Panda et al.'s [11] study. It contains annotations of valence and arousal and the quadrant it belongs to. The dataset consisted of 180 songs that had lyrics annotated while 162 songs had audio annotations. The number of songs that contained both lyrics and audio annotations totaled 133 songs.

2.1.1. Audio files. The dataset contained 158 mp3 files containing 30-second clips each. Each file was converted to wav format with a sample rate of 44100 Hz mono-channel which were used for this study. 126 songs were used for this study as some annotated songs did not have their audio data saved while some songs had low audio quality. The total number of songs in used each quadrant are set as: Quadrant 1 – 38, Quadrant 2 – 36, Quadrant 3 – 27, and Quadrant 4 – 26. The dataset was split into two parts: 66.66% for training and 33.33% for testing. Out of the 33.33% testing data, the test cases contained two parts: 26 songs (20.5%) which only contained audio ground-truth values and 16 songs (12.5%) which contained both audio and lyrical ground-truth data.

2.1.2. Lyrics. The dataset contained 180 songs with links to their lyrics made available. The lyrics were stored in text files and were pre-processed to fit the criteria of the dictionary that was used. The total number of songs in each quadrant are set as: quadrant 1 – 44, quadrant 2 – 41, quadrant 3 – 51 and quadrant 4 – 44. 66.67% out of the 180 lyrics were selected for training and 33.33% were for testing.

2.2. Feature extraction

The features of both audio and lyrical were extracted. Both single (Zero Crossing Rate) and multi-dimensional features (Mel-Frequency Cepstral Coefficient) were used.

2.2.1. Audio extraction. The features were extracted using Giannakopolous' pyAudioAnalysis [2] and an extra feature called tonnetz was extracted from Librosa [3] as tonal features. Table 1 shows the audio features that were used along with their standard deviation. The features selected were not the same for valence and arousal and instead were selected based on Grekow's [4] research. The following features that were used for arousal detection were Energy, Entropy of Energy, Spectral Energy, Spectral Flux, Spectral Roll-off, Beats per Minute, and their standard deviations. Audio features that were used for valence detection were Zero Crossing Rate (ZCR), Energy, Entropy of Energy, the three spectral features, MFCC, Chroma Vector, Chroma Deviation, and their standard deviations. Another feature called tonnetz was extracted from Librosa to have more tonal features present in valence detection.

Table 1. Audio features and their standard deviations were used.

Valence	Arousal
Zero Crossing Rate	Energy
Energy	Entropy of Energy
Entropy of Energy	Spectral Entropy
Spectral Energy	Spectral Flux
Spectral Flux	Spectral Roll-off
Spectral Roll-off	Beats Per Minute
MFCC (13)	
Chroma Vector (12)	
Chroma Deviation	
Tonnetz (6)	

2.2.2. Lyrics extraction. The dataset used links to redirect to the lyrics of each song. The lyrics were stored in text files (collected from the provided links in the dataset [11]. Some of the links provided were unavailable or that the lyrics were written wrong and so the researchers searched the lyrics from other links and corrected them. The resulting lyrics were then saved to text files and were pre-processed to fit the criteria of the dictionary [9] that was used. NLTK was also used to increase pre-processed screening. The lyrics were first stripped of their stop words and punctuations and were finally lemmatized down to its root word. Words with negative prefixes were preserved and words ending with "in'" were corrected as well. The results were saved in text files for the training and testing phase.

2.3. Machine-learning algorithms

The selected machine-learning algorithms, support-vector machine and Naïve Bayes, were used for audio and lyrical classification respectively.

2.3.1. Support-vector machine classifier. The available classifier in Python's sklearn was used with the following parameters. For arousal, the C parameter was set at 150 while valence's C parameter was set at 10^5. The training consists of two models, one for valence and another for arousal.

2.3.2. Naïve Bayes classifier. A modified naïve Bayes classifier that used Warriner et al.'s [20] and the NLTK library in Python to extract and use the pre-processed lyrics was used. Training and testing were split to 66.67% and 33.33%.

3. Results

The SVM classifier was tested with the following results shown on Table 2. Sixteen songs (12.5%) out of the 126 songs were used to test the accuracy of songs where its arousal is predicted with audio features and valence is predicted using lyrics while 26 songs (20.5%) were used for audio only detection. The remaining 84 songs (66.66%) were used for training. Table 3 shows the precision, recall, and f1-score for Naïve Bayes. Tables 4 and 5 show the results of the tenfold cross-validation scores of the audio trained dataset's valence and arousal. The results of the 16 songs used for testing both accuracies for the two classification algorithms are found on Table 6.

Table 2. Results of SVM classifier on the training and testing data

	Training (84) (%) Valence	*Training (84) (%) Arousal*	Testing (26) (%) Valence	*Testing (26) (%) Arousal*	*Testing (16) (%) Valence*	*Testing (16) (%) Arousal*
Precision	88	90	58	100	45	94
Recall	88	89	58	100	44	94
F1-Score	88	89	57	100	44	94

Table 3. Results of Naïve Bayes classifier on the testing data

	Testing (16) (%) -Valence	*Testing (16) (%) +Valence*	Testing (16) (%) *-Arousal*	*Testing (16) (%) +Arousal*	Testing (60) (%) -Valence	*Testing (60) (%) +Valence*	Testing (60) (%) *-Arousal*	*Testing (60) (%) +Arousal*
Precision	90	100	57	78	80	100	76	82
Recall	100	86	67	70	100	62	88	67
F1-Score	95	92	62	74	89	77	82	73

Table 4. Valence Tenfold Cross-Validation Score (%)

Test Case	1	2	3	4	5	6	7	8	9	10
Precision	56	56	59	47	91	91	66	85	89	79
Recall	56	56	56	33	62	62	62	62	25	62
F1-Score	56	56	54	39	69	69	63	64	25	56

Table 5. Arousal Tenfold Cross-Validation Score (%)

Test Case	1	2	3	4	5	6	7	8	9	10
Precision	100	100	91	92	100	92	91	100	94	90
Recall	100	100	89	89	100	88	88	100	88	50
F1-Score	100	100	89	89	100	88	88	100	89	57

3.1. Audio detection

In extracting the audio features for training and testing, a total of 84 songs for the training result showed a (0.88 for the precision, 0.89 for the recall, and 0.88 for the f1-score) of valence while arousal showed (0.90 for precision, recall, and f1-score). Tenfold cross-validation was performed on the training data and showed poor results for valence but high accuracy for arousal. The 26 songs used for testing audio only detection showed (0.58 precision, 0.58 recall, and 0.57 f1-score) for valence and arousal achieved (1.00 precision, recall, and f1-score). The 16 songs that were used for testing resulted in (0.45 precision, 0.44 recall, and 0.44 f1-score) for valence while arousal resulted in (0.94 precision, 0.94 recall, and 0.94 f1-score). The cause of low accuracy on detecting valence has been discussed in the work of Yang, Dong, & Li [20]. It is considered that arousal can easily be distinguished between exciting or calm, with tempo being the key factor in this study. But valence is difficult to distinguish because it is ranked as either positive or negative (pleasant or unpleasant) and people have different opinions towards a song's pleasantness.

3.2. Lyrics Detection

Training data consisted of 120 songs and 60 were used for testing. The classifier achieved an 85% accuracy for valence (51 songs) and 75% accuracy for arousal (45 songs). The confusion matrix at Table 3 shows a high accuracy for valence but arousal predicted poorly this time. The trained model was tested on the 16 selected songs and achieved a high accuracy for valence detection as opposed to the low accuracy found on audio classification by SVM.

Table 6. SVM arousal and NB valence have higher accuracies.

	SVM Arousal Correct (%)	SVM Valence Correct (%)	*NB Arousal Correct (%)*	NB Valence Correct (%)
Quadrant 1	**100**	33.33	33.33	**100**
Quadrant 2	**100**	42.86	85.71	**100**
Quadrant 3	**100**	50	50	**100**
Quadrant 4	**75**	50	75	**72**

4. Conclusion

Arousal detection is highly accurate when used with audio features while valence detection is highly accurate when using lyrics. Arousal is easily distinguishable when listened to since its range would be from high to low. This study focused more on the use of tempo for arousal detection using extracted audio features. Valence detection using lyrics with Naïve Bayes resulted in higher accuracy than the use of audio because it is difficult to distinguish and analyze the tune and the positiveness or negativity of a word as they cannot be distinguished properly [20]. An example would be the difference between Quadrant 3 and Quadrant 4 – the tempo and tone of their songs are mostly alike and are confused with one another. Lyrics, however, uses the meaning of the words to signify the positivity or negativity of a word. This reveals a significant difference between the total positivity or negativity of a song, and can also identify songs representing a happy tone but the meaning of their lyrics is sad.

Future Works

For future tasks, a larger dataset is required for comparison of both audio and lyrical data. Further research also will be focused on valence detection using text classification and how it can be used to increase the accuracy of automatic mood classification problems.

References

[1] Tzanetakis G and Cook P 2002 Musical genre classification of audio signals *IEEE Trans. on Speech and Audio Process.* **10(5)** pp 293-302 doi:10.1109/tsa.2002.800560

[2] Giannakopoullos T 2015 pyAudioAnalysis: an open-source Python library for audio signal analysis *PloS ONE* **12(10)** doi:10.1371/journal.pone.0144610

[3] McFee B, Raffel C, Liang D, Ellis D P W, McVicar M, Battenberg E and Nieto O 2015 Librosa: v0.4.0. *Zenodo 2015* doi:10.5281/zenodo.18369

[4] Grekow J 2018 Audio features dedicated to the detection and tracking of arousal and valence in musical compositions *J. of Inform. and Telecom.* **1(12)** doi:10.1080/24751839.2018.1463749

[5] Raschka S 2014 MusicMood: predicting the mood of music from lyrics using machine learning (Michigan: Michigan State University) *Preprint* arXiv:1611.00138 [cs.LG]

[6] Ascalon E I V and Cabredo R 2015 Lyric-based music mood recognition *DSLU Res. Congress 2015 De La Salle University, Manila, Philippines*

[7] Oh S, Hahn M and Kim J 2013 Music mood classification using intro and refrain parts of lyrics *2013 Int. Conf. on Inform. Sci. and Appl. (ICISA)* pp 1-3 doi:10.1109/ICISA.2013.6579495

[8] Yu L, Lee L, Hao S, Wang J, He Y, Hu J, Lai K R and Zhang X 2016 Building Chinese affective resources in valence-arousal dimensions *Proc. of the 15th Annual Conf. of the North American Chapter of the Association for Computational Linguistics: Human Language Technologies (NAACL-HLT'16) San Diego, CA, USA* doi:10.18653/v1/N16-1066

[9] Warriner A B, Kuperman V and Brysbaert M 2013 Norms of valence, arousal, and dominance for 13,915 English lemmas *Behavior Res. Methods* **45** pp 1191-207

[10] Hu X, Choi K and Downie J S 2016 A framework for evaluating multimodal music mood classification *J. of the Assoc. for Inform. Sci. and Technol.* **68(2)** pp 273-85 doi:10.1002/asi.23649

[11] Malheiro R, Panda R, Gomes P and Paiva R P 2016 Bi-modal music emotion recognition: novel

lyrical features and dataset *9th Int. Work. on Music and Machine Learning – MML'2016 – in conjunction with the European Conf. on Machine Learning and Principles and Practice of Knowledge Discovery in Databases – ECML/PKDD 2016 Riva del Garda Italy*

[12] Yu L, Wang J, Lai K R and Zhang X 2015 Predicting valence-arousal ratings of words using a weighted graph method *Proc. of the 53rd Annual Meeting of the Association for Computational Linguistics and the 7th Int. Joint Conf. on Natural Language Processing* pp 788-93 retrieved from http://www.aclweb.org/anthology/p15-2129

[13] Patra B G, Das D and Bandyopadhyay J 2013 Automatic music mood classification of Hindi songs *3rd Work. on Sentiment Analysis where AI meets Psychology (SAAIP 2013)*

[14] Bhat A S, Amith V S, Prasad N S and Mohan D M 2014 An efficient classification algorithm for music mood detection in Western and Hindi music using audio feature extraction *2013 Fifth Int. Conf. on Signal and Image Processing (ICSIP 2014)* pp 359-64 doi:10.1109/ICSIP.2014.63

[15] Ren J, Wu M and Jang J R 2015 Automatic music mood classification based on timbre and modulation features *IEEE Trans. on Affective Comput.* **6(3)** pp 236-46 doi:10.1009/TAFFC.2015.2427836

[16] Li M, Long Y and Lu Q 2016 A regression approach to valence-arousal ratings of words from word embedding *2016 Int. Conf. on Asian Language Processing (IALP)* pp 120-3 doi:10.1109/IALP.2016.7875949

[17] Romeu A U 2016 Emotion recognition based on the speech using a Naïve Bayes Classifier (Bachelor's thesis) retrieved from https://upcommons.upc.edu/bitstream/handle/2117/98368/Bachelors_thesis_ANGEL_URBANO.pdf?sequence=1&isAllowed=y

[18] Russell J A 1980 A circumplex model of affect *J. of Personality and Soc. Psych.* **39(6)** pp 1161-78 doi:10.1037/h0077714

[19] Nuzzolo M 2015 *Music Mood Classification* (Massachusetts, USA: Tufts University)

[20] Yang X, Dong Y and Li J 2017 Review of data features-based music emotion recognition methods *Multimedia Systems* pp 1-25 doi:10.1007/s00530-017-0559-4

Usability of KITIKIT: a mobile-based app for implementing customizable online product ordering

A Balcita[1], RJ Ventayen[2] and T Palaoag[3]

[1] Don Mariano Marcos Memorial State University
[2]Pangasinan State University
[3]University of the Cordilleras

[1]amypanedabalcita@gmail.com
[2]dayjx@yahoo.com
[3]tpalaoag@gmail.com

Abstract. Pugo, branded as the woodcarving capital of La Union, is well known for woodcarving products. The town has skillful tribes who practice woodcarving. One of these tribes was the Ifugaos, and when they learned that the art of woodcarving could make money for them, a wave of Ifugaos migrated along the Marcos Highway to showcase their hand-carved wood products. The widespread of woodcarving business was due to the availability of the raw materials, tools used and durability of the product. Home decors, furniture and other sculptural forms made out of wood are some of the products that are now distributed in the market. This research was then conceptualized for the benefit of the woodcarving industry *stakeholders. The researchers proposed the KITIKIT, a mobile app to ease the searching for* hand-carved wood products. Usability testing using USE questionnaire had been conducted to find the level of usability of the KITIKIT. The results show that the proposed application is acceptable to users in terms of usefulness, ease of use and satisfaction.

1. Introduction

In the modern-day era, so many things are transformed. The advent of technology has modified the way humans live, work and spends amusement time. Computers are taken into consideration the lifeblood of society; they are everywhere. The advancement in discoveries, each in hardware and software, keep coming, day in and day out. Technology changes and improves at a rapid pace, and business and companies have coped with it. This is because as the ultra-modern know-how of improvement in computer science grows, human's standard increases. They seek and crave for higher and extra environment-friendly approaches to accomplish tedious undertaking [1].

At present, human beings at the back of large and small organization adopt modern technology to survive and compete with market demands. Also, through the use of an online ordering system, different transactions and processing systems have become computerized for greater ease and accuracy. The net and web have been firmly entrenched in modern day commercial enterprise practices. Businesses migrate into the online system for various reasons. One of these is that an online system can decrease operational fee considering that transactions are done through the web. An online ordering system is a powerful tool for business, adapting to these technological advancements is needed for everybody to improve their business internationally or locally and cope up with the problems that come along with it [2].

Rayport mentioned that when ordering online, the transaction is known as online ordering. Online ordering systems help customers to buy goods through the internet. Online ordering has been around over the past two decades, involving many activities associated with the customers. Online ordering has become popular because of the way that it can expand your order and allow your customers a more natural means of placing orders [3]. Samsudin explained that information and communication technology (ICT) had been utilized to automate movement tasks in enterprise transactions such as customer registration system, billing system, and inventory system. However, the evolution of wireless technology and the emergence of mobile gadgets have no longer solely supplied end users enjoyment in using computerized structures but additionally new methods to communicate information. The convergence of wireless and cell applied sciences can facilitate a ubiquitous platform for implementing commercial enterprise applications such as food ordering system [4]. Kasemsap mentioned that businesses nowadays are using an online ordering system to automate transactions. It enables the customers to access the information. Thus, it is observable that companies are dependent on the current high-technologies to improve their operation [5].

The woodcarving industry in Pugo, La Union, is refining its services to its customers and staff through computerized processes and system. Thus, the researchers conducted a study which helps solve the problems in the traditional method. Through this online ordering system, an automated process had been convenient for customers, orders are easier to manage, and attracted more customers because transactions are processed online.

The researchers aimed to develop a mobile app and analyze the usability of the application when being used by stakeholders in searching for hand-carved wood products. This study specifically aimed to identify the information requirements in the development of the KITIKIT, to identify the features of the KITIKIT and finally to identify the level of usability of the KITIKIT. This application got its name from a wood carving tool called KITIKIT, the Iluko term for carving knife that woodcarvers use to pare, cut, form and smooth wood.

2. Literature Review

2.1 Electronic commerce (E-commerce)

E-commerce is rapidly gaining ground as a regularly occurring and used enterprise paradigm. More and more extra commercial enterprise houses are imposing websites and apps imparting functionality for performing business transactions over the web. It is realistic to say that the method of purchasing on the net is becoming commonplace [6]. Devendra and Toral characterized that electronic commerce, commonly known as e-commerce, comprises of the buying and selling of products or services over electronic frameworks such as the web and other computer networks [7]. According to Gangeshwer, the slicing aspect for companies today is e-commerce. It facilitates new types of data based entirely on enterprise strategies for reaching and interacting with customers like online advertising and marketing, online order taking and online purchaser service. It can additionally minimize price in managing orders and communicating with a massive vary of suppliers and trading partners, areas that usually add significant overheads to the cost of products and service [8]. At present, many business establishments, institutions, agencies, and persons tailored the modern approaches of buying and promoting their merchandise or products through internet access. The appreciation of social advertising has shifted and no longer seen as a cutting-edge or fad, having a flexible and well-managed presence in each of the "big three" the Facebook, Twitter, and Google. It has turned out to be a need to for any enterprise searching for to impenetrable a place in both the way of life and digital marketplace.

With the increase in the number of people who participated in shopping online, people and business from Pugo are finding ways to cope with the changes brought about by the digital age. There are many Facebook group pages managed by Pugonians that exists. Users utilize it to post the products they sell, browse through products for sale, and for sellers and buyers to communicate, therefore using Consumer-to-Consumer (C2C) type of transaction. The website is an intermediary, directly just there to suit customers. They do longer check the best of merchandise being offered. By using the Facebook group to shop online, there are many risks and problems that arise, as most of the transactions between the buyers and sellers are completed independently on the website. According to some concerns about the group page, comments of the buyers and sellers are the following the non-delivery of products which have been paid for, delivery of defective products, false reservations of buyers, and also obscure payment and delivery schemes, among others.

The Municipality of Pugo is established as the center of wood trade in Southern part of La Union. The Department of Trade and Industry had approved hand-carved wood products as the *One Town One Product (OTOP)* of the municipality. At present, there are twenty-eight established wood carved shops. The people from nearby towns come into the town of their needs because of the presence of many shops that sell hand-carved wood products. These wood carved shops are somewhat outdated and lagging behind since they are still doing the traditional mode of business. They advertise their products through flyers and during exhibits conducted by the Department of Trade and Industry. Most of the services like record keeping and booking services are done manually. These shops also don't maintain a database for their products and their customers. The sellers even admit that they are knowledgeable of the benefits of online shopping yet their problem is that they don't know how to convert their traditional business into an online store. While their commercial enterprise of in-Store is doing well, the researchers believe that the establishment wishes a new business model which is going online for i. Leveraging its existing in-store unit enterprise and ii. Developing omnichannel to accommodate the change in client behavior that is empowering these clients for purchasing decision making and executing their purchase in any channel as per their convenience [9]. Because of these scenarios, the researchers were inspired to develop a mobile app for the sellers of hand-carved wood products of Pugo, La Union. This can help in fast buying and selling procedure 24/7 as well as easy to find products. Through the proposed KITIKIT it will be simpler for the wood carved shops to pick up clients, promote items or products, low operational costs, way better quality administrations, and services, no require of physical company set-ups, and clients can select items or products from diverse suppliers without moving around physically [10].

2.2 Usability Testing

Usability is commonly comprehended as the methods for improving the ease-of-use during the early design process [11] [12]. Usability testing on the other hand, generally involves the evaluation of the accessibility of an application from the user's perspective with regard to different aspects of the application functions [13] [14]. Aside from the development of the mobile app KITIKIT, another focus of this research is on assessment of the usability of the KITIKIT by using USE questionnaire [15]. Evaluating usability of mobile applications is considered one of the key issues in application development [16] [17]. There are many researches that conduct usability testing from end-users perspective [18] [19] [20] [21]. Usability is one of the key parameters to make a mobile application used and loved by the users, and also enabling users achieving their specific goals. Usefulness, ease of use, and satisfaction are key factors that determine user experience when they are engaging with an application [22].

3. Methodology

3.1 Software Development Methodology

The methodology that was utilized in this study is Agile Methodology which is an iterative aspect development methodology, where both development and testing process are concurrent. Testing is not a separate phase; coding and testing are done interactively and incrementally, resulting in the final stage production, which sports are meeting customer requirements. Further, continuous integration leads to early defect removal and hence time, drive and cost savings [23]. One of the agile strategies is Feature Driven Development (FDD) that is beneficial because it demonstrates that the researchers can focus on domain modeling on an iterative and incremental undertaking and due to the fact it proves that agile-like methodologies can scale. FDD indicates that the researchers can spend a short amount of time at the beginning of the project to establish a clear perception of the area in which they are working and use that appreciation to formulate an ambitious plan besides getting caught in analysis and design paralysis [24].

FDD is a production strategy which highly oriented on ensuring out small blocks of client-valued functionality. This drives builders or developers to come up with working elements once every two weeks usually and it can tune down the project progress with precision FDD, which is one of some agile improvement processes, is an iterative and incremental software improvement method having the main reason of delivering real working software program repeatedly. [25]

Thus, the researchers selected FDD as the methodology for the reason that the method for the motive that its techniques enable and put into effect the repeatable delivery of operating software immediately with correct and significant facts to all vital roles inside and outside the project. FDD methodology helps the researchers to develop and deliver the application in a short period. It can also scale down and still provide value to the development process. The different phases of the FDD are shown in Figure 1.

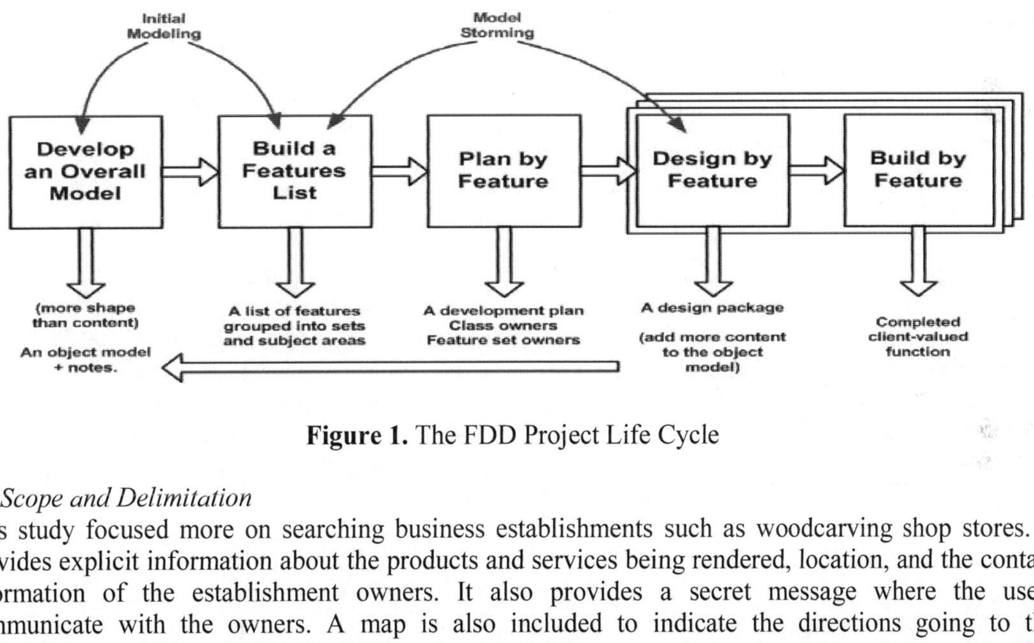

Figure 1. The FDD Project Life Cycle

3.2 Scope and Delimitation

This study focused more on searching business establishments such as woodcarving shop stores. It provides explicit information about the products and services being rendered, location, and the contact information of the establishment owners. It also provides a secret message where the users communicate with the owners. A map is also included to indicate the directions going to the establishments. The mobile application is accessible everywhere; however, the participating establishments are only within Pugo, La Union. It does not have a GPS to detect the location of the user. The app does not cover the online payment. The study is designed to develop an android shopping app that exclusively caters the buy and sell of hand-carved wood products in the Municipality of Pugo, La Union. The creation of a database server, administration website, production of other mobile versions of the app outside of the Android version (e.g., iOS, Windows) will not be covered in this study.

3.3 Data Gathering Techniques

The researchers made use of the following tools and procedures in gathering data for the study:

- Interview

  This type of data gathering technique has been used by the researchers to gather information from the twenty establishment owners. A face-to-face and a survey with guide questions that aims to acquire the most relevant data taking considerations of the stakeholder's opinions, interest, and business workflow.

- Observation
  This method that was used by the researchers is to observe the owners on how they respond to their prospect clients on their struggle in finding quality hand-carved wood products.
- Documentary Analysis
  This is a technique which the researchers used to analyze the given data provided by the establishment owners such as legal papers to prove that the establishment is operating legally.

3.4 Usability Evaluation

This usability evaluation aims to find out how far the stakeholders can interact effectively and how far the application provides user guidance, provide information, and provide consistent interaction [26]. The procedure in conducting a usability evaluation is to use a questionnaire, one of which is the USE questionnaire. The USE questionnaire used 14 questions to evaluate the application from 3 aspects: usefulness, ease of use and satisfaction. The 14 questions were given to 60 respondents as can be seen in Table 1.

Table 1. Number of Respondents

Respondent Types	Number of Respondents
Establishment Owners	20
Clients	20
Product Specialists	20

The researchers surveyed the Municipality of Pugo, La Union using the usability metric based on Measuring Usability with the USE (Usefulness, Satisfaction, and Ease of Use) Questionnaire done by Arnold M. Lund published on October 2001. The questionnaire is constructed with five-point Likert rating scales extending from strongly disagree to Strongly Agree. The respondents tested the proposed system and filled up the questions corresponding to the following categories of usefulness, ease of use and user's satisfaction. To interpret the results, Table 2 presents how the data has been interpreted qualitatively.

Table 2. Data Interpretation.

Point Value	Mean Range	Descriptive Rating
1	1.00 – 1.79	Strongly Disagree
2	1.80 – 2.59	Disagree
3	3.00 – 3.39	Just Right
4	3.40 – 4.19	Agree
5	4.20 – 5.00	Strongly Agree

Table 2 shows the interpretation for each category based on the average results from the survey on the measure of usability of KITIKIT. The information from the table above corresponds to the description of the average score for each category on the study and the total average for the overall usability.

4. Discussion of Findings

4.1 Information Requirements of KITIKIT

Business Workflow. The researchers have interviewed a total of 20 establishment owners regarding the structure and flow of the business. The interviewees have come to discuss first the processes and methods they use to advertise their products and what are the requirements needed to accomplish when the customer would like to reserve for the product. Also is the proof that they are running the business legally like the Bureau of Internal Revenue permit and the business operating permit, the information on the means to contact the establishment owners, the detailed description of the products being advertised together with its price and the exact location of the establishment. From the gathered data, the database of the system was designed using MySQL.

4.2 Process of Information Extraction in KITIKIT.

The system's information extraction for the search function takes place using Entity Extraction of Google's Natural Language Processing (Google NLP) API. The query string is the part of the uniform resource locator (URL) containing information that does not fit helpfully into a hierarchical way structure [27] , these are sent to Google NLP for Entity Extraction and responds with JSON (JavaScript Object Notation is a light-weight data-reciprocation format) formatted payload containing the entities within the transmitted sentence [28]. Entities are notable parts of a sentence that most likely include relevant data concerning the whole context of the passage. The system extracts these entities from the search query and compares each entity against the details related to the establishments. For example, the query "Stores in Marcos Highway with Furniture and Home Decors" contains the following entities: "Stores," "Marcos Highway," "Furniture," "Home Decors." These entities will be queried over the database with each of the establishment detail. If one member of the establishment's information hits the first entity, the establishment will gain a point for the search and another if the second hits and so on. The hits represent the establishment's relevance with the query. Therefore, the establishments with the highest number of a hit are shown first then down to the establishment with only one hit.

4.3 Features of the Proposed Application

Features of the proposes application:

- Establishment Overview
  For a user to look for a product, they must navigate to the app's establishment details screen that presents the establishment's relevant information and documents, its products, the gallery, and its ratings.
- Establishment reviews
  Reviews are essential to help clients create a critical decision in choosing a product. The app lets the previous clients of an establishment give feedback both in words and in a quantifiable rating score that reflects their experience from the establishment.
- Establishment Gallery
  To help the clients get an overview of the establishment without the need to pursue the establishment physically, the researchers considered a gallery feature that will provide a good experience and a clear picture of the prospected products.
- Private Messaging
  The in-application chat enables the establishment representative and the prospect client to share relevant information regarding the transaction of product reservation.
- Security
  It pertains to the safety of every user access levels of user accounts and the administrator account in the system.
- Google Map
  This integrated Google service shows the navigation of the map and how it is used in the application.
- File Maintenance
  In this feature it let the users manage their accounts and update their master files.

4.4 Usability Level of KITIKIT

The following tables show the results of the usability metric interpreted by the researchers for the proposed system. Table 3 shows the effect of the usefulness of KITIKIT. Respondents strongly agree that the system is useful with a highest weighted mean of 4.90. A weighted mean of 3.80 which is the lowest average for the question whether the system saves time when they use it. Hence, the majority of the respondents agree that the system is useful regarding utilizing time, effectiveness and user's expectation.

Table 3. Usefulness of KITIKIT

Indicators	Mean	Descriptive Rating
It helps me be more effective	4.10	Agree
It saves time when I use it	3.80	Agree
It is useful	4.90	Strongly Agree
It does everything I expected it to do	4.00	Agree

In Table 4, the highest weighted mean which is 4.20 which is an evaluation that the system is user-friendly. However, respondents gave the lowest assessment which is 3.70 to the usage of the system. Some of the navigation is easy to use, but the system design should be more simple.

Table 5 shows the satisfaction level of KITIKIT. The respondents were satisfied with how the system works with a highest weighted mean of 4.10. The lowest weighted mean is 3.70 which evaluate the respondents feel that they should have it. The respondents were satisfied with the system, but researchers need to consider the need to have it and the features that will make the system fun to use.

Table 4. Ease of Use of KITIKIT

Indicators	Mean	Descriptive Rating
It is easy to use	4.00	Agree
It is simple to use	3.80	Agree
It is user-friendly	4.20	Strongly Agree
It is flexible	3.90	Agree
Using it is effortless	3.70	Agree
I can use it without written instructions	4.10	Agree

Table 5. Satisfaction of KITIKIT

Indicators	Mean	Descriptive Rating
I am satisfied with it	4.10	Agree
I would recommend it to a friend	4.00	Agree
It is fun to use	3.80	Agree
I feel I need to have it	3.70	Agree

The average of the total weighted mean for each usability was shown in Table 6. The usefulness category has the highest weighted mean of 4.00 which indicates that the user agrees on the usefulness of the system. While the measured usability on the proposed system on ease of use has an average result of 3.90, the result indicates that the users agree to the ease of use. Lastly, the measure of usability on satisfaction has an average of 3.90 which shows that the respondents agree on the satisfaction of the proposed system. Overall, the total average on the measure of usability is 3.90 which means respondents agree on the measure of usability.

Table 6. Usability of KITIKIT

Indicators	Mean	Descriptive Rating
Usefulness	4.00	Agree
Ease of Use	3.90	Agree
Satisfaction	3.90	Agree

5. Conclusions and Recommendations

The information extraction would be a great asset to the proposed application for the search function takes place using the Entity Extraction function of Google's API. The features of the intended application include Establishment Overview, Establishment reviews, Establishment Gallery, Private Messaging, Security and File Maintenance. These features will surely help the stakeholders who are searching for the hand-carved wood products. Based on the evaluation done by the possible users of the system, the system is efficient, useful, and easy to use and the users are satisfied with the overall functionality of the proposed system, but some usability needs improvement. The researchers recommend that information extraction should not only focus on Google's API since the study has NLP it should also include more of its type. The proposed system has already its core functions and services. However, it is recommended that the mode of payment should be handled by the system for more convenience to the client and future researchers to include a feature which allows the filtering of the price range.

References

[1] Visperas A 2010 *Online Ordering System for Silantro-Fil-Mex*
[2] Chavan V, Jadhav P, Korade S and Teli P 2015 implementing customizable online food ordering system using web-based application *Int. J. of Innovative Sci., Eng. & Technol.* vol **2** issue **4** pp 722-7
[3] Rayport J and Jaworski B 2011 *Electronic Commerce* (McGraw-Hill Higher Education)
[4] Samsudin N A 2011 A customizable wireless food ordering system with realtime customer feedback *2011 IEEE Symp. on Wireless Technology and Applications (ISWTA)* pp 186-91
[5] Kasemap K 2010 The role of electronic commerce in the global business environments *Handbook of Research on Interactive Information Quality in Expanding Social Network Communications* ed F V Cipolla-Ficarra (Hershey: IGI Global) pp 304-24
[6] Kodali S 2007 *The Design and Implementation of an E-Commercesite for Online Book Sales* (South Bend: Indiana University South Bend)
[7] Devendra J and Toral J 2012 The electronic commerce act and its implementing rules and regulations *J. Syst. of Inform. Syst.* pp 46-9
[8] Gangeshwer D 2013 E-commerce or internet marketing: A business review from the Indian context *Int. J. of u- and e-Service, Sci. and Technol.* **6(6)** pp 187-94
[9] Retrieved from www.slideshare.net
[10] Retrieved from likemindtech.com
[11] Miller J 2006 Usability testing: a journey, not a destination *Internet Computing* (IEEE) **10 (6)** pp 80-3
[12] Yonglei T 2005 Work in progress-introducing usability concepts in early phases of software development *Proc. Frontiers in Education of 35th Ann. Conf.* **TC4-7-8**
[13] Hussain A, Mkpojiogu E and Hussain Z 2015 Usability evaluation of a web-based health awareness portal on smartphone devices using ISO 9241-11 model *J. Teknologi (Sci. & Eng.)* **77 (4)** pp 1-5
[14] Hussain A and Mkpojiogu E 2015 An application of ISO/IEC 25010 standard in the quality –in-use assessment of an online health awareness system *Jurnal Teknologi, Sci. & Eng.* **77 (5)** pp

9-13
[15] Lund A 2001 Measuring usability with the use questionnaire *STC Usability SIG Newsletter* **8:2**
[16] Hussain A and Mkpojiogu E 2015. The effect of responsive web design on the user experience with laptop and smartphone devices *Jurnal Teknologi, Science & Eng.* **77 (4)** pp 41-4
[17] Hussain A, Mkpojiogu E, Jamaludin and Moh S 2017 A usability evaluation of lazada mobile application *2nd Conf. on Applied Science and Technology*
[18] Az-zahra H, Pinandito A and Tolle H 2015 Usability evaluation of mobile application in culinary recommendation system, wireless and mobile (APWiMob) *IEEE Asia Pacific Conf. on IEEE*
[19] Sia W, Tiu R and Tangsoc J 2017 A User experience evaluation for wendy's online delivery website geared towards improving customer experience *IEEE Int. Conf. on Industrial Engineering & Engineering Management, 2nd Int. Conf.*
[20] Dewi R, Fanani L, Priandani N, Ananta M, Brata K and Tolle H 2018 Usability evaluation of mobile group DSS use for recommendation of culinary micro, small and medium enterprise: enterprise perspective *The 4th Int. Conf. on Computer, Communication and Control Technology*
[21] Hochle H and Viswanath V 2015 Mobile application usability: conceptualization and instrument development *Mis Quarterly* **39.2**
[22] Dewi R, Priandani N, Brata K and Fanafi L 2018 Usability evaluation of mobile-based application for Javanese script learning media *J. of Inform. Technol. and Comp. Sci.* pp 88-93
[23] Tutorialspoint *Agile Testing - Overview* retrieved from https://www.tutorialspoint.com/agile_testing/agile_testing_overview.htm
[24] ProjectConnections *Agile Method Brief – Feature Driven Development (FDD)* retrieved from www.projectconnections.com/templates/detail/agile-techniques-fdd.html
[25] Alivia A and Malana C 2017 *iHome: A Mobile-Based Application for Accommodation Reservation*
[26] Aelani K 2012 Pengukuran usability sistem menggunakan USE questionnaire *Seminar Nasional Aplikasi Teknologi Informasi*
[27] Retrieved from linkerme.info
[28] Retrieved from Mixrank.com

Predicting emotion in music through audio pattern analysis

J M Brotzer[1], E R Mosqueda[2], and K Gorro[3]

[1 2]University of San Carlos, Talamban, Cebu City, Cebu, Philippines
[3]Department of Computer and Information Sciences, University of San Carlos, Talamban, Cebu City, Cebu, Philippines

[1]jomar.brotzer@gmail.com
[2]ethanray.mosqueda@gmail.com
[3]kdgorro@usc.edu.ph

Abstract. Music plays a big part in the development of humanity and has inspired so many people to do great things. Which is why a study was conducted focusing on music and its relation to human emotions. It tackles on how an emotional response is generated from an audio pattern through the use of a model constructed by the researchers. The work done is aimed to generate ratings of the emotion conveyed in a song represented as valence and arousal. Audio feature extraction and Artificial Neural Networks played big parts in the development of the research in creating the regression model and predicting the human emotion being emphasized in an audio pattern. The findings of the research will be used to give meaning to audio patterns and how their structure and composition leads to different emotional states of man.

1. Introduction

Researchers look for different kinds of correlations between music and the feelings people experience. Although ideas have been realized, there is still room to improve in implementations to reach the desired output: emotion in audio. The application of different combinations of machine learning algorithms with the features have yet to be tested and still remain uncertain on which is the best option. This research attempted to determine whether or not using a neural network with regression would yield a better result in determining valence and arousal values as compared to other methods. Valence is defined as the measure of how positive or negative an emotion is while arousal is how calmly or exciting the emotion is felt. In regression, output will be determined using R^2 score, which is a measure in statistics that deals with how close the data points are to the regression line. The research is aimed to construct a model that would predict valence and arousal R^2 scores given an audio sample. Manipulating audio patterns to derive human emotion has proven to be a difficult task. Some methods of classification have been considered to give emotions but only to a certain extent. The study is only limited to the use of the machine learning algorithm Artificial Neural Networks. Furthermore, the type of AI that has been developed in the research will be of a narrow kind for it only outputs mood ratings. The emotion regression was done using the Thayer's model of emotion and the annotations of each song from the dataset.

2. Review of Related Literature

In doing research, it is important to find references and studies about the topic being researched. Previous studies can provide valuable input for researchers especially in terms of methodology. Recently, Neural Networks have shown promising results in complex machine learning which is why the researchers decided to explore neural networks and predicting valence and arousal values in audio patterns. The following are studies related to this research:

2.1. Feature Selection
The study [1] focused on identifying which set of features, when combined, yielded the best prediction results for valence and arousal. This study had a regression approach towards the problem. They used features extracted by Essentia and Marsyas which are both python libraries used for retrieving audio information such as Low-level, Rhythm and Tonal types. The study concluded that the use of several groups of audio features ended up giving better results compared to fewer groups.

2.2. SVM Approach
This study [2] focused more on the concept of Music Emotion Recognition (MER). They approached the problem using regression to predict the values arousal and valence scores directly. The study wished to evaluate the performance of a regression approach to predicting arousal and valence. They used an SVM as a regressor, which in turn yielded them an R^2 score of 58.3% for arousal and 28.1% for valence as their best result.

2.3. Fuzzy Approach
This study [3] focuses on the use of fuzzy logic for music emotion classification. Seeing the subjective nature of how humans perceive music to be, using fuzzy logic allows for a less binary approach to emotion analysis. Fuzzy classifiers allow for a value to be assigned based on how strong an emotion may be perceived. This approach manages to deal with the situation wherein people share different opinions regarding a certain song as well as being able to detect the variations of emotions found in said music.

2.4. Lyrical Analysis
The study [4] makes use of a bi-modal method of audio emotion classification as it deals with both audio and lyrical features, using these as basis for classifying audio to emotion. Selected machine learning algorithms are implemented in classifying to compare the each of their performances on the data.

2.5. Neural Network Architectures
The article [5] provides information on artificial neural networks. Neural networks have different kinds of applications and design architectures. The article mentioned proves to be insightful for researches involving neural networks giving an in depth understanding of the concept.

3. Methodology

3.1. Data Gathering

3.1.1. Selecting Data. The data used in the study was taken from the DEAM (Database for Emotional Analysis in Music) Dataset, consisting of 1802 audio files along with valence and arousal values annotated per second and averaged per song. For the study, the researchers used the annotations averaged per song. These values are real numbers that range from 1 to 9. These values are plotted in a plane using Thayer's Model of Emotion where Valence is the range in x axis and Arousal in the y axis. In Thayer's model, there are four quadrants with different ranges of valence and arousal. Table 1 shows a sample of songs in the dataset and their corresponding placement in the plane.

The emotion ratings were scaled down to -0.2 to 0.2 to have a zero-centred axis and give lenience to error for the model in predicting values. This gives the neural network more room to train and react on the weight changes as the model is subjected to the data. In figure 1, it is clear that the data is condensed at the centre, in the positive and the negative axes. Basing on the Thayer's Model, songs that have mostly negative valence with positive arousal and songs that have positive valence with negative arousal are lacking in numbers. This means songs that are intensely unpleasant and calmingly pleasant aren't prevalent in the dataset.

Table 1. Sample Audio of DEAM Dataset. The table shows the varying of structure depending on where they are located in the Valence and Arousal plane.

Audio Pattern	Information	Averaged Song Ratings
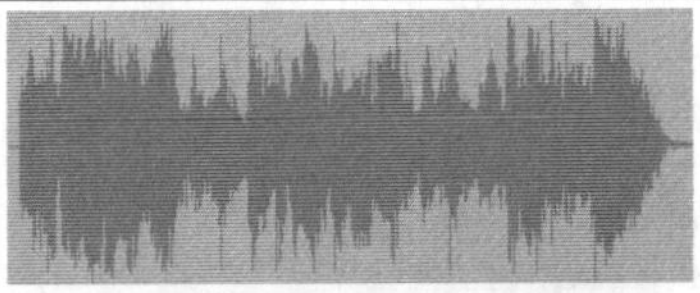	File: 115.mp3 (Quadrant 1) High Valence High Arousal	Valence: 8.4 Arousal: 7.8
	File: 2057.mp3 (Quadrant 2) Low Valence High Arousal	Valence: 3.2 Arousal: 6.8
	File: 488.mp3 (Quadrant 3) Low Valence Low Arousal	Valence: 1.6 Arousal: 3.4
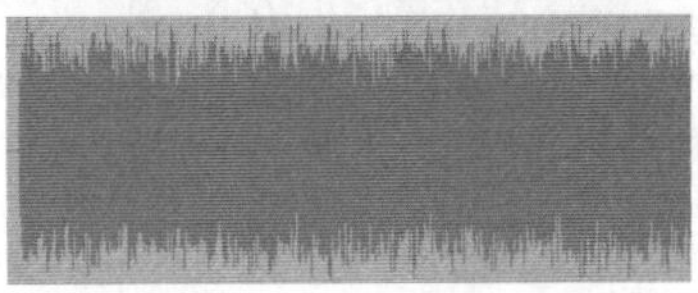	File: 977.mp3 (Quadrant 4) High Valence Low Arousal	Valence: 7.3 Arousal: 3.5

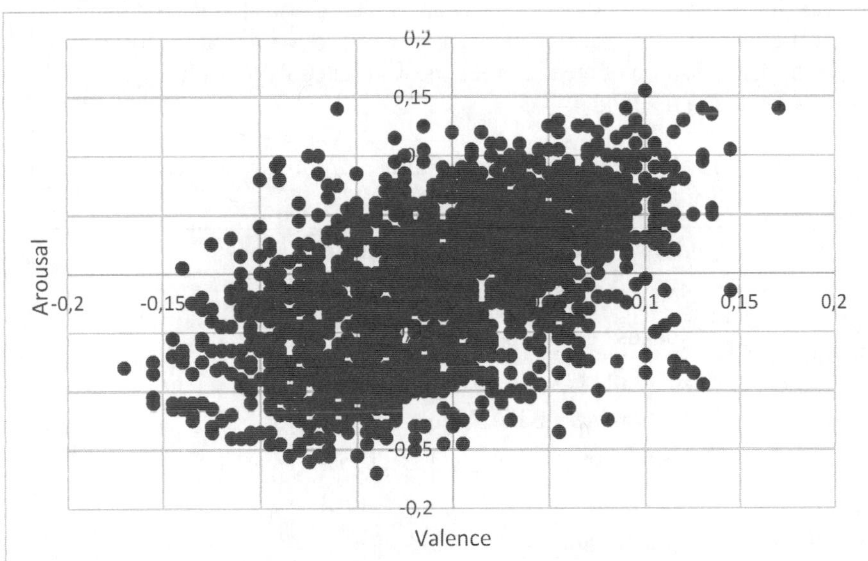

Figure 1. This is a figure showing data points of the valence and arousal labels.

3.1.2. Audio Feature Extraction. For processing the data, the researchers have used LibROSA, a python audio analysis library to extract audio features and NumPy to structure the data into a two-dimensional numpy array of this shape: (1802, 193). The labels of each audio file are likewise stored in the same fashion with a shape: (1802,). These are the selected audio features extracted based on a study on classifying bird songs [6]. The features extracted are already proportioned by LibROSA so that the length of the audio would not affect the shape of the features obtained. Table 2 shows the features extracted with LibROSA.

Table 2. Extracted Audio Features from LibROSA.

Audio Features	Definition	Outputted Shape
Mel-frequency cepstral coefficients (MFCC)	Coefficients that collectively make up a Mel Frequency Cepstrum.	(40,)
Chromagram of a Short Time Fourier Transform	Values that relate to the twelve different pitch classes.	(12,)
Mel-based Power Spectrogram	A Spectrum representation that indicates the fluctuations of the intensity of frequencies	(128,)
Octave-based Spectral Contrast	Decibel difference between peaks and valleys of a spectrum	(7,)
Tonnetz	(German: tone-network) is a conceptual lattice diagram representing tonal space	(6,)
Total Features:		(193,)

Another group of audio features was also taken into consideration from a certain research [2] that gathered information on which audio features contributed greatly to the prediction of valence and arousal. To get the features, the researchers used Essentia to extract the features in table 3 and also shaped them in the same fashion as the audio features extracted from LibROSA apart from the number of features to be inputted to the model

Table 3. Extracted Audio Features from Essentia. (38 features in total)

Audio Features	Feature Type
Melbands Kurtosis	Low-level
Melbands Skewness	Low-level
Spectral Energy	Low-level
Beats Loudness Band Ratio	Rhythm
Key Strength	Tonal
Chords Histogram	Tonal
Chords Strength	Tonal
Harmonic Pitch Class Profile Entropy (HPCP)	Tonal

3.2. Neural Network Construction and Training

Two neural network configurations were created using the libraries: Tensorflow and Keras. One neural network having 1-1-1 NN layer layout (Input, Hidden, and Output) and the other having 1-6-1. The type of neural networks developed were feed-forward; one complex and the other simple. A feed-forward neural network is a kind of neural network where the nodes do not form a cycle and only move in one direction, which is forward and through the nodes. An example of a feed forward neural network is shown in figure 2. The 1-1-1 architecture was to see if a simple NN architecture would be sufficient enough to predict valence and arousal through the use of the different audio features obtained. Otherwise, the researchers would turn to the use of a more complex feed-forward structure.

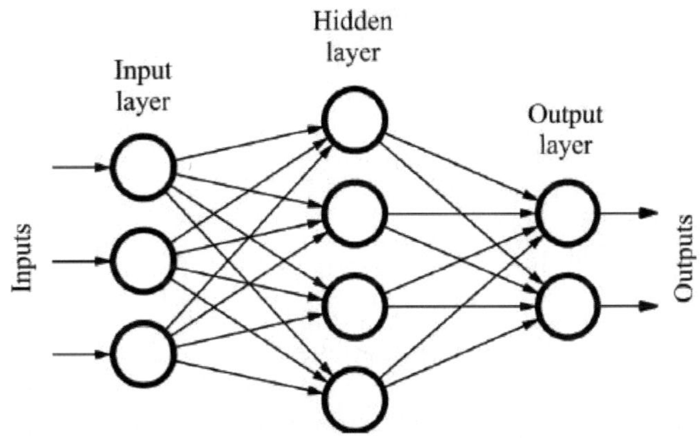

Figure 2. A Simple Feed Forward Neural Network

3.3. Evaluating Models

The models were subjected to a 10-fold cross validation where the dataset was shuffled and per iteration was subjected to training and prediction. Each iteration reported on different regression metrics available from Sci-Kit Learn. This researched used explained variance score, mean absolute error, mean squared error, median absolute error and r^2 score for the verification

4. Analysis of Results

Using the 38 features from Essentia library yielded dissatisfying results when paired with the labels during the train and test process. The 193 features from LibROSA had better results compared to the ones from Essentia but were also ineffective in prediction. So the researchers incrementally added data rows for the train and test process to see if the model would produce better results as the number of data rows increased or decreased. The scaling of the values did not greatly affect the r-squared scores but lessened the mean absolute error. Table 4 displays some of the expected emotion ratings and the predictions made by the model.

Table 4. Arousal and Valence Predictions.

Arousal Predictions		Valence Predictions	
Expected	Predicted	Expected	Predicted
0.0000	0.0166	0.0500	0.0502
-0.1000	-0.0659	-0.0500	-0.0639
0.0500	0.0492	0.0500	0.0421
-0.1000	-0.1058	-0.1000	-0.0886
-0.1000	-0.0562	0.0000	-0.0659

Several number of nodes were tested alongside it but the result from the simple feed forward network that had one hidden layer was not successful compared to the complex network. The best result obtained was from the neural network configuration that had 6 hidden layers each having 500 nodes with rectified linear unit activators and an output node with a linear activation, so the researchers used these values for further testing. Table 5 indicates the best results attained from the model.

Table 5. 10 – Fold Cross Validation Regression Metric Results.

Seed = 7, EPOCHS = 50, Audio Features = 193, Data Rows = 1802.

	Arousal Model	Valence Model
Explained Variance Score	0.3737	0.3285
Mean Absolute Error (MAE)	0.0419	0.0400
Mean Squared Error (MSE)	0.0028	0.0025
Median Absolute Error	0.0355	0.0340
R^2 Score	0.3572	0.3132

5. Conclusion

The neural network architecture developed is experimental and needs refinement. A different type of neural network may prove to be more efficient than a feed-forward network for this kind of problem. A recurrent neural network works well with speech recognition and would be a better option in proceeding with emotion prediction of audio.

The improvement of data would be greatly welcomed as time progresses music continues to evolve. More data rows and different kinds of datasets would better develop the field and lead to better and more accurate models. As of now, the data related to audio sentiment is limited and researchers struggle to get accurate data that would be viable for usage.

It is valid to predict emotion in music because music affects humanity and it is beneficial to know which types of music negatively or positively affects a person. Creating a model that can deduce an audio pattern's emotional rating provides a buffer for people to decide whether or not to listen or create music based on the rating.

References

[1] Grekow J 2017 Audio features dedicated to the detection of arousal and valence in music recordings *IEEE Intl. Conf. on Innov. in Intel. Sys. and App. (INISTA)* pp 40-4

[2] Yang Y, Lin Y, Su Y and Chen H 2008 A regression approach to music emotion recognition *IEEE Transac. on Audio, Speech, and Lang. Proc.* **16(2)** pp 448-57

[3] Yang Y, Chu Liu C and Chen H 2006 Music emotion classification: a fuzzy approach *MM '06 Proc. of the 14th ACM Intl. Conf. on Multi.* pp 81–4

[4] Malheiro R, Panda R, Gomes P and Paiva R 2013 Music emotion recognition from lyrics: a comparative study *6th Intl. Work. on Mach. Learn. and Music, Prague*

[5] Stergiou C and Siganos D 1996 *Neural Networks* online https://www.doc.ic.ac.uk/~nd/surprise_96/journal/vol4/cs11/report.html

[6] Collis J 2016 *Machine Listening: Using Deep Learning to Build an Audio Classifier To Recognise Birdsong*

IOP Conf. Series: Materials Science and Engineering **482** (2019) 012022 doi:10.1088/1757-899X/482/1/012022

InTelect: Interactive Telemedicine Communication Technologies Mobile App

M J C Samonte[1], V Aguilar[1], A K Marino[1] and R Tolentino[1]

[1]School of Information Technology, Mapua University – Makati, 333 Sen. Gil J. Puyat Ave., Makati City, Philippines

Abstract. In over 90 million of the population in the Philippines, for every 10 people, 8 of them reported never having a medical check-up in their lives. Reasons for this are the ongoing economic issues of the country such as poverty, poor infrastructure, and heavy traffic. This paper focuses on developing a system for both medical doctors and patients whose ability and affordability to access healthcare in terms of medical consultations are very limited to almost none. The developed system, a mobile application, provided an environment for the remote communication of doctors and patients utilizing the mobile communication tools for the Information and Communications Technology (ICT) through the concept of telemedicine and the use of Global Positioning System (GPS). Through this research, medical consultations between doctors and patients are made more accessible since both users are able to communicate over long distances, making it possible to look for a doctor especially in the rural areas. The mobile communication tools and the GPS are used to allow patients to search for a specific kind of doctor and consult with them. Patients can also utilize the GPS alone to locate and get directions of the nearest healthcare facilities from their locations. On the other hand, doctors are able to accommodate patients remotely giving them the benefit for comfort and an additional source of income.

1. Introduction

Today, the height of electronic and computer technology can be said to be at its peak, information is obtained so fast and results are conveyed and transported instantly. Communication has become so easy and distance does not matter anymore as compared to years back. Mobility and accessibility are two of the foundations over which information and communications technology (ICT) is built. The emergence of mobile technology is so flexible and versatile that it extends major improvements even to the healthcare industry. It has potentially facilitated and simplified the manner clinical research is conducted and the delivery of disease therapies. Prompt delivery of service and care and effectiveness are vital components that serve as demarcation to patient experience. These are the drivers that healthcare companies are pushing to deliver a much higher level of patient care [1].

The standard way people resort to when they are feeling sick is to consult a family member or a relative or go to a medical doctor for checkup. However, given the situation in the Philippines, consideration should be given to economy-related matters and environmental issues like constant heavy traffic congestion. Most of the time, people deal with one of the country's biggest issues which is heavy traffic. With this condition, people often give up the idea of consulting a doctor or going to a medical facility, more so if they do not feel anything alarming.

The Philippines is just one of the many low and middle income countries (LMIC) where telemedicine networks to aid medical doctors in an environment of inadequate means and supplies. Communication between them has been informal and relatively limited in scope [2] [3]. Though there had been previous researches on teleconsultations, e-mail and SMS were the only means of communication supported by doctors as reference to patients' health conditions. The result of such programs have revealed the capability that telemedicine has on enhancing the delivery of the services that the healthcare sector renders and its positive impact to the patients' wellbeing [4]. Nevertheless, despite the definitive strength of the scheme in dealing with the concerns and issues of the healthcare sector, the extent of its realization is quite poor [5].

The researchers proposed and created a process in line with the goals of the National Telehealth Center (NTHC) of the Philippines which is to engage people to identify problems in the health care system and develop telehealth applications from the people's very own problem-solving contributions. Having stated the problem previously, people's most go-to solution is to browse their mobile devices and search the internet for information regarding their health concerns. However, resorting to such risky self-assumptions only based from personal research concerning the health of a person is not ideally recommended nor is it proven safe. Because majority of the country's population now have the access to mobile technologies, the researchers have created a mobile application which contains the tools for ICTs and GPS functionalities that enables both users, medical doctors and patients, to interact in medical consultations remotely. With the GPS, users are able to track the location of each other before and during the consultation. For the patients, they are able to go to a healthcare facility as they are guided through the GPS in real-time.

1.1. Objective of the Study

The researchers' aim was to improve the conventional way of patients of seeing a medical doctor for health consultations and even doctors accommodating patients allowing speedy delivery of treatment and medication, reduce costs both for patients and providers, and provide additional source of income for doctors. Through remote medical consultations, patients can save both time and expenses finding and travelling to a healthcare facility and looking for a specific doctor according to their medical needs. The system provider can also make use of the application as an entirely new business which can be made into an online healthcare facility assisting patients remotely through employing licensed doctors eligible to participate in the system. To make the system possible, the ICTs and the GPS were integrated to allow the consultation between the doctors and patients remotely.

1.2. Conceptual Framework

Figure 1 shows the conceptual framework which the researchers used in identifying the accessibility features contributing to the efficient implementation of the system. The conceptual framework classifies the traditional and digital processes involved in the development of the system.

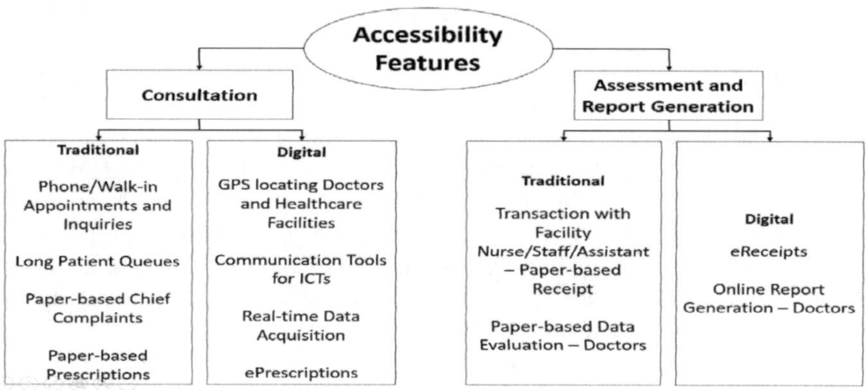

Figure 1. A conceptual framework for accessibility features

1.3. Scope of the Study

The created mobile application is accessible upon download and installation from the Google Play Store. Each of the pages in the application is responsive, meaning it will run in any mobile device as long as it has the Android operating system. The mobile application' is created using HTML5, Bootstrap, and jQuery as its front-end and using Apache Cordova for Visual Studio as its back-end. For monitoring and management purposes, a web application is also created as the server for the entire system. The web application is accessible through its domain name and using the Google Chrome browser. The web application is created using HTML and Bootstrap as its front-end and using PHP as its back-end.

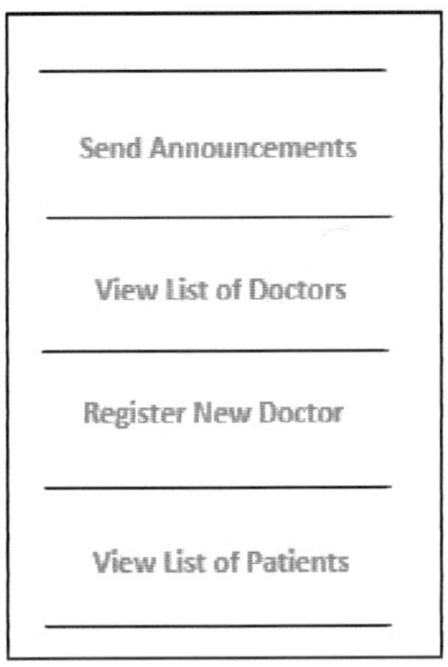

Figure 2. Doctor's Management Module

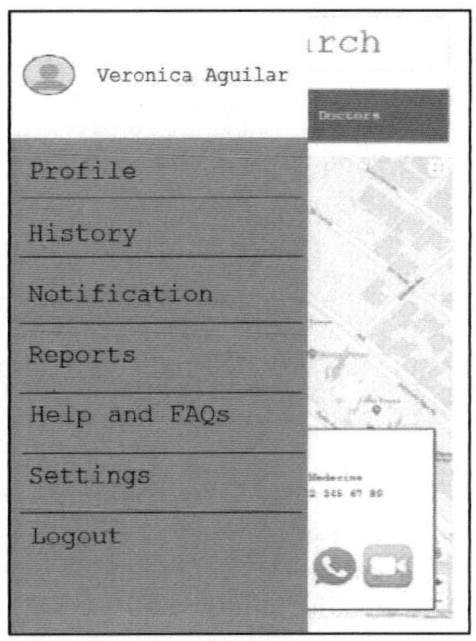

Figure 3. Patient Menu

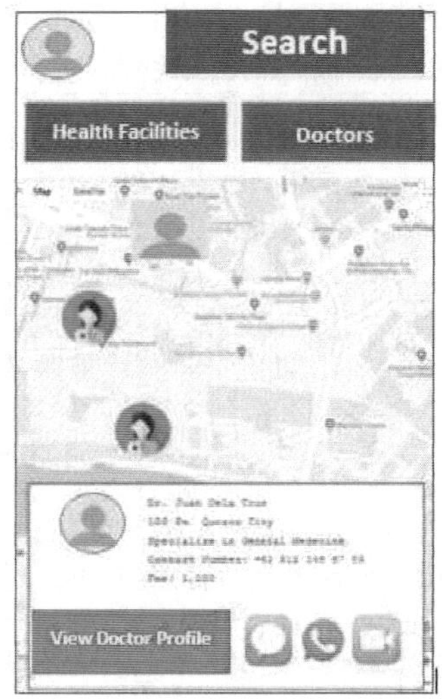

Figure 4. Consultation Module

The web application of the system is for monitoring and user management of the mobile application. The user for the web application is a system administrator. It comprises a user management module which oversees both the doctor and patient users who use the mobile application. Figure 2 shows this feature. In order for a doctor to have an account, the system administrator is the one who is in charge of creating the doctor's account. The NTHC coordinator, Dr. Geohari L. Hamoy, devised with the researchers that this is a necessary requirement to avoid identity theft on the doctors' end.

For the mobile application on the doctors' end, the primary purpose is to assess and accept requests for consultations from the patients. It comprises of a menu module, consultation module, and report module. Figures 3 and 4 show these features. First, the components of the menu module are: Profile, History, Notification, Reports, Help and FAQs, and Settings. This module and its components are for the doctors' information and tracking and evaluation of their performance on the patient consultations they have accommodated and the income they have acquired from it. Second, the components of the consultation module are: Chat, Voice Call, and Video Call Options for the consultation session, Queueing, ePrescription, and eReceipt. This module and its components are for the doctors' consultation session with their patients. They can choose and switch from the provided tools for communication and prescribe their patients. Lastly, the components of the report module is the functionality to generate reports. Here, the doctor is able to assess and evaluate his patient accommodations as well as his income gained on a daily, weekly, and monthly basis.

For the mobile application on the patients' end, the primary purpose is to look for healthcare facilities and doctors utilizing the GPS and getting their directions or consulting with their selected doctors. It comprises of a menu module, consultation module, and feedback module. First, the components of the menu module are: Search Doctor, History, Profile, Notification, and Help and FAQs. This module and its components are for the patients' tracking of their previous in-app medical records. Second, the components of the consultation module are: GPS, Chief Complaint, Payment and eReceipt, Chat, Voice Call, and Video Call Communication Options for the consultation session, Queueing, and ePrescription. This module and its components are for the patients' consultation with their selected doctors. Before proceeding with the consultation session, the patients are required to enter their chief complaint to which the doctors will be able to preview as the patients' concerns. Patients can also interact better through sending images, via capturing a photo or sending an existing image from their devices, as well as attaching file/s and sending them to their doctors. After the consultation is done, the patients receive their ePrescriptions and eReceipts of the consultation that has just ended. Lastly, the components of the feedback module is the functionality for the patient giving feedback to their doctors, right after their consultations have ended.

2. Review of Related Literature

Virtual consultation system plays a significant role in the transformation of the healthcare industry at large as it may reduce the cost of service delivery, give easier and quicker access to both users, bring cost-effective post treatment consultations, conserve time to both users, reduce the costs of travel, and improve the quality and efficiency of medical care.

Health literacy is the degree to which an individual has the capacity to obtain, communicate, process, and understand basic health information and services to make appropriate health decisions. In the Philippines, the low-income group or the lowest class of the society have low level of understanding when it comes to health matters [6].

The doctors are anxious in the increasing mortality rate among the citizenry caused by chronic non-contagious illnesses and lifestyle-related diseases. It's sad to say that patients go to the doctor when their sicknesses are already severe and therefore making the treatment more difficult and costly.

A similar study in the Health Institute of the US found virtual consultation processes are 92% accurate. It concluded that virtual consultations can potentially shorten treatment decision time and enhance clinical outcomes [7].

Over this platform, various conditions can be addressed including cancer, complex orthopedic cases, spine disorders, transplants, congenital heart disorders, joint replacement cases, complex pediatric disorders, certain neurological disorders, genetic diseases, and liver disorders [8].

This changes for the patients to represent advantages for the doctor. More patients mean more revenue, and more revenue allows for more investment in the hospital or practice. Beyond increasing revenue, medical experts are looking to the virtual consultation for their marketing efforts [9].

There are no laws for the security, specifically, between doctors and patients online, only the Data Privacy Act, which is a law that seeks to protect all kinds of information covering both natural and judicial people, serves as a foundation for information in the healthcare industry in the Philippines.

3. Methodology

The researchers utilized the Research and Development methodology for this project. This allowed the researchers to solely refer on the objectives that are applicable for the project as it is very much technology-based. There are three phases in this methodology: basic research, applied research, and the research and development phases. The basic research phase comprises the researchers' studies through continuous meetings with experts on the subject matter such as medical doctors from some healthcare facilities and researchers from the NTHC. Lastly, the researchers conducted user acceptance testing (UAT) and item analysis to gather the primary results needed to formulate the conclusion for the entirety of the research and development stages.

3.1. Accessibility Features

The primary focus of the system lies in its accessibility features through the mobile application which is created varying the doctors' and the patients' end. It is made available in-app that patients are able to locate healthcare facilities and available doctors even without having the ability to consult with a doctor due to reasons of absence of the communication tools on their devices, incapability to register credit card details or pay online, etc. The three main modules addressing accessibility are the menu module, consultation module, and assessment and report module.

3.2. Menu Module
In the menu module within the profile page, the patients and doctors must input their complete details. For both users, they have to fill in their personal information. For the doctors, they have to fill in the work information, as well.

The menu module on the patients' end also comprises of the search doctor page wherein patients can specifically find a doctor. The patient can filter their search in terms of the doctor's name, gender, specialization, and/or location. For both patients and doctors, their menu modules comprise of the history page wherein both users can track and view back their previous consultation records such as ePrescriptions and eReceipts. Figure 5 shows this feature.

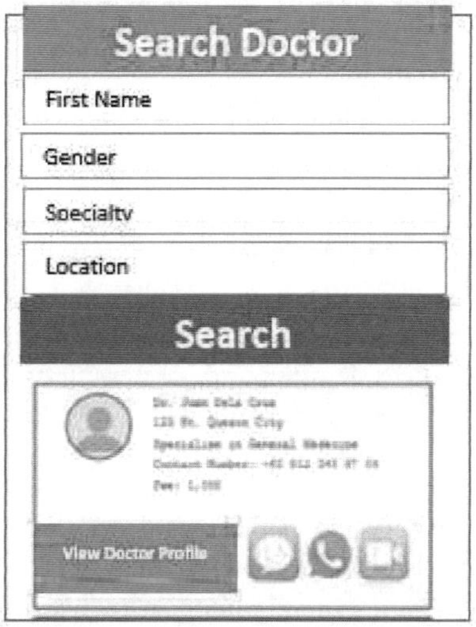

Figure 5. Search doctor page for the patient users

3.3. Consultation Module
In the consultation module, the GPS map is shown and, by default, already shows the healthcare facilities and available doctors for the patients. Selecting a healthcare facility shows its details and an option to which the patient can click to get its directions. Selecting an available doctor shows the doctor's details and consultation fee. It also has an option to view the doctor's profile or proceed to consultation. For the doctors' end, the pending consultations are shown. The doctors may choose to reject or accept the patients' request for consultation, or preview the patients' chief complaint, first. The queueing functionality of the system handles the queues of the accepted requests. Figure 6 shows this feature.

3.4. Assessment and Report Generation Module
The assessment and report generation module is only available on the doctors' end. The doctor can determine his consultation fee within his profile. All transactions made will be stored in his registered bank account. In this module is the report generation page wherein the doctors can assess and view their performance and total income through their patient accommodations from their selected time, for example, 'Weekly'. Upon viewing, the will automatically display the doctor's income breakdown.

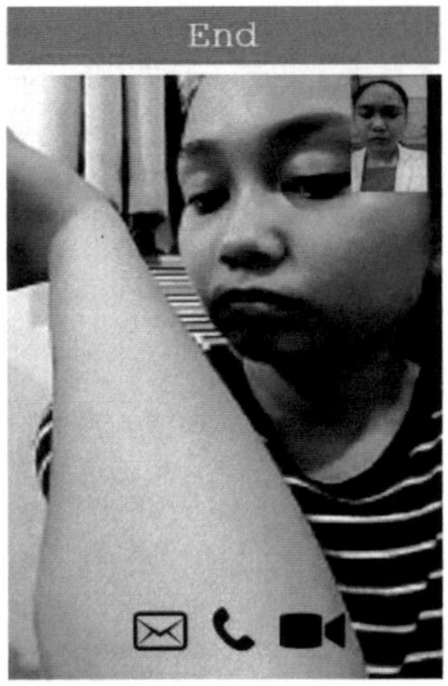

Figure 6. Video call feature on the consultation of patient users

3.5. Monitoring and User Management Module

The monitoring and user management module comprises pages for the system administrator to oversee the performance of the system, in terms of system reports and user ratings and feedback, and also manage all kinds of users in the system, specifically their accounts and legitimacy to proceed in using the application. In this module, the system administrator can add, edit, activate or deactivate, and notify all kinds of users through messaging. User data such as passwords, bank account registrations, in-app medical records, and sessions are kept encrypted to the administrator and is therefore confidential only to every user of the system.

3.5.1. User Access and Login. This component comprises the pages where the system administrator can manage the users of the system according to their roles. The system administrator is also responsible for the registration of the accounts of doctors as it is the only way how doctors are able to use the system.

3.5.2. User Notification. This component comprises the functionality for sending the users a message in terms of system-related messages such as announcements, updates, system maintenance, etc.

3.6. Testing

The researchers conducted UAT, in a form of questionnaires, with 5 medical doctors and 10 patients who registered themselves in the system, located healthcare facilities and doctors, and consulted with doctors, for the patients, and gave prescriptions, for the doctors. Out of 5, the aspects of the system as rated by the doctors are: learnability (4.10), satisfaction (4.47), interface (4.61), and efficiency (4.92). Out of 5, the aspects of the system as rated by the patients are: learnability (4.08), satisfaction (4.32), interface (4.70), and efficiency (4.85).

4. Conclusion

Patients here in the Philippines, especially those residing in the rural areas, find it difficult, time-consuming, and expensive to afford and access their right to quality patient care. Because of this, they resort to their very own ways of finding a solution on how to relieve themselves of their conditions. The project developed raised the possible implementation of a system that utilizes the tools in technology through ICTs. The concept of telemedicine was integrated for the ease of access to healthcare via remote doctor-patient communications addressing the economic issues of the country hindering the people on their right to better health services. The features for accessibility paved the way for the system to be a channel for doctor-patient relationships over distances and for the additional knowledge and research to serve as a feasible project for future advancements.

5. Recommendations

The researchers recommend to add options for the mode of payment of the patients, it can be connected to health insurances and/or Health Maintenance Organizations (HMOs) for cases of patients who use such affiliations for their medical needs. Another recommendation is to include educational materials in-app. Aside from the topic of the research, the future researchers may also integrate the system, whilst possibly already developed, to another system that may be bigger or for the purpose of creating a newer system.

References

[1] Marmot M 2006 *H. Uneq. W.* **368** 2081–94
[2] Bashshur R and Shannon G *Telemedicine as a modality of health care delivery and its implications* Encyclopedia of Cyber Behavior **1** 620–633
[3] Hailey D, Roine R and Ohinmaa A 2002 *Syst. Rev. Evid. Ben. Telemed.* **8** 1–7
[4] Vassallo D, Swinfen P, Swinfen R and Wootton R 2003 *Exp. Low. Telemed. Syst. Thr. Dev. Count.* **33** 33–4
[5] Chanussot-Deprez C and Contreras-Ruiz J 2013 *Telemed. Wou. Car.* **26** 78–82
[6] Broens T, Huis in't Veld R, Vollenbroek-Hutten M, Hermens H, van Halteren A and Nieuwenhuis L 2007 *Deter. Succ. Telemed. Implem. Lit. Stud.* **13** 303–9
[7] Fitzpatrick D, Grabarz D, Wang L, Bezjak, A, Fehlings M, Fosker C and Wong R 2012 *How. Eff. Virt. Cons. Proc. Facil. Multid. Decis. Malig. Epid. Spin. Cor. Comp.* **84** 167–172
[8] Palen T, Price D, Shetterly S and Wallace K 2012 *Com. Virt. Cons. Trad. Cons. Elect. Heal. Rec. Obser. Cas. Cont. Stud.* **12** 2–15
[9] Timpano F, Bonanno L, Bramanti A, Pirrotta F, Spadaro L, Bramanti P and Lanzafame P 2013 *Teleh. Neur. Wh. Poss.* **34** 2263–70

Tagalog text-to-braille translator tactile story board with 3D printing

L A D Arbes, J M J Baybay, J E E Turingan, and M J C Samonte

School of Information Technology Department, Mapua University, Nicanor Garcia Street, Makati City, 1209

Abstract. - New technologies create new challenges, especially for those who cannot read, count, and write. It has been reported that out of 161 million blind people or visually impaired people in the world, 90% of them live in developing communities. However, only 3% are capable to read, write, or count. The Philippines is estimated to have half a million blind Filipinos. National Library's Division for the Blind contained 922 Braille titles, 880 cassette titles, and 462 large print titles. A report concluded that Braille reading materials are lacking in schools and despite the condition of blind Filipinos, many still want to read. This study is focused on developing an assistive system for visually impaired children under kindergarten. The developed system enhanced the existing reading materials that is used by visually impaired students by providing a 3D visual representation of children's book. Significant story lines of a book were translated to Braille text to tactile board then made into a 3D printed product as a learning material presented to visually impaired students in class. The developed system was tested and used by the target users to their full learning potential.

1. Introduction

The overall prevalence of blindness around the world is 0.7 % and mostly 90% of visually impaired lives in developing countries like the Philippines. Prevalence of childhood blindness is about 0.3 in every 1000 children aged 0-15 years old [1].

In accordance to the 2015 annual report of the Resources of the Blind Incorporated, the Braille literacy rate of the Philippines is only 8% and the Braille reading materials that are being produced yearly are only 300,000 – 500,000 pages and most of the said reading materials are mainly for adults. Considering these reports, the researchers came up with the idea of producing a tactile storyboard that will be printed in 3D, giving the children a new approach of learning the Braille and as well as increase the Braille literacy.

Assistive technology is composed of equipment, programs, or product which is used to either increase or improve one's functional capability. In the Philippines, there is an existing law for the disabled persons which is the "Magna Carta for the Disabled Persons (Republic Act no. 7277)", which states that "The state shall adopt polices ensuring the rehabilitation, self-development and self-reliance of the disabled persons. It shall develop their skill and potentials to enable them to compete favorably for available opportunities." This act helps the disabled persons in the advancements of technology. When it comes to blind education, one known method for learning is the Braille [2]. In the Philippines, there is a school that is managed by the government that specializes in teaching visually-impaired students. This school is known as the Philippine National School for the Blind [2].

Braille is the disposition of raised dots that can be read using the fingers of the individuals who have visual impairments. There are 2 different ways to write using Braille, grade 1 or uncontracted and grade 2 or contracted. Most story books for children are written in grade 1, while grade 2 is used for reading materials meant for adults. Contrary to popular belief, Braille is not a dialect. Rather, it is a code by which numerous dialects might be composed and read. Braille code was created by a blind youth named Louis Braille at the age of 15 [3].

Tactile approach is a method on how visually-impaired person read. The current literacy tool that is used for the visually impaired person is embossed braille. In the research "3D Literacy Aids Introduced in Classroom for Blind and Visually Impaired Students," the product of their project is a 3D printed literacy aid that contains the debossed Hangul Characters or Korean alphabet alongside the braille translated character. The children trace the debossed Hangul character allowing them to properly visualize each character. This product has been concluded to have a positive effect on the students' learning on how to read and write Hangul [4].

The goal of the study is to create an assistive technology where tactile story board will be produced by using Braille translator and 3D printing system. The developed system consists of built-in 3D models [5] used by teachers through drag and drop and a canvass for editing and/or designing the storyboard. For every storyboard, it requires the text or story to be inputted by the teacher and the system translates it in Braille.

1.1. Objective of the Study
The research specifically aims to provide a platform that transforms 2D images into 3D models and enable text-to-Braille feature embedded in to the story board. The main goal of the research is to create 3D tactile story board as a supplementary learning material for the visually impaired children and their teachers.

1.2. Scope of the Study
The text-to-Braille translator and 3D modelling software is integrated into one system. The system translates Tagalog and English language into grade 1 Braille. The content of the storyboard will be a short story or riddles. The design of the system has a registration and logging of accounts for the user. It also allows the user to have the option of creating and editing of storyboard. As for the design of the storyboard, it has a default size of 8x6 inches. The target evaluators are kindergarten students and teachers at Philippine National School for the Blind (PNSB). The school entertains students that are partially blind, totally blind, and light perception.

2. Review of Related Literature
There is an increase of new technologies for almost everything in the lives of everyone, especially those who have disabilities. Different assistive technologies or technologies that provide assistance or enhancement to the person using it. Assistive technologies also provides a person to be capable of performing tasks that they cannot normally do [6][7].

3D printers have been a trend from the past few years. These printers were used by different people with different agendas. Some people use it for their hobbies, while some people use it for industrial purposes [8][2]. 3D printers are quite expensive and takes several hours to process and print an object, however, it would still largely benefit people who use it in different fields like in the field of medicine or education.

Braille code is composed of six dots that is read using the fingers of a visually-impaired person [2]. The code is used for reading materials meant for both children and adults. The dots are arranged in three rows of two called a "cell" [5]. These cells translate to a particular character depending on which dots are raised. Braille is very useful for those who want to learn to read even if they are limited because of their disability. A sample illustration of a Braille alphabet can be found on figure 1.

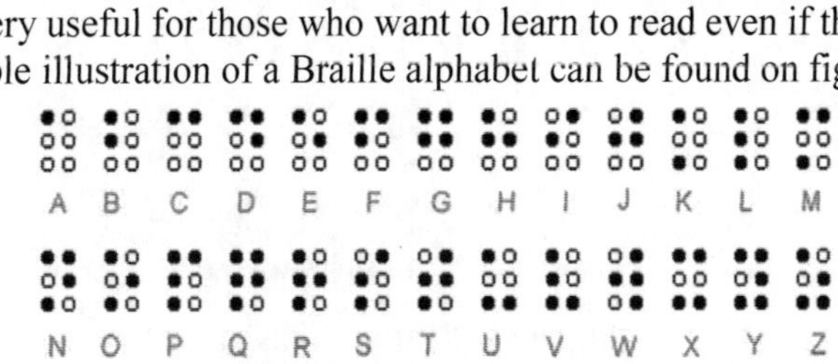

Figure 1. Braille Alphabet

Some 3D modelling systems are too complex for a person to understand. However, new developments made 3D modelling easy because of simple interfaces and designs. Similar studies show that improvements are made, thus, making 3D modelling easier to use, reliable, and responsive. Paper3D is a 3D modelling system. It went through a natural, user-guided evaluation and the feedbacks that were given back were used as basis for the improvement of the system. The modelers that tested the systems said that the application is very mobile in terms of access for it can be used through smartphones that has multi-touch display. Paper3D was also said to be precise and the modelers also wanted more sub-modes or techniques that can be used in the system [5][6][8].

Technologies help people make their lives easier. For the visually impaired people and those who are concerned with them, text to Braille conversion applications benefits them. Fast conversions of text to Braille or vice versa using a software saves more time than converting it traditionally. This will also provide the visually impaired person to learn faster because the materials that they need can be produced quickly [9][10].

Combining the said technologies from different studies, the researchers came up with a 3D Printing System with Text-to-Braille Translator which can produce a 3D printed storyboard for the visually impaired. Similar studies analyses the need for the visually impaired to have reading materials to effectively teach them how to read and even write by tracing the character on the board. The developed technology is a success to improve the literacy rate for the visually impaired [11][12].

3. Methodology

The research is composed of using 3D printers and a translator that converts text to Braille. Both component would work together to form an assistive system that can aid people ease activities. The review of related literature provided information which is used for the study. The researchers used the research and development approach upon developing the study. This approach is composed of six phases; data gathering, analysis, design, implementation, testing, and development and evaluation.

The system is designed to convert text will be translated to braille and allow creation of 3D models that can be used in the 3D storyboard, which will produce a product that is done by a non-visually impaired user. The product allows the visually impaired to visualize the story.

3.1. Designing of 3D models

The researchers used Lithophane Creator for 3D modelling. It is an open source image based 3D modelling software. The researchers used this for the user to input an image with a file format of either JPG or PNG and that will be converted to STL (stereolithography) file format for 3D printing. This allows the creation of 3D models from a 2D image. This tool simply takes any image, calculates the grayscale value for each pixel, and then creates a 3D model that represents the extrusion of those grayscale values. The researchers also considered the use of 'Blender' software to learn more about 3D modelling [13][14].

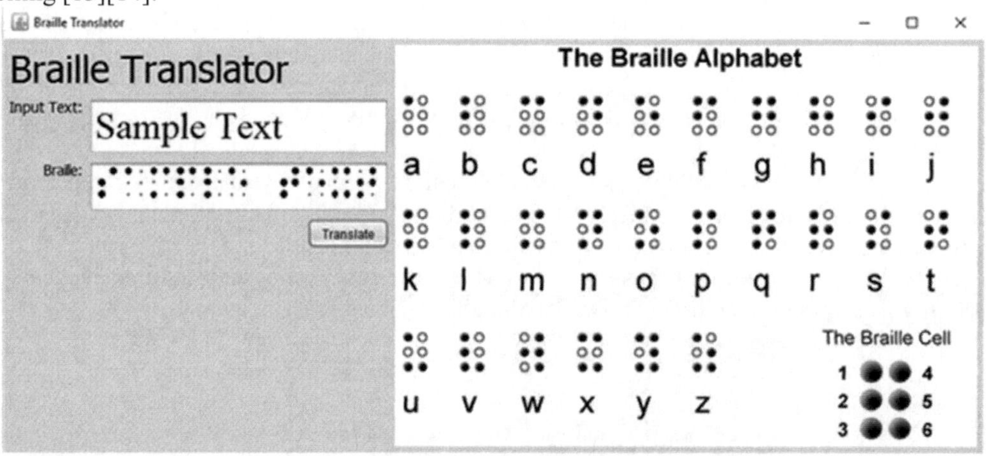

Figure 2. Braille Translator

3.2. Translation of Braille

The system used grade 1 Braille because the product is produced for the academic use of visually-impaired students under kindergarten. Grade 1 Braille is Braille that in every possible arrange of dots corresponds to one letter, punctuation, or number. The teachers of the Philippine National School for the Blind also said in their interview that they use grade 1 Braille for teaching the children. The system focused on translating Tagalog language into Braille but it can also be able to translate English language because it only supports grade 1 Braille.

The Braille translator of the system, which is presented in Figure 2, requires the input of the user in the first textbox. The system will provide the translated Braille in the second textbox once the "Translate" button is clicked. The system also has a Braille code that is mainly for checking.

3.3. *Tactile 3D Printing*

A 3D storyboard will be created through the use of the system. Figures 3 to 7 illustrate the digital 3D story tablet product of the system. These examples illustrate a Tagalog or local Filipino language version of a riddle which is also called *bugtong.*

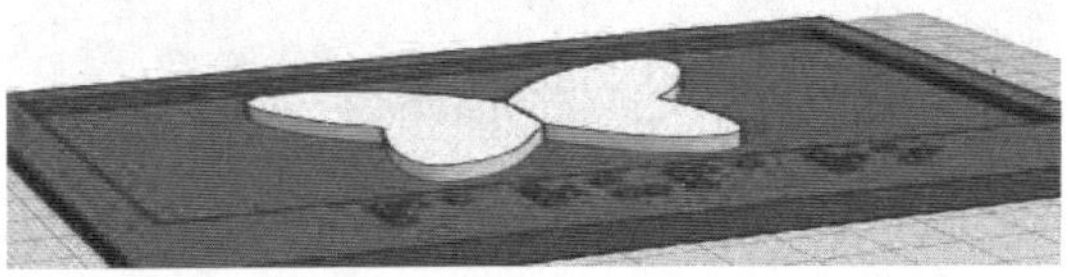

Figure 3. The Riddle in Braille and Butterfly printed in 3D as the answer.

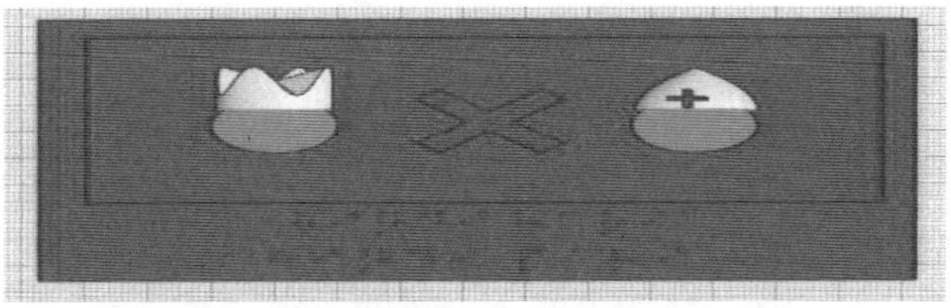

Figure 5. Riddle (Top view)

Figure 4. Example 1 Riddle in Braille -

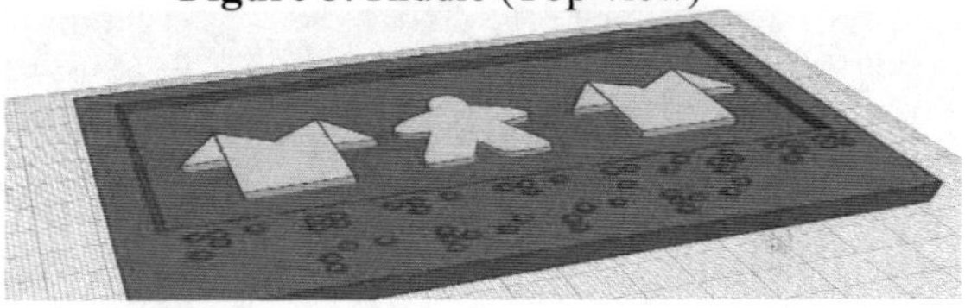

Figure 6. Example 2 - Riddle in Braille

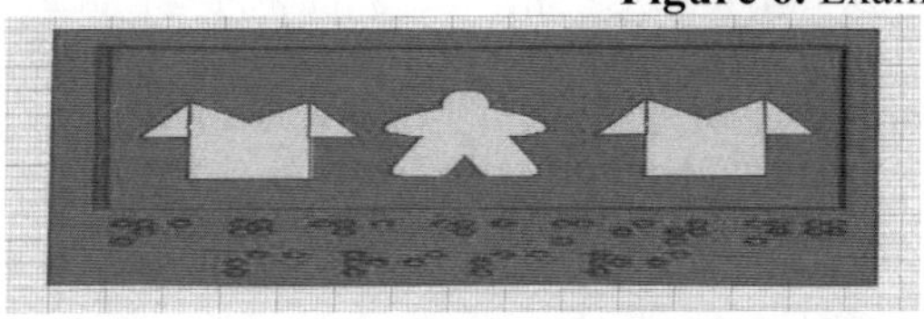

Figure 7. Example 3 - Riddle in Braille (Top View)

3.4. *Databases*

The system has a repository to store unfinished or finished work. This lets the user to save their work and continue whenever they can. The database is only for the storage of the 3D tactile story board created by the user.

4. Result and Discussion

The first phase of developing the system is the gathering of needs of the researchers to create the system. The researchers conducted interviews with teachers at Philippine National School for the Blind and have gathered related literatures that supports the development of the system. These data were assembled and were analyzed by the researchers.

The system is developed using Java language. The windows forms is created using NetBeans IDE and the researchers used open-source APIs for the Braille Translation and 3D Modelling Tool. The system allows users to be able to upload their previous works for them to continue or make changes.

The process flow of the system will start wherein the user will be prompted to create or edit a storyboard and will be required to input a text that will be translated to braille. It will then be connected to a 3D printer to print the product. The system will have three main functions. The creation of tactile story board, text-to-Braille translator, and creation of 3D models from 2D images.

Evaluation of the developed system in terms of functionality and usability were tested by target users who are language teachers of kindergarten level who teaches English, Filipino, and Braille and the product was evaluated by visually impaired children at the Philippine National School for the Blind.

The researchers gathered ten students from Philippines National School for the Blind (PNSB) and assessed their fluency with braille code by giving them an exam. The six students who scored less than 40% were chosen to be part of the user acceptance test was divided into two groups. The first group were taught braille code with the aid of the story tablets and the second group were taught using the traditional method. The story tablets were lent to the language teachers of kindergarten level who teaches English, Filipino, and Braille for them to use as a learning material for a month. The data gathered by the researchers are shown below:

Group #	Student #	Exam #1	Exam#2
Group #1	Student #1	30%	70%
	Student #2	20%	50%
	Student #3	20%	60%
Group #2	Student #4	40%	50%
	Student #5	30%	50%
	Student #5	20%	40%

Figure 8. Result of User Acceptance Test

After a month the participants took the assessment exam again. As shown in Figure 8, the group of students that were taught with the product became more familiar with the braille code and was able to finish the assessment exam faster. The teacher in charge of the students was interviewed about how the students interacted with product. She stated the having a visual representation of the text enhanced the students' interest in reading and fluency with braille code.

5. Conclusion

Technologies can make work easier and faster. Due to the rapid increase in technology, persons with disabilities, especially visually-impaired persons, face struggle in keeping up. The study focused on producing a system that is used as a supplementary academic material for the visually-impaired which is used through tactile approach. The product provided an opportunity for the visually-impaired person to enhance his reading skills at a young age. The product and system undergone a user acceptance test in the Philippine National School for the Blind which proved that product is more efficient and effective as a learning material compared to embossed braille. The researchers also conducted interviews with professionals that specializes in this field which provided an ample information that made the system more reliable.

6. Recommendation

The advancements of technology will continuously incline as years will pass by from now. It will require more research to improve the system especially in the field of tactile education, wherein, it will provide a study of the students that will present where they engage more in learning. It will require a long-term study to provide a more interactive reading material for their academic use.

References

[1] Del Mundo P and Chua C 2015 Causes of blindness and severe visual impairment among children enrolled in an early intervention and preschool program of a school for the blind in the Philippines *Philippine J. of Ophthalmology* **40** p 1

[2] Cruz F and Calimpusan E 2018 Status and challenges of the deaf in one city in the philippines: towards the development of support systems and socio-economic opportunities *Asia Pacific J. of Multidisciplinary Res.* **6(2)**

[3] Morse A 2017 Valentine Haüy and Louis Braille: enabling education for the blind *Foundations of Ophthalmology* (Cham: Springer) pp 45-63

[4] Harianto R, Chen E, Lim Y, Jo W, Moon M and Lee H 2016 3D literacy aids introduced in classroom for blind and visually impaired students *J. of Blindness Innovation & Res.* **6(2)**
[5] Paczkowski P, Dorsey J, Rushmeier H and Kim M 2014 Paper3D: bringing casual 3D modelling to a multi-touch interface *Proc. of the 27th annual ACM symp. on User interface software and technol.* pp 23-32
[6] Damit D, Ani A, Muhamad A, Abbas A and Ali F 2014 Dual braille code translator: basic education tool for visually impaired children *Computer, Communications, and Control Technology (I4CT) 2014 Int. Conf.* pp 399-402
[7] Luo X, Wang J, Liu N, Zhao Z and Zhou Y 2014 Yarep: a personal 3d printing simulator *2014 Int. Conf. on Virtual Reality and Visualization (ICVRV)* pp 408-11
[8] Bernhard J, *et al* 2016 Personalized 3D printed model of kidney and tumor anatomy: a useful tool for patient education *World J. of urology* **34(3)** pp 337-45
[9] Sultana S, Rahman A, Chowdhury F and Zaman H 2017 A novel braille pad with dual text-to-braille and braille-to-text capabilities with an integrated LCD display *2017 Int. Conf. on Intelligent Computing, Instrumentation and Control Technologies (ICICICT)* pp. 195-200
[10] Jariwala N and Patel B 2015 Conversion of Gujarati text into braille: a review *Int. J. of Innovation and Advancement in Comp. Sci.* **4(1)** pp 59-64
[11] Luckner J, Bruce S and Ferrell K 2016 A summary of the communication and literacy evidence-based practices for students who are deaf or hard of hearing, visually impaired, and deafblind *Communication Disorders Quarterly* **37(4)** pp 225-41
[12] Bardot S, Serrano M and Jouffrais C 2016 From tactile to virtual: using a smartwatch to improve spatial map exploration for visually impaired users *Proc. of the 18th Int. Conf. on Human-Computer Interaction with Mobile Devices and Services* pp 100-11
[13] Nielsen J, Eiriksson E, Lyngby R, Wilm J, Jensen J, Aanæs H and Pedersen D 2017 Picprint: embedding pictures in additive manufacturing *Special Interest Group Meeting: Additive Manufacturing*
[14] Pantazis A and Priavolou C 2017 3D printing as a means of learning and communication *Telematics and Informatics* **34(8)** pp 1465-76

IOP Conf. Series: Materials Science and Engineering **482** (2019) 012024 doi:10.1088/1757-899X/482/1/012024

MAR UX Design Principles for Vocational Training

C.V. Dayagdag[1], R. A. Catanghal Jr[2], T. D. Palaoag[3]

[1]Romblon State University, Odiongan, Romblon, Philippines
[2]University of Antique, Sibalon, Antique, Philippines
[3]University of the Cordilleras, Baguio City, Philippines

[1]carlwindayagdag@gmail.com, [2]catanghal.ric@gmail.com, [3]tpalaoag@gmail.com

Abstract. Current scientific research on MAR UX design does not address all facets of user experience in a holistic manner. It only covers usability, value, and desirability of the fundamental elements of UX. No known current published research considers the adoptability design principles of UX in MAR. To address this need, this Paper proposed UX design principles for MAR vocational training application and apply it to the design and development of AR4Juan. AR4Juan is an android based MAR vocational skills training application for Senior High School students specializing in Computer System Servicing National Competency (CSS NC) 2. CSS NC2 is under Technical Vocational Livelihood (TVL) track of the K12 education program in the Philippines. The researchers used MAR UX design principles research process to formulate the proposed UX Design principles. The study found out that other literature's design principles only focus on usefulness, usability, and desirability of UX elements and does not consider the adoptability element. Adoptability principles are essential in ensuring the application's natural users usage increase. Thus, drives business success. AR4Juan is still in its prototype stage and needs to be evaluated in every key UX element to determine the degree of its usefulness, ease of use, user enjoyment and engagement and ease of access. Other researchers could adopt the proposed UX designed principles and improve it iteratively after every evaluation of the MAR application.

1. Introduction

Mobile Augmented Reality (MAR) is an idea developed around the mid-1990s incorporating Augmented Reality (AR) in a portable mobile device such as smartphones. It merges different technology like camera, location-based computing services (LBS), wireless communication and AR to create an interactive environment that allows users to experience an enhanced interactive virtual and physical world [1][2]. When the user aims the camera to the object or a marker, the virtual information is directly superimposed and displayed on the user's screen, and the user can interact with the system through its user interface [3]. MAR is growing rapidly and becoming more mature due to the continued enhancement of the smartphone's embedded technology [4]. In the study, a systematic review of 10 years of AR usability studies 2005 – 2014, nine different AR application area are identified [5] namely collaboration, education, and training, entertainment, and gaming, industrial, interaction, medical, navigation and driving, perception and tourism and exploration.

Vocational Training provides job-specific technical training for work in the industry. This training provides students with hands-on instruction that can lead to certification or diploma. Research on AR in education and training has reported that one of the most important advantages of AR is the increase of

IOP Conf. Series: Materials Science and Engineering **482** (2019) 012024 doi:10.1088/1757-899X/482/1/012024

student motivation, achievement and improve the skill and efficiency in assembly and maintenance tasks [6][7].

MAR applications can create more realistic and innovative experiences. However, when mobile applications and AR technology are being used and developed extensively, the focus is often on technological development, and the UX is not considered [4]. Moreover, Current scientific research on MAR UX design does not address all facets of user experience in a holistic manner. It only covers usability, value, and desirability of the fundamental elements of UX. No known current published research considers the adoptability design principles of UX in MAR. According to Frank Gou [8], adoptability has often neglected aspect of UX strategy. If users do not install the application, all of the product's good features become irrelevant. The main objective of this study is to propose a holistic MAR UX design principles for vocational training application and create an application prototype that applies the proposed design principles.

2. Background of the study

2.1. What is User Experience

UX professionals use the term user experience much more broadly that covers completely ranging from ease of use to user engagement to visual appeal, and it better captures all of the psychological and behavioral aspects of users' interactions with products [8].

2.2. Elements of User Experience

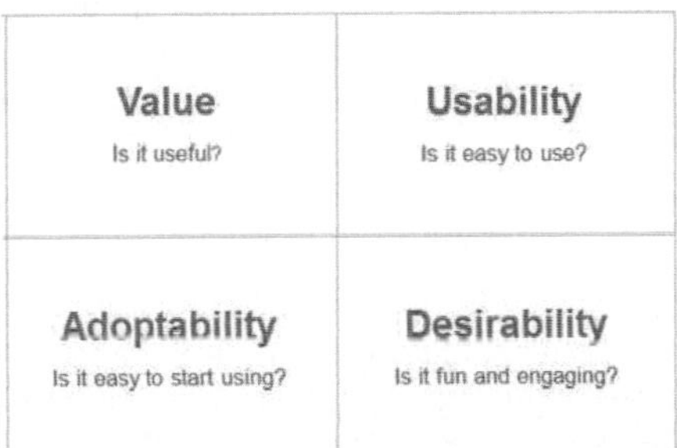

Figure 1. Frank Gou's four elements of UX

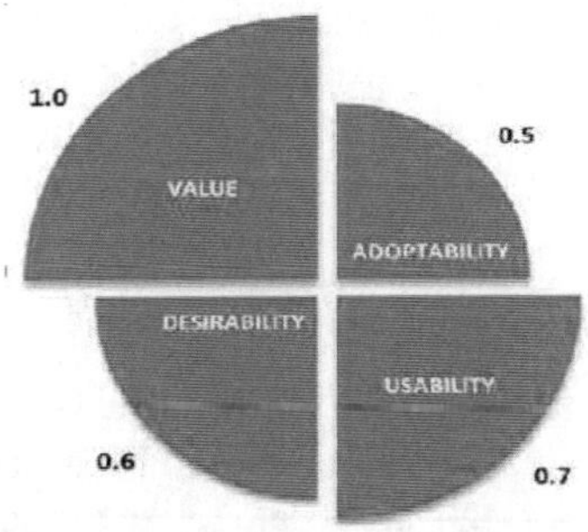

Figure 2. The relative importance of UX elements for MAR Vocational Training Application

In attempting to achieve conceptual simplicity, Frank Gou reduced the many aspects of user experience to the four elements that are the most fundamental namely Value, Usability, Adoptability, and Desirability as shown in figure 1. Usability is about how easily users can complete their intended tasks using a product. Value refers to the usefulness of a product such as software application to the users. Adoptability relates to users' buying, downloading, installing and start to use a product. It should be considered as an integral component of an application design. Desirability relates to user's emotional appeal where the application allows the user to feel fun and encourage interactive engagement in using it [8].

2.3. UX Elements Relative Importance

The four elements of UX do not always have the same importance. Each element's relative importance is identified by the developers based on the type of application being developed and its target users. Understanding the relative importance is critical to correctly prioritizing product design and development efforts. Numeric weightings such .1 to 1.0 are assigned to a particular UX element wherein .1 is the lowest priority while 1.0 is the highest priority. These numeric weightings are used by the developers in assessing which UX element is more or less important in driving great user experience and prioritize the appropriate UX elements when allocating the resources that are available.

3. Research Process

As shown in figure 3, this paper presents the research process that is used in this study.

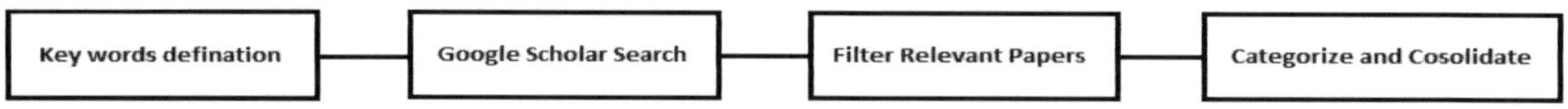

Figure 3. MAR UX Design principles research process

3.1. Keywords definition

Before starting the search, the keywords are first defined, which are necessary for the research in online databases. The researcher used the keywords *"User Experience in Mobile Augmented Reality"*, *"Mobile Augmented Reality in vocational training"* and *"Augmented Reality Design Principles"* for the research.

3.2. Google Scholar Search

Google scholar is the online databased used to search the defined keywords. We used the defined keywords in Google Scholar Search Engine and produced a combined 45 papers.

3.3. Filter Relevant Papers

In the next step, the researchers read all the documents and decided if the paper is relevant to the present study. The paper interest is to discover the holistic MAR UX design guidelines that could be used for vocational training because this is still an unexplored area. Scientific research under medicine, tourism and gaming domain is excluded because this is beyond the scope of this study. The researchers identified six relevant papers, as shown in Table1, focusing on UX in MAR, MAR in Vocational Training and MAR dynamic interactions.

3.4. Categorize and Consolidate

After identifying the relevant papers, the researchers categorize the current work according to holistic UX design elements and consolidate similar principles in the same category into a set of MAR UX design principles for vocational training as shown in Table 2. Design principles whose category belongs to two UX elements are recategorized based on UX elements weights. The final category of the design principle will be in the UX element having the higher weights. For example, the design principle "virtual objects should be aligned appropriately with physical objects and should display continuously over time" is categorized under Value and Desirability elements. In this study, Value elements weight is 1.0 and Desirability only weight 0.6, since Value element has a higher weight than Desirability element, the design principle is finally categorized under Value element.

Since this study is related to produce MAR Vocational Training application prototype, Value element weighs the highest, followed by Usability, then Desirability and lastly Adoptability. Figure 2 shows the researchers assigned weight for every UX element used in the study. Value element weighs the highest because, in the study, MAR vocational training application's usefulness to its intended users is given top priority. Therefore, design principles under this element are prioritized to be applied first and allotted more resources such as time and budget in the design and development of MAR application. Usability element weighs second to the highest because MAR application must be user-friendly in the sense that users need to be able to complete their tasks with efficiency and accuracy. Otherwise, MAR application simply fails to deliver its value. Desirability element weighs third to the highest because the researchers want to ensure that MAR vocational training application must be useful and user-friendly first before making the application fun and encourage interactive user engagement. Finally, Adoptability weighs last because this element provides less effect in delivering effective vocational training but essential in driving MAR application and business success. The researchers of this study make the call in identifying the weigh of each UX elements. The weigh is intended to identify the UX element that will be given higher priority in the design and development and resource allocation of MAR Vocational training application.

Recent research work does not consider adoptability design approach in MAR application design, the researchers looked into the practice of mobile applications in the industry, acquire the adoptability component of it and consolidate it in the proposed MAR UX design principles in this study. These proposed principles could be applied during the development process of MAR application for Vocational Training. This paper aims to propose a holistic UX design principles in designing MAR application for Vocational Training.

4. Recent UX, MAR and AR design principles

Table 1. Recent study on MAR application design principle

Source	Investigation Approach	Key Design Findings
[1]	Literature review on the interaction design of MAR	1) Use the context for providing content, 2) Deliver relevant-to-the-task content, 3) Inform about content privacy, 4) Provide feedback about the infrastructure's behavior and 5) Support procedural and semantic memory
[9]	Interaction Design Evaluation using AR for Assembly and Training Task	1) Simple Design, 2) Feedback, 3) Pre-knowledge, 4) Consistency, 5) User Mode, 6) User Control, 7) Help, Error prevention, Individualization and 8) Help & Documentation:
[7]	Study on Design and Development of MAR in Vocational Education and Training based on Co-creation process and UDL	1) Provide multiple means of presentation, 2) Provide multiple means of expression and 3) Provide multiple means of engagement
[10]	Heuristic evaluation of usability principles on dynamic interaction in MAR	1) Fit with user environment and task, 2) Form communicates function, 3) Minimize distraction and overload, 4) Adaptation to user position and motion, 5) Alignment of physical and virtual worlds, 6) Fit with user's physical abilities, 7) Fit with user's perceptual abilities, 8) Accessibility of off-screen objects and 9) Accounting for hardware capabilities
[4]	Study on demonstrated usability for MAR application	1) User-information, 2) User-cognition, 3) User-support, 4) User-interaction and 5) User-usage
[11]	Study on UX in MAR focusing on Emotions, Challenges, opportunities and proposed best practice	1) Spatial correspondence, 2) Tolerance to movement, 3) Object detail, 4) Object Correspondence, 5) Natural Interaction, 6) Personalized Experience and 7) Emotion-evoking avatar

As shown in Table 1, recent research work on MAR is focused on addressing usability, value and desirability issue of MAR UX design. Thorough evaluation of different area of MAR application and interaction design studies produced generic MAR design approach [1]; MAR application in industry skills training provides design recommendations to strengthen the trainee's procedural skill that is essential for assembly and maintenance task [9]; education and vocational setting designed and developed a MAR application for skills training utilizing co-creation and UDL design approach [7]; Heuristic evaluation of usability principles on dynamic interaction in MAR are performed [10]; usability for MAR application is demonstrated [4]; and in addition best practice in UX MAR design approach is proposed [11]. The recent study does not apply the holistic MAR UX application design approach which constitutes an opportunity for the researcher to discover its deficiency.

5. Proposed MAR UX design principles for vocational training application

The proposed MAR UX design principles for vocational training application are categorized based on four fundamental UX elements namely Value, Adoptability, Desirability, and Usability. Each element has a set of proposed design principles, description and indicative design practice that are consolidated from previous MAR/AR, UX, MAR vocational training related studies and the common practice of mobile applications in the industry.

5.1. Value
Value UX element is given the highest relative importance in driving UX success for this study. The reason is that in Vocational Training, the usefulness of MAR application in complementing the delivery of skills development is given top priority. To help MAR application developer come-up with a useful MAR application training tools, the following seven design principles are proposed: 1) *Collaborative creation process*, MAR developer should apply collaborative creation process (Co-Creation) and Universal Design for Learning (UDL) to identify the requirements needed in producing effective training MAR application. Its indicative practice is to design the MAR Vocational training application with experts in the Vocational Education and Training domain (teachers), software developers, graphic designers, and educational technology experts with an extensive background in UDL; 2) *Mental Model Building*, MAR training application should provide a mental model of an assembly task that describes the internal representation of an entire task. Its indicative design practice is a training task progress bar; 3) *Visual Aids*, MAR training application should present items of illustrative matter, such as a 2d/3d models, videos, images, and text, designed to supplement training written information so that it can be understood more easily. Its indicative design practice is display images and other multimedia objects related to training activity; 4) *Passive Learning*, MAR training application should have an option in the training that trainee is not active and only receives information about the task. Its indicative design practice is video demonstration in assembly training task; 5) *Haptic hints*, MAR training application should trigger the vibratory component of the smartphone or a vibrotactile bracelet that allows giving feedback about current actions and can help to prevent errors. Its indicative design practice is during training MAR application trigger the vibrator in the smartphone when the user commits a mistake; 6) *Fit with user's perceptual abilities*, MAR experience should be within the user's capacity to understand and handle training task. Its indicative design practice is during training; task procedures are displayed one at a time; and 7) *Alignment of physical and virtual objects,* MAR training application's virtual objects should be aligned appropriately with physical objects and should display continuously over time. Its indicative design practice is when the camera is pointed to the target physical object its corresponding virtual object should display next to it accordingly and continuous over time.

5.2. Usability
Value design principles could make Vocational training application useful. To deliver its usefulness, users need to be able to complete their task with ease of use, efficiency, and accuracy. MAR application developer could accomplish the delivery of application's ease of use using the following six proposed Usability design principles: 1) *Match between MAR System and the real world*, MAR application should speak the users' language, with words, phrases, symbols and concepts familiar to the user. One of its indicative design practice is during practical skills training, MAR application training control button resembles the images or icons of a multimedia player such as forward, back, play and pause icon; 2) *Simple Design*, MAR application should have all the required functionalities, not more and not less. Its indicative design practice is MAR system Menu should only contain needed options design to its intended purpose; 3) *Fit with user's physical abilities*, MAR application should allow the user to perform any action easy to perform. Its indicative design practice is when the user's simple hand gesture or voice command, MAR application seamlessly hide or display MAR objects; 4) *Accessibility of off-screen objects*, MAR application's UI control should allow users to easily recall those objects when outside of the field of view. The indicative design practice is a visual hint in MAR application such as an icon showing hidden active training activity objects; 5) Consistency, MAR application should use standards and be consistent in its user interaction while in operation. Its indicative design practice is MAR application produce consistent short tick sound every time the user selects any system controls; and 6) User Control, MAR application should allow the user to go back, forward and to the starting point at

any time in performing training task. Its indicative design practice is MAR application shows visible back, forward, go to begin and go to end button during the training mode.

5.3. Desirability

In general, MAR application is designed to be desirable and engaging. It allows the user to experience fun and encourage interactive engagement in using it. To realize this positivity, eight Desirability principles are proposed for this study. These principles are 1) Accounting for hardware capabilities, MAR experiences should be designed to accommodate for the capabilities & limitations of the hardware platform. Its indicative design practice is MAR Application contains a mechanism to identify smartphones hardware capability and load its settings ideal to the optimal systems operation on application's start-up; 2) System Feedback, MAR application should inform the user about the current user's action status. Its indicative design practice is MAR application provides visual emphasis such as control buttons glows every time user is using it; 3) User Mode, MAR application should have different modes (beginner, intermediate, advanced) to allow advanced users to work faster. Its indicative practice is MAR application provides a different user mode; 4) Error prevention, MAR application should prevent errors. Its indicative design practice is when user attempt or accidental initiate close MAR application, the system display notice that there is unfinished work and ask confirmation to save and exit the current work; 5) System user support, MAR application should give easily understandable help at the right time. Its indicative design practice is straightforward application Help and Documentation option; 6) Good object details, details of MAR objects should be sufficiently high (within the limits of the target hardware) to make them recognizable and appealing. Its indicative design practice is good quality icons, images, animations, and video presentation; 7) Inform about content privacy, MAR application system users information should not go out in the public media without consent. Its indicative design practice is to inform and seek user consent when publishing their skills progress or accomplishment in social media; and 8) Personalize Experience, Making MAR experience personal to each user increases the likelihood for emotional engagement. Its indicative design practice is user has an option to choose and personalized Avatar to assist in training.

5.4. Adoptability

Providing easy access to a product is a design consideration that every MAR application developer should consider. It allows the application to potentially increase the number of its users that could benefit from its features and create good business value. To improve MAR applications adoptability, two Adoptability design principles are proposed: 1) Sharing of MAR training task and accomplishments, MAR applications should allow the user to share task and accomplishments experience to friend's social media, email, IM, and SMS. Its indicative design practice is in every successful training task, the user is asked to share their accomplishments in their Facebook post. Included in the post is the recommendation to download and use the MAR Vocational Training Application; and 2) Share MAR training application, MAR application before closing should provide the user an option to share the application to social media friends. Its indicative design practice is when the user initiates the application's closing; the application encourages the user to share the MAR Vocational Training Application to social media friends.

6. Application of Proposed MAR UX Design principle for AR4Juan application

AR4Juan as shown in Figure 4, is an android based MAR vocational skill training application for Senior High School students specializing in Computer System Servicing National Competency (CSS NC) 2. CSS NC2 is under Technical Vocational Livelihood (TVL) track of the K12 education program in the Philippines. This paper is intended to apply the proposed UX principles in practice and later identify the degree of its usefulness, ease of use, user enjoyment and engagement and ease of access. The application is focused on providing training delivery on Set-up Computer Networks competency particularly in installing network cable of CSS NC2 specialization. The researchers apply the proposed design principles that are group into 4 UX key elements namely Value, makes the system use-full by ensuring that all requirement needed in training are meet through Co-creation process; Usability, make the system user friendly through easy to recall and use system training controls; Desirability, makes the system fun and engaging through personalized avatar and good system to user interaction; and Adoptability, makes the application reach more its potential system user through social media. The following table summarizes the proposed UX design principles practice used in AR4Juan.

Table 2. Proposed UX design principles practice for AR4Juan.

UX Elements	Design Practice
Value	Co-Creation and UDL (perform during MAR Application Pre-Development) 1) Progress Bar 2) images and other multimedia objects related to training activity 3) Only display instructions one at time 4) Physical object's corresponding virtual object should display next to it accordingly and continuous over time
Usability	MAR object could be hidden and reveal easily thru gesture and voice control (incorporated in MAR application), Produce consistent short tick sound every time the user selects any system controls (incorporated in the application) 5) Activity Control Buttons resembles real-world multimedia symbols 6) Display only needed option control and information. 7) Shows visible back, forward, go to begin and go to end button during the training mode
Desirability	Ask the user to confirm current action while ending the training or application (incorporated in the application) 8) good quality icons, images, animations, and video presentation 9) personalized trainer Avatar
Adoptability	MAR user training activity, performance, experience, and application information could be shared using user's social media accounts (incorporated in the application, shows at the end of every training activity)

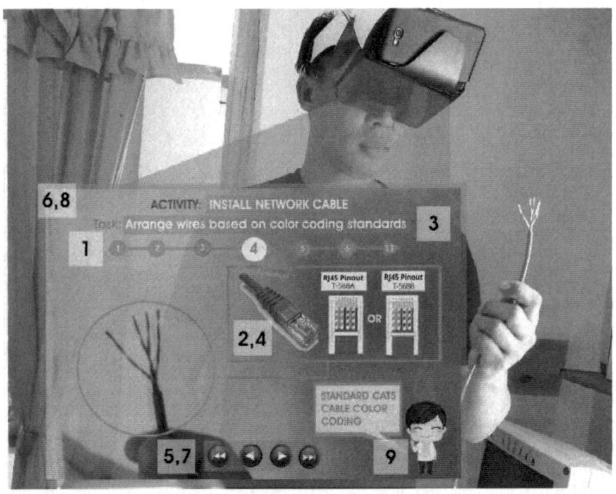

Legend:

1) Progress Bar
2) Images and other multimedia objects related to training activity
3) Only display instructions one at time
4) Physical object's corresponding virtual object should display next to it accordingly and continuous over time
5) Activity Control Buttons resembles real-world multimedia symbols
6) Display only needed option control and information
7) Shows visible back, forward, go to begin and go to end button during the training mode
8) Good quality icons, images, animations, and video presentation
9) Personalized trainer Avatar

Figure 4. AR4Juan Application Prototype

7. Conclusion & future works

The main contributions of this paper are as follows: 1) We proposed wholistic UX design principles for MAR application in vocational training, and 2) Design and Develop AR4Juan MAR application prototype applying the proposed UX design principles. This study is different from other studies reported in the literature regarding the design principle presentation. We classify the proposed design principles based on its corresponding UX key elements. This classification is done to determine the design principles that will confirm each UX element. We also found out that other literature's design principles only focus on usefulness, usability, and desirability of UX elements and does not consider the adoptability element. Adoptability principles are essential in ensuring the application's natural users usage increase. Thus, drives business success. AR4Juan is still in its prototype stage and needs to be evaluated in every key UX element to determine the degree of its usefulness, ease of use, user enjoyment and engagement and ease of access. Other researchers could adopt the proposed UX designed principles and improve it iteratively after every evaluation of MAR application.

References

[1] Kourouthanassis, P E. Et Al. (2013). Demystifying The Design Of Mobile Augmented Reality Applications.
[2] Shukri, S. Et Al. (2017). The Design Guidelines Of Mobile Augmented Reality For Tourism In Malaysia
[3] Tsai, T. Et Al. (2016). Design Of A Mobile Augmented Reality Application: An Example Of Demonstrated Usability.
[4] Liang, S. Et Al. (2015). Design Principles Of Augmented Reality Focusing On The Ageing Population
[5] Dey, A. Et Al. (2018). A Systematic Review Of 10 Years Of Augmented Reality Usability Studies:2005 To 2014
[6] Acosta, J. (2017). Framework For The Design And Development Of AR Learning In Vocational Education And Training.
[7] Bacca, J. Et Al. (2015). Mobile Augmented Reality in Vocational Education and Training.
[8] Gou, F. (2012). https://www.uxmatters.com/mt/archives/2012/04/more-than-usability-the-four-elements-of-user-experience-part-i.php
[9] Werrlich, S. Et Al (2017). An Overview Of Evaluations Using Augmented Reality For Assembly Training Tasks.
[10] Endsley, T. Et Al. (2017). Augmented Reality Design Heuristics: Designing For Dynamic Interactions
[11] Dirin, A. Et Al. (2018). User Experience in Mobile Augmented Reality: Emotions, Challenges, Opportunities and Best Practices

A fuzzy-based career recommender system for senior high school students in K to 12 education

M C B Natividad[1], B D Gerardo[2] and R P Medina[3]

[1]Technological Institute of the Philippines, Quezon City, Philippines
[2]West Visayas State University, Iloilo City, Iloilo, Philippines
[3]Technological Institute of the Philippines, Quezon City, Philippines

[1]marveecheska@gmail.com
[2]bobby.gerardo@gmail.com
[3]ruji_p_medina@yahoo.com

Abstract. In this paper, a proposed career recommender system was presented which aim to help not only the guidance counselor but most especially the senior high school students to guide them in considering numerous factors associated to their decision on what career they will pursue. In dealing with choosing the student attributes from numerous factors, a feature selection technique is appropriate to use to remove irrelevant features that affect the performance of the proposed fuzzy-based system. In this paper, different filter methods are used to select the best attributes then these attributes are used as crisp inputs. The result of the experiment shows reasonable result for making decisions. It is concluded that the proposed career recommender system for the students is very timely and will be one of the significant research work in the new era of education system in the Philippines.

1. Introduction

Life is full of difficult choices like the senior high school students where they need to face a tough decision as to what appropriate career will lead them to a bright future. That is why the education department of the Philippines as they continuously realizing their commitment to give quality education to the nation is now implementing the K to 12 education program that one primary reason is to lengthen the cycle of primary and secondary education from 10 years to 12 years (1). Aside from this, one feature of K-12 is to prepare the students and empower them to confidently join the labor market, so that by the time they graduate, they are already on their legal age and equipped with sufficient skills (2), thus, the Department of Education included career plans in their enhanced curriculum.

Studying career choice is important in employment especially in developing countries (3) like the Philippines as such, course preferences of graduates has a strong relationship with the labor market (4). But nowadays, students are more exposed with various factors that are associated with their career choices which need to be considered like interest in the course (5), situational factors, parental influence, school attended, gender, prestige attachment, and passion which are listed in the study of (6) (7). The ability of the learners to identify their preferred career choice, family, and teachers have significant role to influence the student (8) as well as environmental factors like counselor or other members of the family aside from parents, teachers, and friends, opportunity, their own personality traits (9), and capacity(10).

That is why career counseling has a very important role to help the students to be more aware (11) and not to just pick any course that they want(12) and to address this issue, it is only high time to develop a career recommender system to help the students in determining career tracks suited for them using fuzzy logic since this approach deals with linguistic variables which are words that are less precise than numbers just like in human reasoning.

Since in this study, different factors must first to identify and in dealing with this diverse and vast amount of student data, issues in computational time and complexity must also consider to produce a quality prediction model (13) and to address this, the application of feature selection is needed (14) since it is used in pre-processing step of data by selecting appropriate subset of features as relevant student attributes before they will be used in a career recommender system.

2. Literature Review

2.1. Feature selection

Feature selection techniques are effectively used together with many classification techniques in educational setting like in predicting a student performance model because the learning efficiency and performance accuracy are enhanced especially that the complexity of the learned results reduced (15). Consequently, it gives several benefits such as the removal of irrelevant features (16) and with these limited number of features, still, promising results were achieved (17). In addition, by using feature selection in identifying significant factors as well as the students' characteristics (18) that influence the studying performance of student (19) and to use in a warning system, is very helpful to the learners as well as to the teachers in evaluating academic performance for improvement (20) since this is the primary goal of all educational organizations (21).

There are three feature selection methods namely: filter, wrapper and embedded method, wherein in this study filter methods are used because they are generally functioning in pre-processing as features are selected based on the characteristic of the data (22) as it can easily cope with classification tasks in feature spaces of large dimensionality as found out by (23) in their paper.

2.2. Fuzzy Logic

Fuzzy logic was introduced by Lotfi Zadeh in 1965 as an approach to computing based on degrees of truth rather than the usual true or false of 1 or 0 of Boolean logic. It handles uncertain, imprecise and vague data which are encountered in most real-life problems and make use of them as basis for building fuzzy inference systems(24).

In recommender systems, especially in taking preferences of users, uncertainty cannot be ignored. But, with fuzzy logic, this issue could be handled to support recommender systems in giving accurate recommendation (25) with the effective and efficient result as presented in (26), even in career counseling (27) (28), as well as to analyze students' academic performances (29).

3. Methodology

3.1. Data Collection

The student data were obtained from senior high school students of the two schools division in the province of Isabela where their demographic data are gathered through a structured questionnaire which these attributes were adopted from different literature and with prior approval of the school heads the students' class marks were also collected. From the collected student data, several attributes were drawn as shown in Table 1.

3.2. Pre-processing of Data

The dataset was divided into three sets as sixty percent (60%) of the population were consider as the train set with 429 students, and the remaining forty percent (40%) has been equally divided to validation set as 144 students and test set consisting of 143 students, respectively, with a total of 716 students. All unnecessary spaces are removed and wrong input texts were edited before it converted to a .csv file and Weka 3.8.0 as an open source software and data mining tool was used for feature selection using five filtering methods which are Correlation-based, GainRatio, InfoGain, Relief and Symmetrical, respectively.

Table 1. Description of Student Attributes

Attribute Name	Description	Attribute Name	Description
Gender	Student's gender	**Parent_income**	Parent's monthly income
Age	Age of the student	**Personality**	Perception and Interest
Track	SHS track enrolled	**Parents_influence**	Influence of Parents
Strands	Current strands	**Relatives**	Influence of Relatives
Father_Occupation	Father's Occupation	**Peers**	Peer Influence
Father_educ_attainment	Father's Education	**Socio_economic_status**	Socio-Economic status of the family
Mother_Occupation	Mother's Occupation	**Proximity**	Proximity of School
Mother_educ_attainment	Mother's Education	**Job_opportunities**	Prestige of career
Final_Grade	Cumulative Grade Point Average		

3.3. System architecture

The system architecture of the proposed system as shown in figure 1 presented data from students and other factors associated to their decision as filtered using feature selection techniques to select the best attributes and prepare those data as crisp inputs before inputting in a fuzzy inference system. To make the data become fuzzy inputs, fuzzification process takes place as these inputs transform into grades of membership for linguistic terms of fuzzy sets. Rules now in the fuzzy inference will be evaluated and combined all these rules using fuzzy set operations to obtain a final result and defuzzified to obtain the final crisp output as the appropriate recommendation which the advisor as counselors will give to the student.

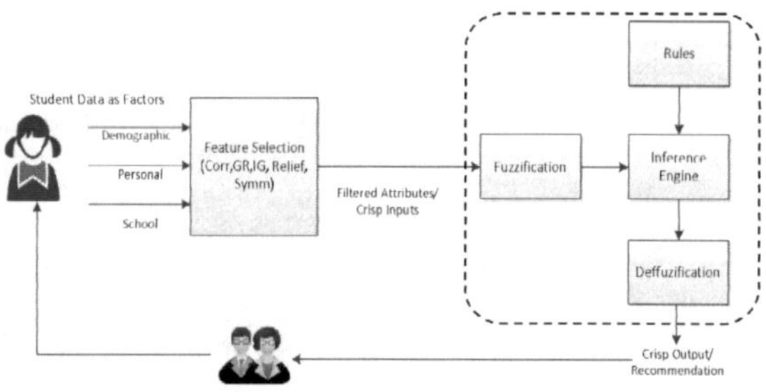

Figure 1. System architecture of the proposed career recommendation system

4. Experimental Results

The selected attribute from the different feature selection techniques were, National Career Assessment Examination (NCAE) result, strands, final grade and interest of the students and these attributes used as variables in developing the fuzzy inference engine in using Matlab as shown in figure 2 consisting of four input variables and one output as result.

The four input variables in figure 3 show the different degree of membership with their respective values and the rule viewer as shown in figure 4 presented 72 rules for the fuzzy model.

The fuzzy inference engine was then evaluated using evalfis and the result was shown in figure 5 where the deffuzification method used is centroid. Here, the four input data are 3 as strands input, 5 for interest, 1 for ncae and 89 as grade which the result as ans is equal to 88.9565 meaning that the student has 88.95% chance of pursuing their chosen career as related to their current track.

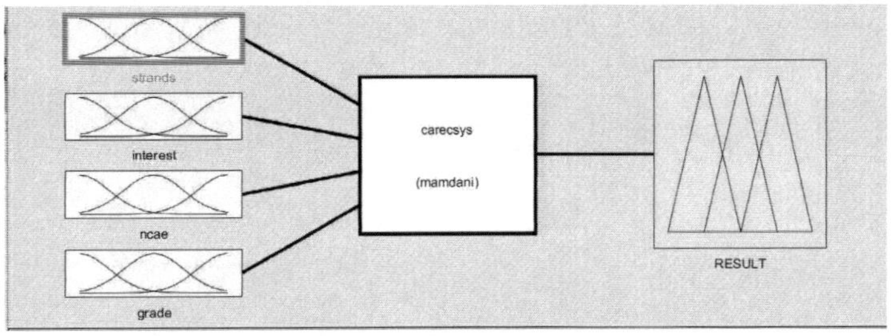

Figure 2. Fuzzy model with Input Variables

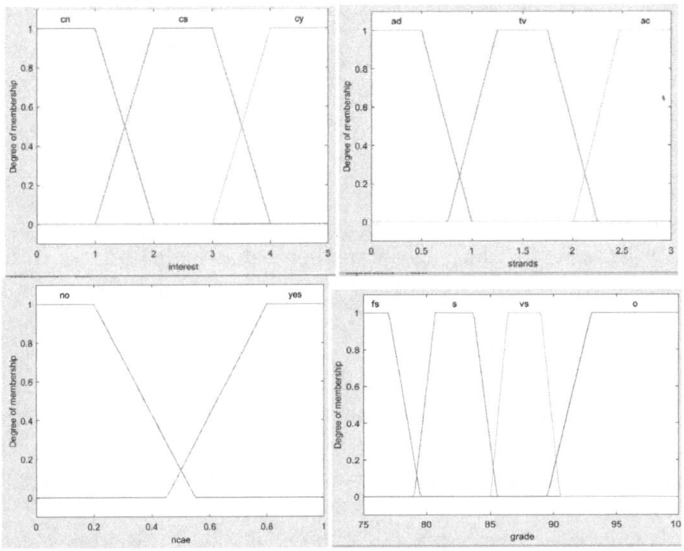

Figure 3. Degree of membership for four input variables

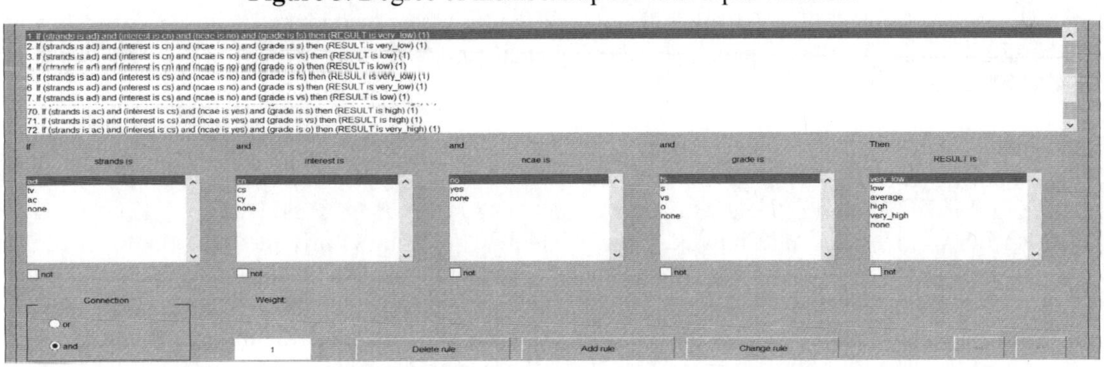

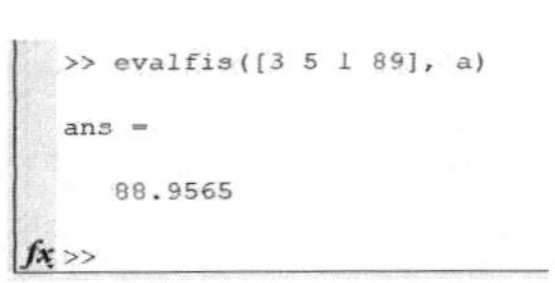

Figure 5. Result of evaluation of the fuzzy inference engine

4.1. Performance Evaluation

The root mean square error (RMSE) and mean absolute error (MAE) are used to measure the performance of the developed fuzzy model as shown in equation (1) and (2).

$$RMSE = \sqrt{\frac{1}{N}\sum_{i=1}^{n} e_i^2} \quad (1)$$

$$MAE = \frac{1}{N}\sum_{i=1}^{n} e_i \quad (2)$$

Table 2. Performance Evaluation of the developed model in the dataset

Dataset	MAE	RMSE
Training Set	2.251	4.482
Validation Set	.1927	3.954
Test Set	.9857	3.861

Table 2 shows that RMSE as a measure of predictive accuracy decreases the value from the training set to the test set which means that a smaller value indicates less difference between the estimated and actual values. Additionally, with the decrease of the RMSE, the predictive accuracy of the model improves.

5. Conclusion and Future Works

It is presented in this study a fuzzy based recommender system that would help senior high school students to consider factors associated to their career decision, but since numerous factors are needed to be considered, feature selection techniques are used to remove irrelevant student attributes so that the fuzzy inference model will generate reasonable results.

For further works, the proposed system needs to be evaluated for more test sets with the different set of criteria since the student attributes selected will be changed as the significant factors from real data will also be changed. It is concluded that a fuzzy based career recommender system is timely and will be useful for helping the senior high school students under the new education program which is the k to 12.

References

[1] SEAMEO INNOTECH 2012 *K to 12 Toolkit*
[2] Yap R H 2011 K to 12: the key to quality education? *Policy Brief* **PB-11-02** pp 1–12
[3] Ng E S W, Gossett C W, Chinyoka S and Obasi I 2016 Public vs private sector employment An exploratory study of career choice among graduate management students in Botswana *Pers Rev.* **45(6)** pp 1367–85
[4] Pitan O S and Adedeji S O 2015 Students' choice of courses : determining factors , sources of information , and relationship with the labour market demands in Nigeria *Africa Educ. Review* **11:3** pp 445-58 DOI: 10.1080/18146627.2014.934997
[5] Laguador J M 2014 Examination of influence and intention towards Lyceum of the Philippines University and career choice of general engineering students *Int. J. Manag. Sci.* **3(11)** pp 847–55
[6] Esther S U and Ayuk A A 2013 Templates that determines the choice of course selection in the university : a case study of University of Calabar Department of Sociology *J. Emerg. Trends Educ. Res. Policy. Stud.* **4(2)** pp 246–51
[7] Eremie M D 2014 Comparative analysis of factors influencing career choices among senior secondary school students in rivers state Nigeria *Arab J. Bus Manag. Rev.* **4(4)** pp 20–5
[8] Shumba A and Naong M 2012 Factors influencing students' career choice and aspirations in

South Africa *J. Soc. Sci.* **33(2)** pp 169–78

[9] Olamide S O and Oluwaseun S 2013 The factors determining the choice of career among secondary *Int. J. Engineeering Sci.* **2(6)** pp 33–44

[10] Gorad N, Zalte I, Nandi A and Nayak D 2017 Career counselling using data mining *Int. J. Innov. Res. Comput. Commun. Eng* **5(4)** pp 0–5

[11] Goyal P and Kukreja T 2015 Narrowing awareness gap by using e-learning tools for counselling university entrants *Int. Conf. Adv. Comput. Eng. Appl.* **11(4)** pp 847–51

[12] Durosaro I A and Nuhu M A 2012 An evaluation of the relevance of career choice to school subject selection among school going adolescents in ondo state *Asian J. Manag. Sci. Educ.* **1(2)** pp 140–5

[13] Shahiri A M, Husain W, Aini N, and Rashid A 2017 A proposed framework on hybrid feature selection techniques for handling high dimensional educational data *AIP Conf. Proc. 1891* 020130

[14] Acharya A and Sinha D 2014 Application of feature selection methods in educational data mining *Int. J. Comput. Appl.* **103(2)** pp 34–38

[15] Anuradha C and Velmurugan T 2016 Feature selection techniques to analyse student acadamic performance using Naïve Bayes student acadamic performance using Naïve *Conf. Proc. Small Mediu Bus* 2016 January 19 pp 345–50

[16] Regha R S and Rani R U 2016 Optimization feature selection for classifying student in educational data mining *Int. J. Innov. Eng. Technol.* **7(4)** pp 490–6

[17] Almayan H and Mayyan W 2016 Improving accuracy of students' final grade prediction model using PSO *6th Int. Conf. Inf. Commun. Manag.* 16 pp 35–9

[18] Maharani K, Adji T B, Setiawan N A, and Hidayah I 2015 Comparison analysis of data mining methodology and student performance improvement influence factors in small data set *Conf. Proc. Int. Conf. Sci. Inf. Technol.* 15 pp 169–74

[19] Saedudin R R and Sutoyo E 2017 Attribute selection on student performance dataset using maximum dependency attribute *5th Int. Conf. Electr. Electron. Inf. Eng.* 17 pp 176–9

[20] Shanmugarajeshwari V and Lawrence R 2016 Analysis of students' performance evaluation using classification techniques *Int. Conf. Comp. Tech. & Intell. Data Engi.* p 16

[21] Zaffar M, Iskander S and Hashmani M A 2018 A study of feature selection algorithms for predicting students academic performance *Int. J. Adv. Comput. Sci. Appl.* **9(5)** pp 541–9

[22] Wang S, Tang J, Liu H and Lansing E 2016 Feature selection *Encycl. Mach. Learn. Data Min. ed* Samut C *et al* (New York City:Springer Science Business) 101–1 pp 1–9

[23] Haindl M, Somol P and Ververidis D 2006 Feature selection based on mutual correlation *Prog. in Patt. Recog., Image Analand Appl.* (Springer Link)

[24] Oladokun V O and Oyewole D I 2015 A Fuzzy inference based decision support system for solving the university-course admission choice problem *Int. J. Comput. Appl.* 112(3) pp 1–7

[25] Samia B, Allel H and Aicha A 2014 Handling preferences under uncertainty in recommender systems *2014 IEEE Int. Conf. Fuzzy. Syst.* 14 pp 2262–9 doi:10.1109/FUZZ-IEEE.2014.6891823

[26] Ari E, Vatansever F and Uzun A 2009 Student professional orientation using fuzzy logic rules and quality *6th Res. Expert Conf. "Quality 2009"* pp 719–24

[27] Gupta M V, Patil P, Deshpande S, Arisetty S and Asthana S 2017 FESCCO: Fuzzy expert system for career counselling *Int. J. Recent Innov. Trends Comput. Commun.* 5(12) pp 239–43

[28] Razak T R, Hashim M A, Noor N M, Hazwam I and Halim A 2014 *Career Path Recommendation System for UiTM Perlis Students Using Fuzzy Logic 14* 10.1109/ICIAS.2014.6869553

[29] Banswal R and Madaan V 2016 SPACS : Students' performance analysis and counseling system using fuzzy logic and association rule mining *Int. J. Comput. Appl.* 134(3) pp 12–7

Automatic wrinkles detection on face image

Izzati Muhimmah[1], Fadhillah Abriyani, Arrie Kurniawardhani

Department of Informatics, Islamic University of Indonesia, Yogyakarta, Indonesia

[1]izzati@uii.ac.id

Abstract. Image processing techniques are beneficial for reconstructing and processing an image in resulting different images. One of its benefits is it can be used for detecting wrinkles on a face. Line characteristic of wrinkles can be detected also with the support of the features that had been provided. The purpose of this research is to design a system that could detect the exact location of the line of wrinkles which occur from one face image. Generally, the step that required for this research is the HSV colour image segmentation step and K-Means Clustering for dividing skin object and non-skin object. The second step is phase image improvement step with Laplacian of Gaussian filter to sharpening line of wrinkles on image face, and the last step is feature extraction which used for taking wrinkles. After all the steps had been done, each of the wrinkle's areas are detected with Canny filter and eliminated based on Area characteristic. From research experiment, the result validation data of system based on an expert are 40.6% for sensitivity, 96.9% for specificity, and 94.2% for accuracy, with the performance of the system 5,247 seconds.

1. Introduction

Human skin will change along with age. The most visible change is the appearance of wrinkles on the hands and face skin. Wrinkles can be a sign of premature aging. Generally, wrinkles can be seen clearly in some areas of the face and palms. Wrinkles are caused by a lack of elasticity of the skin, so the skin loses flexibility [1]. The most severe effect is it can disrupt a person's psychosocial aspects which someone will feel inferior and be ignored by their environment. Therefore, many people start to treat their skin especially for wrinkles. The treatment, however, is quite expensive, because it requires special instruments in the beauty clinic.

Some studies were conducted to identify wrinkle automatically. Batool and Chellappa in their study assumed about the geometrical characteristics and the wrinkles features. They said wrinkles appear in the arches form, continuous, and not intersecting with each other [2]. Wrinkles detection system of Nazre Batool and Rama Chellappa research was carried out using Gabor filters, image morphology, and geometric constraints to detect wrinkles lines on the face. The study can detect wrinkles lines only on the forehead with a deep wrinkles line, not in the whole face.

The second study is conducted by Nur Hayatin, which classifies age based on facial images using important features namely face anthropometric and wrinkles [3]. The wrinkle features calculated are width and depth of wrinkles. The accuracy of that system is 65%, but this study did not show the wrinkles features obtained.

According to the previous studies, wrinkles lines in the facial image can be detected by utilizing image processing techniques. Thus, this study attempts to create a prototype that can identify the presence of wrinkles through facial images automatically without the need to use special instruments owned only by beauty clinics. This study aims to detect the location of wrinkles lines on the face image. The system is expected to help user to detect wrinkle lines on their face. Besides, this system can also be applied as one of the features in beauty product skin advisors and the analysis of this study can be used as a reference for further research.

This journal is described in several sections. Section I describes the introduction and benefits of the system. The methods and data used are explained in section II. The research methodology described in section III. The results and discussion of the research conducted are explained in section IV. Section V explains the conclusions and suggestions from research that has been done.

2. Data and methods

2.1. Wrinkle

Wrinkles are lines or folds which appear on the skin [1]. There are two factors that affect the aging process of the skin, namely intrinsic factors, and extrinsic factors. Intrinsic factors include heredity (genetic), racial, and hormonal. While extrinsic factors include environmental factors such as sunlight, humidity, and free radicals [4].

Wrinkles occur due to reduce of collagen in the skin. Collagen and elastin are the main components of the dermis layer. Loss of these collagen has a negative impact on moisture and elasticity skin which which can cause wrinkles [5]. Skin that lacks collagen can be seen in Figure 1. Wrinkles can be seen clearly in some areas of face, such as forehead, outer corner of the eye, below eyes, cheeks, and between the cheeks and upper lip [6].

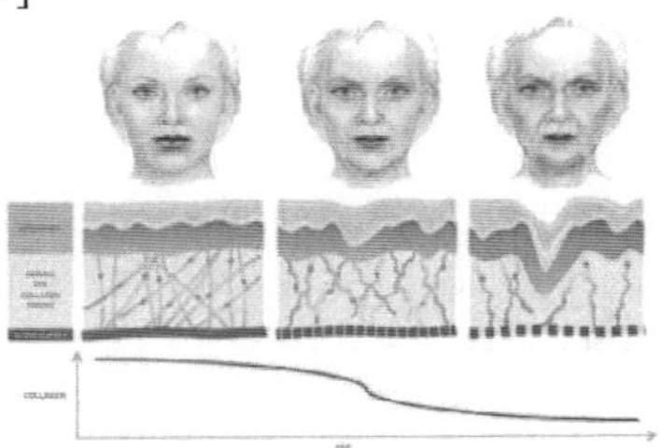

Figure 1. Illustration of facial skin deficiency in collagen [6]

2.2. HSV colour model

HSV color model consist of three layers, Hue, Saturation, and Value. Hue defines true colors such as red, violet and yellow. Saturation states the level of purity of a color, which indicates how much white is given to the color. Value is an attribute that states the amount of light received by the eye regardless of color [7].

2.3. K-means Clustering

K-Means Clustering is an algorithm for grouping data [8]. The number of groups or clusters based on the value k given by user. Data's attributes become a space vector line. The clustering process starts by calculating the distance between each data and each cluster center. The data will be a member of the cluster k_i when the distance of data to center of the cluster k_i has the smallest value compared to the distance to the center of the other clusters. After the distance of all data to all cluster center calculated, the cluster center will be calculated again. Those process will be repeated until there is no change in cluster center.

2.4. Region Filling

Region filling is one of image reconstruction using morphological algorithm [7]. Region filling will remove the holes of an object in the image by filling the hole pixels with the same color to the pixels around the hole.

2.5. Laplacian
Laplacian is one of algorithm for enhancing image. Laplacian filter is used two-dimensional Laplacian operator for sharpening [7]. The filter will be convoluted to the image.

2.6. Canny edge detection
The Canny algorithm approach is carried out by convolution of image functions with Gaussian operators and their derivatives [7]. Canny combine method of smoothness and edge detection into a convolution in one dimension in two different directions, vertical and horizontal. Canny mark all existing edges in accordance with the convolution parameter criteria performed, can give very high flexibility to decide the desired edge thickness, able to localize well, and give a clear response to each edge.

2.7. Validation
System validation test is carried out using the Single Decision Threshold method. This method basically has information used to compare the detection results by the system (predictive value) with the detection results by experts (true value). The true value is generated by experts, skin and genital specialist. There are four terms used, namely:

1. True Positive (TP) is if the true value and predictive value produce positive results, such as, if the true value is "wrinkles lines", then the system decides "wrinkles lines".
2. True Negative (TN) is if the true value and predictive value produce negative results, such as, if the true value is "not a wrinkles line", then the system decides "not a wrinkles line".
3. False Positive (FP) is if the true value is negative, but the system produces positive results. For example, if the true value is "not a wrinkles line", but the system decides "wrinkles lines".
4. False Negative (FN) is if the real value is positive, but the system produces negative results. For example, if the real value is "wrinkles lines", but the system decides "not a wrinkles line".

Sensitivity and specificity can be calculated after the values of TP, TN, FP, and FN are obtained. Sensitivity is a parameter to measure the percentage of positive data that is correctly identified (wrinkles lines detected by the system are the same as experts). Specificity is a parameter to measure the percentage of negative data correctly identified (the system does not detect objects not wrinkles lines of candidates) [9]. The formula of value sensitivity and specificity can be seen in Equations 1 and 2, respectively. Accuracy is a value that describes how accurately the system can detect data correctly [10]. The formula for accuracy can be seen in Equation 3.

$$\boldsymbol{Sensitivity} = \frac{\boldsymbol{TP}}{\boldsymbol{TP + FN}} \quad (1)$$

$$\boldsymbol{Specificity} = \frac{\boldsymbol{TN}}{\boldsymbol{TN + FP}} \quad (2)$$

$$\boldsymbol{Accuracy} = \frac{\boldsymbol{TP + TN}}{\boldsymbol{TP + FP + TN + FN}} \quad (3)$$

2.8. Data
The facial image data used in this study were obtained randomly from internet. The image should be color images (RGB), close-up, and front facing face. The facial image should have wrinkles. The image format used is JPG, PNG, and JPEG. The image should have evenly distributed lighting. There are 39 facial images. Three images were used as training data. Training data are used to identify the facial area and wrinkles features. The results of this identification process would be used as reference data to test other images. Three other images were used as testing data.

3. Methodology
The stages to develop prototype of automatic wrinkles detection on face image are shown in Figure 2.

IOP Conf. Series: Materials Science and Engineering **482** (2019) 012026 doi:10.1088/1757-899X/482/1/012026

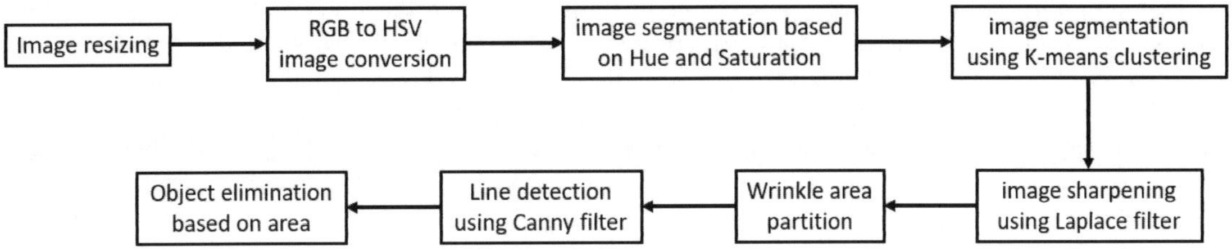

Figure 2. The stages of automatic wrinkles detection on face image

Because the image data are taken from internet randomly, these images have different characteristics and sizes. In order for images to have similar characteristics which are close-up images, the image is cropped manually first. Thus, the largest object in the image is face object. The prototype begins by resizing the image size to 580x580 pixels. Resizing is carried out to equalize the size of the input image in order each process can be implemented for all images.

Second stage is image conversion. The original image is a color image in the Red, Green, and Blue color space (RGB). In segmentation process later, image objects will be separated into skin object and non-skin objects. To recognize the skin color object well, the original image in the RGB color space is converted to the HSV color space (Hue, Saturation, and Value) [11]. The range value on the HSV color model used to recognize the skin object in this study is 0.0-0.25 for H (hue), and 0.15-0.9 for S (Saturation) [12].

The fourth stage is image segmentation. Skin object and non-skin objects is separated using K-means clustering [13]. K-means clustering algorithm will group pixels that have the same color in the image accordance with HSV color model. The K value used in this study for K-means is two that's mean the colors in the image will be grouped into two groups. Group that has the greatest number of pixels will be considered as skin group and vice versa, group that has the least number of pixels will be considered as non-skin group. Therefore, the RGB intensity value of the skin group pixels in the original image will be retained, while the RGB intensity value of the non-skin group pixels will be changed to black. Images that only contain facial skin are obtained. However, in some images, the facial skin image has several holes. Therefore, Region Filling algorithm is implemented in this stage so that the holes in the skin are closed.

The fifth stage is image sharpening. The facial skin image is sharpened using Laplacian filter. Laplacian filter is used to sharpen the face line and wrinkle lines. The Laplacian filter used in this study is 3-by-3 filter and 0.2 for alpha value.

The sixth stage is to take areas of the skin where wrinkles often appear, namely the forehead, the middle-area between the two eyes, the left-side area of the left eye, the right-side area of the right eye, the left eye bag area, and the right eye bag area. In the facial skin image, the entire face skin area is given a percentage value of 100%. The index of initial row, end row, initial column, and end column of the facial skin area are taken. The position of the forehead area is obtained by taking an index value of 20-30% of the rows of facial skin area and 25-70% of columns of the facial skin. The position of the middle area between the two eyes is obtained by taking an index value of 34-43% of the rows of facial skin area and 44-55% of columns of the facial skin. The position of the right-side area of the right eye is obtained by taking an index value of 38-44% of the rows of facial skin area and 15-25% of columns of the facial skin. The position of the left-side area of the left eye is obtained by taking an index value of 38-44% of the rows of facial skin area and 75-85% of columns of the facial skin. The position of the right eye bag area is obtained by taking an index value of 45-55% of the rows of facial skin area and 20-43% of columns of the facial skin. The position of the left eye bag area is obtained by taking an index value of 45-55% of the rows of facial skin area and 55-80% of columns of the facial skin.

After the wrinkles area is obtained, Canny edge detection algorithm is applied in each wrinkle area. in Canny edge detection, there is threshold values used to determine which lines are preserved. The smaller the threshold value is given, the more the edges are detected and vice-versa. Each wrinkle area has their own threshold value because each area has a different edge line characteristic. Threshold value for forehead and the middle-area between the two eyes is 0.08. Threshold value for the left-side area of the left eye and the right-side area of the right eye is 0.06. Threshold value for the left and right eye bag area is 0.1.

The last stage is elimination. Lines detected in each wrinkle area are checked whether they are wrinkled or not. Area information from each line is extracted. In forehead area, lines which area is more than 100 pixels and less than 200 pixels are considered as wrinkles. In the middle-area between the two eyes, the left-side area of the left eye, and the right-side area of the right eyelines, lines which area is more than 50 pixel, or less than 30 pixels are considered as wrinkles. In the left and right eye bag area, lines which area is more than 35 pixels and less than 90 pixels are considered as wrinkles. After wrinkles lines are detected, the Marking process of wrinkle pixel position is carried out on the original image so that the results can be seen by the user.

4. Result and discussion

The results of each step according to Figure 2 are shown in Figure 3. Figure 3.(a) is original image that obtained from internet randomly. Because the largest object in the image is not face object, the original image should be cropped manually before inputted to system. Figure 3.(b) shows cropped image of original image. The cropped images have different sizes, so that the image is resized to 580x580 pixels. The size is chosen in order the image size is not too large which will cause the detection process run too long. Besides, if the size of the image is too small, the wrinkle lines are not clearly visible, thus become difficult to be detected.

Figure 3.(c) shows the result of segmentation process using the K-Means Clustering based on Hue and Saturation value. HSV colour model used in this process because it is the best colour model for detecting skin colour in normal lighting conditions [11]. HSV colour model separates luminance from the colour components while RGB colour model cannot. It must do because there is different lighting condition when the picture is taken.

K-Means Clustering algorithm group pixels according to their colour. Two clusters are selected to separate the pixels of skin colour from pixels of non-skin colours, such as hair colour and background. The pixels in the cluster which has the greatest number of pixels is determined as skin pixels. K-Means clustering algorithm was chosen because it has an accuracy score better than the Decision Tree method [14]. As shown in Figure 3.(c), the segmented image has some holes in skin object. Thus, Region Filling algorithm is needed in this process to fill the holes in skin object. The image result of region filling process is shown in Figure 3.(d).

Figure 3.(e) shows the results of sharpening process using Laplacian filter. Laplacian filter algorithm can enhance the image detail properly without any significant noise interference. The sharpened images can help the wrinkle detection process which wrinkles can be seen more clearly.

Figure 3.(f) and (g) shows results from the process of taking areas where wrinkles often appear, namely the forehead, the middle-area between the two eyes, the left-side area of the left eye, the right-side area of the right eye, the left eye bag area, and the right eye bag area. The method used to get these areas are percentage method. Another method that has been carried out before is Active Contour method. This method can take the wrinkled area but is not perfect because the edge of the wrinkled area is not neat.

Figure 3.(h)-(m) show results of edge detection in forehead, the middle-area between the two eyes, the left-side area of the left eye, the right-side area of the right eye, the left eye bag area, and the right eye bag area using Canny edge detection. Canny edge detection algorithm was chosen because this method is able to detect detail facial lines well compared to other edge detection methods. Canny can localize properly and provide a clear response to fine lines of wrinkles. The next elimination process is carried out based on the area of line. The lines detected in the Canny process image results will be eliminated according to the area of each line. The image result of line elimination based on area is shown in Figure 3.(n).

The evaluation of experiment results was carried out using the Single Decision Threshold method. Test validation is calculated by comparing the results obtained by the system with the doctor's marking results. The experiment is carried out using 36 testing images. From these testing image, the average value of the sensitivity, specificity, and accuracy of the system obtained are 40.6%, 96.9%, and 92.4%, respectively. The low scores of sensitivity are got due to there are the difference in location and length of the wrinkles marking by expert and system. However, based on the specificity scores, the results obtained are very good because system errors in detecting not-wrinkle objects as wrinkle object are rarely found.

When compared with previous studies, this research is quite effective. This study can achieve accuracy score up to more than 90%. Some of the advantages of this study compared to previous study is that the system is able to do segmentation well for Asian and European skin tones. Moreover, the system is able to detect wrinkles lines in the overall facial image in 5.247 seconds for an image.

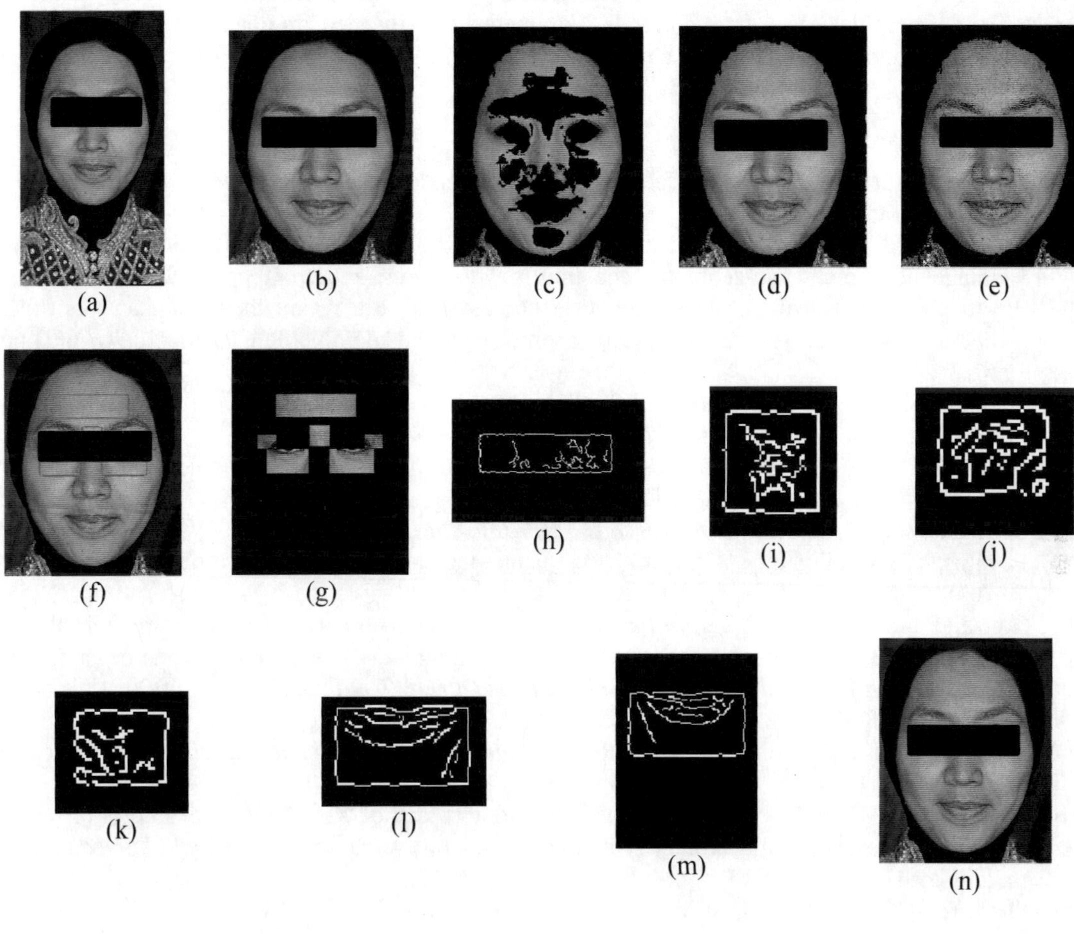

Figure 3. The examples of image result in each process

Not all test images can be processed well by the system, even though the image used is an ideal image that has fulfilled the requirements as an input image. This is caused by several things. First, the system fails to detect facial skin due to the colour similarity between skin objects and non-skin objects. The lighting factor that is too bright on the photo also causes the face to be undetectable. Second, the system fails in the process of taking the wrinkles area appropriately. This is due to the factor for taking an area is still static so that the size of area cannot adapt flexibly. Thus, it causes errors in detecting wrinkle lines. Lastly, the system fails to detect wrinkles lines in the area that has been taken. This is caused the area size limit used is also static.

5. Conclusion

From all the processes that have been carried out, the conclusions obtained from this study are in general testing, the program can achieve a high success rate for images with ideal conditions. The ideal conditions are RGB images contain close-up front facing faces and have a good lighting. The validation is tested

Test validation is calculated by comparing the results obtained by the system with the doctor's marking results. From the test results obtained sensitivity value is 40.6%, the specificity value is 96.9%, and the accuracy value is 94.2%. System performance is quite fast, 5.247 seconds, for the entire process.

For further research, it is better to use images taken by researchers themselves in order the initial characteristic of wrinkles can extract correctly. Parameters for area size are determined more flexible based on facial characteristics. Other characteristics of wrinkles can be proposed to increase the specificity score.

References

[1] Lemperle G *et al* 2015 A classification of facial wrinkles *Plastic and Reconstructive Surgery* **108** pp 1735-50
[2] Batool N and Chellappa R Fast detection of facial wrinkles based on Gabor features using image morphology and geometric constraints *Pattern Recognition* **48** pp 1–32
[3] Hayatin N 2016 Klasifikasi kelompok usia berdasarkan citra wajah menggunakan algoritma neural network dengan fitur face anthropometry dan kedalam kerutan *TEKNOLOGI* **6(2)** pp 61–7
[4] Michniak-Kohn B and Leonardi G R 2017 An overview about oxidation in clinical practice of skin aging *Anais brasileiros de dermatologia* **92(3)** pp 367-74
[5] Geddes K 2016 Age-related changes in the menopausal woman: the role of hormones in the skin *J. of Aesthetic Nursing* **5(3)** pp 120-5
[6] Epilightnewskin 2015 *PDO Threads BEAUTY LIFT - Epilight New Skin Clinic* https://www.epilightnewskin.co.uk/?treatments=facial-lifting-threads retrieved April 15th 2018
[7] Gonzalez and Woods 2002 *Digital Image Processing* 2nd Edition (New York: Prentice Hall)
[8] MacQueen J B 1967 Some methods for classification and analysis of multivariate observations *Proc. of 5th Berkeley Symp. on Math. Stat. and Prob.* (California: University of California Press) pp 281-97
[9] Shortliffe E H and Cimino J J 2017 *Biomedical informatics computer applications in health care dan biomedicine* 3rd Edition (Springer)
[10] Zhang Y N *et al* 2015 Sensitivity, specificity and accuracy of ultrasound in diagnosis of breast cancer metastasis to the axillary lymph nodes in Chinese patients *Ultrasound in medicine & biology* **41(7)** pp 1835-41
[11] Shaik K B, Ganesan P, Kalist V, Sathish B S and Jenitha J M M 2015 Comparative study of skin color detection and segmentation in HSV and YCbCr color space *Procedia Comp. Sci.* **57** pp 41-8
[12] Mujahidin S 2012 *Aplikasi perhitungan jumlah orang dalam satu foto* Bachelor Thesis (Yogyakarta: Jurusan Teknik Informatika, Fakultas Teknologi Industri, Univeristas Islam Indonesia)
[13] Dhanachandra N, Manglem K and Chanu Y J 2015 Image segmentation using K-means clustering algorithm and subtractive clustering algorithm *Procedia Comp. Scie.* **54** pp 764-71
[14] Valentino F, Adji T B and Permanasari A E 2017 *Komparasi metode decision tree dan k-means clustering dalam mengatasi masalah cold-start pengguna baru* pp 268–73

Braille3D: using haptic and voice feedback for braille recognition and 3D printing for the blind

M J Samonte, E D Laurente. K M Magno and C Perez

School of Information Technology, Mapua University – Makati, 333 Sen. Gil J. Puyat Ave., Makati City, Philippines

Abstract. In today's state of technology, people can see the innovations that have evolved throughout time. One of those is assistive technology which exists to help impaired people from doing their daily activities. This technological advancement is made to help people accomplish tasks they could not perform before. Assistive innovations today reside on computers and people, not only the disabled, have become dependent on them as they make life more efficient and convenient. Accessibility has always been a challenge for people with disability. Researchers and inventors resolved this problem by integrating solutions and innovations into society which benefited the disabled community to let them experience a more convenient life, easing their burdens that hinders them from living an enjoyable life. Assistive technology was one of the solutions that innovators eventually came up with. Prosthesis was invented to be an extension to legs, arms, and other body parts for people who lost a limb, wheelchair for those people who were paralyzed or have difficulties in walking. Other assistive technologies were made to accommodate current technology such as accessible keyboards, text-to-speech software, and speech recognition software. With that said, the motivation of this is study is to contribute to the society by helping people with visual impairment to have more access to technology and to give them a better learning environment in which the person can learn the Braille system.

1. Introduction

Today's assistive technology is not only a trend in the medical field, but also in the field of education. This type of technology is vastly used by people who are disabled. Students in classroom take advantage in the uses of this assistive technology. Innovative equipment such as refreshable braille display, screen reader, and such exists to aid students in their classroom lessons. Using a software that translates text into braille are helpful for students who are blind. Text-to-speech applications also can aid them since it converts text into audio so that they could access text on digital devices [1].

Assistive technology is a solution for the blind, mute, deaf, or people who have lost a limb. This type of technology is defined as any item, system, machine, or service that is used to improve the functional disabilities of a person. The most common assistive technology for visually impaired people is the braille. Braille was created to accommodate the blind into reading the text in a specialized paper with little dots that are embossed. Braille has a unique combination that can be characterized as alphabets and numbers [1].

When it comes to the implementation of coded instruction into integrated software or hardware, Computer Aided Instruction (CAI) is generally used. CAI is a simulation of exercises that has the ability to show feedbacks to its user whether the answer is correct or wrong. Interaction is key since the computer will be the proxy of a teacher. Hence, CAI can be called as a supplementary teaching tool that incorporates real training and lessons with interactive feedbacks [1]. The examples of CAI include a combination of infographic texts, audio, and visual exercises that are commonly used as drills for the user. CAI is an instructional program technique that exercises self-learning through interaction of the student for educational purposes. This refer to the utilization of computing materials such as mobile phones and computers as a tool to accelerate the learning process of a student. The researchers have decided to integrate CAI in the project [2].

The proponents are tasked to produce a mobile application that will incorporate braille and 3D printing into a mobile application which will teach the braille characters through sequence of lessons and exercises where assessment modules are assimilated. 3D models will be available in the application for printing to let the student feel the geometry of a specific object. Figure 1 shows the conceptual framework of the accessibility features of the system.

1.1. Objective of the Study

The general objective of this study is to create a mobile learning application incorporating haptic and voice feedback for braille recognition and 3D printing for the blind, specifically for visually impaired kindergarten students. Generally, the goal is to develop an application for the visually impaired with the use of the mobile phones' vibrational and voice capabilities and to promote the geometry of an object through 3D printing.

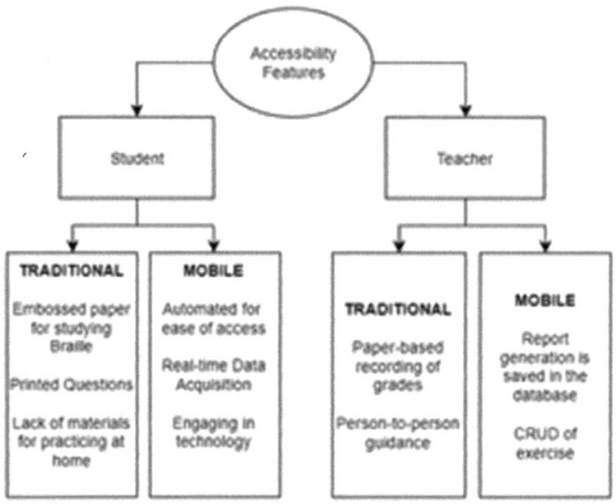

Figure 1. A conceptual framework for accessibility features of the system.

1.2. Scope of the Study

For the learning module, since the target users are visually impaired kindergarten students, the application will feature lessons that will display the basic Braille characters (A to Z, 0 to 9) as well as the Braille spelling for shapes. The user can visit this module to learn and review the pattern for each Braille characters. Each letter shall have example words that the app will read to the student when prompted, as well as an option to 3D print that word example if there is an available 3D model for that word. The learning module for the numbers shall be presented in an orderly manner to also teach the user to count from 0 to 9. The learning module for shapes shall teach the user the spelling of the basic shapes in Braille. Some modules won't be accessible and will be unlocked later based on the learning performance of the user.

The application exercises are to assess the learning progress of the user. Each exercise contains a set of Braille characters. The application will display the dot pattern for the Braille characters and the user will have to identify that character in a multiple-choice format. Choosing the correct answer will earn the user a point and will proceed to the next item, choosing the wrong answer will encourage the user to try again and pick another choice, until the correct answer is selected. After the user answers all the item, the application will assess the user's performance based on the correct answers and retries to pick the correct answer.

There will be another interface for the teacher. The teacher will be the one to view the assessment of the student user. The teacher is also responsible for the Content Management System where they can modify exercises and assign them to the student. The teacher can also manage the lesson modules by adding example words to the Braille characters and also add a 3D model for that word. Unlocking lesson modules for the students is also an option for the teacher user.

Computer Aided Instruction techniques will be followed to make the learning experience of the student more engaging and interactive. The proponents will do this by adding haptic and voice feedback, as well as 3D models for printing 3D objects to make the students more interested [1].

For 3D printing, as have been stated before, some of the word examples in the learning module have 3D models attached to them. 3D models can be printed by connecting the device to a 3D printer via Bluetooth or USB cable, and by using a 3D printing application. Since this process might be too complex for a kindergarten student, the teacher, or someone who has knowledge in 3D printing, must handle the 3D printing process [3].

2. Review of Related Literature

3D printing is considered an aid in literacy for kids in many countries. The study shows that tactile figures like replicas of monuments, maps, and other unfamiliar object result to improved comprehension and literacy skills. [4] Since most of the learning materials for the visually impaired are limited, 3D printed models are a huge boost in the learning capabilities for the blind. Furthermore, 3D printed tactile objects aids the overall cognition of the student of the material itself because physical models generate a sense of familiarization to the student. This gives the student more understandings rather than embossed books. Classes like biology where the student can feel the geometry of a certain organ are proof why 3D printing is needed in the learning materials for the blind. [4] The swift occurrence of technology resulted into the creation of turning materials that are designed from software into a physical manifestation, thus the 3D printing. Since blind people have a difficulty in gathering/accessing information, 3D printing can be of aid to the visual impaired community. Moreover, 3D printed objects give the person the form and structure of the 3D model through the sense of touch. Thus, giving the justification of the relevance of 3D printing in the proposed topic. Other researches were focused on a tactile-based solution to improve touchscreen mobile interface exploration by blind users [5].

The use of mobile in different ways to present the braille. First is the Braille scan where a six-dot Braille layout are placed in a 2 x 3 matrix which are used in mobile phones. People can read the characters by moving the stylus on the device. The dots become available for reading from the moment the stylus was placed on screen until it was lifted off. Second is the Braille Sweep where it uses six-key layout to generate the characters. This method also needs stylus to function properly because it needs to use the sweep method. Vertical and diagonal movements are possible making this method more efficient than the first. It is like Braille Scan only it is laid horizontally. Third is Braille Rhythm which enables the person to read the characters by holding the stylus on the screen allowing a tactile pattern to appear. This method presumably was the most promising method for tactile perception since it allowed the user to navigate and read the characters easier. The study of Braille is to make the methods more effective and to make the representations of Braille to be more practicable to users [6].

Modern smartphones with the use of touch-screen and vibrations introduced for the blind or deaf-blind who mostly relies on their tactile sense to convey and absorb information. In V-Braille, the screen is divided into 6 parts, representing the dots of a Braille character. When any of these 6 cells is touched, the phone vibrates if the cell touched has a raised dot. The screen can be tapped or stroked in different areas, and multi-touch is not supported [7][8].

Computer Aided Instructions is a computer application that was implemented in a coded instruction, it describe also as integration of a software and the hardware. CAI have different type of for, like game, tutorial or simulations. The good thing about CAI is to present the actual feedback responses, and summarizing the result of the performance, it teaches the new skill as supplementary device which associated to the real trainings and lessons. CAI is like a teacher which interact to the user, through CAI the user can achieve information or instructions which can be found in different devices; it is also attractive because of the animations and images [2].

Text to Speech is a type of speech combining an application that would yields to a digitalized spoken sound of the text. Screen Reader usually used, it can read different types files and web pages, or naming the program or folder where the user points on the screen [9].

3. Methodology

The methodology the proponents used for this project is the iterative and incremental method. The project follows an iterative process where each phase is done repeatedly to fully discover and meet the requirement of the system. Each iteration will add an improvement to the system based on the previous phase. This method contains five phases, requirements, analysis and design, implementation, testing, and deployment. The mobile app is created on Android platform using Android Studio, written in Java programming language. The layout of the application is written in XML language. For the backend, Google's Firebase (a backend as a service) platform is utilized, specifically for the authentication, realtime database, and cloud storage.

The accessibility service that the app will use is a built-in function on Android devices called TalkBack. It is a screen reader, it reads not only the texts on the screen but also describes its components (such as buttons and lists). Aside from providing speech, it also provides haptic feedback or vibrations to give the user feedback when some control is being touched. TalkBack also has a setting that will let the user change the speech rate, pitch, or accent of the voice. The app will utilize this function in order for the visually impaired student to navigate and use the application.

Android devices has a built-in tutorial on them on how to use TalkBack, such as the different gestures, how to navigate on certain controls such as scrolling on a list and so on. These tutorials will be shown first to the students to be familiarize with TalkBack before using the application.

The main menu of the application consists of the learn, exercise, my account, sign out as well as instructions button that takes them on a screen that gives them instructions on how to use the app. These functions can be accessed by the student by double tapping on the buttons in the main menu. Computer aided instructions are applied here and throughout the app by filling the content descriptions of each controls with instructions. For example, when the user hovers into the learn button, TalkBack will read a text that will give the user an idea of the next screen. When the learn button is double tapped, it will bring the next screen where student has the option to choose between the four levels. If the user chooses level 1, the user will again choose a letter (A-J) to practice. If the user chooses level 2, the user will again choose a letter (K-T) to practice. If the user chooses level 3, the user will again choose a letter (U-Z) to practice. Lastly If the user chooses level 4, the user will again choose between numbers (0-9). All this levels that contains basic characters are part of the Grade 1 braille.

As shown on figure 2, the screen displays the braille pattern for letter A. This is the screen where the students will familiarize themselves with the patterns and can always come back here for practice. The black dot represents a raised dot and the gray dot represents the flat dot. When the user's finger hovers through the dots, TalkBack will say the dot number and whether it is raised or flat to the user, letting the visually impaired student to be familiar with the position of the dots as well as their state. The user can also choose 3D Model Download button that will download a 3D model of the example word that is stored on the cloud storage. The Example Word button will give the user other example words that starts with the letter that is being practiced.

The students can select which exercises to do which is assigned to them by their teacher. The teacher user is responsible for creating exercises and assigning them to their students. A content management system will be available for the teacher account where they will be able to create exercises and add new students. The exercises for this app are in a simple multiple-choice format. As shown on figure 3, a braille pattern is displayed and three choices are below it. The student will first slide their finger above in which TalkBack will read the exercise number and the item number, so that the student will know which screen is the app currently in. Then, the student will slide over to the braille pattern which is similar with the screen in the learning module except that it does not tell what letter it is. Once the student is able to touch all the dots and gets the pattern, he or she can now choose the answer in the choices below it. If the answer is wrong the app will encourage the student to try again and select the correct answer. Once all the items are answered, the student will be congratulated for answering all and the results (retries for each item and the duration of taking the exercise) will be sent to the teacher's account for assessment.

Figures 4 and 5 shows that the user can also choose 3D Model Download button that will download a 3D model of the example word that is stored on the cloud storage. The user can print the 3D model provided by the app using a 3D printer of their choice. The 3D models are in STL format. For the testing of 3D printing, the researchers used da Vinci 1.0 AiO 3D printer and XYZware 3D slicing and printing software. The process on how the 3D model was printed can be seen in figure 4, a rabbit model is used and the material used for printing the model is a white ABS filament. The total time it took to completely print the rabbit model is 4 hours. Figure 5 displays the finished product.

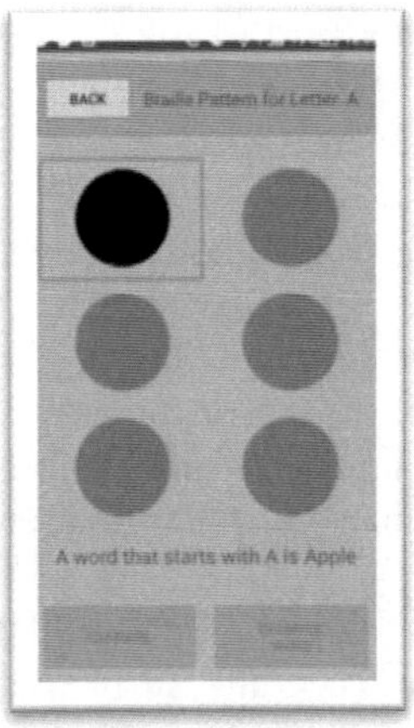

Figure 2. Learning module for letter A

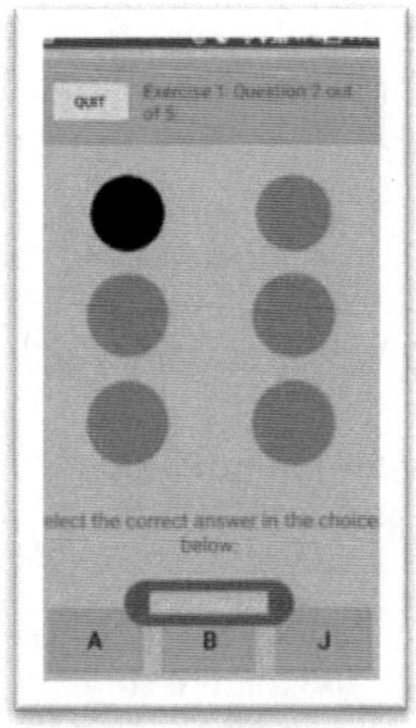

Figure 3. Exercise module

Figure 4. Ongoing 3D printing of the object

Figure 5. Finished product (Rabbit)

4. Results and Discussion

The proponents have gathered two different types of students to test the application namely: totally blind students, and partially blind students. The proponents have tested the application to the blind students at the Philippine National School for the Blind (PNSB). Moreover, the researcher gathered data which are represented below in pie charts:

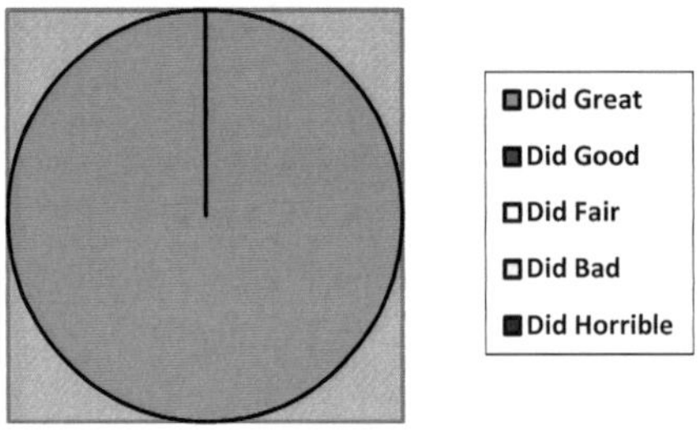

Figure 6. Totally blind students' success rate

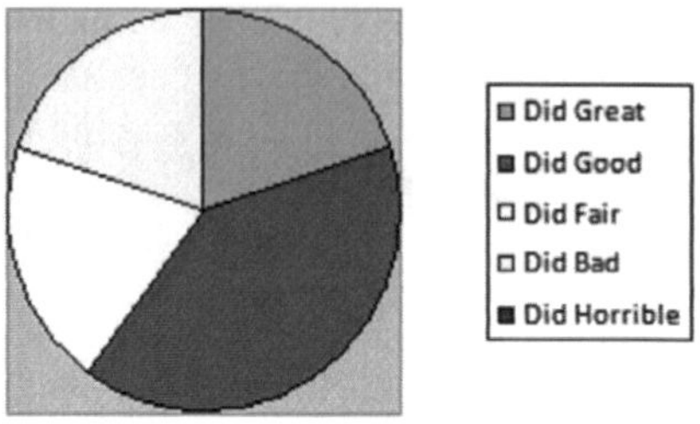

Figure 7. Partially blind students' success rate

It can be seen in figure 6 that the success rate of the application is 100% for the students who did great (did not have a hard time using the application). Meaning, the totally blind student did great on the application test.

For the figure 7, several partially blind students took part in the testing. The student who did great did not really need help in using the mobile phone because the student is already familiar with the Braille patterns. TalkBack was definitely a big help for the students. There are 20% of the student test who did great.

Another student only did good because the student also learned quickly thru the assistance of the TalkBack. The student did not have a hard time using the application and navigating across it. There are 20% of the student test who did good.

The student who did fair started out slowly due because the student struggled first with the navigation. After guiding the student, the student did well on the test. There are 20% of the student test who did fair. The student who did bad had only begun studying Braille. The student had a hard time answering because the student is not fully familiarized with the Braille character. With guidance, the student was able to be familiarized with the navigation and the student was able to answer the test There are 20% of the student test who did bad.

There are 0% of the student test who did horrible. Overall, the success rate of the students who used tested the application is high.

5. Conclusion and Recommendation

The results of this study have shown that usage of Braille3D in a classification of braille characters is advantageous, especially for totally blind students. For the purpose of being portable, the proponents have decided to develop the application on smartphones since hand-held devices are prevalent in today's technological advances. The proponents have noted that the application is exceedingly beneficial to totally blind students because it helped them understand and gain familiarization to the Braille characters faster than the traditional devices they are using.

The proponents recommend to look on more functionalities which can be beneficial to the development of this assistive technology.

References

[1] Lounnas V, Wedler H, Newman T, Black J and Vriend G 2015 Blind and visually impaired students can perform computer-aided molecular design with an assistive molecular fabricator *Int. Conf. on Bioinformatics and Biomedical Eng.* (Cham: Springer) pp 18-29

[2] Morse A 2017 Valentin Haüy and Louis Braille: enabling education for the blind *Found. of Ophthalmology* (Cham: Springer) pp 45-63

[3] Harianto R, Chen E, Lim Y, Jo W, Moon M and Lee H 2016 D literacy aids introduced in classroom for blind and visually impaired students *J. of Blindness Innov. & Res.* **6(2)**

[4] Wilson J and Wooten T 2017 3D printed braille maps—Texas A&M *2017 IEEE Global*

Humanitarian Tech. Conf. (GHTC) pp 1-4

[5] Katzschmann R, Araki B and Rus D 2018 Safe local navigation for visually impaired users with a time-of-flight and haptic feedback device *IEEE Trans. on Neural Syst. and Rehabilitation Eng.* **26(3)** pp 583-93

[6] Zhang X, Tran T, Sun Y, Culhane I, Jain S, Fogarty J and Mankoff J 2018 Interactiles: 3d printed tactile interfaces to enhance mobile touchscreen accessibility *Proc. of the 20th Int. ACM SIGACCESS Conf. on Computers and Accessibility* pp 131-42

[7] Tekli J, Issa Y and Chbeir R 2018 Evaluating touch-screen vibration modality for blind users to access simple shapes and graphics *Int. J. of Human-Comp. Stud.* 110 pp 115-33

[8] Manresa-Yee C, Morrison A, Muntaner J, and Roig-Maimó M 2017 Multi-sensory environmental stimulation for users with multiple disabilities *Recent Adv. in Technol. for Inclusive WellBeing* (Cham: Springer) pp 165-82

[9] Arik S *et al* 2017 Deep voice: Real-time neural text-to-speech *Preprint* arXiv:1702.07825

Virtual reality technology on health intervention for women: a literature review

Rahadian Kurniawan[1], Dhomas Hatta Fudholi[2], Dimas Panji Eka Jalaputra[3], Restu Rakhmawati[4]

Department of Informatics, Universitas Islam Indonesia, Yogyakarta, Indonesia

[1]rahadiankurniawan@uii.ac.id
[2]hatta.fudholi@uii.ac.id
[3]15917208@students.uii.ac.id
[4]16917115@students.uii.ac.id

Abstract. Virtual Reality (VR) technology application has become more affordable, immersive and portable. This is also happening in the health intervention domain. While the usage of the VR applications is still dominated by man, we conduct a literature review study to see the state-of-the-art in the application for recent VR technologies for women's health intervention. This study is done since women are more susceptible to some disease than men, feel pain more easily, more susceptible to autoimmune diseases, and experience anxiety and phobia more easily. It will help future studies to lookup the big impact of VR health intervention for women. The result of the literature review shows that there are three types of VR technology are used in the application (non-immersive, semi-immersive, and full-immersive), there are four category of health interventions done through VR (prevention, diagnosis, treatment, diagnosis-treatment), and the domain of the disorder are focused mostly to psychological and physical aspect.

1. Introduction

In recent years, Virtual Reality (VR) technology has become increasingly affordable, immersive and portable. This allows the use of VR in various environments, including in the domain of health interventions. Some areas of VR use in health intervention domains have promising applications, including anxiety, phobia, obesity, chronic diseases, mental illness, to surgical simulations. VR is able to present subjective experiences that make it an attractive intervention in the realm of health, including preventive, diagnosis and treatment.

Several surveys conducted on VR consumers in America [1] and China [2] showed that the usage of VR was still dominated by men. This shows that men have more alternatives that can be chosen to support health problems and women have fewer opportunities in the experience of using VR as a health intervention. Women have their own problems which in some cases need better health interventions, such as: women are more susceptible to Alzheimer's disease than men [3], women feel pain more easily than men [4], women are more susceptible to autoimmune diseases [5], and women experience anxiety and phobia more easily [6].

In this paper, we conduct a novel review on the studies of VR-based treatments for female patients in global health interventions. This literature study aims to examine VR technology developed over the past few years in dealing with health problems in women. The development of these technologies includes topics of the targeted health issues, types of VR used in health interventions, and VR technology used. In its presentation, this paper divides the research into three technology categories: Full-Immersive; Semi-Immersive; and Non-Immersive VR. According to [7], Full-Immersive VR System is a type of VR that encapsulates audio and visual perceptions of users in cyberspace and cuts all outside information so that the experience is truly immersive. Non-immersive systems are often called a desktop virtual reality (without an input device) and are rely on the displayed screen. This is the window to go into virtual worlds without any additional devices, such as HMD (Head-mounted display). The system is sometimes called Window on World (WoW) systems. A semi-immersive system consists of VR and real-world attributes by realizing computer graphic objects into real scenes. The input to this type of system is entered and controlled by the user such as by using a mouse, keyboard, interaction style, glasses, and joystick. This allows users to interact using hands and sometimes using HMD or DataGloves. The results of this literature study are expected to provide a new knowledge as point of views and reference for the development of VR technology in dealing women's health problems.

2. Methods

2.1. Inclusion and Exclusion Criteria

The chosen research in this study meet the following inclusion and exclusion criteria: (1) The research publication papers are published in or after 2015; (2) Research's respondents are all or dominated by women; (3) The research only discussing women's health problems; (4) The research uses virtual reality technology; (5) There is an explanation of the use of the virtual reality; (6) There is clear data about virtual reality effects for patients. We did not include research that has is aimed for health education, since we are focusing on health service interventions.

2.2. Searching Method

To find out the application of virtual reality technology to deal with women's health problems, a systematic literature review is done. Searching related researches are done by using several appropriate keywords. The used keywords are related to the discussed topic. Keywords are divided into two parts, namely a) "virtual reality" or "health technology" and b) "women" or "female". By using the keywords, we searched for relevant researches using several search engines: Google Scholar, EBSCOhost, Scopus, and ScienceDirect.

2.3. Data Extraction and Analysis

The search results from different search engines are combined and the result duplicates are eliminated. Finally, the result is filtered to match the selection criteria. A spreadsheet tool is used to present the complete attributes, descriptions, and categories of the papers. Research that meets the selection criteria is included in the literature review process. This paper categorizes the review findings based on the type of virtual reality technology and the type of health interventions (preventive, treatment, and diagnosis).

3. Result and Finding

The selection process gives 24 studies as the result, which is then included in the review process (as shown in Table 1). The studies are analyzed and discussed based on two topics: the kind of VR technology used and the kind of health intervention which use the VR technology. Figure 1 shows the finding on the use of VR technology for women's health intervention, based on the two topic given. From the figure, we can see that the three type of VR technology is broadly used on the treatment intervention. Further discussion on the finding is given as follows.

3.1. VR Technology in Women's Health Intervention

The three type of VR technology, namely non-immersive, semi-immersive, and full-immersive, are used in women's health intervention. The following are the finding found.

Table 1. Recent Researches in Virtual Reality Technology on Health Intervention for Women

Year	Ref	VR Type	Health Condition	Purpose	Tools
2015	[19]	Semi immersive	Food craving	Treatment	Notebook with Intel pentium processor, 4 GB RAM, Radeon HD 4570 up to 2301 Hyper Memory graphic card, 15,6 inch monitor; Polarized glasses; Noise cancelling earphone
2015	[8]	Non immersive	Self management strategy	Preventive	XBOX 360; Microsoft Kinect; 1625.6 mm monitor
2015	[23]	Semi immersive	Pain reduction on women who are having their first child (Primiparity).	Preventive	3D Blu-ray/DVD player full HD, BD660 model, connected with video glasses (Wrap 920 system, Vuzix factory, USA) and one 3D movie (IMAX Dolpine and Whales 3D 1080p). External headphones.
2015	[10]	Semi immersive	Urinary Symptoms	Treatment	Wii Fit Plus
2015	[9]	Non immersive	Alzheimer	Treatment	Video game capable computer.
2015	[12]	Semi immersive	Urinary Symptoms	Treatment	Virtual environment with freeware dance game StepMania; Dance mat as motion sensor.
2015	[26]	Full immersive	Fear of crowd	Treatment	300 x 225 cm2 stereoscopic passive screen, CAVE, HMD.
2016	[28]	Full immersive	Eating Disorder; Body Image Disturbance (BID)	Diagnosis; Treatment	Display: nVisor SX111 resolution 1,280x1,024 pixels, 280x1024 with headphone; Advanced Video Control Unit (NVIS, Reston, VA).
2016	[17]	Semi immersive	Social Phobia	Treatment	PC: OS Windows 7, i7, RAM 4GB, NVIDIA 540M; 3D Head-mounted display
2016	[25]	Full immersive	Persecutory Delusions	Treatment	Laboratory room as the environment; Head-mounted display;
2016	[11]	Semi immersive	Urinary Symptoms	Treatment	Wii Fit Plus; Wii Balancing Board
2016	[29]	Full immersive	Eating and weight disorder	Treatment	Head-mounted display; Oculus Rift DK2; Portable computer, intel i7
2016	[15]	Non immersive	Depression	Treatment	Second Life (played by participants)
2016	[22]	Semi immersive	Post stroke	Treatment	27″ monitor, 3D vision glasses, robotic pen
2017	[18]	Semi immersive	Anorexia Nervosa	Treatment	3D Head-mounted display
2017	[20]	Semi immersive	Alzheimer	Diagnosis	2 monitors; Head-mounted display
2017	[31]	Non immersive	Pregnancy; Kinematic motion	Diagnosis; Treatment	Nintendo Wii Fit Plus
2017	[24]	Full immersive	Anterior Cruciate Ligament (ACL) injury	Preventive	10mx15m room for VR environment (TEAM VR); Unity 3D Pro to develop the softwre; Windows 7 NVIDIA GeForce GTX 980; Wireless HMD 1280x720 HD, 60 frames/s
2017	[27]	Full immersive	PTSD	Preventive; Treatment	Head-mounted display (TriVisio VR Vision for 3D visual); Headphone for audio; Olfactor sensor to spray smoke smells through ventilator for multi-sensori stimulation; Joystick for motion control
2017	[30]	Full immersive	Pain management: Ginecology operation	Treatment	VR environment "Enchanted Forest" ; Joystick as a game controller; Patient body sensor; Head-mounted display
2017	[13]	Non immersive	Anorexia Nervosa	Treatment	19" flat panel LCD screen 1280x1024 pixel; 32 bit colour depth
2017	[21]	Semi immersive	Post stroke	Treatment	1 or 2 markerless infrared sensors, workstation touch screen, camera supports, BTS NIRVANA software, and Webcam. The system is connected to a projector or a big screen;
2018	[14]	Non immersive	Diabetes	Preventive	Participant plays Second Life GoWoman
2018	[16]	Semi immersive	Military Sexual Trauma	Treatment	Head-mounted display; Audio

3.1.1. Non-Immersive. Several studies have been conducted using non-immersive VR technology that targets health interventions for elderly women. In [8], the researchers developed the Individualized feedback-based virtual reality (IFVR), a virtual reality which aims to be a strategy for managing health.

This intervention utilizes XBOX 360, Microsoft Kinect, and the monitor screen to train independence in maintaining personal health for women in old age. IFVR interventions have been shown to improve quality of life, mental health, body strength, vitality, emotional health, and general health. Research conducted by [9] raised health interventions for women with Alzheimer's. The therapy is done by inviting women who have Alzheimer's disease and aims to improve their memory. Using VR with the background of a kitchen, the application is run on a laptop. The media of user interaction in the application is by using the mouse. The application was developed using the Unity 3D game engine. Research conducted in [10] involved 46 female respondents to treat Urinary Symptoms. The research uses VR games from Wii Fit Plus. The use of the VR is twice a week for 30 minutes, with a total of 10 sessions. As a result, using the VR program is reducing urinary symptoms in menopausal women. The work in [11] also examined the same thing by inviting 60 women who had experienced menopause. From the results of the interventions carried out with the VR, participants have better muscle endurance. Furthermore, research conducted by Elliot et al. [12] is also related to Urinary Symptoms. The therapy was carried out by combining Pelvic Floor Muscles (PFM) and Virtual Reality Rehabilitation (VRR) exercises. The training is carried out for 12 weeks with assessment done before and after the training. Virtual Reality in this study uses a computer in which free StepMania dance game is installed. Trainees will do the exercises in front of a screen that shows the direction of movement and Dance Mat is used as the motion sensor of the game.

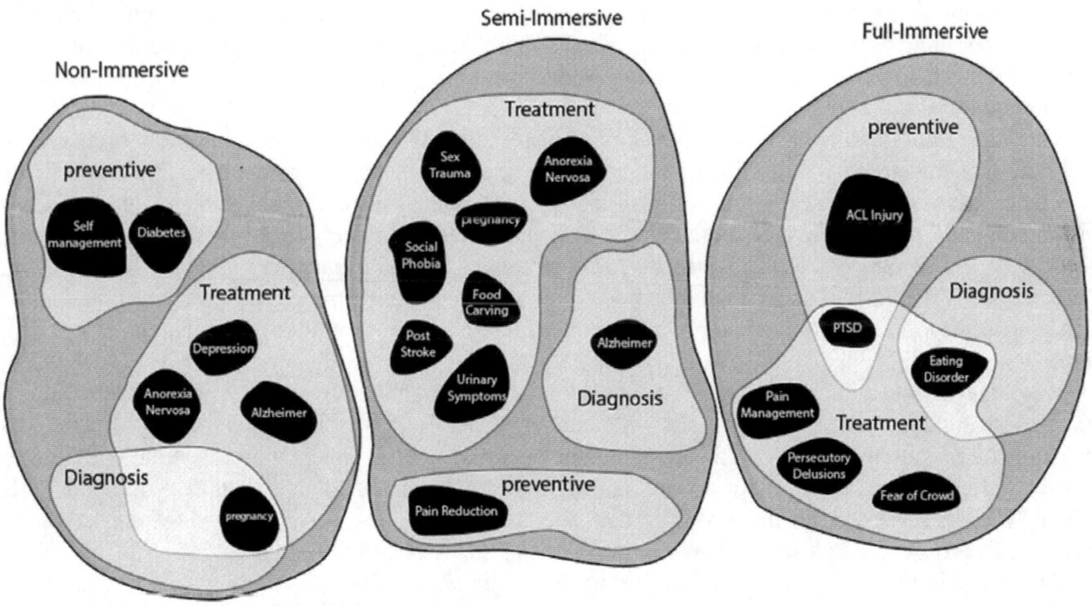

Figure 1. The finding of the use of VR technology for women's health intervention.

Non-immersive virtual reality is also performed on women to increase self-confidence and self-respect. Research conducted by [13], raised the problem of Anorexia Nervosa. Including 15 women who experienced this problem, participants will see their body simulation with VR according to the ideal BMI of each patient. Using VR technology, the participants will be scanned by sensors to measure up their estimated weight and produce their 3D body shapes. Research conducted by [14] developed the GoWoman program based on Second Life virtual reality. GoWoman will help women to maintain their health from the risk of diabetes. Participants will play GoWoman through their respective computers or laptops. Research using Second Life virtual reality was also carried out in [15]. In this study, therapy sessions were conducted for one week using the Second Life application and significantly succeeded in increasing the confidence of participants in depressed women with disability.

The subsequent research conducted in [31] is about diagnosis and therapy for kinematic movements of pregnant women using the Nintendo Wii Fit Plus®. All pregnant women were divided into four groups. The test was carried out in 12 sessions, 30 minutes in a session and three times a week. The results obtained show that the use of Nintendo Wii Fit Plus® is not able to influence the kinematic variables sit-to-stand in the women analyzed. Furthermore, research conducted [17] reported that the use of VR can increase patient motivation to recover from a stroke.

3.1.2. Semi-Immersive. Several studies that use semi-immersive VR technology were conducted by targeting mental health problems. Research conducted in [16] raised the issue of Military Sexual Trauma (MST). This study utilizes HMD and audio stimuli as its virtual technologies. The therapy was carried out for 3 months and has been proven safe and promising for the treatment of MST, especially mental disorders. Furthermore, another study targeting mental health was carried out in [17], which raised the problem of Social Phobia. By using VR technology installed on a computer and 3D glasses as visual media, participants underwent 12 therapy sessions with each 50-minute duration session. The therapy was held for 6 months. Overall participants experienced positive improvement in their phobia symptoms.

Research utilizing semi-immersive VR technology was also conducted to address the problem of self-confidence in body shape and diet. Research [18] utilizes VR to help women who experience problems with Anorexia Nervosa. Participants underwent three-dimensional body scans and conducted treatment sessions using 3D HMD. Participants who experienced Anorexia Nervosa can determine the target weight according to the ideal body weight. Furthermore, research [19] developed a semi-immersive virtual reality to control appetite (cravings). Food craving is a condition where someone wants to consume a food but not based on hunger.

Virtual reality for diagnosis was developed by research [20], namely the use of virtual reality technology to diagnose the inhibition of cognitive function in adult women who are starting to age using two monitor screens and head-mounted displays. 89 participants were divided into two groups. Each group compared its performance on the Three Stroop Modalities, namely multi-item paper-and-pencil Stroop, Single-item Computerized Stroops (no distractions) and Virtual Reality-based Stroops (with Distractions). The results show that the use of Virtual Apartment Stroop Task has the potential to distinguish between inhibition of prepotent response and resistance to inhibition of distractors in aging adult women.

Research of VR which take after stroke domain is carried out by [21]. The research utilizes markerless infrared sensors, touch screen workstations, camera supports, BTS NIRVANA software, and webcams. The system is connected to a projector or a big screen. Nirvana base stations are able to improve attention processes, spatial cognition, and mood in patients. In addition, [22] uses VR tools called RehAtt, succeeded in increasing spatial attention in the activities of daily life of post-stroke patients. RehAtt works by using 27" monitors, 3D vision glasses, and robotic pens.

The study in [23] develops virtual reality to reduce pain postpartum episiotomy. 30 women were divided into two groups. Intervention Group gets regular treatment using VR and Control Group gets treatment without using VR. The result shows that VR can be used as a complementary non-pharmacological method that is effective for reducing pain during episiotomy repair.

3.1.3. Full-Immersive. Research [24] aims to develop virtual reality for female athletes that can reduce the risk of ACL injury. In this intervention, virtual reality was developed in a room measuring 10m x 15m for a VR environment (TEAM VR); Software developed with Unity 3D Pro; Windows 7 NVIDIA GeForce GTX 980; 1280x720 HD HMD wireless, 60 fps. The result shows that virtual reality technology can be used as an immersive and simulation-based exercise scenario to evaluate the patient's biological condition and as a neuromuscular training intervention to reduce the risk of ACL injury.

Several studies have been conducted using VR technology on mental health problems. In [25], the research raised the domain of Persecutory Delusions. The test is carried out for 60-90 minutes each period. The virtual environments are designed with two different environments, namely Subway and Lifts. Therapy is carried out in a virtual lab which use sensors to support the assessment of patient behavior. Patients wear HMD during therapy sessions. For therapy conducted in the virtual subway environment, patients can walk as if they are on a train. Stages of difficulty in the therapy are done by raising an avatar. The number of avatars that appear is a stage of the difficulties faced by patients. Through the results of the therapy obtained, the use of VR technology can prove high effectiveness in treating delusions. Research conducted in [26] shows that VR can be used as a therapeutic tool for phobia patients with crowds. The study utilizes VR technology by displaying visual scenes on a stereoscopic passive screen 300 x 225 cm^2. Users wear polarized stereoscopic glasses. Auditory scenes are presented via Sennheiser HD650 headphones and sound stimuli are processed through binaural rendering using Head Related Transfer Functions (HRTF). The virtual environment reproduces the campus environment of Trinity College Dublin which consists of buildings, alleys and vegetation. Animated virtual individuals, referred to as humanoids, can be placed in this environment. The auditory component of a virtual environment consists of a human voice and an ambient audio environment consisting of bird sounds and urban activities. In virtual scenes used 12 to 96 humanoids can be distributed in virtual environments. Furthermore, Virtual reality used for Anxiety Level diagnosis in PTSD (Post-Traumatic Stress Disorder) patients was developed in [27]. This virtual reality technology uses Head-mounted display Type TriVisio VR Vision for 3D visuals; Headphones for sound; Olfactor sensor to emit smoke odor through a ventilator to ensure multi-sensory stimulation; Joystick for movement control. In this study, 80 participants were divided into 2 groups, High Anxiety (HA) and Low Anxiety (LA). The results show that the HA group manifest higher stress. The VR can be used to prevent stress with interventions that target risk factors for peri-traumatic and post-traumatic in real time.

Several studies have been conducted in the intervention addressing the problem of Eating and Weight Disorder. Research [28] developed a London Bus virtual reality to diagnose Body image disturbance (BID) in women which provide therapy to women who experience BID. The virtual environment used is a 4-minutes trip using the London bus. The simulation environment was developed using the Unity software package. The environment is displayed in color through HMD; the screen used is n SX111 Shield with a resolution of 1,280x1,024 pixels, 280x1024 resolution, and 111 ° field of view. Furthermore, research [29] took 21 female participants with symptoms of Eating and Weight Disorder. Participants will be asked about their ideal body shape. Then they will see the ideal body shape they want through VR. VR technology used is HMDs that are connected to a computer. The computer has the Skinny Belly application which is used to view the participants' virtual body shape. After conducting an experiment using the VR, participants are shown a decrease in the estimation of their ideal body shape and body shape. This application can be developed for women who suffer from weight disorders and appetite.

Finally, a study was carried out in [30], taking a related problem on Pain Management in women who would perform gynecological surgery. VR technology used includes Full-Immersive, using HMD, sensors that are installed in the patient's body and joystick as an interaction tool in the game. As a result, VR patients showed lower pain indicators than patients who did not use VR.

3.2. Health Intervention of VR

Based on the type of the health intervention, we found that there are four types of VR application for women. Three of them are granular intervention: preventive, diagnosis, and treatment. The other one is the combination of diagnosis and treatment. We elaborate them as follows.

3.2.1. Preventive. Several studies have developed virtual reality as a preventive effort for several diseases ([19], [8], [23], [24], and [14]). The preventive interventions given are (i) develops a virtual reality to control appetite (cravings/food craving) to control health [19], (ii) regulates self-health for women in old age to improve the quality of healthy life in mental health, body strength, vitality, emotional health, and general health [8], (iii) reduces pain postpartum episiotomy [23], (iv) evaluates the patient's biological condition and as a neuromuscular training intervention to reduce the risk of ACL injury [24], and (v) maintains the health from the risk of diabetes by taking 16 practice session to experience weight loss, waist circumference, increased physical activity, diet and self-desire in diet [14].

3.2.2. Diagnosis. Virtual reality application for diagnosis can be seen in [20]. The use of virtual reality technology to diagnose inhibition of cognitive function in mature women who are aging. It is proved that the VR has the potential to distinguish between inhibition of prepotent response and resistance to inhibition of distractors in aging adult women.

3.2.3. Treatment. Most of VR applications are developed in treatment intervention. Research [12] combines Pelvic Floor Muscles (PFM) and Virtual Reality Rehabilitation (VRR) exercises. As a result, the incorporation of training using PFM and VRR is acceptable, efficient and satisfying for the elderly with urine problems. The same thing was done in the study [10] using the VR program, that can be promoted in PFM training and reduce urine symptoms in menopausal women. [11] also examined the same thing by inviting 60 women who had experienced menopause. It also shows good results for PFM and can be practiced in a clinical environment. Research conducted in [9] raised Alzheimer's. Therapy is done by inviting a woman who has Alzheimer's disease to be able to improve her memory. With 4 days of therapy with cooking tasks, the result shows that respondents can learn the tasks given in the application and are able to do it in real life stably according to what is done during therapy. Conducting research on Eating and Weight Disorder, the study in [29] took 21 female participants with symptoms of Eating and Weight Disorder to participate. Participants will be asked about their ideal body shape. Then they will see the ideal body shape they want through VR. This can be developed for women who suffer from weight disorders and appetite. Furthermore, in the study [17] raised the topic of Social Phobia therapy with each 50-minute duration session which was lived for 6 months. Overall participants experienced an increase in the value of their phobia symptoms. In [25], the research about Persecutory Delusions is conducted. The test carried out for 60-90 minutes each period. Through the results of the therapy obtained, the use of VR technology can prove high effectiveness in treating delusions. In [30], took the related problems of Pain Management in women who would perform gynecological surgery. In the intraoperative phase, the VR group will use VR as an intervention during this phase. As a result, VR patients showed lower pain indicators than patients who did not use VR. In [13], proposed the problem of Anorexia Nervosa. Using VR technology, the participants will be scanned by sensors to measure shapes up to their estimated weight and produce their 3D body shapes. Patients who underwent therapy showed a decrease in the ideal body shape with their ideal BMI. This shows that VR Environment is able to make patients more realistic in determining diet targets to achieve an ideal body. Another research that develops virtual reality for Anorexia Nervosa is in [18]. After undergoing experiments, the two groups can determine the target weight according to the ideal body weight. In [16], which took the issue of Military Sexual Trauma (MST). respondents were given 6 training sessions for 3 months. Training using VR is carried out for 30-45 minutes within 2 weeks of working days. The results indicate that the Virtual Reality Environment (VRE) provides the delivery of training materials safely and promising for the treatment of MST, especially mental disorders.

3.2.4. Diagnosis and Treatment. Some studies also develop virtual reality that can be used for diagnosis and treatment of a disease. As in the study [28] developing virtual reality diagnoses BID in women and provides therapy to women who experience BID. Subsequent research is about diagnosis and therapy for kinematic movements of women who are pregnant [31]. The test was carried out in 12 sessions and 30 minutes in session and 3 times a week.

4. Discussion

Based on the findings, most virtual reality technologies developed for women health interference were aimed for the treatment of psychological problems and physical exercise. This statement is strengthened using word cloud analysis depicted in Figure 2. The focus of the VR usage in physical treatment is reflected by the high frequency number of the word *body, bmi, weight,* and *physical* (in year 2018). As for psychological problem, the word cloud gives a visualization of the word *trauma, stress, esteem,* and *anxiety*.

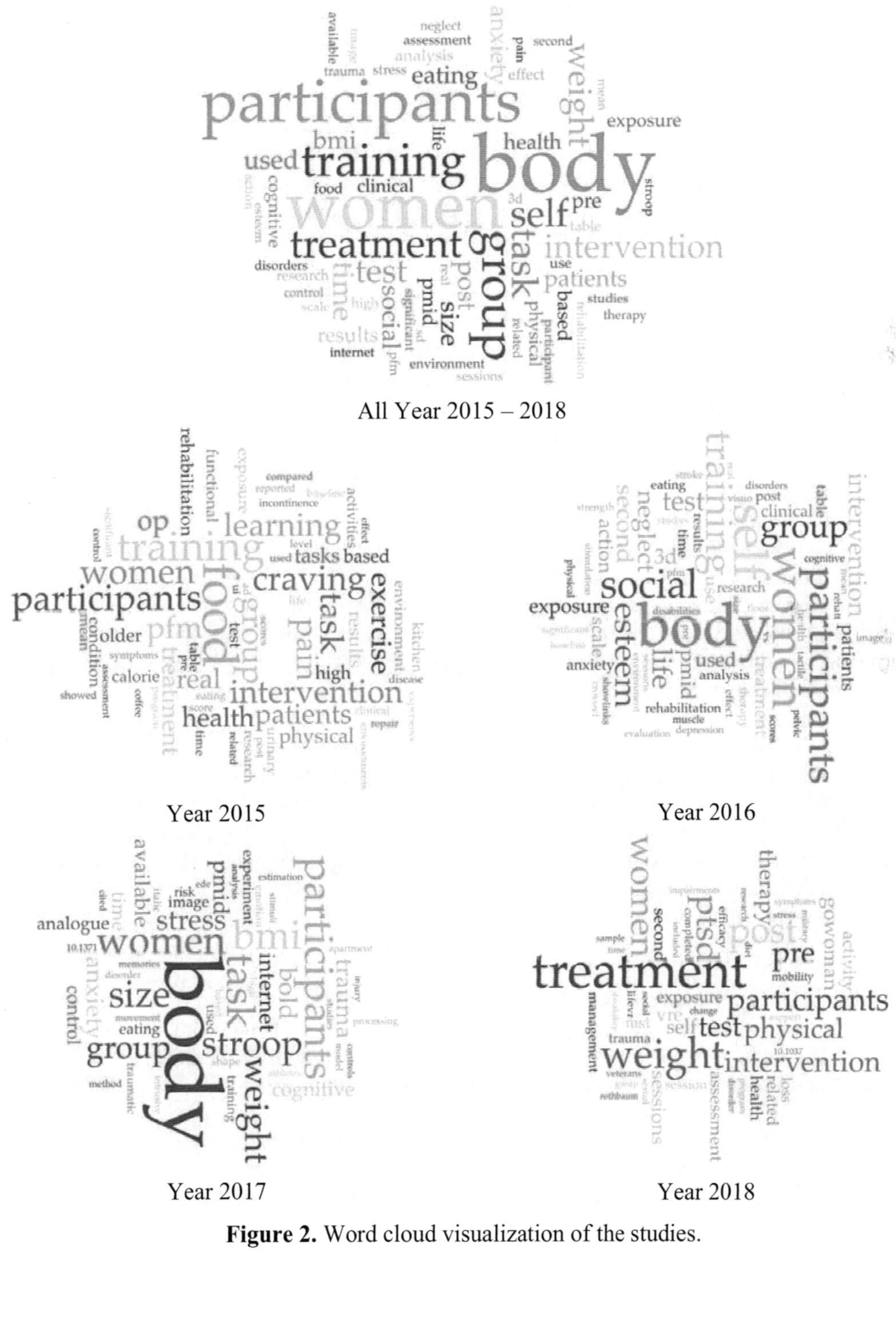

Figure 2. Word cloud visualization of the studies.

Psychology therapy requires a situation that is able to present a real picture of the situation that affects the patient psychologically. As an example, is the treatment of Persecutory Delusions [25]. Persecutory Delusions are psychological problems that make sufferers experience delusions of persecution, so they have excessive fear and alertness when interacting with strangers. Virtual reality environments are designed like underground train stations and elevators. Patients try to play virtual reality as if they were in these two places. Through virtual reality games, patients can practice reducing their anxiety and delusion towards persecution. In addition, in the case of social phobias [17], virtual reality can help people with social phobias to practice interacting with other people with 3-dimensional avatars.

In addition to psychological therapy, virtual reality is also widely developed for physical exercise. With sensors that can record body movements, virtual reality can help patients to do physical exercise therapy according to their needs. The important thing that must be considered during physical exercise is the condition of the patient when doing the exercise and the suitability of the patient in carrying out physical exercise movements. If during the training session there is a clear report on improving the patient's condition and the patient is able to exercise properly, the physical exercise therapy session can be declared successful. Virtual reality technology is able to present a real environment, provide simulations of situations that exist in the real world, and is able to record patient sensors during use.

Trends in the development of virtual reality technology based on the findings will lead to technology for therapy. This can be seen from many studies that develop virtual reality technology that is used as a therapy for health problems, especially psychological problems, and physical exercise. Not only specifically for therapy, but virtual reality technology can also be combined with diagnoses such as in [28] and [31] or combined with preventive as in [27]. In addition, although not many have researched, virtual reality can also be developed into preventive technology.

Virtual reality technology for diagnosis has not been widely developed by researchers. From the search results, only one study developed virtual reality technology for diagnosis. In the case of women's health, the developed virtual reality is a combination of diagnosis and therapy [28], [31]. However, in both studies, the results obtained were not in accordance with the hypothesis. Research [31] is a study that discusses virtual reality technology for the sitting-standing movement of pregnant women. The study was the first study to discuss the effect of virtual reality for diagnosis and sitting-standing therapy for pregnant women. There are several things that affect the research instrument that was recently discovered after the study. One of them is the influence of maternal hormones that change frequently. The hormonal conditions of pregnant women when making a diagnosis with virtual reality compared to their daily hormonal conditions can be different. Likewise, with the state of the hormone when making a diagnosis in each session. In addition to hormonal conditions, physiological changes in pregnant women also affect the results of diagnosis and therapy using virtual reality.

In developing virtual reality for the diagnosis of women's health, it is important to note the side effects that arise during the experiment, such as dizziness, nausea, headaches, eye strain, and the development of responses that are not in accordance with the real world [32]. These side effects can also be affected by the condition of female hormones when conducting experiments. Development of virtual reality technology for women's health in the future must pay attention to the side effects and influences of other conditions that may be experienced by women.

5. Conclusion

Virtual Reality (VR) technology are used broadly in the health intervention domain. This study elaborates the state-of-the-art in the application for recent VR technologies for women's health intervention. The findings of this study show that there are three types of VR technology are used in the application (non-immersive, semi-immersive, and full-immersive). The three technology are used pretty much equally. However, we could see that non-immersive technology is rather be used highly in aged women. From the same findings, from the four category of health interventions done through VR (prevention, diagnosis, treatment, diagnosis-treatment), treatment is one health interventions with the highest number. Finally, the domain of the disorder are focused in disorders related to psychological and physical aspect.

References

[1] Statista 107 *Virtual Reality (VR) and Augmented Reality (AR) Device Ownership and Purchase Intent Among Consumers in The United States as of 1st Quarter 2017 by Gender*

[2] Vive 2017 *China Virtual Reality User Analysis*

[3] Viña J and Lloret A 2010 Why women have more Alzheimer's disease than men: Gender and mitochondrial toxicity of amyloid-β peptide *J. Alzheimer's Dis.* vol **20** no **SUPPL.2** pp S527–33.

[4] Bartley E J and Fillingim R B 2013 Sex differences in pain: a brief review of clinical and experimental findings *Br. J. Anesth* vol **111** no **1** pp 52–8

[5] Quintero O L, Amador-Patarroyo M J, Montoya-Ortiz G, Rojas-Villarraga A and Anaya J M 2012 Autoimmune disease and gender: Plausible mechanisms for the female predominance of autoimmunity *J. Autoimmun* vol **38** no **2–3**

[6] Offord D R, Boyle M H, Campbell D, Goering P, Lin E, Wong M and Racine Y A 1996 One-year prevalence of psychiatric disorder in Ontarians 15 to 64 years of age *Can. J. Psychiatry* vol **41** no **9** pp 559–63

[7] Alqahtani A S, Daghestani L F and Ibrahim L F 2017 Environments and system types of virtual reality technology in stem: a survey *Int. J. Adv. Comput. Sci. Appl.* vol **8** no **6** pp 77–89

[8] Lee M, Son J, Kim J and Yoon B C 2015 Individualized feedback-based virtual reality exercise improves older women's self-perceived health: A randomized controlled trial *Arch. Gerontol. Geriatr* vol **61** no **2** pp 154–60

[9] Foloppe D A, Richard P, Yamaguchi T and Allain P 2015 The potential of virtual reality-based training to enhance the functional autonomy of Alzheimer's disease patients in cooking activities : a single case study *Neuropsychol. Rehabil* vol **2011** no November

[10] Botelho S, Martinho N M, Silva V R, Marques J, Carvalho L C and Riccetto C 2015 Virtual reality: a proposal for pelvic floor muscle training *Int. Urogynecol. J.* vol **26** no **11** pp 1709–12

[11] Martinho N M, Silva V R, Marques J, Carvalho L C, Iunes D H and Botelho S 2016 The effects of training by virtual reality or gym ball on pelvic floor muscle strength in postmenopausal women: a randomized controlled trial *Brazilian J. Phys. Ther.* vol **20** no **3** pp 248–57

[12] Elliott V, Bruin E and Dumoulin C 2015 Virtual reality rehabilitation as a treatment approach for older women with mixed urinary incontinence: a feasibility study *Neurourol. Urodyn* vol **34** no **3** pp 224–30

[13] Cornelissen K K, McCarty K, Cornelissen P L and Tovée M J 2017 Body size estimation in women with anorexia nervosa and healthy controls using 3D avatars *Sci. Rep.* vol **7** no **1**

[14] Nosek M A, Robinson-Whelen S, Ledoux T A, Hughes R B, O'Connor D P, Lee R E, Goe R, Silveira S L, Markley R and Nosek T M 2108 A pilot test of the GoWoman weight management intervention for women with mobility impairments in the online virtual world of Second Life® *Disabil. Rehabil.* pp 1–12

[15] Nosek M A, Robinson-Whelen S, Hughes R B and Nosek T M 2016 An internet-based virtual reality intervention for enhancing self-esteem in women with disabilities: results of a feasibility study *Rehabil. Psychol.* vol **61** no **4** pp 358–70

[16] Loucks L *et al* 2018 You can do that?!: Feasibility of virtual reality exposure therapy in the treatment of PTSD due to military sexual trauma *J. of Anxiety Disorders* (Elsevier) no October pp 0–1

[17] Gebara C M, de Barros-Neto T P, Gertsenchtein L and Lotufo-Neto F 2016 Virtual reality exposure using three-dimensional images for the treatment of social phobia *Rev. Bras. Psiquiatr.* vol **38** no **1** pp 24–9

[18] Mölbert S C, Thaler A, Mohler B J, Streuber S, Romero J, Black M J, Zipfel S, Karnath H O and Giel K E 2017 Assessing body image in anorexia nervosa using biometric self-avatars in virtual reality: attitudinal components rather than visual body size estimation are distorted *Psychological Medicine* pp 1–12

[19] Ferrer-Garcia M, Gutierrez-Maldonado J, Treasure J and Vilalta-Abella F 2015 Craving for food in virtual reality scenarios in non-clinical sample: analysis of its relationship with body mass index and eating disorder symptoms *Eur. Eat. Disord. Rev.* vol **23** no **5** pp 371–8

[20] T D Parsons and M Barnett 2017 Virtual apartment-based stroop for assessing distractor inhibition in healthy aging *Appl. Neuropsychol.* vol **0** no **0** pp 1–11

[21] De-Luca R, Lo-Buono V, Leo A, Russo M, Aragona B, Leonardi S, Buda A, Naro A and Calabrò R S 2017 Use of virtual reality in improving poststroke neglect: Promising neuropsychological and neurophysiological findings from a case study *Appl. Neuropsychol.* pp 1–5

[22] Fordell H, Bodin K, Eklund A and Malm J 2016 RehAtt – Scanning training for neglect enhanced by multi-sensory stimulation in virtual reality *Top. Stroke Rehabil.* vol **23** no **3** pp 191–9

[23] Jahanishoorab N, Zagami S E, Nahvi A, Mazluom S R, Golmakani N, Talebi M and Pabarja F 2015 The effect of virtual reality on pain in primiparity women during episiotomy repair: A randomize clinical trial *Iran. J. Med. Sci.* vol **40** no **3** pp 219–24

[24] Sch A W, Dicesare C, Bonnette S, Kitchen K, Gadd B, Thomas S, Barber-Foss K D, Myer G D, Riley M A and Silva P 2017 Sport-specific virtual reality to identify profiles of anterior cruciate ligament injury risk during unanticipated cutting *Int. Conf. on Virtual Rehabilitation ICVR* 2017 vol June

[25] Freeman D *et al* 2016 Virtual reality in the treatment of persecutory delusions: Randomised controlled experimental study testing how to reduce delusional conviction *Br. J. Psychiatry* vol **209** no **1** pp 62–7

[26] Taffou M, Ondřej J, O'Sullivan C, Warusfel O, Dubal S and Viaud-Delmon I 2015 Auditory-visual virtual environment for the treatment of fear of crowds *Proc. of the 2015 Virtual Reality Int. Conf. on ZZZ - VRIC '15* pp 1–3

[27] Schweizer T, Schmitz J, Plempe L, Sun D, Becker-Asano C, Leonhart R and Tuschen-Caffier B 2017 The impact of pre-existing anxiety on affective and cognitive processing of a Virtual Reality analogue trauma *PLoS One* vol **12** no **12** pp 1–19

[28] Mountford V A, Tchanturia K and Valmaggia L 2016 What are you thinking when you look at me? a pilot study of the use of virtual reality in body image *Cyberpsychology Behav. Soc. Netw.* vol **19** no **2** pp 93–9

[29] Serino S, Pedroli E, Keizer A, Triberti S, Dakanalis A, Pallavicini F, Chirico A and Riva G 2016 Virtual reality body swapping: a tool for modifying the allocentric memory of the body *Cyberpsychology Behav. Soc. Netw.* vol **19** no **2** pp 127–33

[30] Wiederhold B, Vázquez J, Vaca V, Miller I and Wiederhold M 2017 Virtual reality pain distraction during gynecological surgery—a report of 44 cases *Surg. Res. Updat.* vol **5** no **1** pp 12–6

[31] Ribeiro S O, De Sousa V P S and Viana E D S R 2017 Influence of a virtual reality-based exercise protocol on the sit-to-stand activity kinematic variables in pregnant women: A randomized controlled trial *Motriz. Rev. Educ. Fis.* vol **23** no **3** pp 1–8

[32] Srivastava K, Das R C and Chaudhury S 2014 Virtual reality applications in mental health: Challenges and perspectives *Ind. Psychiatry J.* vol **23** no **2** pp 83–5

In the making of effective decision making in public health domain with business intelligence dashboard

MA Setiawan

Department of Informatics, Universitas Islam Indonesia

andri@uii.ac.id

Abstract. The easy access towards public health information system is one of the most important aspects in improving general health public. Today, people have more privilege to travel than a few decades ago as flight cost is getting cheaper and the economic situation for many people are better. However, this also brought some consequences that disease also easily move from one place to another. Hence, for many countries, it is necessary for many public health institutions to be able to check current health information system in more accurate and can answer problems that often happened. One of the approaches that have been used to improve the readability of public health information is through Business Intelligence (BI). BI allows health administration to get a good quality of information that can help them to decide. BI has its role as a decision support system that is based on collected facts. However, there are still lots of challenges in implementing BI for public health information system that needs to be addressed. In Indonesia, public health data are collected, but then some studies have suggested that data that have been collected just sitting idle in the repository, and have not been used effectively to help health administration managing the public health. We proposed a system that can help health administration to decide what needs to be done to improve the public health. This information is able to help health administration to show problems and allows them to assign priority to improve the overall public health condition.

Keywords — public health information system, business intelligence, decision support system

1. Introduction

In the past decade, Indonesia has seen tremendous adaptation in technology. A survey conducted by APJII [1] showed that more than half of Indonesia population have been connected to the Internet. The growth in the internet adoption also increases what it called as *Over-the-top content* (OTT) [2] where users are starting to consume many kinds of things over the internet. New technology also plays part in the healthcare section. Public health is an ongoing issue within the technology adoption spectrum that also needs to be tackled down to improve overall healthcare experiences in Indonesia.

Health care institutions in Indonesia are under a pressure to serve public health well. Even though the number of communal diseases such as dengue fever, tuberculosis, and other diseases has decreased, public health communities are continuously seeking ways to make sure that information is deployed efficiently to improve the quality of patient care. Information in public health is considered to be a very essential in helping authorities to get better approaches [3]. Most importantly, decision-making in public health field requires the timely availability of valid and sound data [4].

Many public health organisations in Indonesia have already implemented information system that records data and information. However, in many occasions, even though the data are in comprehensive collections, but they are only in a raw data form without any specific pattern that users are able to interpret them easily. It is quite often that data is presented in a complex, fragmented, and do not portray a representative look to needs. Other than that, some inconsistencies are often found in how the report is organized and how it affects the availability of valid and consistent data or information [5]. This chaotic organization of public health data will lead to a new level of difficulty for those who have the responsibility to make the decision in public health organizations [6].

Timely and accurate data has an important role as the foundation of the knowledge base for health information system [7]. The knowledge will be used as the source of decision making where the best decision is formed when all relevant data are considered well. One of the technology that is capable to transform data into information that can help decision making process is data warehouse [8]. A data warehouse is a process to extract data periodically and transform them into a certain format and then inserted into a central data storage to create a business intelligence (BI) information system. This data warehouse will be the source of information in a form of ad-hoc report, portal, or a dashboard [9].

The concept of BI has attracted many healthcare organizations and professionals. BI that uses data warehouse approach has been used in many businesses and industries to analyze the data within certain time frame [6]. Recently, data warehouse is generating traction as today information is presented in a more attractive form that is easier to read e.g. via charts and diagrams [10]. Hence, it also makes sense that public health field also uses data warehouse approach to make health data easily accessible in a more effective and efficient ways.

In this paper, we present a practical solution in presenting public health data in community health centre in one of districts in Yogyakarta, Indonesia. The dashboard provides information on national health with the source of the data has been transformed with certain schemes. This dashboard is presented in a way to improve the performance, effectiveness, and the efficiency of public health management.

2. Related Literatures

2.1. Overview of Business Intelligence

With the emerging of Big Data, BI has reserve itself in a nice spot of information technology field [11]. BI itself is a tool that has been used by organizations to make a better understanding of how the activities found from raw data behave, how they run, etc. [12]. BI incorporates many categories of methodologies, applications, and technologies by providing them with an easy interface for faster and more accurate business decisions. BI is often considered as the natural successor of *Decision Support System* (DSS) as it provides similar results from a DSS [13], but with one very distinguished difference which is unstructured data sources. Traditional DSS have limitations with unstructured data sources while BI do well on both structured and unstructured data sources [5], [14].

In the context of public health, BI needs to be introduced as part of improvement effort in healthcare information system [6]. While information system implementation in other fields has reached a mature level, public health information system is often in opposite place. It is found that healthcare information system unsuccessfulness is impeded by the fact that underlying data that construct the information system is not well structured. BI has been seen as a new and useful technology with their ability to help decision makers in healthcare organizations to make better decisions. BI works by extracting set of data and store them in a system called data warehouse. During extraction process, data is transformed and simplified with techniques according to the need of business analysis. The transformation leads to pattern and behavior of the processed data. BI extracts the data sources and presents working data, transactional data, or others into knowledge and has the ability putting the right information at the right time to support decision-making process and to achieve the vision of the organization.

The results of the transformation and simplification are then presented to end users especially those who have the responsibilities to make the decision. The decision is no longer based on intuition and quantitative experiences which often found in many organizations [15], but based on the facts data [16]. BI works by incorporating these steps:

- Identify business problems that need to be tackled with data warehouse and decide which data is needed [17]
- Transform the data from multiple sources into consistent format [18]
- Retrieve the transformed data from centralized database and insert them to data warehouse [19]
- Create systems or applications that will query or analyze transformed data e.g. with online analytical processing (OLAP) [20]
- Visualize the results [21]

3. Methods

3.1. Data Source
Our study uses data from community health centre, at Puskesmas Tempel II, Sleman, Yogyakarta. The information system used by the community health centre collects many kinds of data such as patients' medical records, mothers and babies' health care, etc.

3.2. System Design
In order to have a good outcome of the system, we design our system first with use case diagram, database design, and interface design. Figure 1 shows a use case diagram (in Bahasa Indonesia) which define how users interact with the system. The scenario is, once a user logged in the system, this user will be able to use all information, able to drill-down each report, and can download the raw tables.

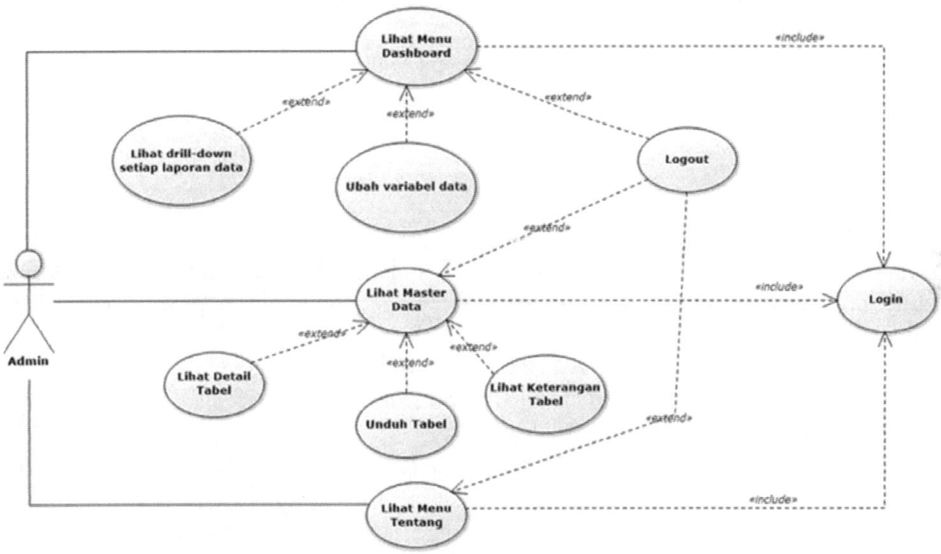

Figure 1. Use Case Diagram

3.3. ETL (Extract, Transform, Load)
ETL is a set of process that related with a way to fetch the data from its sources as presented in Figure 2, which then will be transformed before it then being loaded into data warehouse. Once the data is loaded into the data warehouse, BI tools will use them as needed. An example of ETL is merging two tables or more to create calculation of their aggregates. Varcellis [22] explains the detail of ETL process, and those are:

- *Extraction*

  Extraction is a cleaning up process for data that is gathered from homogenous or heterogeneous sources which then being extracted to make the structures or the patterns of the expected information from data.

- *Transformation*
  Transformation is a process to select, clean up, and add more attributes so that the data that has been extracted can be loaded according to the correct format or after the data warehouse structure.
- *Load*
  Load is a process to insert data into data warehouse after being processed in transformation phase. After the data is formatted and structured according to the structure set in the data warehouse, the loading process begins. Data came from staging area will be moved to the pre-defined site in data warehouse.

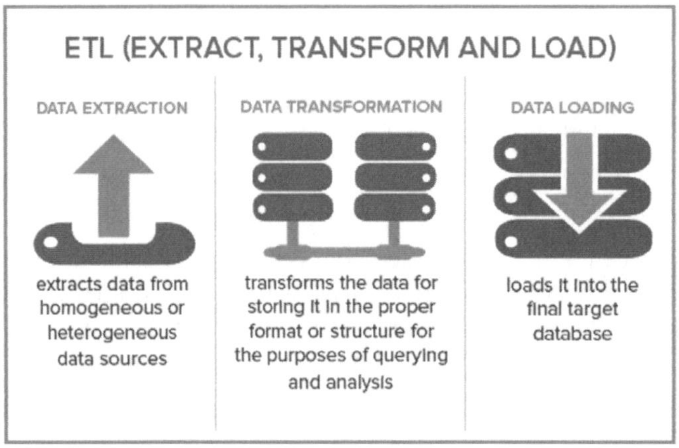

Figure 2. ETL Process

4. Results

4.1. Data Warehouse Architecture

Our data warehouse architecture has the characteristics of object-oriented, integrated, has time variance, and non-volatile (once data is loaded, it's permanent). We use top-down methods where the process is started with data collection from the data sources. Once the data is collected, it will be loaded in staging area (a temporary storage) so that data can be customized and then we generate a sequence of data before being processed and loaded to data warehouse [8] as presented in Figure 3.

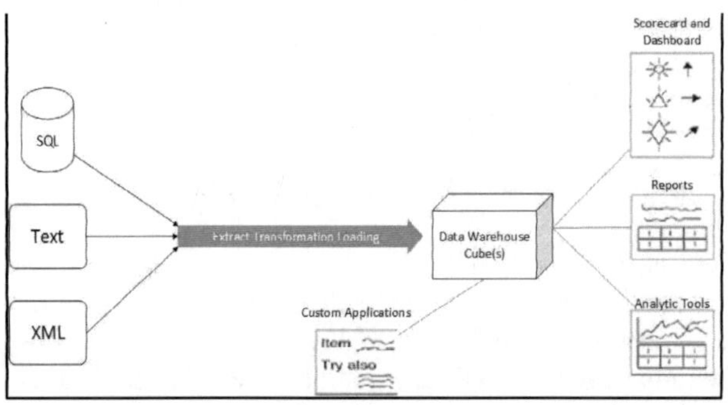

Figure 3. Data Warehouse Architecture

4.2. Data Warehouse Consolidation

Data warehouse consolidation is our effort to cluster data. The consolidation part is designed using OLAP technique that will explain the content of fact table and dimension table from the data sources. There are four fact tables and 23-dimension tables and each of fact tables is created with OLAP Cube techniques. We then assign each four fact tables with few of 23-dimension tables. We use the OLAP Cube techniques to show data in a multi-dimensional form. The OLAP uses star-scheme against 4 fact tables. Once all tables are consolidated, the system then creates a BI dashboard that uses data from the data warehouse. The user who logged into the system will be able to read the information from our prepared scheme in the BI dashboard.

4.3. BI Dashboard

The dashboard presents all facts designed in the OLAP i.e. facts on staffs' load, mothers and babies health data, geographical mapping of the patients, patient's trend, public health screening, etc. In Figure 4 (shown in Bahasa Indonesia), we present how the community health centre targeting surrounding community to have their own initiative in improving health-aware environment. It shows the target, and the achievement so far. From the figure, it was found out that smoking inside a house is commonly found. Hence, the community health centre can decide that they should aim more to alleviate the smoking inside a house problem. Figure 5 presents mother and babies health data. One of the information that can be concluded is that the number of mothers who has been vaccinated with Tetanus Toxoid (TT) vaccine is still low. This can pose serious problem in the future.

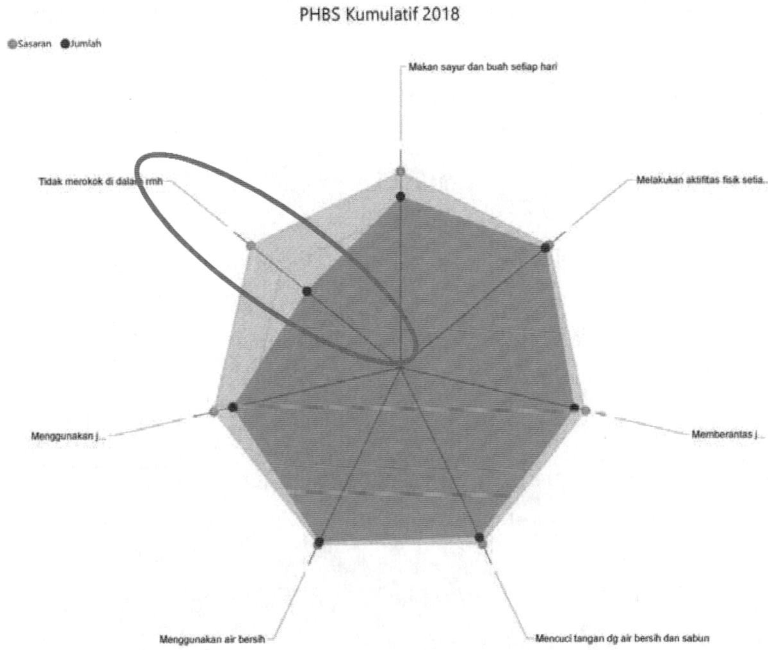

Figure 4. Community Improvement Public Health

Figure 6 provides information on where are the patients from. This information is really important for community health-centre decision makers. They can see where are the blank spot of the patients, where they can promote more. Figure 7 presents staffs' load distribution. This can help management people to decide how the organization is going to be run.

All facts are presented as in those above dashboards (not all are shown here due to page limitation of the paper). All information is shown in a form of pie charts, maps, bar diagrams, and others which can help decision makers to decide and set strategy for future development and set up priority for public health in the district.

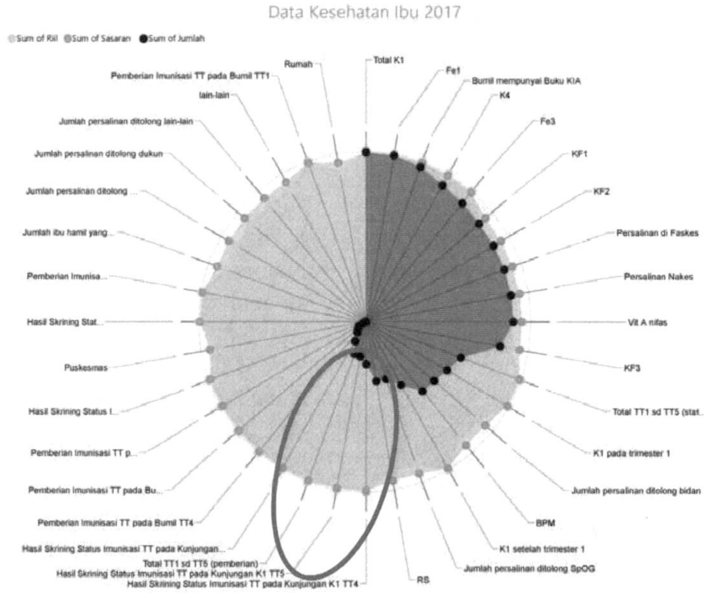

Figure 5. Mother and Babies Health Data

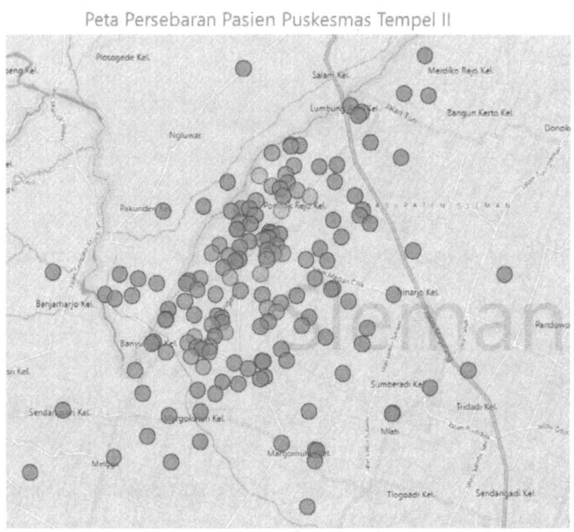

Figure 6. Where are patients from

5. Conclusion

We have successfully created a BI system that complement a public health information system. As all information is now presented in a graph and chart form along with simple and short analysis, people who have responsibilities on public health decision making will have a better understanding on how public health in Indonesia is running. Charts are easier to read than tables. Moreover, we no longer see the unstructured data, instead user is presented with nice and sleek graphics which even allow user to understand of what is happening in the public health field at a glance. Difficulties to understand how public health run in the district of Tempel, Yogyakarta, is gone. BI technology serves as an assisting framework in decision making process. Hence a right approach in timely manner is expected with the use of BI systems. Our study is without limitation. The system needs to enrich itself by incorporating from multiple related sources. The national bureau of statistics actually provides private data that is more comprehensive than its public version. We also preparing our system to be able to look up for disease trajectory. This will be particularly very important to prevent the disease from spreading further or to prevent it from happening as we can learn the pattern from collected data in the past.

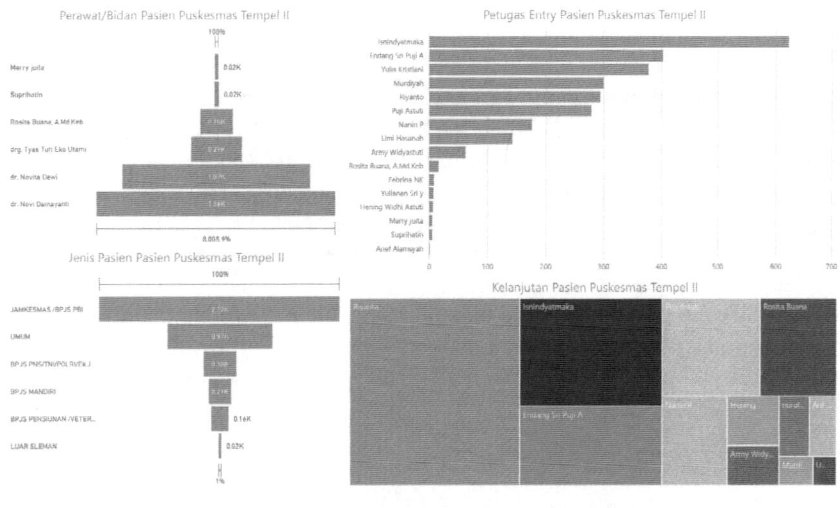

Figure 7. Staffs' load distribution

Acknowledgment
Authors would like to thank the Ministry of Research and Higher Education for the generous support in the research funding.

References

[1] APJII 2016 *Penetrasi & Perilaku Pengguna Internet Indonesia* Survey
[2] Green W, Lancaster B and Sladek J 2006 *Over the Top Services* (LTC Int.)
[3] Foshay N and Kuziemsky C 2014 Towards an implementation framework for business intelligence in healthcare *Int. J. Inf. Manag.* vol **34** no **1** pp 20–7
[4] C of the E Communities 2007 Together for health: a strategic approach for the EU 2008-2013 *European Commission* (Brussels)
[5] Baars H and Kemper H-G 2008 Management support with structured and unstructured data—an integrated business intelligence framework *Inf. Syst. Manag.* vol **25** no **2** pp 132–48
[6] Bonney W 2013 Applicability of business intelligence in electronic health record *Procedia-Soc. Behav. Sci.* vol **73** pp 257–62
[7] Heeks R 2006 Health information systems: failure, success and improvisation *Int. J. Med. Inf.* vol **75** no **2** pp 125–37
[8] Inmon W H 2005 *Building the Data Warehouse* (John wiley & sons)
[9] Guerra J and Andrews D 2011 Why you need a data *Copyr. Andrews Consult. Group Inc* pp 1–8
[10] Chaudhuri S, Dayal U and Narasayya V 2011 An overview of business intelligence technology *Commun. ACM* vol **54** no **8** pp 88–98
[11] Chen H, Chiang R H, and Storey V C 2012 Business intelligence and analytics: From big data to big impact *MIS Q* vol **36** no 4

[12] De-Voe L and Neal K 2005 When business intelligence equals business value *Bus. Intell. J.* vol **10** no **3** p 57
[13] Power D J and Sharda R 2009 Decision support *Springer Handbook of Automation* (Springer) pp 1539–48
[14] Brannon N 2010 Business intelligence and e-discovery *Intellect. Prop. Technol. Law J.* vol **22** no **7** p 1
[15] McAfee A *et al* 2012 Big data: the management revolution *Harv Bus. Rev.* vol **90** no **10** pp 60–8
[16] Laudon K C *et al* 2011 Essentials of management information systems *Pearson Upper Saddle River*
[17] Grigori D, Casati F, Dayal U and Shan M-C 2001 Improving business process quality through exception understanding, prediction, and prevention *VLDB* vol **1** pp 159–68
[18] Gangadharan G R and Swami S N 2004 Business intelligence systems: design and implementation *26th Int. Conf. on Information Technol. Interfaces 2004* pp 139–44
[19] Kimball R and Caserta J 2011 *The Data Warehouse? ETL Toolkit: Practical Techniques for Extracting, Cleaning, Conforming, and Delivering Data* (John Wiley & Sons)
[20] Chaudhuri S and Dayal U 1997 An overview of data warehousing and OLAP *ACM Sigmod Rec.* vol **26** no **1** pp 65–74
[21] Aronson J E, Liang T-P and Turban E 2005 *Decision Support Systems and Intelligent Systems* (Pearson Prentice-Hall)
[22] Vercellis C 2011 *Business Intelligence: Data Mining and Optimization for Decision Making* (John Wiley & Sons)

Assessment of the effectiveness of learning theories using gamified android app in teaching C programming

J A Talingdan[1] and C R Llanda[2]

Abra State Institute of Science and Technology, Lagangilang, Abra, Philippines

[1]janelyn.ambre@gmail.com, [2]john_llanda@yahoo.com

Abstract. New knowledge of learners can be easily attained by applying different strategies in the classroom instructions. In the pedagogical domain, learning theories had been enforced and embedded to all instructional materials especially on gamification. Gamification is a concept of game-based approach in the teaching and learning process by offering challenges, tasks, rewards, badges and assessment. An android-based application called C-Rocks had been used in this study which comprises three learning theories: constructivism, behaviourism and instructivism. A pre-test was conducted to the fifty first year IT students who learned C programming in a traditional way of teaching to determine their prior knowledge. After that, the C-rocks was used by the students to facilitate the learning of C programming. A post-test was conducted to determine the post-knowledge gained after using the android app. The students were clustered using the k-means clustering algorithm of educational data mining to good, medium, and poor standard students. It was found out that most students excelled in variable topic while they performed least in the control statement topic. A T-test was also utilized and showed that the result of pre-test and post-test is highly significant with a t-value of 7.236 and a probability of 0.000. With the result established, the C-android app containing three learning theories is an effective Gamified tool in learning C programming as students can achieve greater learning.

Keywords – Learning theory, gamification, programming,

1. Introduction

Learning process requires of acquiring and modifying knowledge, strategies, skills, attitudes, beliefs and behaviours. Learners can gain variety of forms of knowledge such as cognitive, linguistic, motor, and social skills [1]. Learner's experience with the interaction to the environment or discovering scenarios of the world provides huge factor to the development change of human performance. This learning concepts covers the attributes commonly associated with behaviourism, instructivism, and constructivism [2]. Behaviourism is a learning process based on objectively observable changes in behaviour. The learning begins when a cue or stimulus from the environment is presented and the learner reacts to the stimulus with some type of response [3]. Learning through instructivism is the controlled display of text or graphical instructions for the user to follow [4]. Constructivism promotes active assimilation and accommodation of new information to existing cognitive structures. A discovery by learners [5].

With the increased availability of mobile devices, educational games have the potential to become a major tool for learning. One of the emerging fields is using gamification in education which is more fun and facilitates information keeping. Gamification means the use of game thinking, game elements, game techniques, game methodologies, and game framework in a non-game context in order to motivate users, solve problems, increase user experience, and encourage good desired behaviours [6]. Gamification increases level of engagement and motivation in classrooms which are directly needed by students [7]. There was a study that used gamification in teaching C programming language to evaluate learning effectiveness and engagement. The study was conducted using the mixed-method sequential explanatory protocol. A gamification framework is prepared to enable students to ask questions and learn subjects related to C programming language using basic elements of game mechanics. The results show that most students continued working even after earning the maximum amount of grade points to collect all the badges and to keep learning. The author mentioned that gamification is successful and engagement is a valuable indicator of students' academic achievements [8].

Teaching programming requires a systematic approach that involves a combination of activities like planning, designing, testing and debugging [9]. To learn on how to develop, student needs to understand the syntax of programming language. Programming is a technical course when if not properly handled, students will find frustrated and tend to be unmotivated due to its nature of complexity [10]. But with the aid of gamified application, learning programming is becoming more fun and engaging. There is the incorporation of different learning theories to make learning more comprehensive [11].

C rocks is an android app that provides C programming instructions, simulations, and tutorials of syntax, semantics and assessment. The gamified android application is incorporated with several pedagogical approaches to ensure the student's learning can be easily grasp on the idea of basic C programming [9]. The C-rocks android application comprises of several learning styles which aims to arouse the learning process of students. The gamified application is embedded with constructivism, behaviourism and instructivism learning styles. Constructivism is applied in the android apps by influencing the design and world appearance of the game. Drawing the theories of constructivism, the game offers challenges and tasks for each level which provides opportunities to the user to explore and make decisions about their learning and the content they want to find out. On the other hand, behaviourism was integrated by using cue or stimulus on which the user can react to the stimulus with some type of response. Learning through instructivism was applied by displaying text or graphical instructions for the user to follow. This approach is necessary for the progress of the student, especially as the game will be used by students with only simple programming knowledge being taught to them. However C Rocks does not contain all the knowledge about C programming, a feedback feature is given to encourage the user to look at other sources to further their knowledge.

With the learning theories integrated in C Rocks, the researchers prompted to conduct a study that assessed its effectiveness on the learning process of the students. The researchers administered a pre-test to assess first the prior knowledge of students who learned C programming in the traditional way of teaching in the first three topics namely, keywords, variables and conditional statements. Eventually, the C-Rocks android app was utilized by the students as a game learning tool in learning C programming. The students were engaged in exploring and discovering on how to operate and traversing every level of the app. After using the app, the researchers conducted a post-test to assess the gained knowledge using the gamified learning tool. Clustering method of educational data mining using k-means algorithm was applied to group the students to good, medium and poor classification of students. And, T test was used to determine the significant difference of pre-test and post-test.

2. Conceptual Framework

The overall process in evaluating the student's concepts in C programming using Android Gamified Application comprises with several steps with series of activities. Pre-test was used to determine the pre-knowledge of learner regarding C programming concepts on the three topics which are keywords, variables and conditional statement. A paper and pen assessment was conducted and it was performed by fifty first year IT students for thirty questions items in the three selected C programming topics.

To test if the learning theories applied in the development of C Rocks android app is debunk or not, the system logs were gathered. After using the C-Rocks game based learning, post-test was administered to evaluate the level of improvement on C programming skills on the three topics namely keywords, variables, and conditional statements. After which a T-Test instrument was used to determine the significant difference of pre-test result and post-test result. T-test outcome was the basis of evaluation on programming skills of first year IT students from its pre-knowledge in C programming to post-knowledge using C-rocks android app (see Figure 1.)

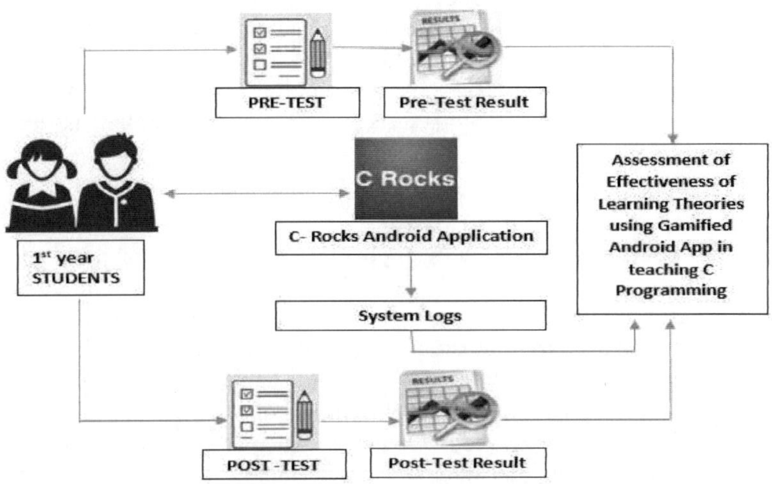

Figure 1. Conceptual Framework of the study

3. Methodology

3.1 Research Design

The study used the descriptive type of research wherein the researchers conducted a pre-test to test the prior knowledge of the BSIT students who learned C programming in traditional way of teaching. Post-test was administered after C-Rocks was introduced and used by the students to learn C programming. This is to assess the effectiveness of learning theories integrated in the android application. Likewise, progress reports logs were also gathered to determine the learning progress of students using the application.

3.2 Sources of Data

The fifty first year students enrolled in Bachelor of Science in Information Technology in the Abra State Institute of Sciences and Technology served as the sources of data in the study. C programming is an introductory programming course for BSIT program that must be enrolled in the second semester and for this, first year IT students were selected as respondents or learner to this research. On the other hand, the progress report logs of students in using the C Rocks android application were also recorded.

3.3 Data Instrumentation

For objective 1, a pre-test was conducted to test the pre-knowledge of students in C programming skills in the three topics namely the keywords, variable, and conditional statements. The test was composed of 30 items covering the three topics. Post-test was also administered to determine their post knowledge after letting the students used the C-Rocks application on their learning process.

The pre-assessment and post-assessment content comprises of three major topics namely keywords, variables and control statement. Each topic contains ten questions with a total of thirty items. Questions in the first topic emphasizes on keywords as an introductory part on learning C programming. It includes C language keywords such as void, int, printf, scanf, etc. After the introductory part is variable which contains int, char, float, etc. The third topic is control statement which contains if and if else statement.

For objective 2, a gamified android app called C-Rocks was utilized by the first year IT students to facilitate the learning process of students in C programming for three topics: keywords, variables, and conditional statements. The game-based android apps was developed through the incorporation of different learning theories such as behaviourism, instructivism and constructivism to make the application as a tool in learning C programming. The progress report logs were recorded such as how many attempts the students answered the teacher question and end level questions and the knowledge score the students acquired while playing the android app.

For objective 3, the results of pre-test and post-test were evaluated using T-test to determine the significant difference of knowledge level of students on their programming skills before and after using the Gamified android app.

3.4 Data Analysis

The clustering method of educational data mining applied in this study was the k-means algorithm. This algorithm was used to choose the best cluster center to be the centroid. It is one of the simplest unsupervised learning algorithms used for clustering. Initially the students are all in one group then the researchers segmented the students into three groups based on their rating for each topic in the post-test. The clustered groups were labeled as good, medium and poor standard students. Through this, it is easy to determine the topic that need improvement and the group of learners that need extra effort in the teaching process. In addition, the progress reports logs were analyzed using frequency count and mean. On the other hand, T-test using two tailed was used to determine the significant difference of pre-test and post-test results and the level of significant used was 0.01.

4. Results and Discussion

This context shows the result and discussion of the study. The 30 items exam that was given to the first year BSIT students was focused on the three topics of C programming such as keywords, variables and control statements. In the pre-test and post-test assessment, it was found out that post-test is higher with an average of 77.33 compared to the 70.08 average of the pre-test. The result implies that the students gained more knowledge after using the C- Rocks android application in learning C programming.

Furthermore, it can be inferred in figure 2 the k-means clustering centroid plot which describes the performance of the students during the assessment of their knowledge in C programming. The students were clustered into three based on their performance and categorized as poor standard students, medium standard learners, and good standard learners. Therefore students belong to Cluster 1 (green line) had mediate scores and they excelled more in keywords but they got a very low performance in control statement. This group is called medium standard students. Students who belong to cluster 2 (red line) had the best performance in all topics. Students belong to this group are called good standard students. On the other hand, students belong to cluster 0 (blue line) got the lowest score on the first two topics keywords and variables but they excelled in control statement. Students belong to this group can be labelled as poor standard students. The result implies that most of the students did not excel in the control statement topic this may be due of having complex syntax compared to the two topics. Furthermore, most students excelled in the variable topic this may be due of greater understanding about variables since they are already using Python which variables have almost same function and scope in C programming.

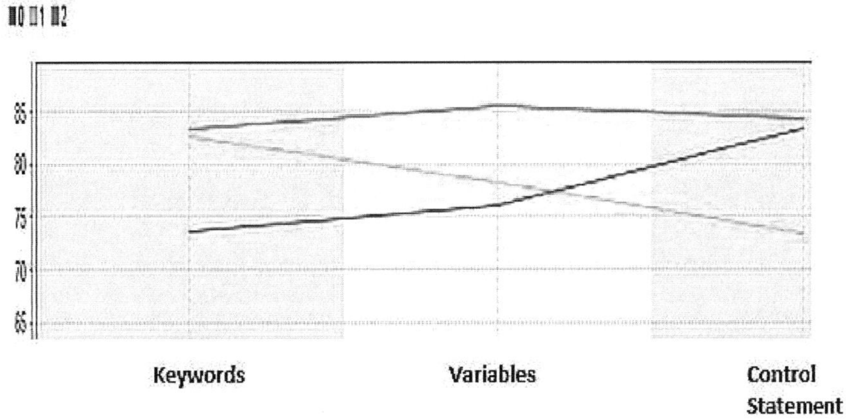

Figure 2. K - means Clustering Centroid Plot

Table 1 shows the average result of progress report logs of students in using C- Rocks application. In the Teacher Question (TQ), it can be inferred that students has lesser attempts in answering the teacher question in 2nd try with a general average of 1.43 compared to the average of 1st try which is 1.94. In the End Level Question (ELQ), it also shows that on the second try the student has lesser attempts with a general average of 1.34 compared to 1.84 general average of 1st try. Furthermore, the table shows that students has higher knowledge score on their second try in Knowledge Gained (KG) with a general average of 8.67 compared to 7.03 general average of 1st try. The result implies that students learned more and the learning theories constructivist, behaviourist and instructivist are effective as students' attempts of answering questions decreased, the knowledge score increased.

Table 1. Progress Report Logs

Level	T Q (1st try)	TQ (2nd try)	ELQ (1st try)	ELQ (2nd try)	KG (1st try)	KG (2nd try)
1	1.84	1.3	1.76	1.24	7.64	9.42
2	1.94	1.46	1.78	1.32	6.88	8.32
3	2.04	1.54	1.98	1.48	6.58	8.28
Gen. Avg	1.94	1.43	1.84	1.34	7.03	8.67

However, in the T-test result of pre-test and post-test, the derived T-test value was 7.236 with a probability of 0.000 therefore the null hypothesis is rejected and it is highly significant. The result implies that the learning theories applied in the development of C-Rocks android app has a great impact to students learning because they learned more when they used the app as their learning tool in learning C programming.

5. Conclusion

The intervention of C-rocks android app improves the knowledge of students in C programming as the t-test result revealed that it has significant difference. Likewise, on the progress report logs it attest that as the students used the android apps they are learning more as their attempts in answering teacher question and end level questions is decreasing while knowledge score is increasing. Therefore, the incorporation of learning theories such as behaviourism, constructivism and instructivism in the C-Rocks android application as tool in learning C-programming is effective as students can achieve greater learning.

References

[1] Cranton P 1994 *Understanding and Promoting Transformative Learning: A Guide for Educators of Adults* (Jossey-Bass Higher and Adult Education)
[2] Siemens G 2014 *Connectivism: A Learning Theory for The Digital Age*
[3] "Learning Theories", Available: http://www.thepeakperformancecenter.com/educational-learning/learning/theories/ Last Accessed: 27th February 2017
[4] Dybvig T 2010 *Learning Styles* Available: http://www.teresadybvig.com/learnsty.htm Last Accessed: 19th January 2017.
[5] Perks S 2014 Game based learning of C programming on the android platform: pedagogical approaches and learning styles *Proc. 13th School Conf. for Annual Research Projects*
[6] Duggal K, Srivastav A and Kaur S 2014 Gamified approach to database normalization *Int. J. of Comp. Applic.* vol **93** pp 47-53
[7] Elshiekh R and Butgerit L 2017 Using gamification to teach students programming concepts *OALibJ* vol **4** no **8**
[8] Ibáñez M-B, Di-Serio Á and Delgado-Kloos C 2014 gamification for engaging computer science students in learning activities: a case study *IEEE Transactions on Learning Technologies* vol **7** Issue **3** pp 291-301
[9] Tom M 2015 Five C Framework: A student-centered approach for teaching programming courses to students with diverse disciplinary background *J. of Learning Design* vol **8** no **1** p 23
[10] Salleh S M, Shukur Z and Judi H M 2013 Analysis of research in programming teaching tools: An initial review *Procedia-Social and Behavioral Sci.* vol **103** pp 127-35
[11] McKenna P and Laycock B 2004 Constructivist or instructivist: pedagogical concepts practically applied to a computer learning environment *SIGCSE Bull.* vol **36** no **3** pp 166-70

IOP Conf. Series: Materials Science and Engineering **482** (2019) 012031 doi:10.1088/1757-899X/482/1/012031

Apiculturists' issues and challenges: a basis for invoking IT model in beekeeping industry

S Kupsch[1], **T D Palaoag**[2], **A P Balcita**[1]

[1]Don Mariano Marcos Memorial State University, Philippines
[2]University of the Cordilleras, Gov. Pack Rd, Baguio, Benguet, Philippines

skupsch@dmmmsu.edu.ph, tdpalaoag@gmail.com, amypanedabalcita@gmail.com

Abstract. The most significant activity of honey bees, as far as advantages to people, is their pollination of natural vegetation. Bees and other pollinators seem to be declining globally. Beekeeping in the Philippines is a thriving industry that perfectly matches the natural landscape of the country. The goal of this study was to define the challenges and issues of Apiculturists in the Philippines where-in an IT approach was formulated to address these issues. Up to now, there is no initiative Information Technology approach in the beekeeping industry in the Philippines. Design thinking has been used as methodology in this study. Several challenges and issues were determined, and it is stated that the main problem to be addressed in order for the bee colonies to have a longer lifespan is its Apiary Location, an IT solution via Drone Technology with a Computer Vision approach was proposed named as Apiary Locator. A proposed framework of the Apiary Locator has been prepared. It is then recommended for the Apiary Locator to be developed as it shall benefit the apiculturists of the Philippines. Also, it will be a basis for developing IT solutions for the Beekeeping industry.

1. Introduction

One bite in in three of the sustenance you eat specifically or indirectly relies upon fertilization by honey bees. The most significant activity of honey bees, as far as advantages to people, is their pollination of natural vegetation. In calm atmospheres, most woodland trees are wind-pollinated, however numerous sorts of shrubs, little trees, and herbaceous plants, including numerous wildflowers, are honey bee pollinated. Protection of numerous habitats thus relies on the conservation of honey bee populaces, for if the honey bees vanish, proliferation of huge components of the vegetation might be seriously scarced [1].

The wonders of Honey Bees and its advantages to humanity but with its declining in production triggered the author to conduct this study so as to heighten the production as well as to save the bees. The study is also vertically aligned towards the Sustainable Development Goals of the United Nations specifically Goal 12: Responsible Consumption and Production [2].

Up to now, there is no initiative Information Technology approach in the beekeeping industry in the Philippines. This is the initial phase of creating a well-needed technology for the beekeeping industry in the country.

The goal of this study was to define the challenges and issues of Apiculturists in the Philippines where-in an IT approach shall be developed to address these challenges and issues.

2. Beekeeping Status Quo

Beekeeping is a well-paid trade even utilizing straightforward management strategies techniques however needs to think about nearby culture and economy for it to be fruitful. Beekeeping as an enterprise fits in exceptionally well with little scale ranchers' occupations. It isn't intrusive; honey bees work along the common examples of nearby agro-environmental zones and give beneficial outcomes to the fauna and verdure found inside [3].

Honey bees and different pollinators – both common and managed – appear to decrease all around, especially in North America and Europe. The nonattendance of energetic provincial or universal ventures expected to screen the present status and examples of pollinators suggests there is a noteworthy powerlessness in the scale and level of this decrease [4].

Concern over the pollinator declines followed by the sudden disappearance of honey bee colonies is well documented in North America and Europe. The dramatic decline in by and large pollinator populaces is a fundamental issue for agribusiness production [5].

"The Food and Agriculture Organisation of the United Nations (FAO) estimates that out of some 100 crop species which provide 90% of food worldwide, 71 of these are bee-pollinated. In Europe alone, 84% of the 264 crop species are animal pollinated, and 4 000 vegetable varieties exist thanks to pollination by bees." – UNEP, 2010 Beekeepers first sounded the caution about vanishing honey bees in the United States in 2006. Apparently solid honey bees were simply forsaking their hives as once huge mob, never to return [6]. The known misfortunes alone are striking. Honey bee colony mortality in Europe has arrived at the midpoint of around 20% (with an extensive variety of 1.8% to 53% between nations) [4].

Beekeeping in the Philippines is a thriving industry that perfectly matches the natural landscape of the country. All types of honey bees utilized for beekeeping exist in the Philippines. These are the European bumble bees (Apis mellifera), Asian bumble bees (Apis Cerana), and stingless honey bees (Trigona spp.).Queen quality, biological conditions, flower arrangement, sorts of innovation and asset administration are among the noteworthy elements influencing the benefit of beekeeping enterprises. In addition, colony quality, kinds of hives utilized, the age of the queen, swarming of colonies and bumble bee management hones are additionally central point affecting the productivity of beekeeping organizations. The two essential difficulties of beekeeping in the Philippines are the absence of nature of Apis mellifera and varroa bugs pervasion. Apis mellifera queens are transported in by thousands consistently from USA and Australia. Medications for varroa vermin are expensive and have a tendency to inoculate the parasites following a couple of years. The National Apiculture Research Training and Development Institute (NARTDI) together with the University of the Philippines Bee Program have just conceived an approach to limit, if not take out these difficulties by utilizing queen instrumental insemination innovation. Through this innovation, queens will be accessible all year in the nation. This innovation will likewise deliver honey bees impervious to varroa vermin invasion. In any case, the essential test is government's nonappearance of assistance for beekeeping progression contrasted with developed countries; their legislatures are at the front line of beekeeping attempts. The world supply of pure honey is declining. China, a noteworthy provider of bumble bee items, is restricted from trading to USA and other European nations since its honey bee items are debased with prohibited antibiotics. This would be an open door for the Philippines to fill the void caused by this ban. The Philippines is in the situation to pitch natural honey bee items to the world market.

Several challenges pose a threat to the industry such as climate change that severely affects the health of bees, entry of fake and adulterated honey, and competition from imported bee products. Further, the difficulty of beekeepers in the evaluation, status monitoring, communication, and inventory has affected the efficiency of honey production. The need to increase production requires training for more beekeepers and funding from the government. The magnitude for the beekeeping industry provides a tremendous market opportunity for beekeeping technology to be introduced, grow, and develop. In the Philippines, particularly in the region, no apiary uses this any kind of technology in their operation. Thus demand for such technology will be advantageous.

3. Methodology

The author used Design Thinking as an innovative approach in this study.

Establishments and associations must advance in response to the rivalry and quickly changing business sector demands. For this intention, Design Thinking is seen as an unfaltering methodology for an extent of business challenges that should be looked for after by both designers and nondesigners [7]. Figure 1 illustrates the Design Thinking process.

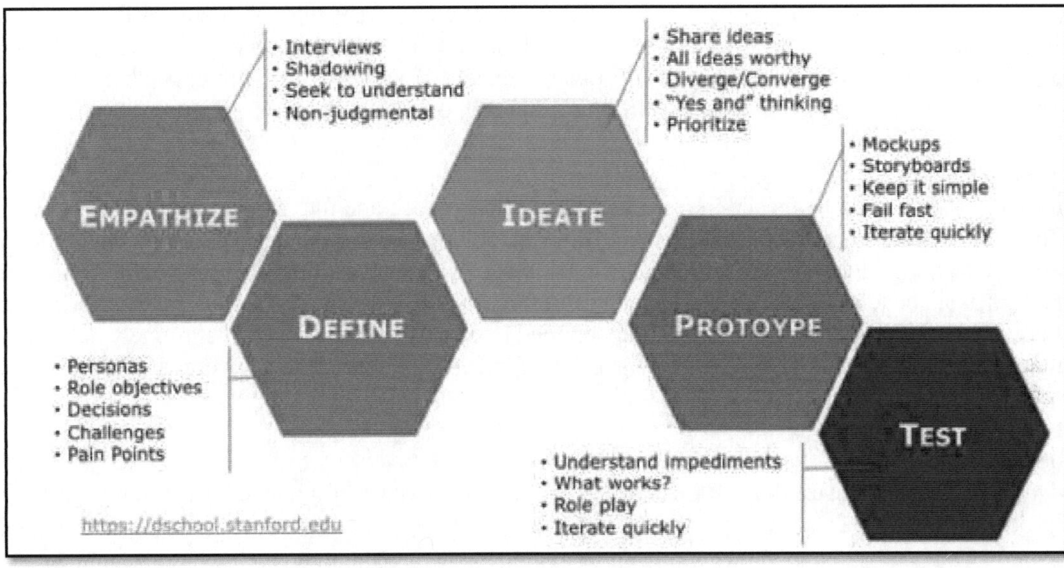

Figure 1. Design Thinking Process

3.1 Empathize.

The initial period of the Design Thinking process is to get an empathic cognizance of the issue you are endeavoring to determine. This incorporates advising experts to find more about the zone of worry through witnessing, drawing in and relating to understand their experiences and motivations, and furthermore submerging yourself in the physical condition to have a more significant individual appreciation of the issues included [8]. Empathy is the establishment of a human-focused outline process where you observe and connect with clients and submerge yourself to reveal their requirements [9].

Interview with various registered beekeepers and apiculturists has been conducted by the author at DMMMSU-NARTDI (Don Mariano Marcos Memorial State University - National Apiculture Research Training and Development Institute) [10].

3.2 Define.

Thru the Define stage, we set up together the data we have made and accumulated amid the Empathize stage. We will break down the perceptions and incorporate them so as to characterize the center issues that have been recognized as yet [8]. The define mode is the point at which you unload and incorporate your empathy findings into convincing needs and bits of knowledge, and degree a particular and vital challenge. It's critical to the design process since it unequivocally communicates the issue you are endeavoring to address through your endeavors [11].

Based on the interview conducted during the first stage of Design Thinking, a persona with the name Condoriano has been established. In this study, Condoriano represents the Beekeepers and Apiculturists in the Philippines. Condoriano's problems in beekeeping are listed as the difficulty of hive evaluation, the difficulty of hive status monitoring, the difficulty of communication between beekeepers, the difficulty of inventory, the difficulty of locating healthy habitat. As emphasized by Condoriano, his main problem is on how to locate a healthy habitat wherein he can deploy his apiary. The apiary must be at least 100 meters away from the forest where-in the bees would harvest their nectar. This study focuses on addressing that problem.

3.3 Ideate.
Ideate is the technique for your strategy in which you mean to create radical outline options. Objectively it addresses a strategy of "going wide" in regards to thoughts and results – it is a technique of "flaring" instead of “focus” [9].

Ideas were discussed with the registered beekeepers and apiculturists through focus-group. It is then conferred that the idea in using Drone Technology (Unmanned Aerial Systems - UAS) in scouting a land area is a more feasible solution to the problem hence it would give real-time results of the land survey. And with the data given by the UAS, this will produce a more accurate Apiary location.

3.4 Prototype.
Prototyping is getting thoughts and investigations out of your head and into the physical world [11].

Computer Vision will be adapted. Support vector machines or SVM is a broadly utilized method in supervised learning. Including a non-probabilistic binary linear classifier, SVM is appropriate for characterizing labeled images in computer vision. Additionally, as an essential image processing technique, adaptive thresholding is extensively applied to edge detection, which is extremely helpful in evaluating land textures. The prototype will be implemented in Python and leverages the OpenCV library [11].

3.5 Test.
Designers or evaluators carefully test the entire item utilizing the finest arrangements distinguished amid the prototyping stage. Last stage of the 5-stage model, in any case, in an iterative procedure, the outcomes produced amid the testing stage are frequently used to rethink at least one issues and advise the comprehension of the clients, the states of utilization, how individuals think, behave, and feel, and to empathise. Notwithstanding amid this stage, modifications and refinements are made so as to preclude issue solutions and determine as profound comprehension of the item and its clients as would be prudent [8].

4. Findings
After numerous interviews to beekeepers and apiculturists, the challenges and issues of Condoriano are summarized as follows. (1) Difficulty of hive evaluation. Condoriano still uses a paper-based method in recording and evaluating each hive. (2) Difficulty of hive status monitoring. Condoriano must visit the apiary and inspect each Hive too often wherein he is losing too much time. (3) Difficulty of communication between beekeepers. During dearth times, the drought of honey, honey or sugar should be supplied to the colony. When this happens, Condoriano should be able to contact the nearest beekeeper in order to ask for a colony rescue. (4) Difficulty of inventory. Where-in Condoriano is still using the traditional paper-based method in recording data. Written notes often get lost or become unreadable. (5) Difficulty of locating a healthy apiary habitat. The location is suitable for beekeeping if the vegetation provides ample sources of nectar and pollen. Also, the location should be at least 100 meters away from the forest. Condoriano is having a hard time in evaluating areas all by himself because he should scout the area manually. Another, there are areas for evaluation that Condoriano cannot reach via human-foot like mountain areas. Condoriano’s problems in beekeeping are listed in Figure 2.

Condoriano has given emphasis that his main problem is number 5, difficulty of locating a good apiary location. For, an excellent location is one in which forage is abundant throughout the year. Also, high areas are hard to reach, roads to an area may also be rough and may not be accessible. Hence, a proposed IT solution that may be seen in Figure 3 has been prepared.

Phase 1 would let Condoriano capture a real-time image of the current condition of the target location. While on Phase two, the image will be evaluated according to the characteristics of a healthy apiary location, e.g. sources of nectar and pollen. Lastly, findings will display the location that is best suited for the Apiary.

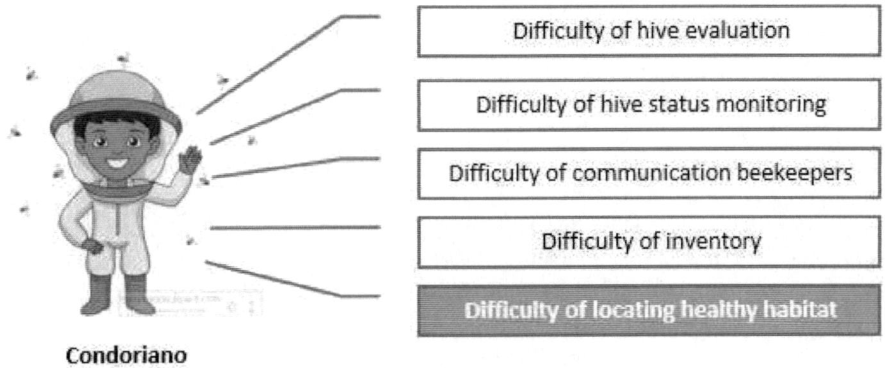

Figure 2. Persona of a beekeeper

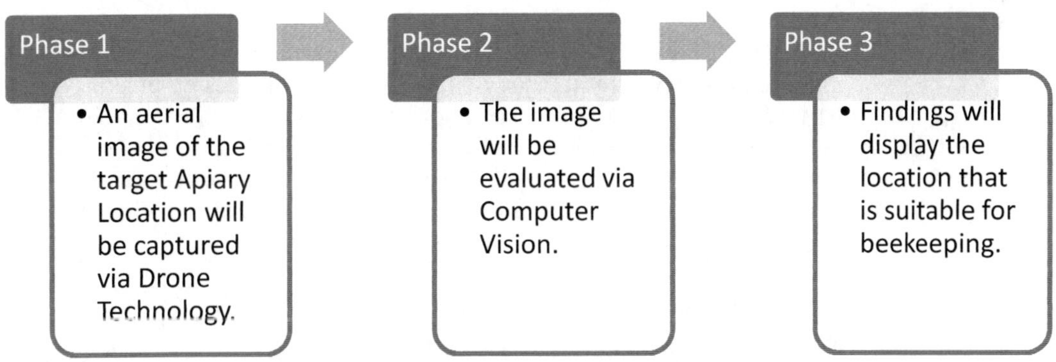

Figure 3. Proposed Framework of the Apiary Locator.

5. Conclusions and Recommendations.

As a result of this study, several challenges and issues were determined as Apiculturists has difficulties in hive evaluation, hive status monitoring, communication between apiculturists, inventory and locating a healthy apiary habitat.

It is mentioned that the main problem to be addressed in order for the bee colonies to have a longer lifespan is its Apiary Location. An excellent location is one in which forage is abundant throughout the year.

In order to address that main problem, a solution was formulated. It was then proposed to develop an IT Solution via Drone Technology with a Computer Vision approach. An Apiary Locator.

Thus, it is then recommended for the Apiary Locator to be developed as it will benefit the apiculturists of the Philippines. Also, it will be a basis for developing IT solutions for the Beekeeping industry further.

6. Future Works

In this study, using the Design Thinking methodology, the authors have reached the Ideate stage. Future researchers may use the proposed framework that can be seen in Figure 3 to address the main problem of Condoriano. The authors expect to complete the stages of the mentioned methodology in the continuance of this research since the primary purpose of this study is to provide an IT framework. Future researchers may continue through Prototype stage and Test stage based on this study.

As seen in Figure 2, there are other difficulties that Condoriano is facing which needs to have a solution. From this, future researchers may take this study to a broader benefit for Condoriano.

References

[1] Michener C D 2007 *The Bees of the World* (Baltimore: Johns Hopkins University Press)
[2] The United Nations (n.d.) *Sustainable Consumption and Production* retrieved February 21, 2018, from http://www.un.org/sustainabledevelopment/sustainable-consumption-production/
[3] Hilmi M, Bradbear N and Mejia D 2011 *Beekeeping and Sustainable Livelihoods* (Rome: Rural Infrastructure and Agro-Industries Division, Food and Agriculture Organization of the United Nations)
[4] Tirado R, Simon G and Johnston P 2013 *Bees in Decline: A Review of Factors That Put Pollinators and Agriculture in Europe at Risk* (Amsterdam: Greenpeace International)
[5] Pannure A 2016 Bee pollinators decline: perspectives from India *Int. Res. J. of Natural and Applied Sci. (IRJNAS)* vol 3 Issue 5 ISSN: (2349 - 4077)
[6] Natural Resources Defense Council 2011 *Why We Need Bees: Nature's Tiny Workers Put Food on Our Tables* retrieved February 28, 2018 from https://www.nrdc.org/sites/default/files/bees.pdf
[7] Seidel V P and Fixson S K 2013 Adopting design thinking in novice multidisciplinary teams: the application and limits of design methods and reflexive practices *J. Prod. Innov. Manage* vol 30 no 1 pp 19–33
[8] Dam R and Siang T 2018 5 *Stages in the Design Thinking Process* retrieved March 23, 2018 from https://www.interaction-design.org/literature/article/5-stages-in-the-design-thinking-process
[9] Terrar D 2018 *What is Design Thinking?* retrieved March 23, 2018 from https://www.enterpriseirregulars.com/125085/what-is-design-thinking/
[10] Anonymous 2018 __ retrieved March 23, 2018 from http://nartdi.dmmmsu.edu.ph/
[11] OpenCV 2016 Open Source Computer *Vision Library* http://opencv.org/about.html

IOP Conf. Series: Materials Science and Engineering **482** (2019) 012032 doi:10.1088/1757-899X/482/1/012032

Workspace layouts for communication quality in Scrum teams

Kholid Haryono[1], Za Idatin Nikmah[2]

Informatics Department, Faculty of Industrial Technology, Universitas Islam Indonesia, Jl. Kaliurang KM. 14.5 Sleman Yogyakarta Indonesia
[1]Corresponding author: kholid.haryono@uii.ac.id

Abstract. Purpose of study to show the influence of workspace design on the quality of communication, especially on the Scrum team. The motivation is to find out the effect of changes in layout of the workspace on the quality of communication. Communication quality is an important concern about the performance of software development in the Scrum team. Problem: How does the layout change affect the Scrum team's workspace on the quality of communication between members on a team; team members with other teams, and differences in the effect of changing layouts with old layouts. Methodology: Research steps use Participatory Action Research (PAR). This method has four main steps, namely planning; acting; observing; and reflecting. Data retrieval uses survey method by distributing questionnaires, conducting daily observations, and Forum Group Discussion (FGD). Findings: The Development Team Cell (DTC) workspace model is better than the Conference Cell (CC) model. This finding strengthens the previous research conducted by Scott and Pawel Role. Conclusion: There are three conclusions. a) changes in layout can improve the quality of individual communication. b) changes in layout have a positive effect on the quality of communication between team members and outside parties. c) changes in layout have a positive effect on team members' understanding of the work and methods used.

1. Introduction

Effective communication is one of the keys to success on the Scrum team [1][2]. Scrum is a product-oriented project management approach. This approach emphasizes product release rather than development documents [3], so that communication between members in the team will be more intensive than communication with project planning documents. Many project failures are caused by poor quality communication. The most frequent failure is due to incompatibility between requirements and results. In addition, the software delivered is outdated, and it is not in accordance with the needs of stakeholders.

Scott W Ambler, in 2002-2005 explained about the most effective communication model, namely face to face with whiteboard, face to face conversation, video conversation, and phone conversation [1]. These communication models can be applied one way by redesigning the development team's work space. The principles expressed by Scott are important considerations in developing them. Workspace design for this team can improve the communication quality of team members both internally and externally [4]. Workspace barriers that are set improperly can hinder team performance. Performance that does not match expectations both from results and processes has become a clear consequence of communication barriers in the team.

Purpose of this study will show the influence of workspace design on the quality of communication, especially on the Scrum team. Case studies were taken from the journey of the Information Systems Agency (BSI) at a higher education institution in Indonesia. BSI is the unit responsible for the development and service of information technology and systems. They are overhauling the technology to adjust to current needs. As a unit that is trying to rebuild its information system after the old system is declared out of date, the number of systems to be rebuilt, and the large number of teams, BSI adopts the Scrum approach. This approach was chosen because it has been proven to be able to develop software in an agile and effective manner [5][6][7][3].

One of the problems encountered in the initial stages is the difficulty in integrating the work of each team members. From the seven teams that were formed often found miscommunication and had not cooperated well. Even each members of team also could not cooperate with other members, so that they complete their own work. This study will examine the results of research conducted by Pawel Rola about workspace that can improve the quality of communication in teams by implementing the design at BSI. The contribution of this research provides more important lessons and findings of information that are useful for business people, especially organizations that use the Scrum approach.

2. Previous research

There have been several previous studies related to work space and communication quality on development teams that use the Agile (Scrum) methodology. Some of them are:

Pawel Rola (2016) examines the concept of a workspace model for the Scrum team. The purpose of this study is to provide support to the project team so that they are able to complete the mission effectively. The results of this study indicate that changes in workspace design have a positive effect on the performance of the Scrum team. There are at least three effects achieved, namely (a) a good workspace can maintain the team to continue to improve work in accordance with the Scrum framework; (b) change workspace to improve team performance; and (c) the attention of team members increases especially to leaders, Agile trainers, and co-workers who work in teams [4].

Betteke Van Ruler (2014) examined the effectiveness of Scrum team communication in the public sector. The case study was carried out in the Netherlands. The results of the study indicate that the involvement of communication theory is still needed to support effective communication, especially with external parties [2]. Nils BredeMoe, et al (2010) examined the team model to improve understanding of the Agile (Scrum) method. The aim is to provide an understanding of the agility of small teams in managing themselves. In addition, it also shows the challenges faced by the team in the Scrum project. They find problems in team orientation, team leadership, and coordination. Furthermore, highly specialized skills and appropriate job sharing are important obstacles to achieving team effectiveness. [6].

Based on these studies, the position of this study will take the results of Pawel Rola (2016) about the layout he offered. The design of the recommendations will be implemented at BSI. The effect will be measured by asking feedback from team members regarding the changes. At BSI, testing was carried out at the development unit. The method used is Participatory Action Research (PAR) with college case studies. This is important because the cultural complexity in universities also influences team performance. The involvement of the team and all stakeholders in designing work, communication, and achieving development goals requires good collaboration as required by the Agile Manifesto [3].

3. Research Methodology

This study will test the quality of communication from the workspace design team implementation using the Scrum approach. Measurements are made before and after design changes. Before the change in layout will be revealed the initial conditions, both physical layout and communication conditions of team members. This communication includes internal, communication with other team members, and with external parties. The implementation stages of workspace change use the Participatory Action Research (PAR) approach [8]. There are four steps in PAR approach : Planning, Acting, Observing, and Reflecting. All of step will discuss in the bellow.

Analysis of the initial conditions and implementation of PAR was carried out with three main methods, namely Forum Group Discussion (FGD), observation, and questionnaire. The FGD was conducted by five people, namely three field heads and two development team managers. Observation is used to observe teamwork by sitting together when the team activities while recording the events that caused miscommunication. While the questionnaire to get the team members' responses about two things. First, does the design change have a positive effect on improving the quality of communication? and second what influences the quality of communication in addition to workspace design.

3.1. Initial Conditions
This section shows the initial conditions of BSI before doing the workspace design and adopting the Scrum approach. This is important to provide convenience in understanding the results of this study. Initial conditions include initial layout design, team managerial approach used, and team conditions.

The initial condition of workspace has three forms. First, the closed cubicle cell (CCC) model. This model provides more privacy to team members. Second, open cubicle cell (OCC). This model is used by teams from third parties on duty at BSI. They are students who are in the final project, contract employees, and apprentices. Third, the conference cell (CC) model with a table in the middle and participants in a circular shape. These workspace forms are shown in Figure 1

A. CCC

B. OCC

C. CC

Figure 1. Initial condition workspace

Management approaches are still conventional. Organizations use command or structural systems with clear job descriptions. Structurally there are five divisions. Namely, the planning division, development division, operational division, service division, and the Administration and Project Management (AMP) division. The total number of members is 42 people. The division with the most members is the development division which is 18 people and the operational division is 15 people. They consist of engineers (develompment, DB, QA, Devop, Network, and Security), Customer Support, and administration. One characteristic of a conventional approach is giving a clear job description to team members. This can close the creativity of each member, because they are working on something as assigned.

This research we focused for team who involved in the development process directly. Therefore, total of members who we got to be respondent from all members of development and part of members in the operation division. Total respondents are 18 from members of development division and others are members of operation division dan service division. So, total respondent are 21-23.

In early 2017, BSI adopted the Agile approach, Scrum. This method has proven effective in organizing employees, especially in the software development environment [7]. Among the steps in this approach is to build a smaller team, consisting of 5-9 people per team [3]. The formation of the team was successfully carried out with the composition in the development division and operating division with the most members. The development division was formed into 4 teams and operational divisions were formed by 3 teams. In fact, communication is still not optimal. One reason is the position of the workplace that uses closed cubicles so that privacy still dominates each member. Initial attempts to put the team in the meeting room, the results were good enough even though there were still some miscommunications in several places.

3.2. Participatory Action Research
This stage implements the workspace design to see the effect on the quality of communication for each member. The approach taken is Participatory Action Research (PAR) [8]. Workspace implementation starts from planning to reflecting results taken from the conditions after implementation. The steps include four stages: Planning → Acting → Observing → Reflecting.

3.2.1. Planning.

This stage is a crucial part because it will affect the validity and quality of research. At this stage includes several steps. First, look for an ideal workspace design for the Scrum team based on previous research. Second, look for workspace designs from best practices that have been used by organizations that apply the same approach.

Based on literature study and review of previous research, researchers found articles published by Pawel Rola [4] about workspace designs that are most suitable to be implemented. At least inspired by the five main forms of work space, namely conference cell, social kitchen cell, chill out cell, development team cell, and product owner cell.

Workspace design, especially seating composition, is a concern for the Scrum team. The need to improve team communication is to bring distance between members, even though they have to sit back and forth. This is shown in the design of the Team Development Cell workspace. The product owner's space is close to the team for intensive inspection needs. Social cells are useful for teams to brainstorm and gather ideas. Conference cells can be used for releases and meetings with stakeholders.

Workspace reviews from the best practices that used by startup companies and also using the agile-Scrum approach include Kudo Technology Indonesia. The main form of workspace is made together in a large space with a table and chair arrangement like a conference room. The principle of laptops were facing outwards with chairs close together as a way of bringing communication closer between team members and the ease of management in carrying out inspections of the course of teamwork. Stay supported with chill space, social room, and even a refreshing place and ping pong table. A similar workspace design is used by Agate Studio. This company is one of the game development companies in Indonesia located in Bandung.

3.2.2. Acting.

The workspace model that supports the quality of good communication in the Scrum team in the planning phase is Development Team Cell (DTC). In this acting phase. The focus at this stage is to change the form of two existing conference cells to DTCs. The first step is to ensure that all teams have space permanently because it will be used to store all notes on the blackboard and note board. The space provided is 4 spaces for four teams. Called teams 1, 2, 3, and 4. After implementation of space, the form of workspace used is shown in Figure 2.

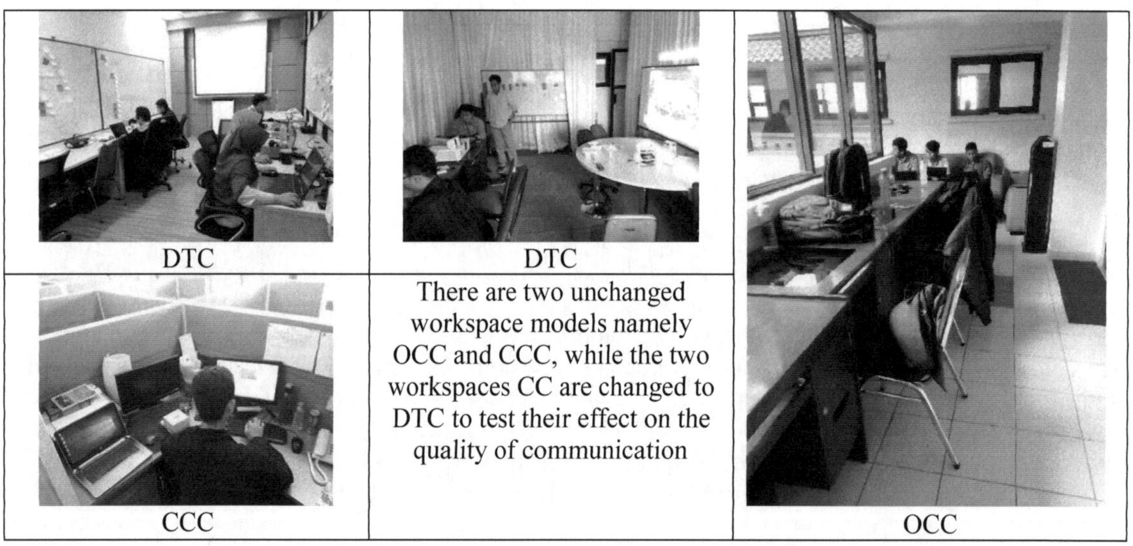

Figure 2. Workspace design after changes

In principle, the layout of seats is physically close to a position that facilitates team members in communicating with each member. The member's computer screen is facing out to facilitate discussion of various issues.

3.2.3. Observe.
This stage is used to retrieve data from workspace changes. Some of the activities carried out were: conducting daily observations, holding meetings (FGD) with management at the operational level held every week, and distributing questionnaires to team members to confirm what they felt after the implementation of the layout design.

Daily observations focus on several behaviors. That is, a) communication behavior of members in one team; b) mobility / movement of people to approach each other because of communication needs; and c) ease of communication between members of a team and other team members. Using the role of the product owner and Scrum master coordinator, this observation shows positive results. Among the space design models that produce the most effective behavior in communication is the model development team. This model allows the mobility of someone who is quite agile, especially supported by a blackboard as a means of communication.

The FGD was conducted with two division heads namely the planning division and development division and involved two managers. The FGD was conducted in conjunction with regular meetings conducted by operational management at BSI. This meeting is usually used for evaluating information system development process. Some notes summarized from several FGDs regarding the implementation of workspace changes include: a) workspace changes can improve the quality of communication between members in one team and between teams; b) the new layout of workspace has provided convenience in carrying out Scrum events such as sprint planning and daily Scrum; c) there are still some problems in communication but not related to the workspace, first, the problem of maturity adoption of the young Scrum approach. second, changes in team members are too fast. During the implementation of workspace changes, team members have changed twice. In fact, this change causes the quality of communication between members to decline.

The feedback from the team was carried out through questionnaires. Team members were asked to provide an assessment of a layout they had felt before and after the change. There are four layout models assessed, namely: conference cell (CC); development team cell (DTC); open cubicle cell (OCC); and covered cubicle cell (CCC). Each form of layout was submitted 16 questions. The weighting of each statement uses a Likert scale that is between 1 and 5. This method is also used in the CMMI method [9]. High scores indicate more significant. The results of the assessment are shown in Table 1.

Table 1. Results of respondents' assessment

No	Question Is ...?	CC		DTC		OCC		CCC		Max Score
		R	S	R	S	R	S	R	S	
Q1	The Layouts make communication between team members effective	23	4.39	23	3.87	22	2.36	22	1.64	4.39
Q2	The layout supports mobility of team members	23	3.65	23	4.22	22	2.91	22	1.68	4.22
Q3	The availability of the blackboard and its position	23	4.30	23	3.87	22	1.95	22	1.45	4.30
Q4	The position of the LCD screen is very appropriate for coordination	23	4.52	23	3.96	22	1.36	22	1.32	4.52
Q5	The layouts support communication with PO, SM, and Stakeholders	23	4.04	23	3.39	21	2.10	22	1.45	4.04
Q6	The layout supports communication with other teams	23	3.43	23	3.61	22	2.27	22	1.50	3.61
Q7	Seating position is appropriate	23	3.65	23	3.57	22	2.36	22	2.05	3.65
Q8	The availability of the road to pass through is adequate	23	2.74	23	4.17	22	2.91	22	2.36	4.17
Q9	The space area for the daily Scrum is fulfilled	23	3.17	23	4.00	22	2.23	22	1.59	4.00
Q10	The space area for review and retrospection is sufficient	23	3.74	23	3.78	22	1.95	22	1.55	3.78
Q11	The layout supports videoconferencing activities	23	4.17	23	3.65	22	2.18	22	2.41	4.17
Q12	The layout supports telephone conversation	23	3.70	23	3.61	22	2.23	22	2.77	3.70
Q13	The layout supports online/chat conversation	23	4.00	23	3.70	22	2.64	22	2.86	4.00
Q14	The layout supports overview document	23	4.04	23	3.61	22	2.18	22	2.09	4.04
Q15	The layout supports video recording	23	3.65	23	3.57	22	2.36	22	2.59	3.65
Q16	The layout supports audio recording	23	3.70	23	3.57	22	2.27	22	2.73	3.70

Some of the questions in Table 1 are taken from Scott Ambler, owner of Agile Modeling [1]. In each column the layout model (CC, DTC, OCC, and CCC) consists of two columns. Column R (Respondent) is the number of respondents who fill. Column S (Score) is an assessment score. The score is calculated by multiplying the number of respondents and the weight of the score divided by the number of questions so the results are in the range of 1 to 5. For example, the data in the CC layout model contains 23 respondents with a score of 4.39, numerical calculations are shown in Table 2.

Table 2. Calculation of scores for one of the questions in the Conference Cell model

No	High numbers are increasingly significant	Conference Cell					Number of Respondents' Answers
	Significance weight	1	2	3	4	5	
1	The Layouts make communication between team members effective	0	0	2	10	11	23
	Total number of answers * weight	0	0	6	40	55	101
	Score = (number of answers * weight) / total answers						4.39

Calculation of scores for all items in each layout model in Table 1 uses the same method.

3.2.4. Reflecting.

The activity of this section focuses on the reflection of each individual team member; reflection of personal relationships with team members; and structural reflection which includes an individual's understanding of the subject under study [8]. All three were analyzed between before and after the change in layout so that conclusions could be drawn. Table 1 column of the CC model is the layout before the change, while the DTC model after the change. Other models, the CCC and OCC models are not discussed because of the old layout and only show that the two models are not good for Scrum team communication. This is indicated by the small score of the two models.

First, reflection is felt and observed in each individual. Table 1 shows Q2 (seat mobility), Q3 (screen or whiteboard position with seating), and Q7 (seat position). CC scores are greater than DTC in Q3 and Q7 with a difference in scores of 0.43 and 0.09, while the Q2 is greater in DTC with a difference in the score higher than 0.57. This indicates that the influence of individuals on the quality of communication has no effect before and after change. It is only 0.01 bigger than DTC.

Second, a reflection of the relationship between members is shown in Q1 (communication between team members), Q6 (communication with other teams), and Q7 (availability of space for roads). CC score is greater than DTC in Q1 with a difference in score of 0.52, while Q6 and Q7 are greater in DTC with differences in scores of 0.17 and 1.43. This indicates that the effect of the layout on communication between members and between teams has a positive effect with a difference of 0.36 greater DTC.

Third, structural reflection of individual understanding of layout is shown in Q4 (communication to LCD position), Q5 (supports communication with PO, SM, and stakeholders), Q9 (space area for daily Scrum), and Q10 (space area for review). CC scores are greater than DTC in Q4 and Q5 with differences in scores of 0.57 and 0.65, respectively. Whereas Q9 and Q10 are larger DTC with a difference in scores of 0.83 and 0.04. This indicates that the effect of layouts on members' understanding of workspace changes is better before being changed by a total difference of 0.09.

4. Finding

Findings from testing the effect of the implementation of workspace design in improving communication of the Scrum team stated that the CC and DTC models occupy the highest position. This reinforces the publication by Scott [1] This reinforcement the publication by Scott that face to face with a whiteboard is the most effective way of communication. DTC in this study previously came from the design of the conference model workspace. The findings that can be drawn from the analysis of the field data are shown in Figure 3.

DTC as a space that was changed from the CC model and tested in this study shows a positive influence on the quality of communication on the Scrum team. At least the influence occurs in Q2, Q6, Q8, Q9, and Q10. These five attributes are the core of the Scrum team's main activity so that the DTC layout model can be recommended as an ideal layout to improve the communication of Scrum team members.

5. Conclusion

After testing the changes in the workspace model on the Scrum team and observing the effect on the quality of the communication that occurred, several notes can be drawn that conclude in this paper. a) changes in layout can improve the quality of individual communication. b) changes in layout have a positive effect on the quality of communication between team members and outside parties. These parties include PO, SM, and stakeholders, and c) changes in layout have a positive effect on team members' understanding of the work and methods used.

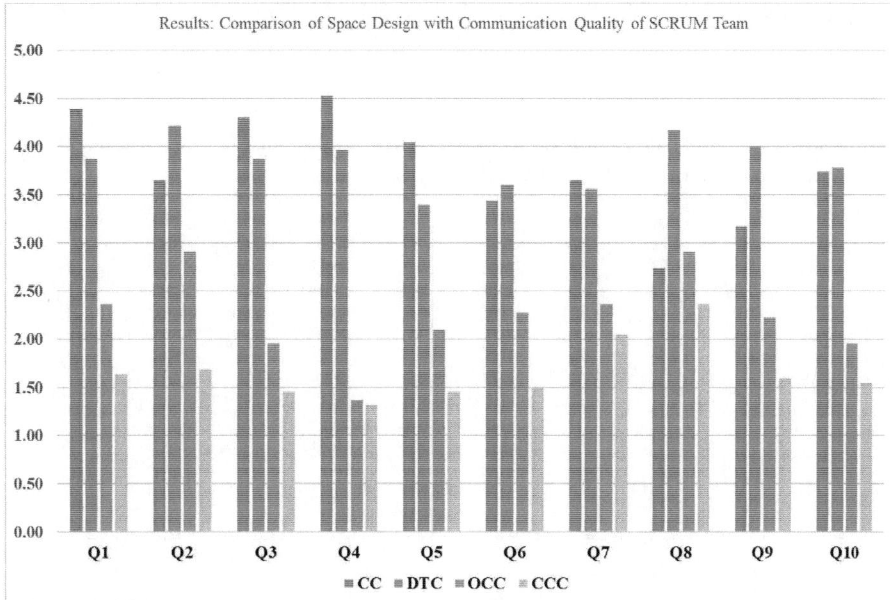

Figure 3. Comparison of Space Design with Communication Quality of Scrum team

6. Future Work

Research opportunities related to effective communication in Scrum implementation are still broad. Especially about workspace and layout for management team in Scrum approach. Future research can attempt to propose about effective collaboration in the team Scrum related workspace. This issue is significant which collaboration be more important to achieve development successfully.

In addition to workspace, an important issue that can be developed is the communication design that is influenced by the type of leader, the adoption of the method that is half-way (as best as possible), the change of the team in the middle of the road, and the main psychological aspects of team happiness. Happiness is an important issue because it affects performance and productivity. Not all happiness is judged from the material. It can also be seen from the point of view: feeling safe, comfortable, and positioned as an important person in the team.

References

[1] Ambler S 2011 Communication on agile software teams *Agile Modeling [Online]* Available: http://agilemodeling.com/essays/communication.htm.
[2] Van Ruler B 2015 Public relations review agile public relations planning: the reflective communication Scrum *Public Relat. Rev.* vol **41** no **2** pp 187–94
[3] Sutherland J 2014 *SCRUM: The Art of Doing Twice the Work in Half the Time* (Crown Business)
[4] Rola P, Kuchta D and Kopczyk D 2016 Conceptual model of working space for Agile (Scrum) project team *J. Syst. Softw.* vol **118** pp 49–63
[5] Martin R C 2002 *Agile Software Development Principles, Patterns, and Practices* (New Jersey: Prentice Hall)
[6] Moe N B, Dingsoyr T and Dyba T 2010 A teamwork model for understanding an agile team: A case study of a Scrum project *Information and Soft. Technol.* vol **52** no **5** pp 480–91
[7] Partogi J 2015 *Manajemen Modern dengan SCRUM* (Yogyakarta: Andi Offset)
[8] Bergold J and Thomas S Participatory research methods: a methodological approach in motion *Forum Qual. Soc. Res.* vol **13** no **1** p 30

[9] White S K 2018 What is CMMI? A model for optimizing development processes *CIO From IDG [Online]* available: https://www.cio.com/article/2437864/developer/process-improvement-capability-maturity-model-integration-cmmi-definition-and-solutions.html.

Acknowledgments
Our thanks to Informatics Department of Universitas Islam Indonesia for allowing us to use facilities, academic atmosphere, and funding support in this study.

Topic modeling of weather and climate condition on twitter using latent dirichlet allocation (LDA)

Ahmad Fathan Hidayatullah[1], Silfa Kurnia Aditya[2], Karimah[3], Syifa Tri Gardini[4]

Department of Informatics, Universitas Islam Indonesia
Jalan Kaliurang KM 14,5 Sleman Yogyakarta Indonesia

[1]fathan@uii.ac.id, [2]silfa.kurnia.aditya@gmail.com, [3]rima.alkatiri@gmail.com, [4]syifatrigardini@gmail.com

Abstract. This study aims to apply topic modeling approach using LDA for Twitter dataset shared by the official Twitter account of BMKG in Java Island. The topic model result can be seen as the representation about what kind of information that posted by BMKG through Twitter. In addition, it can also illustrate the weather, climate, and disaster trends that occurred in Java Island. Based on the topic modeling result, we found five notable topics from BMKG's Twitter accounts. As for the topics discussed, BMKG's Twitter accounts disseminate the information about weather Information and weather forecast; latest weather information in Yogyakarta Region; weather forecast and warning in Central Java and West Java; earthquake information; latest articles and calendar cropping information. In this study, we also illustrated the trends about weather and disaster by analyzing the most frequent words from each region in Java.

1. Introduction

With the emergence of social media, the dissemination of public information is spreading rapidly. Social media that provides flexibility and simplicity, makes people easier to obtain and share information with others. Therefore, social media is a powerful tool that provides accuracy and immediacy of information happening in real time [1]. Twitter is a good example of social media to spread vital and valid information that currently happening. A lot of organizations around the world, including government organizations in Indonesia, utilize Twitter as one of the media to disseminate important information to audiences [2].

Along with the use of social media among the government agencies in Indonesia, there is an agency called BMKG (Badan Meteorologi Klimatologi dan Geofisika) or the Indonesian Agency for Meteorology, Climatology and Geophysics. BMKG has utilized Twitter to inform the current climate condition and natural disaster to the community. BMKG is one of the government agencies that responsible to monitor and also disseminate the information about climate and natural disaster in Indonesia. The information about weather, climate, and disaster are the most important report to be shared as a warning or a reminder to the public. That information plays an important role in various fields such as transportation safety, flight activity, and disaster information. With the availability of accurate information about weather, climate, and disaster, unwanted events can be anticipated earlier and faster.

This study aims to apply topic modeling approach using LDA for Twitter dataset shared by the official Twitter account of BMKG in Java Island. The topic model result can be seen as the illustration of what information provided by BMKG through Twitter. Furthermore, the result can also represent the weather, climate, and disaster trends that occurred in Java Island.

The remainder of this paper is organized into the following structure. Section 2 describes the related work. Section 3 presents the research methodology. In section 4, the results and discussions are explained. Finally, section 5 describes the conclusion of our research.

2. Related Work

2.1. Latent Dirichlet Allocation
Topic modeling is an unsupervised machine learning method that applies clustering to discover latent variables from a large amount of text data. The most popular method for topic modeling task is Latent Dirichlet Allocation (LDA) which introduced by Blei and Jordan (2003) [3]. LDA is explained as a generative probabilistic model for finding the semantic structure of a corpus based on hierarchical Bayesian analysis of the texts [4]. LDA represents a collection of documents as blend topics which contain words with certain probabilities. Figure 1 illustrates the flow of LDA algorithm.

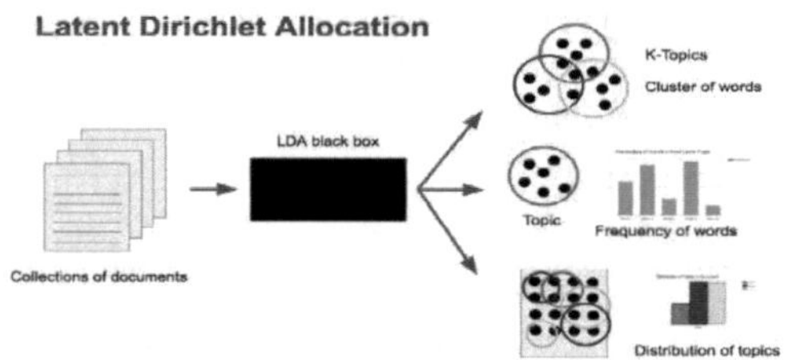

Figure 1. The flow of LDA
(Retrieved from: https://toolbox.kurio.co.id/topic-modeling-696d7ba2592f)

The procedure how LDA works is explained as follows:
1 Initialize some parameters, including the number of documents, the number of topics, and the number of iterations. In LDA, the most important parameter is the number of topics k.
2 Assign each word to a particular topic randomly according to a dirichlet distribution.
3 Repeat the steps for all words in the corpus.

2.2. Topic Modeling using LDA
Since the emergence of topic models, researchers have introduced this approach into the various fields by applying Latent Dirichlet Allocation (LDA) method to build the topic models. LDA has been applied to illustrate the topic regarding traffic information in Java that posted by the Traffic Management Center via Twitter [2]. Wang, et al. (2014) applied topic modeling using LDA on health topic to detect health-related topics discussed in Chinese social media [5]. Liu, et al. (2016) studied the application of topic modeling using LDA and PLSA (Probabilistic Latent Semantic Analysis) in the bio-informatics domains [6]. Neill, et al. (2017) focused on the use of topic model for legislative texts in Britain that could assist in easier tracking of important domain terms that correspond to compliance related issues [7]. LDA has applied to help the historians to identify the most important and interesting topics in newspaper during a specific time period [8].

3. Research Methodology

3.1. Data Retrieval
The data set was retrieved from the official Twitter account of BMKG in Java Island. In this research, we did not collect tweet from all official Twitter account of BMKG because there are several BMKG Twitter accounts that no longer active post the tweets. Therefore, we only grab tweets from the official Twitter account of BMKG that still active to post the tweets until April 2018. The tweet was collected using GET status/user_timeline method on Twitter API v1.1. The user_timeline method provided by Twitter, can reach until 3200 tweets from an account. We also use an open source library in Python

called tweepy to access the Twitter API. Total 26,420 tweets were collected from total 10 official Twitter account of BMKG in Java Island.

3.2. Data Pre-Processing

Pre-processing Twitter data is necessary due to the unstructured text in the tweets [2]. Twitter texts also have some unique characteristics such as character limitation (no more than 140 characters), hashtags (#), RT (retweet), and @username. Therefore, those characters will be removed from the tweets because they do not have any impact to the desired result. Overall, the pre-processing steps in this study follow the previous research [9]. There are some tasks that carried out in pre-processing such as case folding, tokenization, removing URLs, removing punctuations, removing symbols, and removing stop words.

Moreover, we also identify the occurrence of word pairs by joining two words into one term/token because this research will model the topic from a token. To find the word pairs, we use bigrams, a library provided by the Natural Language Toolkit (NLTK) in Python. We then analyze and select which pair of words is an important phrase that often appears in the corpus. For example, the occurrence of the word *"prakiraan"* (forecasts) and *"cuaca"* (weather) arise together 8925 times in the clean tweets. Therefore, we join both words into one token *"prakiraancuaca"* (weather forecasts). Table 1 shows a sample of 10 major pairs of two words from the tweets.

Table 1. Sample of 10 Major Word Pair

No	Word Pair	Number	No	Word Pair	Number
1	*prakiraan cuaca* (weather forecasts)	8925	6	cerah berawan (partly cloudy)	957
2	*citra radar* (radar image)	3204	7	kalender tanam (planting calendar)	920
3	*peringatan cuaca* (weather warning)	2463	8	katam terpadu (integrated planting calendar)	917
4	Jawa Tengah (Central Java)	2024	9	hujan ringan (light rain)	847
5	*info gempa* (earthquake info)	973	10	Jawa Barat (West Java)	785

3.3. Topic Modeling

The process of topic modeling on the whole tweet data is performed by the Latent Dirichlet Allocation (LDA) method. Moreover, this research uses LDA from the Gensim library in the Python programming language. These are the topic modeling steps that conducted in our research:

1 Building term dictionary and corpus.
Dictionary is created by assigning each unique token to an index. The dictionary result will then be used to create a corpus by converting the document list into a document term matrix.

2 Building LDA model object.
The LDA model object is created with the Gensim library and then runs it by training the LDA model on the document term matrix and define the number of topic as a parameter.

3 Visualization.
We use PyLDAvis library to visualize and analyze the topic result. It provides interactive visualization for topic modeling that makes it easy to interpret the generated topic model [10].

4. Result and Discussion

4.1. Topic Modeling

This research grouped tweet topics into 5 topics. The visualization of topical modeling results is shown in figure 2. The left panel illustrates the distance between topics from the generated topic model. Based on the five pieces of the resulting topics, it appears that the topics 1 and topic 3 are mutually interconnected. It shows that there are similarities and closeness of the topic objects discussed between the two groups of topics. As for other topics, the topics 2, 4, and 5 seem to have a considerable distance from each other indicating that among the three groups of topics there is no topic closeness and similarity.

In addition, the right panel illustrates the 30 important words that appear in the corpus. The right panel illustrates about the dominant words discussed topic from the dataset. Based on the right panel visualization, the appearance of the term *'prakiraancuaca'* (weather forecasts) becomes the most frequent word appear in the corpus. Therefore, we infer that the official BMKG Twitter accounts are very often to spread information concerning weather forecast to the society. There are also some other terms that related to weather information for example *"cuaca"* (weather), *"peringatancuaca"* (weather warning), *"infocuacajogja"* (Jogja weather information), *"citraradar"* (radar image), *"kelembaban"* (humidity), *"suhu"* (temperature), and *"angin"* (wind).

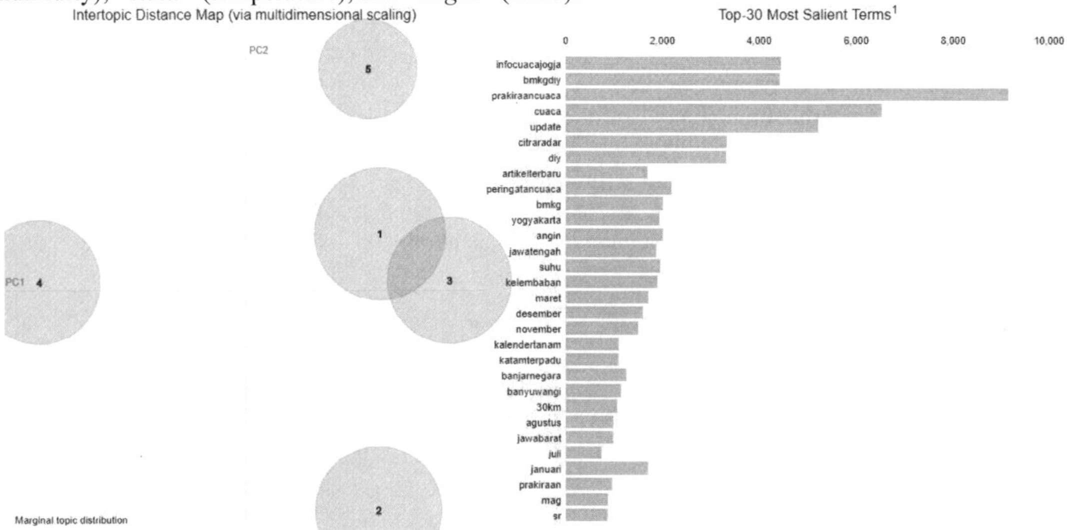

Figure 2. PyLDAvis Visualization

Moreover, we discuss about the interpretation of the meaning of the ten models of topics that have been obtained. The interpretation of the meaning of a topic is viewed based on the 10 most commonly occurring and dominating words in each topic. In addition, we also view the interrelationship between the words of the 10 most frequently appeared from a topic as shown in Table 2.

Table 2. Topic Model Result

Topics	Terms
Topic #1: Weather forecasts	*cuaca* (weather), *prakiraancuaca* (weather forecasts), *angin* (wind), *suhu* (temperature), *kelembaban* (humidity), *banyuwangi*, 30km, *cerahberawan* (partly cloudy), *timur* (east), *hujanringan* (light rain)
Topic #2: Weather information update in the Special Region of Yogyakarta	*infocuacajogja* (jogja weather information), *bmkgdiy* (BMKG of special region of Yogyakarta), *update*, *cuaca* (weather), *citraradar* (radar image), *diy* (Daerah Istimewa Yogyakarta/Special region of Yogyakarta), *Yogyakarta* (Yogyakarta), *januari* (January), *oktober* (October), *februari* (February)
Topic #3: Weather warning for Central Java and West Java Region in a few months	*Prakiraancuaca* (weather forecasts), *peringatancuaca* (weather warning), *jawatengah* (Central Java), *maret* (March), *desember* (December), *November* (November), *update*, *januari* (January), *agustus* (August), *jawabarat* (West Java)
Topic #4: Earthquake	*bmkg* (BMKG), *banjarnegara*, *mag* (Magnitude), *sr* (Richter scale), *stasiunmeteorologi* (Meteorological Station), *infogempa* (earthquake info), *lok* (location), *bt* (east longitude), *ls* (south latitude), *jateng* (Central Java)
Topic #5: Recent articles, crop calendars, and weather forecasts	*artikelterbaru* (latest article), *kalandertanam* (planting calendar), *katamterpadu* (integrated planting calendar), *prakiraancuaca* (weather forecasts), *juli* (July), *september* (September), *juni* (June), *mei* (May), *musimkemarau* (dry season), *terimakasih* (thank you)

The most suitable topic for topic #1 is weather forecasts. It can be inferred from the term probability as shown by column number 1 in Table 2. There are two most salient terms of the term probability, *'cuaca'* (weather) and *'prakiraancuaca'* (weather forecast). Other words, such as *'angin'* (wind), *'suhu'* (temperature), *'kelembaban'* (humidity), *'cerahberawan'* (partly cloudy), and *'hujanringan'* (light rain), describe the terminologies that reinforce information about weather forecasts.

Topic #2 illustrates about the weather information update in the Special Region of Yogyakarta. On this topic, the terms that illustrate about weather in Yogyakarta appear frequently, such as *infocuacajogja'* (Jogja weather information), *'bmkgdiy'* (BMKG of Special region of Yogyakarta), *'update'*, *'cuaca'* (weather), *'citraradar'* (radar image), *'diy'* (Daerah Istimewa Yogyakarta/Special Region of Yogyakarta), and *'yogyakarta'*. Moreover, there are few words that illustrate regarding the month of the weather information, for instance *'januari'* (January), *'oktober'* (October), and *'februari'* (February).

The appearance of the term *'prakiraancuaca'* (weather forecast) in topic #3 is very high. Therefore, the most dominant information for this group is about weather forecasts. Besides, there are some terms, such as *'peringatancuaca'* (weather warning), *'jawatengah* (Central Java), *'jawabarat'* (West Java), *'maret'* (March), and *'desember'* (December), which illustrate the weather warning for Central Java and West Java Region in a few months.

The term *'bmkg'* has the most probability in topic #4. The words, such as *'mag'* (magnitude), *'sr'* (*skala Richter*/Richter scale), *'infogempa'* (earthquake information), *'lok'* (lokasi/location), *'bt'* (*bujur timur*/east longitude), and *'ls'* (*lintang selatan*/south latitude) illustrate the information about earthquakes. Two other words, Banjarnegara and Jateng (Central Java) represent the earthquake location.

Finally, topic #5 illustrates several topics, including recent articles, crop calendars, and weather forecasts. The occurrence of the term *'artikelterbaru'* (recent articles), is quite dominating on topic 5. The terminology *'kalendertanam'* and *'katamterpadu'* represent the information about cropping calendar. The term *'prakiraancuaca'* illustrates the information about weather forecasts.

4.2. Weather and Disaster Trends

Furthermore, we investigate the information from each BMKG's Twitter accounts to see the brief illustration of the weather and disaster information from each region. The investigation is observed based on the occurrence of the terms which related to weather and disaster. From our dataset, there are twelve terms which indicate the weather and disaster information, such as *angin kencang* (strong winds), *badai* (storm), *banjir* (flood), *berawan* (cloudy), *cerah* (bright), *cerah berawan* (bright cloudy), *gelombang tinggi* (high waves), *gempa* (earthquake), *hujan* (rain), *hujan lebat* (light rain), *hujan ringan* (heavy rain), *kilat* (lightning), *petir* (lightning), *siklon* (cyclone), and *tsunami*.

Figure 3 shows that the BMKG of Banyuwangi and Jatiwangi have the most varied weather and disaster information among the BMKG's Twitter accounts. However, BMKG of Jatiwangi more frequently shared about bright cloudy and light rain. In addition, the appearance of the word *hujan* (rain) is quite evenly in all regions except Surabaya (BMKG of Juanda). The most frequent words appeared from BMKG of Juanda is *gempa* (earthquake). Moreover, the information about earthquakes is also frequently shared by BMKG of Bandung Twitter's account.

5. Conclusion

This study has implemented topic modeling using LDA on Indonesian Twitter messages posted by the official Twitter account of BMKG in Java. LDA has promised a good algorithm in topic modeling research. We decided to choose 5 topics as a parameter for LDA method. Based on our study, it can be concluded that discussed topics are about general weather Information and weather forecast; latest weather information in Yogyakarta Region; weather forecast and warning in Central Java and West Java; earthquake information; latest articles and calendar cropping information. From the topic modeling result, BMKG Twitter accounts in Java were mostly used to provide information to the public about the weather condition, which the primary functionality of BMKG itself. The rest topics were discussed about the earthquake information, recent articles, and calendar cropping information. In addition, we also illustrated the trends about weather and disaster by analyzing the most frequent words from each region in Java. In the future works, it will be interesting if we can use the topic modeling result to predict the weather condition and disaster. This will be useful for the community as an early warning and precaution. In addition, further study can be held by using the improvement of LDA or other topic modeling method in to obtain better topic modeling results.

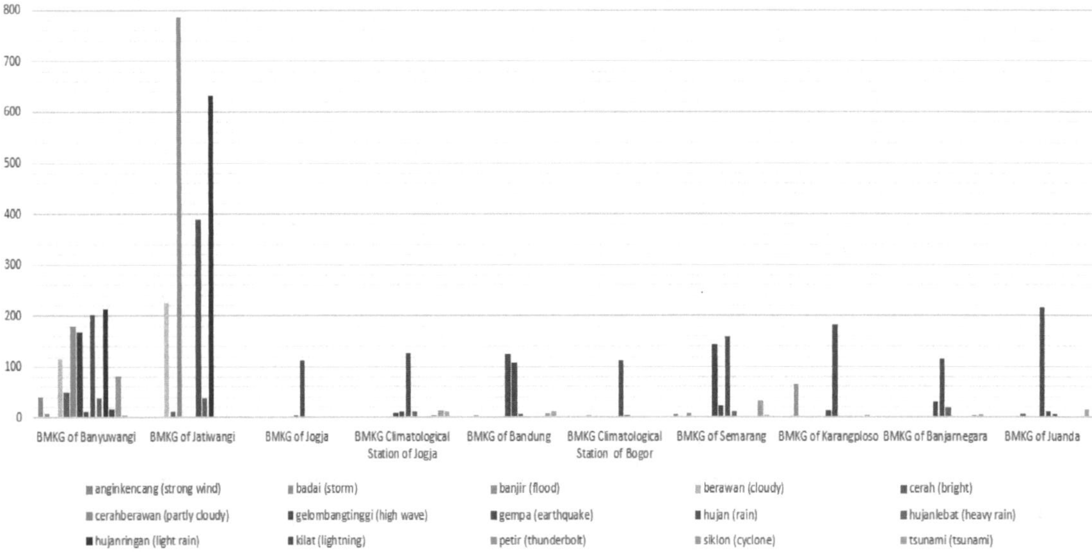

Figure 3. The Illustration of Weather Condition in Each Region

6. Acknowledgement

The authors would like to thank the Department of Informatics, Universitas Islam Indonesia, for technical and financial support.

References

[1] Palen L and Vieweg S 2008 The emergence of online widescale interaction in unexpected events *Proc. of the ACM 2008 Conf. on Computer Supported Cooperative work* pp 117–26
[2] Hidayatullah AF and Ma'arif MR 2017 Road traffic topic modeling on twitter using latent dirichlet allocation *Int. Conf. on Sustainable Information Engineering and Technology (SIET)* pp 47–52
[3] Blei DM, Ng AY and Jordan MI 2003 Latent dirichlet allocation *J. Mach. Learn. Res.* vol **3** pp 993–1022
[4] Blei DM, Edu BB, Ng AY, Edu AS, Jordan MI and Edu JB 2003 Latent dirichlet allocation *J. Mach. Learn. Res.* vol **3** pp 993–1022
[5] Wang S, Paul MJ and Dredze M 2014 Exploring health topics in Chinese social media: An analysis of Sina Weibo *AAAI Workshop on the World Wide Web and Public Health Intelligence* vol **31** pp 20–3
[6] Liu L, Tang L, Dong W, Yao S and Zhou W 2016 An overview of topic modeling and its current applications in bioinformatics *Springerplus* vol **5** no **1** p 1608
[7] O'Neill J, Robin C, O'Brien L and Buitelaar P 2017 *An Analysis of Topic Modelling for Legislative Texts ASAIL*
[8] Yang TI, Torget AJ and Mihalcea R 2011 Topic modeling on historical newspapers *Proc. of the 5th ACL-HLT Workshop on Language Technology for Cultural Heritage, Social Sciences, and Humanities* pp 96–104
[9] Hidayatullah AF and Ma'arif MR 2017 Pre-processing tasks in indonesian twitter messages *IOP Conf. Series: J. of Phys.* 801
[10] Sievert C and Shirley K 2014 LDAvis: A method for visualizing and interpreting topics *Proc. of*

the Workshop on Interactive Language Learning, Visualization, and Interfaces pp 63–70

Online corpus of spoken Ilokano language

F Apostol [1] and A Malicdem [2]

[1] Mariano Marcos State University, Ilocos Norte, Philippines
[2] Don Mariano Marcos Memorial State University, La Union, Philippines

[1]franklinrapostol@gmail.com, [2]amalicdem@dmmmsu.edu.ph

Abstract. There has been a great effort in the collection of different languages in the past years all over the world, and the development of online corpus outside the country brought new possibilities in the Philippines. However, there is a limited resource for the Ilokano Language. This paper introduces the Corpus of Spoken Ilokano Language, an online repository of spoken Ilokano in the Philippines specifically in region 1. The main component of this study is spoken Ilokano. It has been specifically built for natural language processing. It shows the difference of Ilokano language as spoken by Ilokanos in the region. The database consists of 160 speakers, 40 speakers in each province of the region, each speaking about 74 statements. Spoken Ilokano language was audio recorded and transcribed. A web application has been developed making the dataset available online. The corpus was validated to provide a useful resource of data that can be used for automatic speech recognition models.

1. Introduction

With the various language across countries, different corpora have been persistently developed and have become a fertile ground for investigation. The studies of [1][2][3][4] have outlined the development of a repository of British, English, German, Italian and Hindi language in the past years. [5] also illustrated the evolution of corpus specifically for use in the telephone. The birth of the world wide web made these corpuses available over the internet. [6] says that the world wide web is not only a tool for information retrieval and exchange but also a massive repository of authentic data, "a self-renewing linguistic resource" offering "a freshness and topicality unmatched by fixed corpora." However, with all the researches that have been done, [7] says that the use of corpus by different language service providers and language professional remains limited due to the existence of computing resources that are likely to be perceived as less demanding regarding time and effort required to obtain them.

Corpus, as defined by [8], is a collection of writings, conversations, and speeches in a particular language or languages that are to be stored, managed and analyzed in digital form. The growth of the corpus outside the country has led the birth of the Philippine language corpus [8]. PALITO is a collection of written texts of religion and literature and Filipino Sign Language focusing on the four most commonly used dialects in the Philippines namely Tagalog, Cebuano, Hiligaynon, and Ilokano. There has been a great effort in the collection of Philippine language and making it available online in the past years, however, due to the time, it takes to collect such data, [8] strongly recommend the building of a corpus of spoken Philippine language first in Cebuano, Tagalog, Hiligaynon or Ilokano.

With this, it is believed that the building of an online corpus of spoken Ilokano language, the data collection and keeping them track will leverage the quality of Philippine corpus.

Ilokano is an Austronesian language of the Philippine type spoken by about nine million people [9]. It is a member of the cordilleran language family which comprises many language of Northern Luzon Island, Philippines. It is ranked third in terms of its number of mother-tongue speakers (probably over 7.7 million native speakers or 10.1 % of the total population). Ilokano is also considered as regional lingua franca of the Filipinos residing in the Northern Luzon. A very small minority speaks it in the Mountain Province and Davao and Bukidnon in Mindanao. It is also spoken as a second language by the Ilokano people who have migrated to other parts of the Philippines and of the world. It is spoken by the Ilokanos residing in Metro Manila, Central Luzon and as far as Mindanao. It is used by Ilokanos residing in Hong Kong, Saudi Arabia, European countries and United States of America; where the largest concentration is found in California, Alaska and Hawaii. Aside from the high number of first and second language speakers of Ilokano, it has an immense volume of literature; next in number to Tagalog literature.

While it is true that Ilokano is considered as lingua franca in the region, people in different municipality speak the same word differently. The original Ilokano speaking areas include the present-day provinces of Ilocos Norte and Ilocos Sur [9]. People from these places speak the purest form of the language, however, due to the migration of the Ilokanos southward and eastward, much of northern Luzon is heavily influenced by Ilokanos language and culture. The provinces of La Union and Pangasinan are dominated in most areas by Ilokano speakers, speaking the southern dialect which has minimal lexical differences from the northern one, but a significant phonological difference.

This building of an online corpus of spoken Ilokano language will be built specifically for natural language processing. It aims to provide an excellent resource of data to be used for automatic speech recognition models and therefore, hope to enrich the quality of Philippine corpus. Likewise, the results may provide possible implications for Ilokano language teaching in the Philippines and outside. For researchers and scholars who are interested in Ilokano language, the study can provide additional input to their studies thereby contributing to the continuing evolution of studies in Ilokano language. For teachers of Ilokano, this can also provide data for activities such as spoken practices and exploration in their use of Ilokano.

With the existing infrastructure as provided by the World Wide Web and the Internet that virtually connects people from various physical locations, contributing to the development of such a collection of spoken language is now a reality.

The primary objective of this study is to build an online corpus of spoken Ilokano language particularly the language used by native speakers in Region 1. Specifically, the study aimed to: (1) Determine the difference of Ilokano language as spoken by Ilokanos in Region 1, (2) Build the corpus of spoken Ilokano language by: (a) recording voices of native citizens of Ilocos Norte, Ilocos Sur, La Union and Pangasinan, (b) transcribing the audio recordings, (c) creating a database of spoken Ilokano language, (3) Develop a web application for the corpus; and (4) Determine the validity of the corpus for speech recognition.

2. Methodology

The researcher used the descriptive and developmental research methods to organize the presentation, description, and interpretation of data gathered. The study was descriptive in distinguishing the differences of Ilokano language as spoken by Ilokanos in region 1, the collection of data and in the selection of respondents. The developmental research was considered the most suitable since the study dealt with the development of Online Corpus of Ilokano Language.

2.1. Voice Recording

With the use of purposive sampling, participants were selected and limited to those: a) who were born and raised in their home provinces; b) whose first language is Ilokano; c) who were educated in their local schools, colleges, and universities; and d) who used Ilokano language in their conversation. This was to ensure that the participants represent the genuine Ilokano speakers of Ilokano language. Furthermore, the number of respondents was determined using the central limit theorem. The number of participants was limited to 160, with 40 participants from each province, a number which was well within the recommendation of the theorem. The study adapted the age labels used in DARPA Switchboard Corpus for the classification of age groups of Ilokanos in the region. Perceived age was grouped into four main categories: Youth (16-25), Intermediate Adult (26-35), Adult (36-50), Senior (51 and above) with 5 males and 5 females in each group. Table 1 shows the distribution of respondents with equal participants from each province and age group. Equal distribution was employed to ensure objectivity and reliability in the gathered data.

Table 1. Distribution of Participants Per Province.

Age Group	Ilocos Norte N-593,081	Ilocos Sur N-689,668	La Union N-786,653	Pangasinan N-1,146,456
A1 (16-25)	10	10	10	10
A2 (26-35)	10	10	10	10
A3 (35-50)	10	10	10	10
A4 (51 Above)	10	10	10	10
Total	40	40	40	40

Legend: N – Population; n - sample

Data were gathered from the Ilokano-speaking provinces of Region 1. The provinces were limited to Ilocos Norte, Ilocos Sur, La Union, and Pangasinan as these are the homelands of genuine Ilokano-speaking people. Furthermore, data were limited from the recorded oral reading of a short story from BANNAWAG magazine, the umbrella organization of Ilokano writers in the Philippines and other countries. This magazine contains the purest Ilokano language needed in the study. The selected article was a folktale story entitled "Ti Ari ken ti Kawitan" written by Juan S.P. Hidalgo, Jr, issue of March 28, 2011. It contains the four types of sentences (Imperative, Declarative, Interrogative, Exclamatory). These types of sentences were used in the determination of the different pitches of Ilokanos. The article is composed of 74 sentences, 970 words, including the title of the story. The instrument that was used in data gathering for this study aimed to elicit the participants' personal profiles and speech samples which were audio recorded. The participants were requested to read the article written in Ilokano language in an audible manner and normal voice. Each participant's oral reading was audio-recorded. Efforts were made to take advantage of the diverse population in schools, local government units, local congregation, and some barangays. Speakers were recruited by advertising throughout the said environment. The materials used in the recording were ZOOM H1 and BY-M1. ZOOM H1 is a handy recorder for audio and video that runs in a 24 bit and 96 Khz audio, with Lo-cut filter and auto level controls for reducing noise and a X/Y microphone design that captures perfect stereo image. It also includes microphone function for external microphone. BY-M1 is a clip-on lavalier microphone micro-cravate used for audio recorder. A 8 gigabyte of micro SDHC was used as storage of the recorded audio. The recordings were distributed in home, office, streets, rooms as recommended in LREC 2004 Workshop: Speech Corpus Production and Validation. Participants were given instruction prior to the recording. The speaker was seated in a mono block chair in one corner of a closed room. The microphone was covered with foam windscreen for filtering noises and was clipped 5 inches below the chin. Air condition, electric fans and other appliances were turned off to lessen the noise of the environment. During the recording, no one was allowed to speak aside from the speaker to avoid distraction.

2.2. Transcription of Audio Recordings

In the transcription of the audio recording, the software audacity was used. Audio clips were selected from the speakers recorded oral reading as samples for transcription and interpretation. The four types of sentences, imperative, declarative, interrogative and exclamatory were applied as criteria in selecting the samples. The researcher selected 4 sentences from the story, one for every type of sentence. The spectrum of these audio clips was plot for analysis using a mathematical algorithm known as *Fast Fourier Transform (FFT)*. Seven points, from 1 through 7, in the spectrum were chosen for each type of sentences in the determination of the pitch pattern. Point 1 represented the beginning of the statement,

4 represented the middle and 7 represented the end of the statement. The frequency of each points were recorded. The mean of these seven points in every three statements in each sentences was computed in terms of frequency and were presented digitally through a chart and a table.

2.3. Creation of Database

With regards to the process of the creation of a databank of spoken Ilokano language, cloud computing was utilized as the online storage of the data. The data was stored in the local server with 1 Gigabyte Random Access Memory with unlimited bandwidth for the server to cater the functions of the web application that is not supported by a third party server. In the development of the Web application used in the Corpus, the Model/View/Controller (See Figure 1) design pattern was used to have a systematic approach in development. PHP, HTML, CSS were used to design the Application Program Interface.

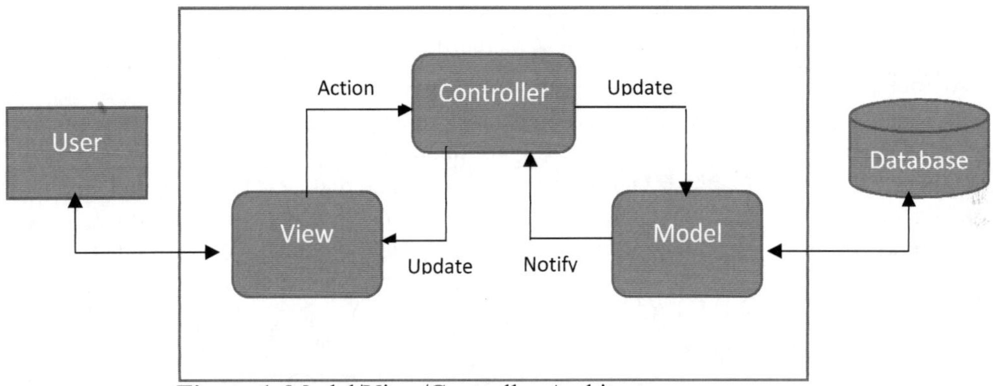

Figure 1. Model/View/Controller Architecture

***2.4. Data** Analysis*

Validation was made to ensure the reliability of the data provided in such a corpus. The validation covered the voice transcription, digital representation of pitch pattern, and the analyses and interpretation of the data being presented from the selected samples of speakers. Seven validators that were expert in Ilokano language were requested to validate the corpus. Every validator indicated their grade to the description related to the speaker. They evaluated the description and interpretation of the digital representation of voice to see if it matched to the audio recording. This was done to determine whether the audio came from a genuine Ilokano speaker or not. The validation was interpreted using the 3-point Likert scale.

$$Weighted\ Mean = \frac{(Number\ of\ Validator)x\ (Weigh\ \ of\ 1st\ Column)}{Total\ Number\ of\ Validator} + \frac{(Number\ of\ Validator)x\ (Weight\ of\ 2nd\ Column)}{Total\ Number\ of\ Validator} + \frac{(Number\ of\ Validator)x(Weigh\ \ of\ 3rd\ Column)}{Total\ Number\ of\ Validator}$$

The scale and their equivalent that were used to interpret the weighted mean are as follows:

Point	Range	Descriptive Equivalent	Descriptive Interpretation
1	1.00 – 1.66	Poor	Not Valid
2	1.67 – 2.33	Good	Valid
3	2.34 – 3.00	Excellent	Valid

Weighted mean with average ranging from 1.67 to 3.0 was interpreted as valid which means spoken data comes from genuine Ilokano speakers, and 1.00 – 1.66 was interpreted as not valid and means speakers were not genuine Ilokano, which specifically needs to be changed for a better corpus of Ilokano Language.

3. Results and Discussion

3.1. Difference of Ilokano Language as Spoken by Ilokanos in Region 1
Based on the use of e/ə, Ilokano may be divided into two dialects: the front medial dialect (e) and the high back unrounded (ə). [9] pointed the "Southern Dialect" to La Union and Pangasinan which use both the high back unrounded (ə) and front medial dialect (e) and the "Northern Dialect" to Ilocos Sur and Ilocos Norte which use the front medial dialect (e) only. Ilokano has six contrastive vowels (five in northern dialect) represented in orthography by five letters /a,e,i,o,u/. The orthographic symbol 'e' in Ilokano constitutes two separate vowel sounds in southern dialect and one sound in northern dialect. Table 2 shows the summary on the difference of Ilokano language as spoken by Ilokanos in Region 1.

Table 2. Summary on the difference of Ilokano language in Region 1.

Province	Difference of Ilokano language in Region 1		
	Dialectal Background	Use of e/ə in Pronunciation	Ilokano Vowels
Ilocos Norte	Northern Dialect	front medial dialect (e)	a,e,i,o,u
Ilocos Sur	Northern Dialect	front medial dialect (e)	a,e,i,o,u
La Union	Southern Dialect	high back unrounded (ə) and front medial dialect (e)	a,e,i,o,u,ə
Pangasinan	Southern Dialect	high back unrounded (ə) and front medial dialect (e)	a,e,i,o,u,ə

3.2. Building the corpus of spoken Ilokano language

3.2.1. Recording of Voices Respondents were requested to read a folktale story aloud while it was being recorded. The place of recording was distributed in home, schools, office, streets, and rooms as recommended in LREC 2004 Workshop: Speech Corpus Production and Validation. The dimension of the room was 4 x 4 meter, the ceiling was 4 m high. During the recording in the street and school community, the microphone was placed 2 inches below the chin. This was to ensure that the voice was captured and at the same time lessen the noise of the environment captured by the microphone. Teachers, employees of Local Government Units, Barangay Officials and students were also requested to become part of the data gathering.

3.2.2. Transcription of audio recordings Audio clips were selected from the speakers recorded oral reading as the samples for transcription and interpretation. The four types of sentences, imperative, declarative, interrogative and exclamatory were used as criteria in selecting the audio clips. These were believed to give the necessary data to be used in determining the pitch pattern of speakers in the region. The spectrum of the selected audio clip was plot for analysis. See Figure 2. Seven points, from 1 through 7, in the spectrum were chosen for each type of sentences in the determination of the pitch pattern. Point 1 represented the beginning, point 4 represented the middle and point 7 represented the end of the statement. The points were where the pitch in the spectrum fades in and fades out. The frequency of each points were recorded. The mean of these seven points in every three statements in each sentences was computed in terms of frequency and were presented digitally through a chart and a table. Three statements for each type of sentences were analyzed in the determination of the pitch pattern of speakers. Seven points in the spectrum were obtained for each type of sentences to see the pattern of pitch. The mean of the seven points in every three statements in each sentences was computed in terms of frequency and was presented digitally. Audio clips from all of the 40 speakers in each province were plot and analyzed to come up with a single pitch pattern of speakers in each province. The average was computed and combined into single pitch patterns. Based on the computed means of the four types of sentences, interpretations were drawn.

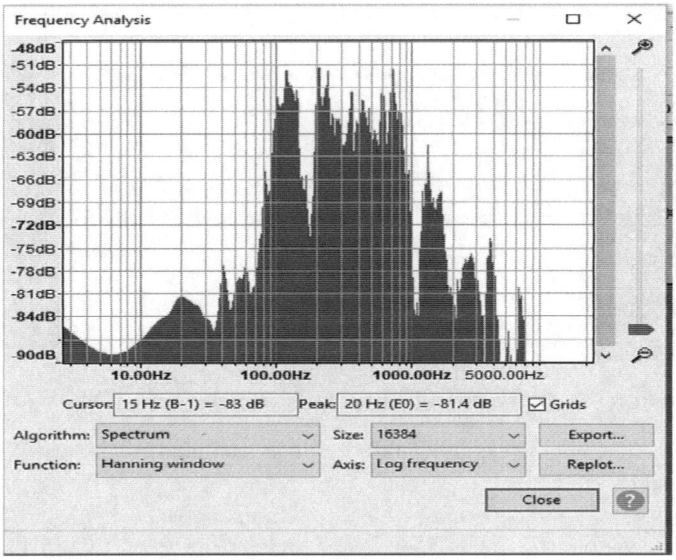

Figure 2. Screenshot of Audacity Spectrum

3.2.3. Creation of Database of Spoken Ilokano Language This study used the cloud computing environment for the online banking of spoken Ilokano language. Databases were made for the audio, respondents, province, users, and for those who will request to become contributors. Entities were constructed for each database. Attributes were also defined. Primary keys and foreign keys were also assigned. Relationships were established between these databases. The data was stored in the local server with 1 Gigabyte Random Access Memory with unlimited bandwidth for the server. The data was stored in the local server to cater the functions of the web application that is not supported by a third party storage.

3.2.4. Development of Web Application In the development process, sublime was used as the text editor and Node.js was the platform. In designing the Graphical User Interface and other functions of the web, both for the client and the admin, Angular JS was used as a framework. Yeoman was used as scaffolding of the web app. Bootstrap was used in the styling of the website. Bower and NPM was utilized to manage the frameworks, libraries, utilities and node.js packages. GruntJS was used as a task runner. The domain name used for the corpus was 'ilokanocorpus.net'. Figure 3 shows the front page of the developed web application for the corpus.

3.2.5. Validity of the Corpus To ensure the reliability, validation was made for the corpus. The validation was made by experts in Ilokano language in the region. Seven validators were requested to evaluate the corpus against specified requirements. The validation covered the voice transcription, digital representation of pitch pattern, and the analyses and interpretation of the data being presented from the selected samples. Pitch pattern was described based on the digital representation of the four types of sentences and compared as to who has the highest and lowest pitch among the speakers in the region. Every validator indicated their grade to the description related to the speaker. They evaluated the description and interpretation of the digital representation of voice to see if it matched to the audio recording. This was done to determine whether the audio came from a genuine Ilokano speaker or not.

Using the formula for weighted mean, the weighted mean in each province was computed and presented in Table 3.

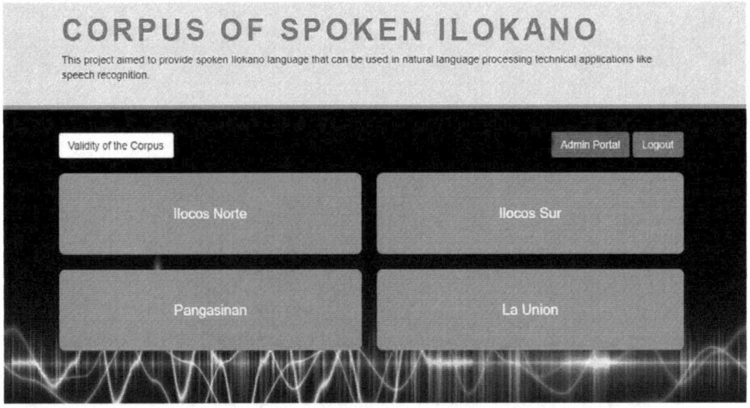

Figure 3. Screen Shot of Corpus of Spoken Ilokano Front Page

Table 3. Validation Report

Speaker	Mean	Descriptive Equivalent	Descriptive Interpretation
Ilocos Norte	2.29	Good	Valid
Ilocos Sur	2.14	Good	Valid
La Union	2.14	Good	Valid
Pangasinan	2.14	Good	Valid
Grand Mean	2.17	Good	Valid

Referencing to the scale and their equivalent presented in the data analysis, the grand mean, 2.17, was interpreted as valid. The voice recording perfectly matches to the digital representation and interpretation of the pitch pattern of Ilokano in Region 1. Thus, the voice recordings came from a genuine Ilokano speaker. With these results, the overall quality of the data in the corpus was good and the corpus was valid and there should be no problem in using it for its intended purposes and other applications.

4. Summary, Conclusions and Recommendations

In this paper we have discussed the difference of Ilokano language as spoken by Ilokanos specifically in region 1. The use of high back unrounded (ə) and front medial dialect (e) makes a difference in the pronunciation of Ilokano language. The challenges of building the corpus has been elaborated. Some of these challenges were in terms of the selection of native Ilokano speakers, methods of data collection, and the transcription of collected data. We have also shown the process of the development of the web application. The web application was developed in a way that could help interested parties access the data. Finally, the validity of the corpus has been presented. Validators deem that the corpus was valid for its intended purposes for it showed the grand mean of 2.17.

In the future we hope more and more affiliations could participate in this work. Only in this way, can the corpus be constructed more efficiently and be used or shared more easily.

References

[1] Grefenstette G and Kilgarriff A 2003 Introduction to the special issue on the web as corpus *Assoc. for Computational Linguistics* vol **29** no **3** pp 333-47

[2] Baroni M, Bernardini S, Ferraresi A and Zanchetta E 2009 The WaCky wide web: a collection of very large linguistically processed web-crawled corpora *Language Resources and Evaluation* vol **43** no **3** pp 209-26

[3] Belhe S, Dubey M, Kulkarni M, Magdum D, Patil T and Shah R 2015 Methodology for designing and creating Hindi speech corpus *2015 Int. Conf. on Signal Processing and Communication Engineering System* pp 336-9
[4] Yin Z and Li A 2017 A standardization program of speech corpus collection *20th Conf. of the Oriental Chapter of the Int. Coordinating Committee on Speech Databases and Speech I/O Systems and Assessment (O-COCOSDA)* pp 1-5
[5] Zhigang Y 2016 How to make a telephone speech corpus *Proc. of the 2nd Int. Conf. on Communication and Information Processing* pp 174-8
[6] Mautner G 2005 Time to get wired: Using web-based corpora in critical discourse analysis *Discourse & Society* vol **16** no **6** pp 809-28
[7] Bernardini S and Ferraresi A 2013 Old needs, new solutions: comparable corpora for language professionals *Building and Using Comparable Corpora* pp 303-19
[8] Boutaba R, Cheng L and Zhang Q 2010 Cloud computing: state-of-the-art and research challenges *J. of Internet Services and Applications* vol **1** no **1** pp 7-18
[9] Burnard L 1995 *British National Corpus: Users Reference Guide British National Corpus Version 1.0* (Computing Service, Oxford University)
[10] Dita SN, Roxas RE, Inventado P 2009 Building online corpora of Philippine languages *Proc. of the 23rd Pacific Asia Conf. on Language, Information and Computation* vol **2**
[11] Rubino CR 2000 *Ilocano Dictionary and Grammar: Ilocano-English, English-Ilocano* (University of Hawaii Press)

Frequency of compromised passwords used by students and staff of Asia Pacific College: an analysis using NIST SP 800-63B and pwned passwords

JV Roig[1], J de la Cuesta[2], J Castillo[3], J Cabardo[2], E Casiño[2], E Salalima[3] and M Sanchez[2]
[1]Advanced Research & Consulting, Asia Pacific College, Philippines
[2]School of Computing and Information Technology, Asia Pacific College, Philippines
[3]Information Technology Resources Office, Asia Pacific College, Philippines

[1]jvr@apc.edu.ph, jvroig@gmail.com

Abstract. The National Institute of Standards and Technology (NIST) released new guidelines in June of 2017 that recommended new standards for managing and accepting user passwords. Among the new guidelines is a requirement that verifiers should check if a user's supplied password is compromised – that is, already listed in previous breach corpuses. Using a corpus of 320 million breached passwords, the researchers conducted an experiment to gauge what percentage of the population of their home institution, Asia Pacific College, use compromised passwords. The study found that 16.72% overall – or 1 in 6 people – were using passwords that are part of the 320M breach corpus. This paper also provides a methodology that other institutions and companies can use to conduct the same analysis in order to gather data specific to their population that can guide the improvement of their password policies and related IT security services.

1. Introduction

The National Institute of Standards and Technology (NIST) published new recommendations for verifiers (organizations that maintain username-password credentials) for managing user accounts and passwords [1]. Included in their latest publication is a recommendation that user passwords be checked against a list of known passwords obtained from past exploits and breaches [2]. If the checked password is found in such a list, the verifier should alert the user that her password is unsafe and should select a new one.

While this is a very practical advice, the difficulty is in obtaining a corpus of such breached passwords. These breached passwords are actual passwords that have been obtained throughout the years by black hats who have exploited many different services and websites, and eventually publicly dumped into various nooks and crannies of the internet. Fortunately, someone in the security community stepped up to the plate to help institutions get hold of such a breach corpus. Troy Hunt, owner and maintainer of "Have I Been Pwned" (https://haveibeenpwned.com), released his own breach corpus for public download called "Pwned Passwords" – the same breach corpus that powers his online service. This initial release consisted of 320M passwords found in previous data dumps from hacks and breaches of various black hats, and can enable institutions to efficiently comply with NIST's latest guideline regarding checking of passwords.

With this breach corpus, the IT team of Asia Pacific College (APC; the home institution of the authors of this report) embarked on the implementation of this new guideline. Along the way, the authors collected data about how many faculty, staff, and students used passwords that were found in the list of 320M breached passwords from Troy Hunt, and the results of analysis of that collected data is reported in this paper.

2. Background and Related Literature

It has been long known that the password hygiene of users is generally terrible. Without special guidance, users will generally fall into usage of insecure passwords due to convenience and the properties and limitations of human memory [3][4]. Bryant and Campbell (2006) concluded that a contributing factor in this lack of care for passwords is the belief that they, the users, are not personally at risk [5]. Shay et al (2010) measured user sentiment regarding password policies (based on older NIST password guidelines, circa 2007), and found that respondents felt that the new policies were annoying, and felt generally neutral when asked if it was worth the effort to comply with the new guidelines, which was the same sentiment when asked if they would rather switch back to the old, less stringent policies [6]. Florencio and Herley (2007) found that the average password reuse is 5.67 sites per password per user [7]. They also found that user type an average of 8 passwords per day – no wonder they'd rather reuse passwords (and use very easy to remember and type, i.e., insecure passwords).

Users in general can't be bothered to keep up or comply with stringent password policies. When forced to do so by systems that strictly enforce complex password rules (say, forcing a combination of upper-case, lower-case, numbers and special characters along with a minimum length, and forcing frequent password changes), the security of individual passwords do increase, but users inevitably cope with negative behavior, such as writing down passwords [8].

There are other purely technology-based solutions that aim to improve the security of passwords, such as the usage of password strength meters. This has been shown to increase the password strength of users [9], but this isn't a full victory – forcing or nudging users to create a strong, complex password is one thing, making sure users can remember them is another. Forcing a complex password, other than leading to written-down passwords, also lead to forgotten passwords, which increase the workload of IT support people who have to answer an "I-forgot-my-password" call and reset the user's password. This also means an inconvenience on the part of the user, which motivates her to finally write it down somewhere or make it predictable for her, both of which are password habits that are not supposed to be encouraged.

NIST's latest guidelines take a more human-oriented approach, instead of the traditional tech-centric policies that has led to the various negative user sentiments and coping mechanisms described above. This report focuses on one specific guideline – checking passwords against a list of known passwords – to measure how many users fail this check within the setting of an educational institution.

3. Experiment Design and Methodology

A list of 320 million breached passwords was obtained through security researcher Troy Hunt of HaveIBeenPwned.com. Hunt publicly released this list after the NIST's 2017 guidelines was published, specifically to enable institutions to adopt one of the latest guidelines recommended by NIST. This list of passwords does not contain plaintext entries of passwords – instead, every entry is the SHA1 hash of the plaintext passwords. The list also does not contain matched usernames. The list is entirely SHA1 hashes. (https://www.troyhunt.com/introducing-306-million-freely-downloadable-pwned-passwords/)

The research team, enabled by the assistance of APC's in-house software development team, inserted hooks in APC's own internal school management system so that the passwords of each user logging in can be compared to the corpus of breached passwords, determining whether the password being used is compromised or not, in accordance with the NIST's latest guidance.

This special software infrastructure, composed of short PHP codes and an SQLite3 database, was left running and recording results for 3 months, capturing peak periods of system activity – the midterms and the finals. This experiment period spanned most of the second trimester of school year 2017-2018 (September – December 2017).

Since the corpus of breached passwords contain only SHA1 hashes, the passwords of users are similarly hashed before searching through the corpus for a match. Only the hashes – not the plaintext password – is ever used and recorded by the research team. At no point does the research team ever use or store the actual plaintext password of any user, beyond using them initially in order to produce the SHA1 hash.

Whether a match is found or not, the user is recorded into a special database table created solely for this research. If a match is found, then that user's record will reflect that his password is compromised. If no match is found, then that user's record will reflect that his password is not compromised.

Users are not recorded twice. If a previously recorded user logs in again (an expected outcome, since users will log in several times during the experiment period), the software simply updates the user's existing record.

After 3 months, 88.33% of every APC user was recorded by this special infrastructure: 84.62% of employees, 91.52% of College students, and 84.64% of Senior High students.

4. Analysis and Results

4.1. Overall Results

After the experiment period, the researchers recorded a total of 2,249 unique user logins. 376 of these users have passwords that were in the breach corpus from Troy Hunt - meaning 16.72% of APC's population (employees and students) are using very unsafe passwords. That's 1 in 6 people using passwords that require minimal effort for black hats to compromise.

4.2. Population Breakdown

The collected data can be broken down into three different user categories – Employee, College Student, and Senior High School. Table 1 shows the breakdown per user category, and compares the recorded users vs the total active users. Overall, almost 90% of the active APC population during the experiment period were successfully recorded by the experiment infrastructure; the unrecorded are those that did not use the school's integrated information system during the 3-month experiment period (e.g., students that did not bother to check their grades online, or employees that had no need to file their leaves online).

Table 1. Percentage of recorded data vs total population size per user category.

User Type	Recorded Users	Actual Population Size	% of Population Recorded
Employee	220	260	84.62%
College Student	1,252	1,368	91.52%
Senior High	777	918	84.64%
TOTAL	**2,249**	**2,546**	**88.33%**

4.3. Compromised Passwords Per User Category

Table 2 compares the percentage of accounts per user category that were found to use compromised passwords. As described in Section 3, "compromised passwords" refer to passwords that are in the breach corpus of 320M passwords.

14% of Employee users were found to use compromised passwords, slightly below the combined (all user types) percentage of 16.72%. Employees, however, were also the smallest group of users, comprising only 10% of the recorded data. College students, the largest group in the data set (comprising 55% of the total recorded users), had a slightly higher percentage, at 17.17% found to be using compromised passwords. Senior High School students, at 16.73% with compromised passwords, is almost exactly in line with the overall percentage of 16.72%. They comprise about 35% of the recorded users. The researchers note that while employees aren't too far behind the overall average of users with compromised passwords, it is slightly reassuring that employees seem to have better password hygiene than students.

Table 2. Usage of compromised passwords per user category.

User Type	Recorded Users	Users with Compromised Passwords	% of Users with Compromised Passwords
Employee	220	31	14.09%
College Student	1,252	215	17.17%
Senior High	777	130	16.73%
TOTAL	**2,249**	**376**	**16.72%**

4.4. Other Interesting Results

We present in this section some statistics derived from auxiliary data.

4.4.1. Gender. Table 3 presents the percentage of compromised passwords per gender. Male users were found to be slightly worse than female users, with almost 18% of male users (higher than the population average of 16.72%) using compromised passwords. Female users were slightly better than the average, with about 15% using compromised passwords. (The lead author would note here that his male brethren has truly disappointed him.) There are four recorded users classified as "unknown" - this means the experiment apparatus was not able to properly retrieve gender information for four users, as their gender data was missing from APC's information system. Of the four, two were found to use compromised passwords. Whether these four are all male, all female, or a mix, is mostly irrelevant now. Even if the 2 unknown accounts that did not have compromised passwords were male, and the 2 unknown accounts were female, the percentages will not change much. In that sort of best-case scenario for male users, the percentages would become 17.84% (male compromised) vs 15.32% (female compromised).

Table 3. Usage of compromised passwords per gender.

Gender	Recorded Users	Users With Compromised Passwords	% of Users with Compromised Passwords
Male	1,248	223	17.87%
Female	997	151	15.15%
Unknown	4	2	50.00%
TOTAL	**2,249**	**376**	**16.72%**

Do female users just naturally have better password hygiene? That's an interesting question, but, aside from simply sharing our raw data, outside of the scope of this report.

4.4.2. Password Length. Majority of the APC population use passwords with 8-12 characters. 72% of the population are found within this range, with each length comprising over 10% of the population: 12% have 8 characters, 16% have 9 characters, 17% have 10 characters, 14% have 11, and 13% have passwords that are 12 characters long. The mean password length is 11.2 characters, while the mode is 10 characters. Only 1% of the population use less than 8 characters: 24 users have 7-character passwords, and 1 user has only 4 characters. Over 25% of users have 13 characters or more. 1% of the population have 18-24 characters in their passwords. These facts and figures are shown in Table 4.

5. Conclusion and Recommendations

5.1. Conclusion

This report details the results of an empirical study made in APC to measure the usage of known unsafe passwords found in breach corpuses. A custom-designed infrastructure that hooks into the existing school information system collected properly anonymized data (non-confidential and non-personally-identifiable) regarding password usage of APC users (employees, college students, and senior high students). 88.33% of the active population was successfully recorded during the experiment period, which spanned most of the second trimester of school year 2017-2018.

Table 4. Password length distribution of APC users.

Password Length	Number of Users	% of Total Population	Running % of Total Population
4	1	0.04%	0.04%
5	0	0.00%	0.04%
6	0	0.00%	0.04%
7	24	1.07%	1.11%
8	275	12.23%	13.34%
9	353	15.70%	29.04%
10	377	16.76%	45.80%
11	321	14.27%	60.07%
12	295	13.12%	73.19%
13	195	8.67%	81.86%
14	166	7.38%	89.24%
15	99	4.40%	93.64%
16	82	3.65%	97.29%
17	22	0.98%	98.27%
18	15	0.67%	98.93%
19	7	0.31%	99.24%
20	10	0.44%	99.69%
21	4	0.18%	99.87%
22	0	0.00%	99.87%
23	0	0.00%	99.87%
24	3	0.13%	100.00%

It was found that 1 in 6 users (16.72%) were using passwords found in the Pwned Passwords breach corpus provided by Troy Hunt. Students fared slightly worse than employees – only 14.09% of employee users used these unsafe passwords, compared to 17.17% for college students and 16.73% for senior high. Categorizing the users per gender also drew an interesting comparison – male users fared slightly worse than female users (17.87% vs 15.15%). Whether this means female users are inherently better is an interesting topic to explore, but outside the scope of this study.

5.2. Limitations of This Study

The current experiment infrastructure used is based solely on Troy Hunt's trove of 320M breached passwords. It lacks commonsensical checks such as commonly-known passwords used in Filipino (the native language in the research environment of this study). It also lacks checks against patterns that are specific to the organization itself (for example, using "APC" in simple patterns). This means a black hat tasked with cracking passwords will likely have better a success rate than just 1 in 6 passwords. So, while a statistic of 1 in 6 unsafe passwords is already bad, the actual figure is most likely higher. It also bears noting that this study isn't directly a measure of the prevalence of weak passwords. Unlike most previous password weakness studies, this study only measures inclusion in the Pwned Passwords list. There are certainly weak passwords that won't necessarily be in that list (such as localized common words and organization-specific terminology or patterns).

5.3. Recommendations and Future Direction

The researchers encourage other organizations to conduct similar experiments and share their results to the community at large. Not only will this enable other organizations to be compliant with the newer NIST guidelines, this can also help shape commonsensical password policies. Creating a repository of unsafe passwords (with which to check user passwords against) based on Troy Hunt's password list is a good start, but we recommend supplementing these with localized bad passwords (e.g., common Filipino passwords), as well as organization-specific bad passwords (i.e., passwords based on your organization or system name).

Will improved password policies and an aggressive information campaign result in minimizing the use of unsafe passwords found in breach corpuses? Going forward, the researchers plan to conduct another measurement during the following trimester (third trimester, school year 2017-2018) after conducting appropriate information campaigns regarding better password habits and developing newer password policies based on the latest recommendations of NIST. Stay tuned!

References

[1] Grassi P A, Garcia M E and Fenton J L 2017 *NIST SP 800-63-3: Digital Identity Guidelines* https://doi.org/10.6028/NIST.SP.800-63-3

[2] Grassi P, Fenton J, Newton E, Perlner R, Regenscheid A, Burr W, Richer J, Lefkovitz N, Danker J, Choong Y, Greene K and Theofanos M 2017 *NIST SP 800-63B: Digital Identity Guidelines: Authentication and Lifecycle Management* https://doi.org/10.6028/NIST.SP.800-63b

[3] Yan J, Blackwell A, Anderson R and Grant A 2004 Password memorability and security: empirical results *IEEE Security & Privacy* vol **2** no **5** pp 25-31

[4] Adams A, Sasse M A and Lunt P 1997 Making passwords secure and usable *People and Computers XII* ed H Thimbleby et al (London: Springer)

[5] Bryant K and Campbell J 2006 User behaviours associated with password security and management *Australasian J. of Information Syst.* vol **14** no **1** pp 81-100

[6] Shay R, Komanduri S, Kelley P, Leon P, Mazurek M, Bauer L, Christin N and Cranor L 2010 Encountering stronger password requirements: user attitudes and behaviors *Proc. of the Sixth Symposium on Usable Privacy and Security 2010* p 2

[7] Florencio D and Herley C 2007 A large-scale study of web password habits *Proc. of the 16th Int, Conf. on World Wide Web* pp 657-66

[8] Stanton J, Stam K, Mastrangelo P and Jolton J 2005 Analysis of end user security behaviors *Computers & Security* vol **24** no **2** pp 124-33

[9] Egelman S, Sotirakopoulos A, Muslukhov I, Beznosov K and Herley C 2013 Does my password go up to eleven?: the impact of password meters on password selection *Proc. of the SIGCHI Conf. on Human Factors in Computing Systems* pp 2379-88

IOP Conf. Series: Materials Science and Engineering **482** (2019) 012036 doi:10.1088/1757-899X/482/1/012036

Bias detection in Philippine political news articles using SentiWordNet and inverse reinforcement model

T A Quijote[1], A D Zamoras[2] and A Ceniza[3]

Department of Computer and Information Sciences, School of Arts and Sciences, University of San Carlos, Cebu City, Philippines

[1] thessaquijote@gmail.com, [2] allenazamoras@gmail.com, [3] amceniza@usc.edu.ph

Abstract. Not all information posted on the internet is deemed 'trustworthy.' Some articles, especially those related to politics, seem to display traces of bias, whether they be for or against the Philippine administration. This research aims to determine if a news article—and by extension, a news outlet—is biased based on its sentiments and use of lexica. Data were harvested from chosen websites and news outlets provided by Alexa. These data underwent pre-processing and were scored based on their sentiments with the use of SentiWordNet. The resulting scores were then fed into the Inverse Reinforcement Model, which determined whether an article is biased or not. With the use of Inquirer, Philstar, Manila Bulletin, The Manila Times, and Journal Online news articles, the system was able to detect bias with an accuracy rating of 0.89, precision of 1, recall of 0.60 and F-Measure of 0.75.

1. Introduction

Nowadays, more and more people turn to the internet as their primary source of news and information, but with this lies a problem: the internet is not as credible as some may think. Because of its vast and open nature, the internet is riddled with biased and deceitful content posted by non-credible sources, causing the spread of misinformation and pretenses.

According to Parker and Berman, bias means that observations are constantly influenced away from the truth and swayed towards one direction [1]. It poses a potential problem not only in clinical research, but also in articles of a political nature. Since people rely on news articles to deliver informative and accurate reports on current events, objectivity is a must when it comes to reporting the news.

It is important then to detect whether biases exist in the news articles we read, most especially if these biases present themselves subtly. Multiple solutions to this problem have been devised with the capacity to detect bias, especially with the use of sentiment analysis and its multiple approaches. In their comparative study, Collomb et al. [2] pointed out the differences among the approaches. The lexicon-based approach in particular is a famous one, and is known to work well and produce good accuracy ratings. This approach is usually implemented using lexicon detection, followed by sentiment strength measurement, such as context-based sentiment analysis. However, the integration of a different algorithm such as machine learning would achieve higher accuracy and boost its performance [3].

The employment of the Inverse Reinforcement Model, then, is unconventional. On its own, Inverse Reinforcement is more than capable of detecting bias and controversy [4], but combining it with lexicon-based sentiment analysis has not been documented before. The aim then is to detect bias in online news articles using SentiWordNet and Inverse Reinforcement.

2. Review of Related Literature

The following are related studies reviewed in this research.

2.1. Sentiment analysis in bias detection

In a study by Enevoldsen and Hansen [5], sentiment analysis was used to obtain more objective and quantitative measures of the political biases in newspapers. Their study focused on identifying how the two newspaper outlets, Berlingske and Information, portrayed two political parties, Alternativet and Liberal Alliance. The results showed that Information's articles about Alternativet were more positively worded than articles about Liberal Alliance, while Berlingske showed no significant difference in their portrayal of the two parties.

Another study by Zhang et al. [6] focused on the sentiment aspect of news articles to develop a system which detected and visualized sentiment tendencies of different websites. Their system would detect a website's sentiment bias when its tendency was different from the others. Further study using sentiment analysis to detect bias was conducted by Lazaridou and Krestel [7], wherein they aimed to detect the expressed sentiment an outlet will show about a politician which they captured in a signed graph.

2.2. Lexicon-based approach

Lexicon-based sentiment analysis is one of the most popular approaches to sentiment analysis. With this approach, a dictionary of words is required where each word is assigned a positive or negative sentiment value [8].

One study improved upon the traditional sentiment analysis (SA) approach—assigning text to a specific opinion category—by making use of the HK method. The main advantage of this method is the implementation of an automated analysis since it is a supervised SA. A dictionary has to be defined, though, which leads to its major drawback: the difficulty in classifying opinions expressed through ironic or paradoxical sentences, or in appreciating all the nuances of language, such as specific jargons [9].

2.3. Inverse reinforcement model in bias detection

The Inverse Reinforcement Model was developed in 2006 by Lauw, Lim, and Wang, where they observed that bias and controversy are mutually dependent, and proposed a reinforcement model to address this mutual dependency. They tested the model on product reviews where the reviewer assigned a 0 to 5 rating to an object. The reviewers were placed in ranked lists in descending order according to their bias values computed by Naive and Inverse Reinforcement. The results showed that there were significant differences between the ranked lists due to cases where Inverse Reinforcement disagreed with Naive by taking into account the mutual dependency between bias and controversy. These results were deemed to be encouraging [10].

Semine and Tumulak [4] developed a judging and scoring system in which the Inverse Reinforcement Model was embedded in order to detect bias and controversy. After a series of tests and validations, they concluded that the Inverse Reinforcement Model can be well implemented in an automated judging and scoring system.

3. Methodology

This section discusses the methodology employed by the research in order to develop a system that can effectively detect bias in Philippine news articles found online. Figure 1 illustrates the conceptual framework of the research.

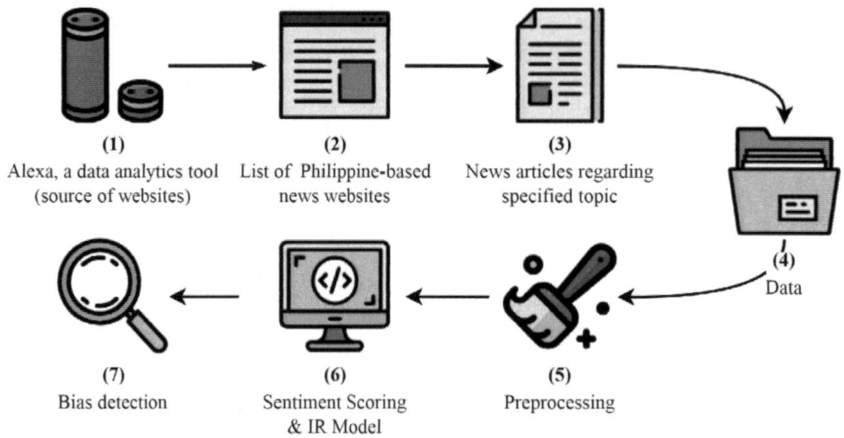

Figure 1. Conceptual framework of the research

3.1. Gathering of data

Alexa, a web traffic and data analytics tool from Alexa Internet, Inc., served as the source for this research (1). It ranked multiple news outlets in the Philippines based on popularity, and was used to identify which news websites should be considered to gather data from. Under the category 'News and Media > Newspapers', the news outlets used are: Inquirer.net, Philstar, Manila Bulletin, The Manila Times, and Journal Online (2).

Articles used for the research were manually picked out by the researchers. A topic was chosen, and articles related to the topic from each aforementioned news outlet was selected for scraping (3). A web scraper was used to extract the body of the articles, then exported into CSV files, which served as the data for this research (4). A total of 22,825 words were gathered.

3.2. Pre-processing of data

The collected data underwent pre-processing (5). Every character in the article was converted to lower case to obtain uniform casing. The resulting text was then tokenized into words. Stop-words such as "am", "the", "she", "to", etc. were removed from the document. All punctuation marks, except for the hyphen, were then removed from the remaining text. 13,126 words were left for processing.

3.3. Scoring with SentiWordNet

After pre-processing, all words in the document were assigned a raw score (6). SentiWordNet, a lexical resource for supporting sentiment analysis, was used to assign three scores: positivity, negativity, and objectivity score. The assigned raw scores were within the range 0 to 1, where the sum of the scores was 1 [11]. The highest score was identified and served as the word's raw score.

Afterwards, each document was classified into either positive or negative [12]. The total score for the positive and negative words were calculated separately, then normalized by calculating the ratio of each score to the sum of both scores. If the normalized positive score was greater than the normalized negative score then the document was labeled positive. Otherwise, negative.

3.4. Inverse reinforcement model

The Inverse Reinforcement Model was employed to detect bias (6), which adopted an iterative process. Initial data were taken from the tabulated scores produced from sentiment scoring. The model then computed for the deviation values for each outlet on each article they published using the formula for deviation, Equation 1. d_{ij} is the average distance between r_i's score and the score of each co-outlet r_k, where $m_j > 1$ denotes the number of co-outlets, e_{kj} is the score of r_k, and e_{ij} is the score of r_i.

$$d_{ij} = \frac{1}{m_j - 1} \sum_{r_k \neq r_i} \left| e_{kj} - e_{ij} \right| \tag{1}$$

$$b_i = Avg_j\, d_{ij} \cdot \left(1 - c_j\right) = \frac{\sum_{j=1}^{n} d_{ij} \cdot (1 - c_j)}{n_i} \quad (2)$$

$$c_j = Avg_i\, d_{ij} \cdot (1 - b_i) = \frac{\sum_{i=1}^{m} d_{ij} \cdot (1 - b_i)}{m_j} \quad (3)$$

n is the total number of articles, m is the total number of outlets, n_i is the number of articles evaluated by r_i, and m_j is the number of outlets evaluating article o_j.

From the computed deviation values, the initial bias values b_i were calculated using the formula in Equation 2, with the average deviation values for each outlet as the initial controversy values c_j. The controversy values c_j were then computed using the formula in Equation 3, taking the computed initial bias values b_i and the deviation values as inputs. The bias values were computed again using the formula in Equation 2, now using the deviation values and the newly computed controversy values as inputs. The whole process of computing the bias values and controversy values were repeated using each other's values and deviation values as inputs, thus, applying an iterative process. The iterations stopped only if they reached convergence, which was when the bias values do not change [10].

The Inverse Reinforcement Model assumed that all outlets are biased, but differed in their level. Bias levels were categorized into four according to the level of bias which covered outlets, as shown in Table 1.

Table 1. Four levels of bias.

Category	Bias Values	Level of Bias
Level 1	Lower than 0.1	Allowable bias
Level 2	Above 0.1 but lower than 0.2	Relatively considerable bias
Level 3	Above 0.2 but less than 0.25	Biased
Level 4	Above 0.25	Extremely biased

To properly detect bias, the system considered an allowable bias of not more than 0.19 (7). This allowed the system to detect relatively biased outlets and articles [4].

4. Results

Before feeding the actual gathered data into the system, testing was conducted to verify the validity of its results. The system was first tested on synthetic data wherein an outlet's articles simulated bias sentiments. The second test involved analyzing unbiased articles. The third test was a combination of biased and unbiased articles to test the application's accuracy in detecting both. After passing all three tests, the application was used to determine bias in actual data. The system's accuracy, precision, recall, and F1 score were calculated based on the following formulas:

$$Accuracy = \frac{TP + TN}{TP + FP + FN + TN} \quad (4)$$

$$Precision = \frac{TP}{TP + FP} \quad (5)$$

$$Recall = \frac{TP}{TP + FN} \quad (6)$$

$$F1 = \frac{2 \cdot (Recall \cdot Precision)}{(Recall + Precision)} \quad (7)$$

where TP is the number of correctly predicted biased articles, TN is the number of correctly predicted unbiased articles, FP is the number of wrongly predicted unbiased articles, and FN is the number of wrongly predicted biased articles. A threshold of 80% served as the basis for success.

Table 2. Test 1 with simulated bias sentiments.

Outlet	Sentiment Values[a]	Deviation Values[a]	Bias Values
1	0.44	0.12	0.10
2	0.44	0.13	0.11
3	0.41	0.14	0.11
4	0.43	0.12	0.10
5	0.85	0.42	0.35

[a] Values shown are averages of the values of the 'articles' in each outlet.

For test 1, outlet 5 was purposely given biased scores. Results in Table 2 showed that outlet 5 was correctly predicted as biased, while the rest were correctly predicted as unbiased.

Table 3. Test 2 (a) with unbiased articles and Test 3 (b) with a combination of biased and unbiased articles.

(a)				(b)			
Article	Sentiment Values	Deviation Values	Bias Values	Article	Sentiment Values	Deviation Values	Bias Values
1	0.16	0.20	0.17	**1**	0.20	0.32	0.27
2	-0.17	0.19	0.16	**2**	-0.17	0.16	0.14
3	0.17	0.21	0.18	**3**	-0.22	0.19	0.15
4	-0.13	0.15	0.13	**4**	-0.27	0.23	0.19
5	0.14	0.19	0.16	**5**	0.17	0.29	0.24
6	-0.11	0.15	0.12	**6**	-0.11	0.15	0.13
7	-0.11	0.15	0.12	**7**	-0.11	0.15	0.13

The system correctly predicted all unbiased inputs as unbiased for the second test. For the third test, articles 1 and 5 were correctly predicted as biased; 2, 6 and 7 were correctly predicted as unbiased; and articles 3 and 4 were falsely predicted as unbiased. Combining all TP, TN, FP, and FN values across all three tests yielded an accuracy of 0.89, precision of 1, 0.60 recall, and 0.75 F1.

Table 4 presents the results produced by the system after feeding it with the actual gathered data.

Table 4. Results of the bias detection system with actual data.

News Outlet	Sentiment Values[a]	Deviation Values[a]	Bias Values
Inquirer	-0.14	0.19	0.16
Philstar	-0.18	0.15	0.13
Manila Bulletin	-0.05	0.22	0.18
The Manila Times	-0.08	0.23	0.19
Journal Online	-0.10	0.18	0.15

[a] Values shown are averages of the values of the articles in each outlet.

According to the results, articles from The Manila Times were considered to be biased, and by extension, the outlet as well.

5. Conclusions

This research was able to detect bias from online news articles by implementing the Inverse Reinforcement Model with SentiWordNet. Using SentiWordNet, the general sentiment of articles were obtained and were later used as input for the Inverse Reinforcement Model. The model calculated the deviation values and controversy values of the articles to determine the bias values of the outlets. After a series of tests and validations, the system was able to determine that articles from leading Philippine-based news outlets were unbiased, except for those from the Manila Times. The Manila Times was then deemed to be biased, with an accuracy rating of 89%.

The results produced by the research are known to be significant since they provide concrete values and figures–i.e. accuracy, precision, recall, and F1–that describe how successful the system is. Other related studies and approaches have failed to state how successful their methods were in terms of statistical approaches. In addition, this research focused on the news articles themselves and how effective the IR Model was on them, seeing as how it was not used on this type of data before. Consequently, this research did not dwell on the political parties that news outlets sided with.

It is important to note that the current system only determines the magnitude of bias. For our future work, we wish to include identifying the direction of bias to further enhance our system's capability. Furthermore, we would also like to consider the incorporation of other sentiment analysis techniques.

6. Acknowledgements

The researchers would like to extend their gratitude to the Department of Computer and Information Sciences and the University of San Carlos for supporting this research endeavour.

References

[1] Parker R A and Berman N G 2016 *Planning Clinical Research* (Cambridge University Press) pp 203–16
[2] Collomb A, Costea C, Joyeux D, Hasan O and Brunie L 2014 A Study and Comparison of Sentiment Analysis Methods for Reputation Evaluation
[3] Mudinas A, Zhang D and Levene M 2012 Combining lexicon and learning based approaches for concept-level sentiment analysis *Proc. of the First Int. Workshop on Issues of Sentiment Discovery And Opinion Mining* p 5
[4] Semine R V and Tumulak J 2010 Automated bias detection judging and scoring system using inverse reinforcement (IR) model *Philippine Information Technol. J.* vol **4** no **1** pp 3-10
[5] Enevoldsen K C and Hansen L 2017 Analysing political biases in danish newspapers using sentiment analysis *J. of Language Works* **2(2)**
[6] Zhang J, Kawai Y, Nakajima S, Matsumoto Y and Tanaka K 2011 Sentiment bias detection in support of news credibility judgment *44th Hawaii Int. Conf. on System Sciences* pp 1–10
[7] Lazaridou K and Krestel R 2016 Identifying political bias in news articles *Bulletin of the IEEE TCDL* **12**
[8] Jurek A, Mulvenna M D and Bi Y 2015 Improved lexicon-based sentiment analysis for social media analytics *Security Informatics* **4(1)**
[9] Ceron A, Curini L, Iacus S M and Porro G 2014 Every tweet counts? how sentiment analysis of social media can improve our knowledge of citizens' political preferences with an application to italy and france *New Media & Society* **16(2)** pp 340–58
[10] Lauw H W, Lim E and Wang K 2006 Bias and controversy: beyond the statistical deviation *Proc. of the 12th ACM SIGKDD Int. Conf. on Knowledge Discovery and Data Mining* pp 625–30
[11] Baccianella S, Esuli A and Sebastiani F 2010 Sentiwordnet 3.0: an enhanced lexical resource for sentiment analysis and opinion mining *Lrec* **10** 2010 pp 2200–4

[12] Sharma S, Chakraverty S, Sharma A and Kaur J 2017 A context-based algorithm for sentiment analysis *Int. J. Computational Vision and Robotics* **7**

IOP Conf. Series: Materials Science and Engineering **482** (2019) 012037 doi:10.1088/1757-899X/482/1/012037

Optimizing Git for handling database backups: findings and insights from another 6,000 hours of experimentation

JV Roig
Advanced Research & Consulting, Asia Pacific College, Philippines

E-mail: jvr@apc.edu.ph; jvroig@gmail.com

Abstract. This is the second in a series of papers detailing results from on-going investigations of using Git - a distributed source control system - as a database backup and disaster recovery tool. The first paper, published early 2017 after 2,000 hours of data processing and simulations, warns that the default configuration of Git is unsuitable for database backups, but also shares configuration tweaks that make it safe and more space-efficient than zipped .sql archives - as low as only 1% of the storage space needed by .sql archives. This second paper focuses on optimization, looking specifically to identify which configuration tweaks and factors affect the reliability, scalability and efficiency of a Git-based database backup and disaster recovery system. Critical insights regarding the reliability and scalability of a Git-based database backup system were found after 6,000 hours of experimentation, centered around the problem of configuring Git to be able to handle .sql files measured in gigabytes.

1. Introduction

This is the second in a series of papers that share the results of my on-going experiments in deploying Git – one of the most popular source control systems in use today – as a database backup and disaster recovery system. For most companies, there is an obvious benefit to be gained (if using Git proves to be reliable for database backup and disaster recovery): Being one of the most popular source control systems, Git know-how is likely already a skill that exists within a company's technology team, so using the same tool for backups and disaster recovery means being able to leverage existing skillsets. Git is also free and open-source software, so it has the further advantage of being free (zero cost) to deploy within the organization for any use.

This paper shares the results of thousands of hours of experimentation aimed at finding out how to properly optimize Git – that is, tweak relevant or critical configuration settings of Git – in order to maximize the reliability and scalability of Git for use as a database backup and disaster recovery system.

2. Background and Related Literature

The first paper in this series (published early 2017) dealt with the initial investigation of feasibility: "Could Git be a reasonable, cheap option for managing database backups?" [1] The results found in this study was promising:

- Git can be safely used for database backups as long as the configuration *windowMemory* and *threads* are set to avoid exhausting all RAM, and *bigFileThreshold* is set high enough to be able to delta-compress your biggest expected .sql file size.
- Delta compression used by Git resulted in >99% storage space savings compared to a collection of zipped .sql files.

There has been little published information about investigative experiments or results regarding the use of Git for database backups. This was true when I first started the experiments in late 2016 and had it published early 2017. This scarcity of data persists today. The interest in using Git for database backups, however, is common from developers of all kinds, as can be seen from company blogs, personal blogs, and even in popular developer-oriented forums like *StackOverflow*. [2] [3] [4] [5] Part of this is due to Git's lower storage space usage compared to traditional source control systems. [6][7]

While the current definitive text available in public literature for experimental data in using Git for database backups remains my original paper, it has a couple of limitations:

- It does not investigate factors that may affect the reliability and scalability of Git.
- The individual file sizes handled in those experiments were under 2GB, with most being under 1GB, as those were the available backup data on hand.

The follow-up experiments described in this paper fill those gaps. Whereas the first paper was based on 2,000 hours of processing and simulations, this second paper is based on an additional 6,000 hours.

3. Experiment Design and Methodology

6,000 hours of data processing and simulations were done using 4 research servers (specs detailed in Table 1) handling 1,000 zip files. These zip files contain .sql files that are 1.5 – 2.5 GB each. These .sql files are historical backup data from an actual production system that I manage. Each server processes the thousand .sql files using different combinations of different Git settings in order to figure out which settings – or combination of settings – yield the lowest backup footprint (lowest storage space needed for the resulting Git repository) and which settings affect the speed of checking out a specific snapshot (i.e, how fast it takes for Git to retrieve a specific SQL file from a specific backup date).

In general, the experiment design and methodology used has not diverged much from the experimental design described in the first paper. An automated script, with only minor modifications from the original script used in the first paper, carries out the same steps as before per experiment:

1. A Git repository is initialized inside the main experiment folder. The configuration file of Git is also automatically tweaked according to the desired settings being for this experiment.
2. A zip file is selected (in chronological order according to the timestamp in the filename) from a folder that contains all 1,000 zipped .sql files the experiment will process.
3. This zip file is copied inside the Git repository, any existing .sql files are deleted, the contents of the zip file are extracted, then the zip file is deleted. Only the new .sql files remain in the working directory of the Git repository after this step.
4. The changes are staged and committed into the Git repository.
5. Depending on the current experiment being run, the script will trigger Git's garbage collection at a certain loop number. In some experiments, this may only be at the very last commit; in others, this may be every *n* commits.
6. The script loops back to step #2 until all 1,000 zip files have been processed and their contents committed to the Git repository.

Once the script finishes one entire run, the resulting Git repository has simulated what it would have looked like if the database backups were done using Git: it now contains every backup file, in the correct, chronological order. The 1,000 zip files represent 500 days of backups; the simulation done by the script takes only 1 or 2 days, depending on the specific settings of the experiment.

There is also a separate script that does automated checkouts, retrieving each commit one at a time. This tests how fast or slow retrieval becomes based on various Git settings used to create the repository.

The experiments are focused on testing the effects and contributions of the following Git settings:

- *threads* – how many threads should Git use for our particular use case?
- *bigFileThreshold* – this setting limits the biggest file size that Git will consider for delta compression. Given the larger file sizes in these new sets of experiments, how much should this setting be increased?
- *windowMemory* – this setting limits how much memory Git will use during packing. How far should this setting be increased without risking exhausting all RAM and swapping, effectively killing the commit process, and therefore the backup process?
- *depth* – how long should the delta chains be, and how does it affect space usage and retrieval times?
- *GCMOD* - not an internal Git setting itself, but an important configuration tweak for the entire backup process itself: how often to trigger Git's garbage collection (the *git gc* command). The name comes from "garbage collection modulo", the operation that computes whether the script should trigger *git gc* or not.

Each experiment is a combination of various values for the above settings. The raw results of each (processing time, final repository size, and checkout/retrieval speed) are logged by the script itself, which I then collected afterwards to process and summarize. The volume of experiments done is summarized in Table 2.

Table 1. Specifications of the research servers used. They are all identical to minimize variance.

Server Name	CPU	RAM	OS
Galaxy	Intel i7-6700 (Skylake)	32GB DDR4	Fedora 25
Constitution	Intel i7-6700 (Skylake)	32GB DDR4	Fedora 25
Sovereign	Intel i7-6700 (Skylake)	32GB DDR4	Fedora 25
Intrepid	Intel i7-6700 (Skylake)	32GB DDR4	Fedora 25

Table 2. Experimentation volume summary.

Server Name	# of Commit Experiments	# of Checkout Experiments	Total Experiment Time (hours)
Galaxy	88	160	2,570.14
Constitution	80	152	2,506.92
Sovereign	32	44	618.87
Intrepid	17	29	367.93
TOTAL	**217**	**385**	**6,083.86**

Table 3 shows a partial list of the experiments done in the "Galaxy" server. To make sure that results are consistent and replicable, most experiments were duplicated in a separate server. Almost the entire experiment list of "Galaxy", for example, was also done in the "Constitution" server. The entirety of the work done by all servers, including the full list of experiment configurations, raw results, and the automated scripts themselves are available as supplementary information from the dedicated research website for this experiment: https://research.jvroig.com/git_research_2.

4. Analysis and Results

After 6,000 hours spread out over 600 total experiments, critical insights regarding the reliability and scalability of a Git-based backup system were found. By *reliability*, I refer to the characteristic of the backup process not failing (i.e., the .sql file is always successfully committed into the Git repository).

By *scalability*, I refer to the characteristic of being able to handle bigger backup sizes and more volume, and also retaining an acceptable retrieval speed.

Table 3. Partial list of experiments in "Galaxy" server to illustrate the exploration of different Git settings. The columns "GCMOD" to "BigFileThreshold" refer to configuration values of those Git settings for the experiment. These Git settings were described earlier in section 3.

Exp #	GCMOD	Depth	Threads	WindowMemory	BigFileThreshold
1	5	50	1	8G	6G
2	5	200	1	8G	6G
3	5	300	1	8G	6G
4	200	50	1	8G	6G
5	200	200	1	8G	6G
6	200	300	1	8G	6G
7	100	250	2	4G	2G
8	100	250	2	4G	6G
9	4	250	2	4G	2G
10	4	250	2	4G	6G

4.1. Threads

Increasing the *threads* setting showed no improvement in repository size. And since repository size was unaffected, the speed of retrieval was likewise unaffected. There is no upside in making Git use more than 1 CPU thread. There is a huge downside: Git's memory consumption is multiplied by the number of threads being used. From a reliability perspective, this means using more than one thread will greatly increase the chances of Git using more memory and exhausting RAM. From a scalability perspective, this greatly limits the maximum file size you can successfully back up. With no upside but severe downsides in both scalability and reliability, Git's *threads* setting should always be explicitly set to 1.

4.2. BigFileThreshold and WindowMemory

These two settings were found to be linked. Increasing only one or the other does nothing, if the other setting is a bottleneck (i.e., changing *bigFileThreshold* to 4G while retaining *windowMemory* at 2G is no better than having both at 2G). For maximizing reliability, the experimental results suggest that both settings should be set to twice the size of the largest .sql file size that you expect to handle. For example, if your backups routinely handle up to a maximum of 4G, they should both be set to 8G. A reminder of a critical finding from the original paper is warranted: the backup process will fail if Git exhausts the available RAM. For this reason, these settings assume that you have enough RAM in the backup machine that uses Git. Ideally, the machine should have more RAM installed than your highest setting (e.g., if *windowMemory* is set to 8G, it is recommended that your machine has more than 8GB installed, to take into account all other processes that also use RAM, including the OS itself.)

4.3. Depth

Git's *depth* setting controls how long the delta chain is, i.e., how often until a full copy of a file is stored instead of a delta. The longer the delta chain (the higher the *depth* setting), the smaller the repository size becomes since fewer full copies of files are stored. However, this has a more significant effect in retrieval times, as having a longer delta chain means any commit to retrieve will take Git longer to process. Experimental results found that retrieval times increase much faster than the decrease in storage space. Using *depth=250* makes the repository size only ~75% compared to the default of *depth=50* (0.875GB vs 1.126GB); however, average checkout times are more than doubled, taking >250% of the original average checkout time (61.19 seconds vs 22.11 seconds). Even the slow checkout time is still in the ballpark of a minute, though, so it can be a reasonable trade-off to increase *depth* from its default value to maximize storage space. From a reliability and scalability perspective, there is no significant downside, so this is more of a specific optimization to fit the organization's particular use case. Whether to do so would mostly depend on any service-level targets that may exist in your organization for recovery time commitments.

4.4. GCMOD

Increasing *GCMOD* means triggering Git's garbage collection less frequently (the command *git gc*). In the experiments, a *GCMOD* of 5 means *git gc* is triggered every 5 commits, a *GCMOD* of 20 means the trigger is every 20 commits, and so on. Similar to the *depth* setting, the repository size also decreases as *GCMOD* increases. However, unlike *depth, GCMOD* does not make retrieval times lag as much. From *GCMOD=20* to *GCMOD=200,* the repository size decreased by 35% (1.7GB down to 1.1GB), while retrieval times only got slower by 54% (14.36 seconds to 22.11 seconds).

More frequent garbage collection (lower *GCMOD*) does have the effect of lowering the peak size a Git repository will grow to, although the garbage-collected repository size (i.e., effective size right after garbage collection) is always bigger. Delaying garbage collection for a month, for example, will yield the smallest repository size right after garbage collection. However, the repository size will increase until then. This has no effect from a reliability perspective, but it may impact scalability. If you have a maximum limit to what the Git repository size should peak at, then you will have to adjust *GCMOD* to be more frequent (for example, if available space in your backup machine can only accommodate a 100GB repository, then you should trigger *git gc* before this limit is reached, so that the repository size shrinks back to ~1GB). In the absence of this concern, garbage collection can be set to a monthly schedule, such as triggering *git gc* through a cron job every beginning of the month.

4.5. Storage Efficiency

Similar to the original paper's findings, Git shows extreme storage efficiency for .sql files. The data set is composed of 1,000 backup files, zipped, weighing in at 121GB. Unzipped to raw .sql files, these backup files weigh in at approximately 1.2 – 1.5 TB. Experimental results showed that Git can store all these in a repository that is less than 1GB (875MB), shown in Table 4 along with Git's comparative % efficiency. Retrieval times average only a minute for this best space-optimized configuration. This best space-optimized result used the following settings: *Threads=1, BigFileThreshold=6G, WindowMemory=8G, Depth=300* and *GCMOD=200.*

Depth=300 lowers file size by making the delta chain longer, while *GCMOD=200* maximizes the number of commits before a delta chain is created. These two settings are the primary drivers of the vastly reduced file size (0.875 GB). The *Threads* setting, as noted previously, should always be 1. *BigFileThreshold* and *WindowMemory* only need to be set to be at least twice the expected largest inidividual file size. For this experiment, the largest file size is 2.5GB as noted in Section 3, so any setting that is at least 5G will suffice.

Table 4. Comparing the most space-efficient experiment result against the actual backup data.

Raw .SQL Files (GB) (estimate)	**Zipped Archives (GB)**	**Git Repo Size (GB)**	**Git % vs Raw .SQL files**	**Git % vs Zipped Archives**
1,300	121	0.875	0.06731%	0.72314%

5. Conclusion and Applications

Small and medium enterprises can use insights in this paper to setup a free but extremely efficient database backup and recovery system using Git, since it is free and open-source software. This is compatible with practically all database products, as it only ever deals with dumped .sql files, and dumped plain text .sql files are a near-universal-feature for almost all database products. Even large enterprises may be able to significantly reduce their backup costs by switching to a Git-based backup system from a proprietary, licensed system. For large enterprises that handle different database products (e.g., having SQL Server, Oracle, and Postgres in one data center), switching to a Git-based backup system can unify the database backup process and potentially lower operating costs.

This study has revealed that the biggest bottleneck for a Git-based backup & disaster recovery system isn't the total database size. Rather, Git is only affected by the largest individual file size it ever needs to process – this is the basis for adjusting *bigFileThreshold* and *windowMemory*, as noted in section 4.2. To scale Git towards handling databases that are tens of gigabytes without demanding a server with extreme RAM requirements, one possible approach is to dump the database into separate .sql files per table, as opposed to one .sql file containing the entire database. Instead of being bottlenecked by the total size of the database, Git's bottleneck would then shift to whatever the size of the largest table is. Testing this scenario is the recommended next step in this series of experiments, to further stretch the limits of Git scaling.

References

[1] Roig JV 2017 2,000 hours of git and databases: raising the bar for the minimum standard of database backups in the enterprise *Proc. of the 17th Philippine Computing Science Congress* pp 17-23
[2] Eisinger D 2009 Backup your databases in git *Viget* https://www.viget.com/articles/backup-your-database-in-git
[3] Kulbertis B 2011 Synchronizing a MySQL database with git and git hooks *Ben Kulbertis* http://ben.kulbertis.org/2011/10/synchronizing-a-mysql-database-with-git-and-git-hooks/
[4] Rafique J 2016 Automatic database backup using git (bitbucket/gitlab/github) *Jabran Rafique* http://jabran.me/articles/automatic-database-backup-using-git-hosting/
[5] Wobbily_col 2014 Is backing up a MySQL database in git a good idea? *Software Engineering Stack Exchange* http://softwareengineering.stackexchange.com/questions/241109/is-backing-up-a-mysql-database-in-git-a-good-idea
[6] Pearce S 2013 GitSvnComparison *The Linux Kernel Organization Git Wiki* https://git.wiki.kernel.org/index.php/GitSvnComparison
[7] Packard K 2007 Repository formats matter *Keith Packards's Blog* https://keithp.com/blog/Repository_Formats_Matter/

Pairing clients and psychologists using stable marriage problem approach

S N Huda

Department of Informatics, Universitas Islam Indonesia
Yogyakarta, Indonesia

sheila@uii.ac.id

Abstract. Health is an important thing in human life, it includes not only physical health, but also mental health. Choosing the right psychologist will help individuals with mental problem to overcome the problem. Choosing a psychologist is important, because the psychologist's personality will influence the process of psychotherapy. This research paper will describe our approach in building a Decision Support System model that will compute the pairing of clients and psychologists. In this study, Weighted Sum Model is adopted to calculate the preference value of each individual, both psychologists and clients. Then based on this preference value, the client-psychologist pairing will be calculated using the Stable Marriage Problem approach. The Gale-Shapley Algorithm is used in this system to solve the Stable Marriage Problem and has been able to produce a stable matching between one set of Clients and one set of Psychologists, and there is no blocking pair from the results of this algorithm.

1. Introduction

Health is an important thing in human life, it includes not only physical health, but also mental health. Mental health can be disrupted with a mental disorder, such as depression. As an example, based on [1], in the UK around 18% of the population in working age experience mental health problems.

Depression and anxiety have a correlation with hopelessness [2]. Hopelessness is defined as a symptom that originates from negative experiences and leads to a perception that the person is powerless and have no future [3]. Hopelessness also has a relationship with suicidal behavior [4]. Therefore, it is very important for someone with a mental disorder to get treatment from experts, such as psychologists and psychiatrists.

There is difference between psychologist and psychiatrist, the basic thing is that a psychiatrist is basically a doctor who studies mental disorders. As a doctor, a psychiatrist can prescribe medications which will work chemically to stabilize hormones in the body. Psychologist is an expert who studies human's psychology. Psychology emphasizes on the pattern or behavior of human life along with the reasons compared to the biological side as in the psychiatrist. Solutions offered by psychologist are more likely about mindset and lifestyle changes.

However, based on WHO estimation in developing countries, the proportion of psychologists in lower to middle income countries is 2-14 psychologists per 1,000,000 people, while in middle to higher income countries, this proportion increases to 147-379 per 1,000,000 people [5][6]. Even in developed countries like US, it is estimated that only 20.9% individuals with 12-month major depressive disorder get adequate handling [7].

High pressure in life can also results in depression. Furthermore, if depression conditions are not handled properly, they will also affect work status [8]. This becomes a chain of circles that needs to be disconnected. By knowing where to look for help, someone with psychological problems can be courageous to get handling from experts. One of them is through a psychological counseling bureau which is a company where a group of psychologists open counseling therapies (psychotherapy).

Choosing the right psychologist will help individuals feel comfortable. Choosing a psychologist is important, because the psychologist's personality will influence the process of psychotherapy [9]. According to [9], there are several things that become factors which affects the pairing of therapists to maximize the progress of psychotherapy. One of the consideration is a pairing based on therapists' field of interest (A-B dichotomy). Suppose therapist A is more successful in dealing with the problem of chronic depression, while therapist B is more successful at handling family and child problems, or therapist C is more successful at handling problems with addiction. In addition, similarity in race and social class between clients and therapists also often the criteria for selecting client-therapist partners, but there is no evidence that this variable has a positive influence on successful psychotherapy. Other factor that influences the pairing of clients and psychologists is gender. For example, the majority of clients from male therapists are women. Female therapist has a tendency to be more empathic and active, whereas male therapist tends to be more confrontational.

In this paper, we choose one psychology counseling bureau as a case study, a group of psychologists opened a joint practice in Yogyakarta. In the bureau, there is one person who work as the customer service and first interviewer. Client who wants to counsel needs to register first, then the client will be "interviewed" about the general idea of the problem by the customer service. The customer service will then choose the suitable psychologist for the client and arrange the consultation schedule. Through this mechanism, pairing client-psychologist tends to be a tacit knowledge, and if the person who is in charge of interviewing the initial client changes (for example, if the previous customer service stop working), this tacit knowledge will be lost along with previous customer service. In addition, this pairing can also be subjective with the officer tendency and can result in an imbalance load of the counseling among psychologists. It would be better if the clients can provide criteria of their desired psychologist, and the psychologists also give the client preference they want to handle, then some computation is done to pair clients and psychologists that will allow for equal distribution of counseling load and satisfy both clients and psychologists preferences.

This research paper will describe our approach in building a Decision Support System model that will compute the pairing of clients and psychologists. This DSS model uses Stable Marriage Problem/Stable Matching Problem (SMP) approach. Stable Marriage Problem is a problem to find a stable pairing between two sets of the same size given the order of preference of each element. The order of preference for each individual is computed using Weighted Sum Model. With this DSS model, it is expected that the pairing of client-psychologist becomes more appropriate to each preference and balances the workload of psychologists in the counseling bureau.

2. Related Works

Problems with pairing or matching have many applications in the real world. For example, [10] discussed about matching children who graduated from elementary school in Singapore to junior high schools. This matching takes into account the preferences of these children with school preferences in the form of grades of primary school graduation (Primary School Leaving Examination).

In other case, matching issue also exists in the economics domain, for example in [11] where a group of firms had a requirement for a group of workers candidate, while a group of workers candidate had a set of firms that they wanted to go to. This problem was also solved using stable matching approach.

While in the health domain, one of the most famous research in stable matching implementation is the study of [12] to pair the resident doctor to their first hospital, it took into account the pairing from two sides, both from the resident's preferences and hospital's preferences.

IOP Conf. Series: Materials Science and Engineering **482** (2019) 012038 doi:10.1088/1757-899X/482/1/012038

3. Methodology

3.1. Weighted Sum Model

Weighted sum model [13] is a widely used method for multi-criteria decision making. This method was introduced more than 50 years ago, but remains as a fundamental and most popular method for multi-criteria decision making [14]. In this study, WSM is adopted to calculate the preference value of each individual, both psychologists and clients. Then based on this preference value, the client-psychologist pairing will be calculated using the Stable Marriage Problem approach.

The value of each alternative is calculated from the sum of the values for each criterion multiplied by the weight of the criterion. The optimal value is taken based on the minimum value if the criterion is a constraint, or based on the maximum value if the criterion is profit, in accordance with Equations (1) and (2).

$$x = \min_{(x_1, \dots, x_n) \in D} \sum_{i=1}^{m} w_i a_i \quad (1)$$

Or

$$x = \max_{(x_1, \dots, x_n) \in D} \sum_{i=1}^{m} w_i a_i \quad (2)$$

Where x is the optimal value achieved, which is the minimum value (if the criterion is a constraint or a cost) or the maximum value (if the criterion is an advantage) of n alternatives available in domain D, where the value of each alternative i is the sum of each criterion from m number of criterion alternative multiplied with criterion weight w.

3.2. Stable Marriage Problem

Stable Marriage Problem is a well-known problem in the game theory with broad applications covering the fields of mathematics, economics and computer science. Stable Marriage Problem is a problem in finding a stable matching between two sets with the same size, given a sequence of preferences for each element. Matching is a one-to-one mapping from elements in one set to the elements on another set. A matching is said to be unstable if:

a. There is an element A in the first set that prefers an element B in the second set than the A's paired element, and
b. B also prefers element A compared to the element which become B's partner

This problem is generally called a Stable Matching Problem, where specifically called a stable marriage problem when the number of elements in the first group is the same as the number of elements in the second group. A Stable Marriage Problem involves two separate sets with size n, namely the set of men and $women$. Each individual has associated preference list of all members in the opponent set. For example, individual p prefers q than r, where q and r are members of the opponent set, and if and only if q precedes r from the list of property preferences of p [15].

For this example, a matching M is a one-to-one correspondence between the set of men and women. If men m and women w are a pair in M, then m and w are called $partner$ in M, and is written as Equation (3) below.

$$m = PM(w), w = PM(m) \quad (3)$$

Where:
$PM(m)$ is $M - partner$ of m;
$PM(w)$ is $M - partner$ of w.

Men m and women w are said to be a blocking pair of M when m and w are not a pair in M, but m prefers w than $PM(m)$, and w prefers m than $PM(w)$. A matching with at least one blocking pair is called unstable, whereas otherwise it is called a stable matching.

3.3. Gale Shapley Algorithm
Gale and Shapley [16] formulated an algorithm to obtain a stable matching solution of SMP. In the first iteration, every man from the set of Men submits a proposal to the woman he likes most. Every woman who receives a proposal will say "maybe" to the man she likes most from the entire sender of the proposal, while she will say "no" to the other man. By saying "maybe", this woman and the man have been "temporarily bond". In the next iteration, men who do not have a temporary partner submit a proposal to the woman who has not refused him according with his preference list, then every woman who receives the proposal, if she has been "temporarily bond" and accepts proposals from other men who she prefer than her temporary partner, she may break the temporary bond and say "maybe" to the new person whom she prefer. This process is repeated until there is no individual without partner. When all individuals have been paired, the problem is resolved and a stable matching is obtained.

3.4. System Architecture
The decision support system in this paper consists of 4 modules. The first module is the client registration module. This module consists of a registration page or login page, as well as input page for client data and preferences. The second module is the admin module, which contains data management pages, including psychologist data, and client data. The third module is a module for calculating the order of clients preferences for psychologists and psychologists preferences for clients. This module uses the Weighted Sum Model to calculate the order of preferences. The fourth module is the calculation of Stable Marriage Problem solutions using the Gale-Shapley algorithm. The general architecture of this system can be seen in Figure 1.

Weighted Sum Model is used to calculate client's preference towards psychologists with the attributes such as field of expertise from psychologists, race origin of psychologist, and psychologist's gender, and age. The same algorithm is used to calculate psychologist's preference towards client with the attributes such as client field problem, and client's gender.

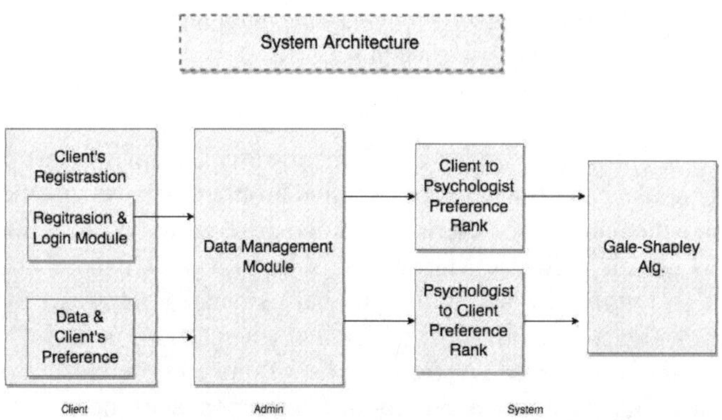

Figure 1. System architecture

4. Results
This application is created using the PHP programming language and MySQL database. The main page for visitor is a one landing page with several sections. In the Client's Registration Module, new client is required to complete his/her personal data and enter psychologist preference data which they prefer. The process of entering the client's personal data and preferences is shown in Figure 2.

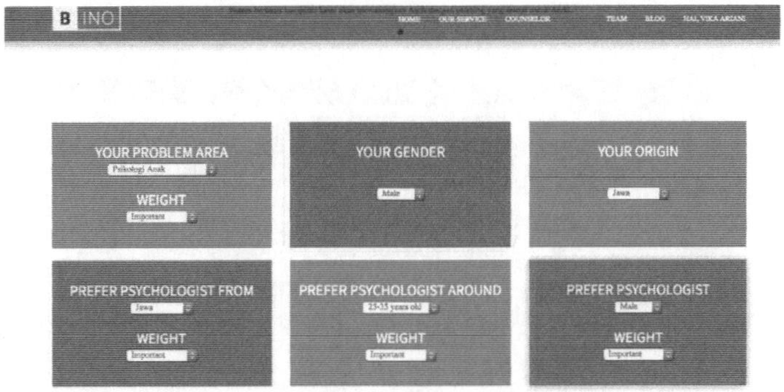

Figure 2. Client registration module

In the Data Management Module, system admin can manage the Psychologists and Clients data. Example of Psychologist data management is shown in Figure 3.

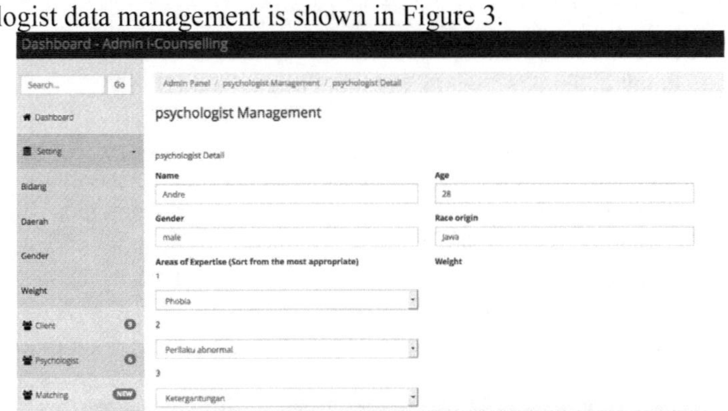

Figure 3. Data management module

Module 3 computes the preference rank from Clients to Psychologists and preference rank from Psychologists to Clients. Weighted Sum Model is employed to compute the clients' preference order with several attributes, such as field of expertise from psychologists, race origin of psychologist, and psychologist's gender, and age. Each attribute is given a proper weight based on the client's choice in the Client Registration Module. For example, a client named "Kiky" has depression problem, and she prefers female psychologist with race from Kalimantan and around 45-55 years old. This client feels that Psychologist's field suitability with her problem is high important while the suitability of gender and age is medium important and the suitability of race origin is low important. Using the Weighted Sum Model as in Equation (2), Kiky's preference order towards all psychologists is obtained. The same Model is used to compute the psychologists' preference order but with limited preferences. Psychologist preferences toward clients are computed using variable such as the suitability of client's problem with his/her expertise, and psychologists can state their preference for client's gender. For example, a psychologist named "Sarah" is a 30 years old female psychologist originated from Sumatera. Sarah has expertise in children psychology, teenager psychology, marriage psychology, depression, addiction, abnormal behavior, and phobia, with the order is taken into account. She prefer female clients over male clients. The suitability of client's problem with her expertise has more influence than the suitability in client's gender. Using the Weighted Sum Model, Sarah's preference order towards all clients is obtained.

Result from Module 3 is passed to Module 4 to compute the stable pairing between Clients and Psychologist. Given the preference order for both of Psychologists to Clients and Clients to Psychologists, the next step is to obtain a stable matching from this two sets using Gale-Shapley Algorithm. Example result of these module is shown in Figure 4. This calculation is also verified using manual calculation and resulted in the same pairing. The manual calculation is not written in this paper due to the page limitation. The result is also verified as a stable matching, since no blocking pair is appeared in the result.

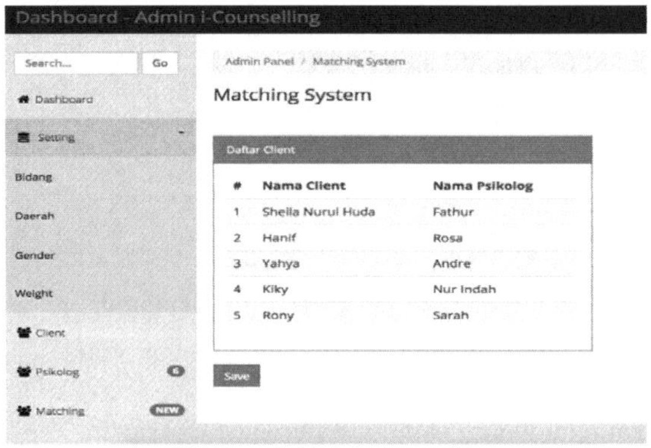

Figure 4. Pairing results

5. Conclusion

Decision Support System Model using the Weighted Sum Algorithm has resulted in Clients preference order for the Psychologists and the Psychologists preference order for the Clients with the Psychologist's attributes include expertise field suitability and gender preference, and attributes for the Client include expertise field suitability, gender preference, age preference and the preference of the race origin of the psychologist. The Gale-Shapley Algorithm used in this system has been able to produce a stable matching between one set of Clients and one set of Psychologists, and there is no blocking pair from the results of this algorithm.

In the future, the algorithm for calculating the order of preference can be made more complex, for example by using a combination of several methods other than simple Weighted Sum Model.

References

[1] van Stolk C, Hofman J, Hafner M and Janta B 2014 *Psychological Wellbeing and Work*
[2] Hope V and Henderson M 2014 Medical student depression, anxiety and distress outside North America: a systematic review *Med. Educ.* **48** 963–79
[3] Oyekcin D G, Sahin E M and Aldemir E 2017 Mental health, suicidality and hopelessness among university students in turkey *Asian J. of Psychiatry* **29** 185-9
[4] Bagge C L, Lamis D A, Nadorff M and Osman A 2014 Relations between hopelessness, depressive symptoms and suicidality: mediation by reasons for living *J. Clin. Psychol.* **70** 18–31
[5] Kakuma R, Minas H, van GInneken N, Dal Poz M R, Desiraju K and Morris J E 2011 Human resources for mental health care: current situation and strategies for action *The Lancet* **378**
[6] World Health Organization 2014 *Mental Health Atlas*
[7] Kessler R C, Merikangas K R and Wang P S 2007 Prevalence, comorbidity, and service utilization for mood disorders in the United States at the beginning of the twenty-first century *Annual Review of Clinical Psychology* **3** 137–58
[8] Tefft N 2012 Mental health and employment: the sad story *Economics and Human Biology* **10** 242-55
[9] Anderson L and Carter J H 1982 Psychotherapy: patient-therapist matching reconsidered *J. of The National Medical Assoc.* vol **74** no **5** p 461
[10] Teo C P, Sethuraman J and Tan W P 1999 Gale-shapley stable marriage problem revisited: strategic issues and applications *Integer Programming and Combinatorial Optimization IPCO 1999 (Lecture Notes in Computer Science)* vol **1610** ed G Cornuéjols et al pp 429-38
[11] Sotomayor M 2004 Implementation in the many-to-many matching market *Games and Economics Behaviour* vol **46** no **1** pp 199-212

[12] Roth A E 1986 On the allocation of residents to rural hospitals: e general property of two-sided matching markets *Econometrica: J. of the Econometric Society* vol **54** no **2** pp 425-7
[13] Zadeh L 1963 Optimality and non-scalar-valued performance criteria *IEEE Transactions on Automatic Control* vol **8** no **1** pp 59–60
[14] Filatovas E and Kurasova O 2011 A decision support system for solving multiple criteria optimization problems *Informatics in Educ.* vol **10** no **2** pp 213–24
[15] Gusfield D and Irving R W 1989 *The Stable Marriage Problem, Structure and Algorithms* (The MIT Press)
[16] Gale D and Shapley L S 1962 College admissions and the stability of marriage *American Mathematical Monthly* vol **69** pp 9-15

Mobile games interaction design for people with visual impairment using participatory design approach

Galang Prihadi Mahardhika[1], Arrie Kurniawardhani[2], Debby Yolhanda[3]

[1,2,3]Department of Informatics, Universitas Islam Indonesia
Jl. Kaliurang km 14,5, Yogyakarta, Indonesia 55584

[1]galang.prihadi@uii.ac.id
[2]arrie.kurniawardhani@uii.ac.id

Abstract. Mobile games are one of the technologies that can change the way people express and interact with their environment. In certain development, mobile games can also be used as medium to support the learning process. One of the main elements of mobile games is accessibility. Accessibility can be illustrated by the ease of interaction in the mobile games itself. There are not many mobile games developed for players with special needs, such as people with visual impairment. This research was conducted to develop alternative interaction design on mobile device that can be used by people with visual impairment. The development of alternative design was conducted by using participatory design approach. The participatory design approach is considered to be able to create interaction design which is appropriate with the participants wish. This study involved 10 visual impairment people as participants in the participatory design process. The final result showed that the alternative design produced were in accordance with participants desire.

Keywords: mobile games, accessibility, interactions, visual impairment, participatory design

1. Introduction

Computer game is a type of game that is highly developed today [1]. The development of computer games was driven by the development of commercial markets for computer-based games and other multimedia products. This type of game has become something that can change the way people express, learn, and interact with the environment around them [2]. The advantages of gim computer have made this type of game got new challenges where computer-based games are not only made for entertainment purposes. Currently computer game has also been developed for educational purposes, such as mathematics learning [3][4], for medical purposes, such as the introduction of nutritional needs [5] and others. One type of game that is growing rapidly today is a game built on mobile devices or better known as mobile games. Mobile games become popular today because of their ease of being able to be played anytime and anywhere [6].

One of the important elements in mobile game is accessibility that focuses on the ease of users to access all game contents. The accessibility development on computer game is more difficult than the development of accessibility on a general computer application [7], especially in the development of gim computer accessibility for people with special needs. Most computer game were not developed to be played by people with special needs [8], this is because people with special needs cannot access general input device and graphical interface perfectly [9].

On the other hand, the Special Interest Group on Game Accessibility (SIG GA) under the Independent Game Developers Association (IGDA) recommends several guidelines that can be used as a reference for developing the accessibility of computer games [10]. The recommendation is proposed for applications used by visually impaired (blind) and deaf (hearing impaired) users. However, in some cases, the guidelines cannot be applied properly [2]. To anticipate the needs that are not facilitated properly when developing computer game for blind people, the game development process needs to be carried out through collaboration between developers (designers) and users [11]. One of approaches that can be used to realize the collaboration is participatory design.

There are several software design approaches that are commonly used, namely user centered design and participatory design. On the user centered design, the research subjects have a role to provide information on the needs of the system, while on the participatory design the research subjects will contribute to make the system as participant. To get the original results, the participatory design will be used in this study. Participatory design is a software design approach that involves all actors (application developers, application users, etc.) at the design process stage [12]. The main challenge in implementing participatory design is to motivate users (as design actors) so that they are able to express creative and innovative ideas during the design development process [13].

This research was conducted to find out the types of interactions on mobile devices that are suitable for blind people. In this study, the participatory design approach will be carried out using cognitive walkthrough method. In cognitive walkthrough method, the design development will be conducted through several iterations until the final design agreed can be realized [14].

2. Related Works

Efforts to utilize multimedia technology were attempted by Miller through the development of an audio-based computer game or better known as audiogame [8]. The game developed in that study is played by playing a sound on a particular pattern, the player is asked to follow the sound pattern by pressing several characters on the input device. Audiogames make totally blind people possible to access the game.

In other studies, Yuan & Folmer tried to develop an interaction device that can adapt to a particular computer game [15]. The game chosen was "Guitar Hero", one of the instrument simulation type games that was not designed for blind users. In the study, Yuan & Folmer used haptic feedback that was attached to a certain glove. Through this feedback, the players can find out the right time to press the button on the controller device that is used as the interaction media in the game.

This research was conducted to develop an interaction design by utilizing multimedia elements such as sound, text, and color as output media and touch screen as input media. In addition, this research will be developed in accordance with the characteristics of blind people in Indonesia, especially in one of the extraordinary school in the Yogyakarta Special Region province. The challenge is the differences in acceptance related to technology, especially on technology of gim computer with mobile device-based.

3. Method

This research was conducted in 3 main steps. First, identifying the participants habits while using mobile device (Mobile Device Usage). Second, developing a participatory design scenario based on the elements needed in the development of interaction design (Participatory Design Scenario). Third, doing the participatory design process based on the scenario that has been made. The final result of the participatory design is an interaction design recommendation resulting.

3.1. Mobile Device Usage

Based on observation result at one of the extraordinary school in Yogyakarta (Indonesia), there were 10 people who can participate (4 totally blind people and 6 low vision people). The participants age varied between 4 years old and 6 years old.

The use of mobile device technology can be well received by students. About 80% of students can use the device. Students can take action such as press the button (tap), press and hold the button (hold), and move the screen (swipe). The use of Android-based mobile device has been generally used for learning (accessing information) and communicating. The use of mobile device has been more useful because it has various forms of interaction and relatively easy to operate than the use of computer technology such as PCs (personal computers) or laptops that need to keyboards and pointers.

The use of color combinations that are suitable for blind people with low vision category must have high contrast values. The colour combination commonly used are black-white, red-white, or blue-white.

3.2. Participatory Design Scenario

Participatory design in this study will be conducted in several processes (iterations) in accordance with the element that has been developed. Each iteration must be completed in sequence, therefore the completion of an iteration must be done before moving on the next iteration. The term iteration of the participatory design process refers to the workshop action with the participants. The element that developed is output elements (color, text, and sound) and input elements.

3.3. Participatory Design Process

Participatory Design was conducted in accordance with a predetermined scenario. The iteration that has been carried out in the participatory design process consist of color determination, text determination, sound determination, and determination of the form of input action.

3.4. Color Selection

The colors that can be seen clearly by the participants are high contrast colors. The colors used consist of 3 colors (1 basic color, and 2 accent colors). The use of basic colors must be used on each display, while accent colors can be applied either one color or two combination color but cannot overlap. The alternative color combinations used in this study are white-red-blue, yellow-red-blue, and black-white-yellow, with hexadecimal values #FFFFFF (white), #E30000 (red), #002B82 (blue), #FFFF00 (yellow), and #000000 (black). Alternative color design can be seen in Figure 1 (points a, b, and c).

The process of determining the color was conducted in 2 stages. The first stage was done to determine the color combination that has been selected. The second stage was done to determine the clarity level of color that has been used. Color determination was only involved low vision participants (6 people).

From the iteration process, it was found that 4 of the 6 participants had chosen the first design (white-red-blue), while the other 2 have chosen the second design (yellow-red-blue). In addition, there were 2 participants who saw blue as purple in the second design, and 1 participant who saw yellow as green in the third design. From these results it was agreed that the color used was a combination of white-red-blue (first design).

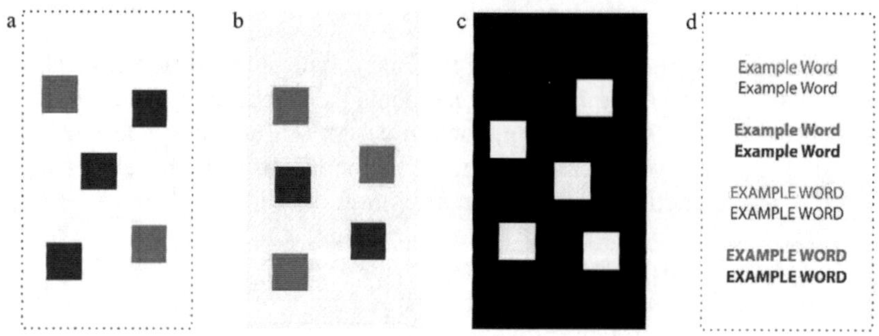

Figure 1. Screenshot of the color alternative design (a. white-red-blue, b. yellow-red-blue, c. black-white-yellow) and the font combination alternative design (d)

3.5. Text Selection

The type of font that has been recommended for blind people in the low vision category is Sans Serif [16]. Sans serif font has good readability, and clear simple form. The font used in this study is called 'Myriad Pro', which is one of types of Sans serif font that was commonly used.

The process of determining text is done to determine the combination of text capitalization that can be read easily by participants. The combination of capitalization text that has been used is capitalize each word (Example Word), Capitalize each word bold (**Example Word**), uppercase (EXAMPLE WORD), and uppercase bold (**EXAMPLE WORD**). This process only involved low vision participants (6 people). Alternative design for text combination can be seen in Figure 1 (point d).

Iteration result showed that 5 out of 6 participants have chosen the fourth combination (uppercase bold), while one participant did not choose because he cannot read. There was only one participant of 6 partisipants who can read all types of combinations clearly. Based on this result, the combination of text used is uppercase bold.

3.6. Sound Selection

In the interaction design that has been developed, sound has a function as an output medium that has been used to convey information related to action that must be conducted by the users. The process of selecting the type of sound was focused on determining how fast a sentence must be read. The sentence used for the process of determining the sound is "apakah suara saya terdengar dengan jelas?" (native language of Indonesia), which means "is my voice heard clearly?" There were 3 types of alternative speeds, namely fast (0.15 second/syllable), medium (0.4 second/syllable), and slow (0.7 second/syllable).

Design testing was conducted by make the partisipant listen to each sound that has been produced using headphone. The participants were then asked to choose one of the three types of sounds that have been played. The process of determining sounds involves all participants (low vision and totally blind participants).

From the process that has been conducted, there were 7 participants choosing medium speed, while 3 other participants have chosen fast speed. There was no participant who has chosen a slow speed, it was because reading at a slow speed can weaken the capture of the meaning of the sentence that has been read. From the results, the speed of sentence reading used is medium speed (0.4 second/syllable).

3.7. Determination of Input Action

Input action has been developed by referring to the action of pressing a button (tap), pressing and holding the button (hold), and shifting the screen (swipe) on the mobile device. The input action that has been developed must be able to accept at least 2 input options.

There were four alternative designs developed for the determination of input action design, namely 2 answers swipe (portrait), 4 answers swipe (portrait), 2 answers hold (landscape), and 3 answers hold (landscape). The action of pressing (tap) button was used as a triger to activate/repeat commands (questions). To test the suitability of orders and actions, simple mathematical questions were used in alternative design. Alternative design for input action can be seen in **Figure 2**.

The process of selecting the input action showed that the the input action design that most participant can answer math problems correctly is the hold interaction with both, 2 answers and 3 answers. However, most partisipants chose 2 answer swipe for the input action design. In selecting the best input action design, weight are given for each choice from the participants (3 points for the first choice, 2 points for the second choice, 1 point for the third choice, and 0 points for the last choice).

4. Results

The final result in the process of determining the input action can be seen in **Table 1**. Based on the participatory design rules, the alternative mobile game interaction design that has been recommended is 2 answers swipe. The interface in the interaction has been built with a combination of white, red and blue. The capitalization text used is uppercase bold with a reading speed of medium (0.4 second/syllable).

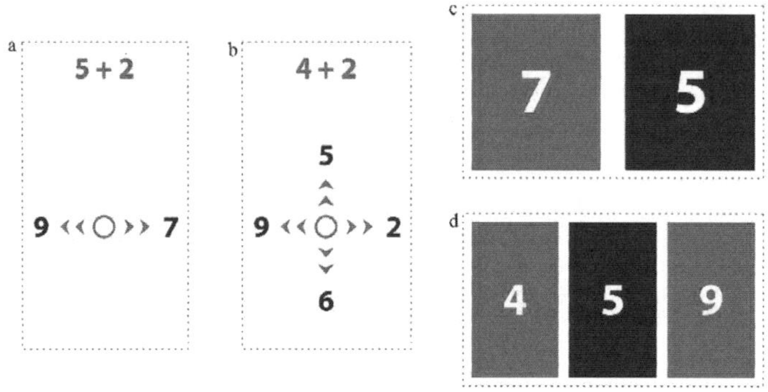

Figure 2. Screenshot of the alternative input action (a. 2 answers swipe, b. 4 answers swipe, c. 2 answers hold, d. 4 answers hold)

Table 1. Determination result of input action

No	Input Action	Results of The Answer Test			Results of The Selected Design	
		Avg of Correct Answers	Avg of Incorrect Answers	Percentage of Correct Answers	Weight of The Selected Design	Percentage of The Selected Design
1	2 answers swipe	8	2	80%	24	40%
2	4 answers swipe	8.6	1.4	86%	12	20%
3	2 answers hold	9	1	90%	18	30%
4	3 answers hold	9	1	90%	6	10%

5. Conclusions

The process of participatory design in this study was carried out by developing several alternative designs which were then selected and developed by participants. Alternative designs are used to provide stimuli to participants so they can immediately generate their ideas. In this study, alternative designs are submitted to participants in stages through 5 iterations until the desired model can be achieved.

Mobile game interactions design that is suitable for people with visual impairment can built with a combination of white (background), red (foreground), and blue (foreground). Type of font that can be used is 'sans serif', with 'uppcercase bold' capitalization type. To accommodate users in the totally blind category, the questions/instructions will be read at medium speed (0.4 second / syllable). The interaction model chosen by the participants is the 'two answers swipe' model.

6. Discussion

There were limitations of participants to be able to describe the design they expected. To anticipate this, we conducted the process of collecting design based on paticipants suggestions. From these suggestions, several design alternatives are made so that participants can choose the best design that meets their needs. In this way, we have assumed that the resulting design is the original result of the participants' ideas.

From the result of selecting input action design, the form of interaction with swipe 2 answers has become an alternative form of interaction that has been widely chosen. However, this form of interaction design has also become a form of interaction with the lowest level of answer's truth. This was probably caused by a type of problem that was not identical question. To get a certain answer, the

IOP Conf. Series: Materials Science and Engineering **482** (2019) 012039 doi:10.1088/1757-899X/482/1/012039

two alternatives (swipe 2 answers and hold 2 answers) must be compared with the appropriate method in the next research.

References

[1] Mahmoudi H, Koushafar M, Amani J and Pashavi G 2015 The effect of computer games on speed, attention and consistency of learning mathematics among students *Procedia – Soc. and Behavior. Sci.* **176** pp 419–24

[2] Yuan B, Folmer E and Harris F C 2011 Game Accessibility : a survey *Universal Access in the Inform. Society* **10(1)** pp 81–100

[3] Cheng H N H, Lin Y J, Wang M and Chan T W 2015 Math detective: digital game-based mathematical error dectection, correction and explanation *Proc. - IEEE 15th Int. Conf. on Advanced Learning Technologies: Advanced Technologies for Supporting Open Access to Formal and Informal Learning ICALT 2015* vol **15** pp 122–6

[4] Mahardhika G P 2016 Math Ship: digital game based learning to improve students' mathematics skill *Adv. Sci. Letters* **22(10)** pp 3181–4

[5] Ledoux T *et al* 2016 An educational video game for nutrition of young people : theory and design *Simulation & Gamming* **47(6)** pp 399–402

[6] Jeong E J and Kim D J 2008 Definitions, key characteristics, and generations of mobile games *Mobile Computing: Concepts, Methodologies, Tools, and Applications* pp 289–95

[7] Archambault D, Ossmann R, Gaudy T and Miesenberger K 2007 Computer games and visually impaired people *Upgrade 8(November 2006)* pp 1–21

[8] Miller D, Parecki A and Douglas S A 2007 Finger dance : a sound game for blind people *Proc. of the 9th Int. ACM SIGACCESS Conference on Computers and Accessibility* pp 253–4

[9] Buaud A, Svensson H, Archambault D and Burger D 2002 Multimedia games for visually impaired children *LNCS* **2398** pp 173–80

[10] Bierre K, Hinn M, Martin T and McIntosh M 2004 *Accessibility in Games: Motivations and Approaches* White Paper (International Game Developers Association)

[11] Triantafyllakos G, Palaigeorgiou G and Tsoukalas I A 2011 Designing educational software with students through collaborative design games: The We!Design&Play framework *Comp. and Edu.* **56(1)** 227–42

[12] Kensing F and Jeanette B 1998 Participatory design: Issues and concerns *Comp. Supported Cooperative Work (CSCW)* **7(3–4)** pp 167–85

[13] Lindberg S 2013 Participatory design workshops with children with cancer: lessons learned *Proc. of the 12th Int. Conf. on Interaction Design and Children* pp 332–5

[14] Mahatody T, Sagar M and Kolski C 2010 State of the art on the cognitive walkthrough method, its variants and evolutions state of the art on the cognitive walkthrough method, its variants and evolutions *Int. J. of Human-Comp. Interac.* **26(8)** 741–85

[15] Yuan B and Folmer E 2008 Blind hero: enabling guitar hero for the visually impaired *Proc. of the 10th Int. ACM SIGACCESS Conf. on Computers and Accessibility* pp 169–76

[16] Russell-Minda E, Jutai J W, Strong J G, Campbell K A, Gold D, Pretty L and Wilmot L 2007 The legibility of typefaces for readers with low vision: a research review *J. of Visual Impairment & Blindness* **101(7)** pp 402–15

Engineering kids health monitoring system in Indonesia

Irving V Paputungan[1,2], Hari Setiaji[3]

[1,3]Informatics Department, Universitas Islam Indonesia
[2]Universiti Teknologi Petronas, Malaysia

[1]irving@uii.ac.id
[3]hari.setiaji@uii.ac.id

Abstract. There is a healthcare program in Indonesia called Pos Pelayanan Terpadu (Posyandu). Services such as nutritional counselling, immunization service, health education and birth preparedness are conducted in the program. All the information collected form such program is still in paper-based document, called Kartu Menuju Sehat (KMS). A development of an electronic KMS (e-KMS) is proposed. This paper presents several related works on e-KMS for Indonesia. This paper derives requirements towards the most workable features for e-KMS based on the literatures.

Keywords: healthcare, posyandu, kartu menuju sehat, e-kms

1. Introduction

Healthy kids are a dream of every parent. Thus, it becomes very important for every parent to ensure that their kids are healthy. However, in Indonesia, large number of kids is still suffered malnutrition. Based on the survey by TNP2K [1], 27.5% of kids in Indonesia, mostly in east Java, are in nutritional deficiencies. Such number of stunting is not only because the Kid healthcare service is poor, but also lack of parental awareness in kids health as they grow.

Indonesian government has a program to ensure better health-care for toddler and pregnant mother. The program is called Posyandu (Pos Pelayanan Terpadu). It normally conducts important services such as nutritional counselling and immunization service. In some cases, Posyandu also provides health education and birth preparedness. Nutritional counselling is an assessment to analyse diet and exercise. Immunization is the process of making someone is immune to infection. Health education helps people to improve their health by increasing knowledge about health. Birth preparedness is an advance planning and preparation for baby delivery.

The children information collected from such program is currently in paper-based document. The document is called Kartu Menuju Sehat (KMS). Using KMS, parents can acknowledge the status of their children in terms of condition from time to time graphically. They can also react to any recorded irregularities. However, the use of paper-based KMS, to some extent, is less reliable. Paper-based document is easy to lost or misplaced and not efficient. In some cases, parents keep forgetting to bring the KMS during the program. It becomes difficult for the officer to record and retrieve the children data regularly. Furthermore, due to different handwriting types, reading handwritten health record also becomes an issue [2].

This paper presents the need towards the development of Electronics KMS (e-KMS). Such model of KMS is proposed to digitalize children health record. In next section several previous works related to digital KMS are presented. Such works are needed to derive some requirements for e-KMS. Section 3 presents the requirements. The summary is then shown in the final paragraph of this paper.

2. Related Work

Several studies have been performed related to the development of digital KMS. A multi-user information system has been developed by Nuraeni (2011) [3]. It is a desktop-based system that is installed in a health centre for patient data including new-born data. A web-based system was then proposed by Lestari et. al (2012) [2] to support online system. However, both implementations focus more on the general patient data.

A database system to manage children health data in healthcare centre was proposed by Indrijani (2013) [4]. The database covers all the records obtained during Posyandu. Searching patient data and reporting become easier using such database. An e-KMS was initially developed by Priskila and Wibowo (2013) [5]. There are 4 (four) forms created to input the data into KMS. Although such system is useful, graphical result and some health indicators are not supported. Patient identity is also missing in such development. Maulidia et al. (2015) proposed similar e-KMS with body weight chart as its additional features [6]. Such chart called anthropometric, is shown in Figure 1, shows children health status based on their weightage and height. It is also proposed by Sholihah and Kusumadewi (2013) [7] and Setyarini (2013) [8]. Radio Frequency Identification (RFID) is proposed by Eridani and Widianto (2014) as an identity for each patient [9]. Digital KMS that can be used on mobile platform is proposed by Windasari and Yana (2016) [10]. They developed an android app, called M-KMS, to keep the kids health records and to ease parents when accessing their kids' data. The app has anthropometric feature and immunization history. In 2013, Bhaskoro et. al (2013) proposed posyandu cloud to provide software as a service for posyandu [11].

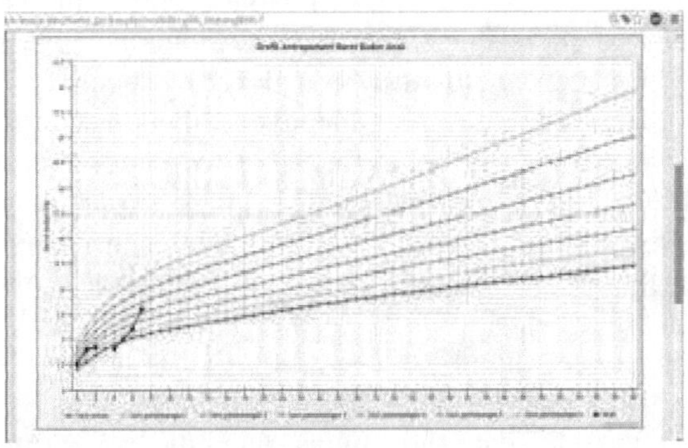

Figure 1. KMS chart for kids

There are also other works that proposed online KMS. KMS-Online (http://kms-online.web.id) by Javakedaton Indonesia and PrimaKu (http://www.primaku.com) by Ikatan Dokter Anak Indonesia (IDAI). Both implementations are a breakthrough in the area of children health, because it helps parents to promptly monitor their kids' condition via Internet. They also provide some useful health articles for guiding parents when take care the children. It, thus, makes online health monitoring becomes necessary. Figure 2 shows the trend of KMS development based on the literatures.

3. Findings: Requirements for e-KMS

Based on the aforementioned literature review, there are some important points that can be derived as requirements for digital KMS or e-KMS in Indonesia. They are:

a) Unique ID for each kid.

Digital identification is a basic item in any application. Identity plays a role for an application, even beyond the login screen. In e-KMS, identity is needed for authentication and user profile. Authentication dictates the way a user logs into the e-KMS. Previously, a simple username and password are enough; currently the use of ID number is more feasible for authentication. In addition, in Indonesia, kids are now can have an ID card. By having such identification, kids profile can also be retrieved.

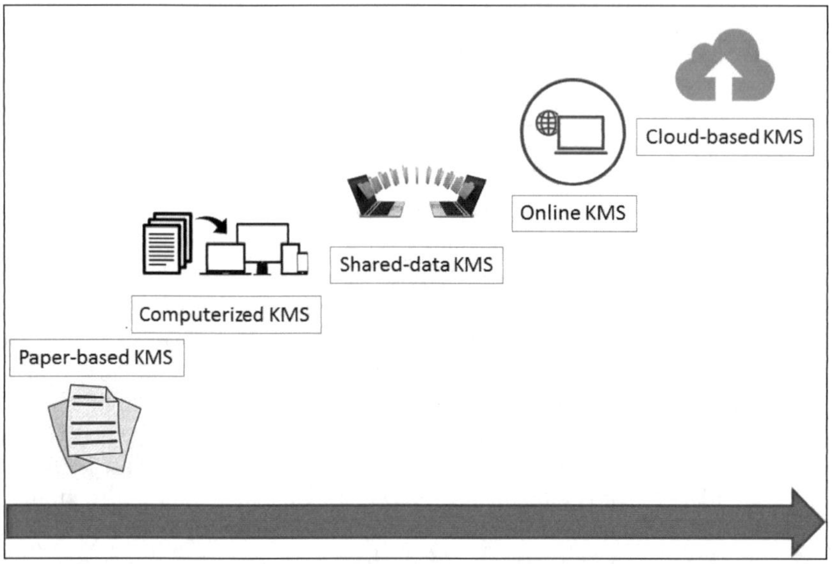

Figure 2. Graph of KMS trend based on literature

b) Cloud-based data management system

Continuous monitoring causes the need of large storage for the captured data. Yet, the number of children that are taken care in the posyandu increases. Thus, the data needs to be managed properly and the system must be maintained regularly. The emerging technology called cloud computing enables users to have computing resources off premises. The adoption of cloud technology is believably able to reduce the cost of having electronics healthcare records [14]. Cloud-based storage [12] makes users easy to deal with growing file size and cloud-based system [13] will have less maintenance stage from users' perspective. Furthermore, managing e-KMS and its data is important especially when it comes to distributed system. Since e-KMS will be used regularly in several places, e-KMS is better to be developed as a service. Cloud-based e-KMS, that is accessible via web or mobile, is a feasible option to simplify the system and data management. By only using mobile apps, posyandu staffs can input the data and parents can immediately read it.

c) Intelligent anomaly detection.

Having a regular data from the kids makes anomaly detection or estimation becomes feasible task to do. Monthly weight and height data, immunization data, and other additional data that are collected during posyandu will led to a good analysis when it comes to anomaly detection and estimation. Thus, any worsened situation can be prevented earlier.

d) Smart reminder.

Immunization date is crucial to the kids, but parents often forgetting due to their daily business. SMS is currently popular and most widely accepted mode of reminder. There is strong consistent evidence that reminder improved rate of attendance compared with no reminders. The use of reminder, either by SMS or any type of reminder, for parents is very helpful in order to keep alerted that immunization must be performed.

e) Specific article for each developmental milestone.

When reaching different stage of growth, kids will behave accordingly. Behaviours such as taking a first step, smiling for the first time, and waving "bye-bye" are called developmental milestones. Children reach milestones in how they play, learn, speak, behave, and move (for example, crawling and walking). Parents, especially who just had their first child, do not have any idea what to do to their kids when reaching certain milestone. Thus, providing some articles in the e-KMS can be a good option.

Figure 3 shows the abstract diagram of proposed e-KMS. Children data will be recorded by the Posyandu staffs to the cloud server using KMS app or Web-based KMS. Parents can view such data from web-based KMS or KMS app.

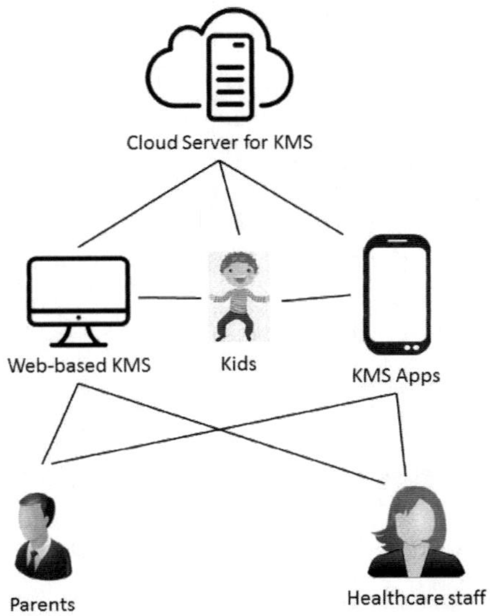

Figure 3. Abstract diagram of proposed e-KMS

There are other thoughts on how the e-KMS in Indonesia might be extended, they are:

a) **Healthcare logistic prediction.**
From the healthcare point of view, logistic becomes another issue when the posyandu program is conducted in remote area. There is only little number of staffs in the posyandu to cater a number of kids during such program. They also sometimes travel only by city car which has small luggage space. It means that they have to carry the logistic optimally. Predicting the quantity of logistics using Artificial Intelligent (AI) method, thus, should be performed.

b) **Internet of Things (IoT).**
Monitoring is one of the most important factors of child's growth and development. New technology called IoT enables new possibilities in the area of health monitoring. It normally uses wearable sensor technology to obtain the kids physical condition (e.g body temperature) real time. Parents will then be informed regularly about their kids either via SMS or Web-based system.

4. Summary

KMS in Indonesia is a good approach for monitoring children health. Cloud-based KMS that has kids' identification, able to detect certain anomaly, has smart reminder, and provides some articles related to kids healthcare, is arguably able to provide better children health monitoring. Parents and healthcare officer can obtain some huge advantages from such e-KMS. In addition, the use of AI and IoT can be a good option for future e-KMS.

References

[1] Jawa-Pos 2017 *PressReader.com - Connecting People through News* https://www.pressreader.com/indonesia/jawa-pos/20170713/282029032269956 retrieved June 2nd 2018

[2] Lestari Y K, Khoiri A and Sandra C 2012 Evaluasi terhadap pelaksanaan rujukan berjenjang kasus kegawatdarutan maternal dan neonatal pada Program Jampersal di Puskesmas Kencong Tahun 2012 *Jurnal Universitas Jember* **1** pp 1-126

[3] Nuraeni A 2011 *Sistem Informasi Kesehatan Ibu dan Bayi di Puskesmas* Bachelor Thesis (Jakarta: Universitas Indonesia)

[4] Indrijani 2013 Membangun basis data posyandu menuju Indonesia sehat sentosa *ComTech* vol **4** no **2** pp 618-26

[5] Priskila O and Wibowo A 2013 Efektivitas penggunaan kartu menuju sehat (KMS) elektronik untuk meningkatkan kecepatan pelayanan, mempermudah pendataan dan pengambilan keputusan status kesehatan di Posyandu *Jurnal Biometrika dan Kependudukan* vol **2** pp 27–32

[6] Maulidia, Nyoto R D, Sukamto AS 2015 Sistem informasi KMS (kartu menuju sehat) (studi kasus: UPTD Puskesmas Kecamatan Pontianak Barat) *Jurnal Sistem dan Teknologi Informasi (JUSTIN)* vol **1** no **1**

[7] Sholihah N and Kusumadewi S 2013 Sistem informasi posyandu kesehatan ibu dan anak *Prosiding SNATIF ke-2*

[8] Setyarini I 2016 *Perancangan Sistem Informasi Posyandu Guna Mendukung Pelaporan Data Perkembangan Bayi dan Balita* Bachelor Thesis (Kediri: Universitas Nusantara PGRI)

[9] Eridani D and Widianto E D 2014 Simulasi aplikasi posyandu berdasarkan konsep RFID *Sistem Komputer* vol **4** pp 37–41

[10] Windasari I P and Yana R R 2016 Aplikasi mobile kartu menuju sehat (M-KMS) *Jurnal Sistem Komputer* vol **6** no 2

[11] Bhaskoro S B, Supangkat S H and Nugraha I G B B 2013 Posyandu cloud: Healthcare networks to support user-oriented (maternal and neonatal) in Indonesia *Int. Conf. on ICT for Smart Society*

[12] Camarasu-Pop S, Cervenansky F, Cardenas Y, Nief J Y and Benoit-Cattin H 2010 Overview of medical data management solutions for research communities *IEEE/ACM Int. Conf. on Cluster, Cloud and Grid Computing*

[13] Rosenthal A, Mork* P, Li M H, Stanford J, Koester D and Reynolds P 2009 Cloud computing: A new business paradigm for biomedical information sharing *J. of Biomedical Informatics* **43** pp 342–53

[14] Taylor N F, Bottrell J, Lawler K and Benjamin D 2011 Mobile telephone short message service reminders can reduce nonattendance in physical therapy outpatient clinics: a randomized controlled trial *Arch. Phys. Med. Rehabil. 2012* **93** pp 21–6 (10.1016/j.apmr.2011.08.007.s)

Classification of project management tool reviews using machine learning-based sentiment analysis

R A Baro[1], M V Pagudpud[2], L M Padirayon[3] and R E Dilan[4]

[1]Sapilang, Bacnotan, La Union 2515 Philippines
[2]Magsaysay, Saguday, Quirino 3400 Philippines
[3]Namuac, Sanchez Mira, Cagayan 3518 Philippines
[4]Consolacion, Agoo, La Union 2504 Philippines

[1]rufo.baro@gmail.com, [2]mvpagudpud@gmail.com, [3]desmpadirayon@gmail.com, [4]raymund.dilan@gmail.com

Abstract. Managing the daily responsibilities in an organization is a great task for company administrators. Employing a dedicated project management tool is a great aid in all the phases of project management. With a variety of project management tools available in the market, it is a demanding task to look for a tool appropriate to the needs of the organization. It is an observation by scholars and suggests project management tool users tend to read first the reviews and comments of the product before deciding on the tool they are going to implement. This study facilitates sentiment analysis which can help identify effective and efficient project management tool. Microsoft Office Project is one of the most popular and most reviewed tools for more efficient project and portfolio management. The researchers manually extracted data from various websites containing comments and reviews from different users or reviewers. Machine learning-based-approach using RapidMiner was utilized to analyze the collected data from the web reviews. Sentiment analysis from MS Project reviews was applied using supervised learning K-nearest neighbor (KNN). The first level of classification involved classifying statements as "satisfied" and "dissatisfied." The second level of classification involved clustering sentiments as cost, experience, task, support and interface. Statements classified as satisfied is greater than dissatisfied. The best features to apply when classifying the PM tool reviews dataset are using stopwords, length, stemming and trigram.

1. Introduction

Many information technology (IT) projects are commenced with bright concepts and enormous investments in various resources [1]. Moreover, IT project could be a challenging task which gears ITs, time-bound and within budget requirements, bound to reach a declared business profit and can be used in a fast changing and increasing range of tools used to hold out the management of operations in a company [2]. Project management (PM) is one of the fastest emerging disciplines in organizations today which is used in various purposes, like product development and/or improvement, system deployment, process building, process re-engineering, new service initiation, marketing campaigns and software development [3]. The implementation of project management throughout an organization supports in creating planned value chain which enables corporations a lead over their rivals [4]. Information technology project management has changed significantly over the previous decade from the traditional scheduling and resource management to a complete system that sustains the whole project programs, life-cycle of projects, and project portfolios [5]. As a result, many companies are reaping their rewards from their investments including: lesser budget requirements, better efficiencies, enhanced satisfaction among consumers and stakeholders, and an edge over their rival companies [4]. These days, there are numerous selection of project management suites offered in the market. However, despite the claims and implementations of PM software tools to project management of different organizations, most of the projects do not achieve much success [6]. The online product reviews are a strong influence of users' purchase and usage decisions because it helps lessen the doubts on the product quality and aligned to the requirements of users [7][8]. Many software users take advantage of the internet which can be accessed anywhere around the globe to read online software product reviews before they do buy the product.

MS Office Project, commonly known as MS Project, is a product that is developed and distributed by Microsoft. It is a range of applications that can make project management easier and well-organized. Thousands of reviews were posted online regarding the said software product. Reviews posted were of the different level of opinion and area of focus, hence, the conduct of this study. The researchers thought of conducting a sentiment analysis regarding MS project to identify further the strength and weaknesses regarding cost, experience, tasks, support, and interface of the PM tool in advantage to the project manager who wants to use the said PM tool.

Sentiment analysis (SA) is a tool used in analyzing peoples' sentiments, views, ideas, attitudes, among others towards different issues and matters of interest the company is concerned [9]. It is also known as opinion mining and extraction, sentiment extraction, effect analysis and emotion analysis. Furthermore, sentiment analysis is usually applied in advertising, movie reviews and other purposes which captures customers' feedback [10]. One of the three top approaches to perform sentiment analysis is the machine learning-based approach which requires a corpus containing tagged examples. The K-nearest neighbour (KNN) was used in this study. KNN is easy to understand and implement. It is also scalable to new modifications as it is possible to eliminate many of the stored data objects, but still retain the classification accuracy of the KNN classifier [11].

1.1 Objectives of the study

This study was intended to classify sentiments about the MS Project PM tool. Specifically, this study aimed to provide analysis on the:

1. categories to be used in classifying the sentiments;
2. feature selection;
3. training and testing of the classifier; and
4. overall sentiment towards the PM tool.

1.2 Related Literature

1.2.1 Sentiment analysis

Sentiment analysis is a task of natural language processing and information extraction that intends to acquire reviewers' comments and queries by evaluating a huge number of documents [12]. Sentiment analysis has gained numerous usage for the past years as internet usage and exchange of public opinion increases and becomes the driving force of this discipline [13]. Sentiment analysis is geared towards establishing common grounds on varying thoughts of people concerning a certain topic. People can post their sentiments through different media, such as product reviews, online forums, blogs and social media sites. Several software packages can be availed as a tool for sentiment analysis. Examples are General Architecture for Text Engineering (GATE), KH Coder, Coding Analysis Toolkit, TAMS, QDA Miner Lite, VisualText, Datumbox and the RapidMiner which was used in this study. Each of the said tools claims best features such as ease of use and topic clustering ability.

1.2.2 Supervised Classification Algorithms

To calculate discrete classes, an algorithm which 'learn' the patterns of data should be applied. This is where supervised classification algorithms comes in. They are valuable because of its flexibility in statistical prediction techniques collectively known as machine learning techniques [14]. K-nearest neighbor (KNN) supervised classification technique was tested in this study.

1.2.2.1 K-Nearest Neighbor

[15] discussed that KNN is a type of instance-based learning or lazy learning which function is only approximated locally and all calculation is deferred in anticipation of the classification. It is non parametric technique used for classification or regression. It is called non parametric because it does not assume a functional form. In industries, however, KNN is widely used in classification problems. Generally, there are 3 important aspects to evaluate classification techniques which are: a) ease to interpret output, b) calculation time, and c) predictive power. KNN is used widely for its easy of interpretation and low calculation time.

1.2.3 Product Reviews

Nowadays, growing numbers of e-commerce platforms offer product reviews [16]. Product reviews are a great source of first hand experiences by users of a product. Satisfaction and dissatisfaction ratings enables consumers make wiser decision in purchasing the product. On the other hand, product owners can use the reviews to improve or enhance their product when the need arises.

2. Methodology

The study utilized the methodology as shown in Figure 1. The steps involved are data extraction from the internet, data cleaning, data pre-processing, feature selection and classification.

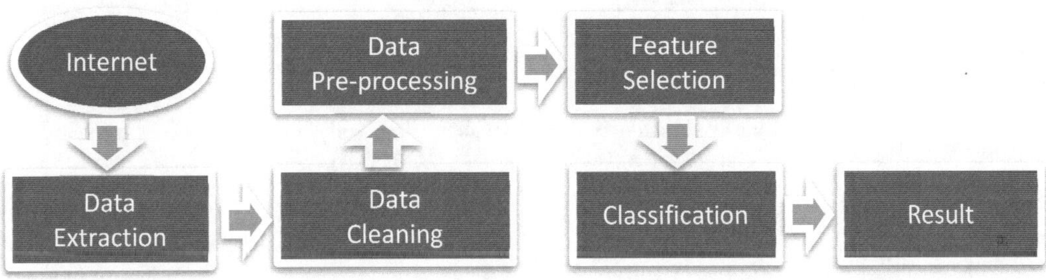

Figure 1. The methodology of the study

Data extraction is the procedure of collecting data from a source. In this study, data extraction is performed manually. The following are three among the other web pages that were considered in gathering data:

- https://www.getapp.com/project-management-planning-software/a/Microsoft-project/reviews/
- https://www.capterra.com/p/1419/MS-Project/
- http://technologyadvice.com/products/project-pro-reviews/

The data gathered from these three sites composed the majority of the data considered in the sentiment analysis. Overall, there were 1242 collected software product reviews. Prior to exporting to CSV Excel file, mined data was saved in local database. The 1242 reviews collected underwent data cleansing. The paragraphs were broken down into sentences since this study focuses on sentiment analysis at the sentence level. There were 1564 sentences produced and all data collected were English words. Non-English words were removed. An annotation scheme was developed for the analysis of the texts in the corpus. The annotator was tasked with the identification of text categories related to the reviews. The categories were presented later in this paper. Another category used by the annotator was to classify the reviews if it denotes satisfaction or dissatisfaction.

In this study, the software RapidMiner was used. RapidMiner [17] is a software tool for data preparation, deep learning, text mining, machine learning, and predictive analytics. Using the said software, the MS Project software product sentiment analysis run through this process. The operator 'Retrieve MSProject_Reviews' is utilized to read the database where the PM tool reviews dataset was saved. The operator's output is the metadata derived from transforming the loaded data by RapidMiner in the database. The 'Cross Validation' operator was used to train and test the different algorithms using 10 fold cross-validation. k- Nearest Neighbor (KNN) was the algorithm considered in training and testing the classifier.

3. Result and Discussion

The classification scheme used to categorize the data is composed of five categories as follows:

- Cost – These are sentiments about the price of the MS Project were categorized here.
- Experience – These are sentiments about the experience, occurrence, and familiarity of users towards MS Project
- Tasks – These are sentiments regarding project management tasks, activities, and features.
- Support – These are sentiments regarding online community support, manual and references.
- Interface – These are sentiments of the users regarding the user interface of the tool.

The categories were derived from the themes of the software product reviews. Table 1 shows examples of statements with annotation based on the annotation scheme used in this study.

Table 1. Sample classification scheme applied to the statements

Category	Satisfied	Dissatisfied
Cost	*I feel that MS Project is cost-effective to manage simple projects*	*Expensive license cost, not cloud-based.*
Experience	*It works well as a teaching tool because the students tend to understand quickly and easily.*	*If a group is not tech savvy, it takes some time in getting used to it.*
Tasks	*Fast data entry*	*Formatting the header to accommodate various icons or logos is not easy*
Support	*Easy to find the correct help video*	*Support is limited to forums and a 500$ per hour fee*
Interface	*I like that the interface allows me to enter the data about the project into what looks like a spreadsheet.*	*The interface is clunky at times and not always intuitive.*

Table 2 shows the data distribution as annotated based on the scheme. Of the 1564 sentiments considered in this study, 833 imposed satisfaction and 731 were found to be dissatisfied as perceived by the annotators.

Table 2. Data distribution for the text annotation related to MS Project PM Tool

Category	Satisfied	Dissatisfied
Cost	18	126
Experience	199	163
Tasks	413	212
Support	125	142
Interface	78	88
Total	**833**	**731**

For feature selection, the algorithm was tested using a 10-fold cross-validation. Features were added after running the algorithm. Table 3 shows the result of the 10-fold cross validation without any features added. The result shows minimum accuracy and f scores.

Table 3. The 10-fold cross-validation of the proposed sentiment classifier

Features	F Score										
	Cost		Experience		Tasks		Support		Interface		Accuracy
	S	*D*	*S*	*D*	*S*	*D*	*S*	*D*	*S*	*D*	
K-NN (3)	0.39	0.42	0.48	0.50	0.36	0.42	0.53	0.50	0.43	0.45	52.08 %
Baseline – K-NN (3)	0.51	0.46	0.66	0.64	0.65	0.66	0.64	0.65	0.63	0.57	64.87 %
Baseline + n-gram (3) – K-NN (3)	0.69	0.71	0.74	0.68	0.69	0.67	0.70	**0.71**	0.75	0.69	70.83 %

Legend: Baseline = filter_stopwords + filter_Length + Stem

A slight increase can be seen when features such as filter_stopwords, filter length, and stem were added. Further, when n-gram (3) was applied to the baseline, the result shows that there is an increase in all the f-score and accuracy. It is noteworthy that all the features in all their respective f scores conforms to the 10-fold cross-validation (see Table 2) except the "interface" category where dissatisfaction is higher than satisfaction. The same category gave the highest f score of 0.75 on "satisfaction"

Table 4 shows the result of the classification using the K-NN algorithm. It shows the accuracy, precision, recall and F-measure score.

Table 4. Result of the fine-grained classification

Algorithm	Accuracy	Precision	Recall	F-Measure
KNN (3)	70.83 %	0.688	0.720	0.781

4. Conclusion

On this study, machine learning using RapidMiner was performed. The following are the conclusions derived from the experiment: The first level of classification involved classifying statements as "satisfied" and "dissatisfied." The second level of classification involved clustering sentiments as cost, experience, task, support and interface. Statements classified as "satisfied" is greater than "dissatisfied." Statements are mostly task related. In addition, "stopwords" removal, filtering of tokens length, stem and n-gram are the features used in this study. The best features to apply when classifying the PM tool reviews dataset are using "stopwords," length, stemming and trigram. The said features, when added to the classifier, increase the f-score and accuracy which means that the features enhanced the classifier. Even so, all values for accuracy obtained during testing were above 50%, which means a relatively accurate classification. The algorithm used manifested an accuracy of 70.83%.

5. References

[1] Mirza MN, Pourzolfaghar Z and Shahnazari M 2013 Significance of scope in project success *Procedia Technol.* vol **9** pp 722-9

[2] Bennington P and Baccarini D 2004 Project benefits management in IT projects—an Australian perspective *Project Management J.* vol **35** no **2** pp 20-30

[3] Shenhar A and Dvir D 2007 Project management research—the challenge and opportunity *Project Management J.* vol **38** no **2** pp 93-9

[4] PMI 2010 *The Value of Project Management* (Project Management Institute, Inc.)

[5] Ahlemann F 2009 Towards a conceptual reference model for project management information systems *Int. J. of Project Management* vol **27** no **1** pp 19-30 ISSN: 0263-7863

[6] Alexander M 2017 *How to choose the right project management tools* (IDG Communications, Inc.)

[7] Kwark Y, Chen J and Raghunathan S 2014 Online product reviews: implications for retailers and competing manufacturers *Information Systems Research* vol **25** no **1** pp 93-110 ISSN 1047-7047

[8] Li M, Huang L, Tan C and Wei K 2013 Helpfulness of online product reviews as seen by consumers: source and content features *Int. J. of Electronic Commerce* vol **17** no **4** pp 101-36
[9] Liu B 2012 *Sentiment Analysis and Opinion Mining* (Morgan and Claypool Publishers)
[10] Vaitheeswaran G and Arockiam L 2016 Machine learning based approach to enhance the accuracy of sentiment analysis on tweets *Int. J. of Adv. Res. in Comp. Sci. and Management Studies* vol **4** Issue **5** ISSN: 2321-7782 [Online] Available: http://www.ijarcsms.com/docs/paper/volume4/issue5/v415-0036.pdf
[11] Sharma D 2016 Experimental analysis of KNN with Naive Bayes, SVM and Naive Bayes Algorithms for spam Mail detection *Int. J. of Comp. Sci. and Technol.* vol **7** Issue **4** ISSN: 0976-8491 (online) | ISSN : 2229-4333 (print)
[12] Mukherjee S 2012 Sentiment Analysis: A Literature Survey *Indian Institute of Technology, Bombay Department of Computer Science and Engineering* Retrieved on December 1, 2017, Retrieved from https://arxiv.org/ftp/arxiv/papers/1304/1304.4520.pdf
[13] Fang X and Zhan J 2015 Sentiment analysis using product review data *Journal of Big Data* vol **2** DOI 10.1186/s40537-015-0015-2
[14] Stephens D and Diesing M 2014 A Comparison of Supervised Classification Methods for the Prediction of Substrate Type Using Multibeam Acoustic and Legacy Grain-Size Data *PLoS ONE* vol **9** no **4** e93950 https://doi.org/10.1371/journal.pone.0093950, 2014
[15] Dey L, Chakraborty S, Biswas A, Bose B and Tiwari S 2016 Sentiment Analysis of Review Datasets Using Naïve Bayes' and K-NN Classifier *International Journal of Information Engineering and Electronic Business* (IJIEEB) vol **8** no **4** pp 54-62 DOI: 10.5815/ijieeb.2016.04.07
[16] Lackermair1 G, Kailer D and Kanmaz K 2013 Importance of Online Product Reviews from a Consumer's Perspective *Advances in Economics and Business* vol **1** no **1** pp 1-5 DOI: 10.13189/aeb.2013.010101 (Horizon Research Publishing)
[17] Hofmann M and Klinkenberg R 2013 RapidMiner: Data Mining Use Cases and Business Analytics Applications *Chapman & Hall/CRC Data Mining and Knowledge Discovery Series* (CRC Press)

IOP Conf. Series: Materials Science and Engineering **482** (2019) 012042 doi:10.1088/1757-899X/482/1/012042

Classifying soil texture images using transfer learning

E P B Guidang

Abra State Institute of Sciences and Technology, Lagangilang, Abra, Philippines

rieann12@gmail.com

Abstract. Transfer learning is a machine learning technique which makes use of a pre-trained neural network to classify new objects. This study was conducted to evaluate the performance of Inception-v3 in classifying Soil texture images on different conditions. Specifically it achieved the following objectives 1) Identified the features of Inception-v3 and 2) Classified Soil Texture images using Inception-v3. The study used literature review to identify the features of Inception-v3. The study found that Transfer Learning comprises of two portions: a) feature mining and b) classification. Moreover, Inception-v3 highest prediction rating of a Soil texture image is 98% and 86% as the lowest. The study concludes that Transfer Learning method through the use of Inception-v3 can be used to classify Soil texture images.

1. Introduction

Today's image recognition models have millions of factors. Training models from scratch requires a lot of computing power and labeling them could be tedious. Transfer Learning (TL) is a method that cut short the processes by learning from a model that was previously trained on one task and apply the same on a new task [1]. It is a machine learning technique which makes use of a pre-trained neural network. It is a procedure where a model trains on a set of task and re-configured on another set of task [2]. Moreover, it allows a rapid performance when applying on a related task [3].

The purpose of Transfer Learning is to improve learning by maximizing information from the source task. Transfer might improve learning with the following measures: a) Use of transferred knowledge before any further learning; b) Amount of time spent on the transferred knowledge versus the time spent in learning knowledge from scratch and c; Final performance level attainable in the target task versus the final level from scratch [4].

Today, "TL" is common in predictive modeling problems using image data as input.

The said image could either be a photograph or video. For these types of problems, a deep learning model pre-trained for ImageNet Large Scale Visual Recognition Challenge (ILSVRC) can be used. ILSVRC is a classification and detection on hundreds of object categories and millions of images standard [5].

Oftentimes, the research organizations releases for reuse the final model of the ILSVRC under a permissive license. There are three models of this type that can be downloaded namely: Oxford VGG Model, Google Inception Model, and Microsoft ResNet Model.

The common version released by Google is the Inception-v3 model. The said model is an image classifier [6]. It is said to be trained for the ILSVRC using the data from 2012. It is a standard job in computer vision [1]. It is a widely-used image recognition model because of its 78.1% accuracy in prediction [6].

This study was conducted to evaluate the performance of Inception-v3 in classifying Soil texture images on different conditions. Specifically it achieved the following objectives:

1) Identified the features of Inception-v3
2) Classified Soil Texture using Inception-v3

2. Methodology

To identify the features of Inception-v3, the author conducted a literature review by of twenty (20) Transfer Learning articles written by enthusiasts, data scientist and the tutorial developed by TensorFlow.

To classify Soil texture images, the directories were labeled as Clay, Sand and Silt. Next, images which were highly recognizable and intentionally blurred using Gaussian and Motion were placed into the directories manually as shown in figure 1, 2 and 3. 70% of data were used for training while 30% percent of the data were used for testing. Testing data were blurred using Gaussian, Pixel and Motion.

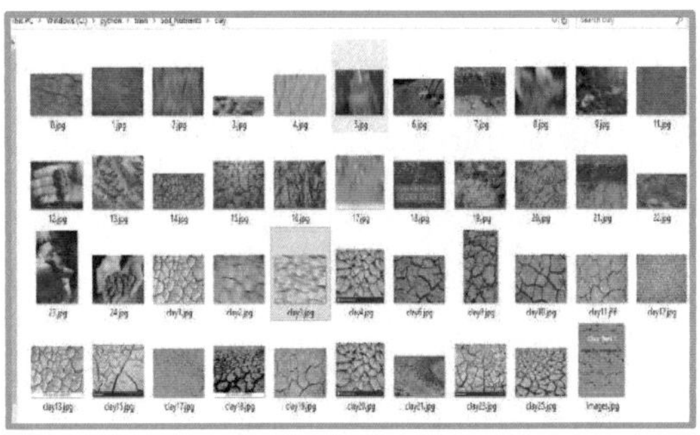

Figure 1. Clay images

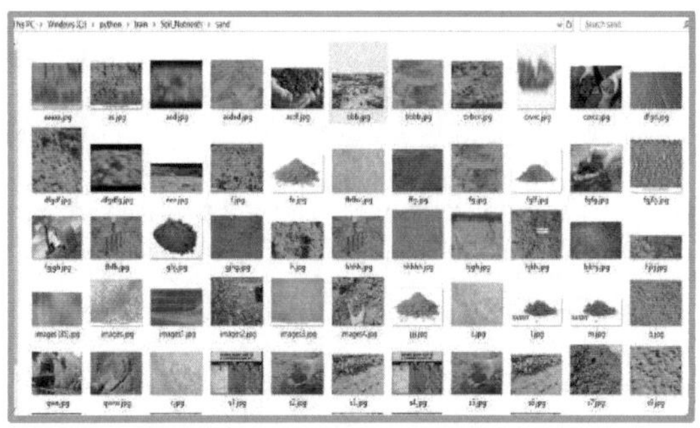

Figure 2. Sand images

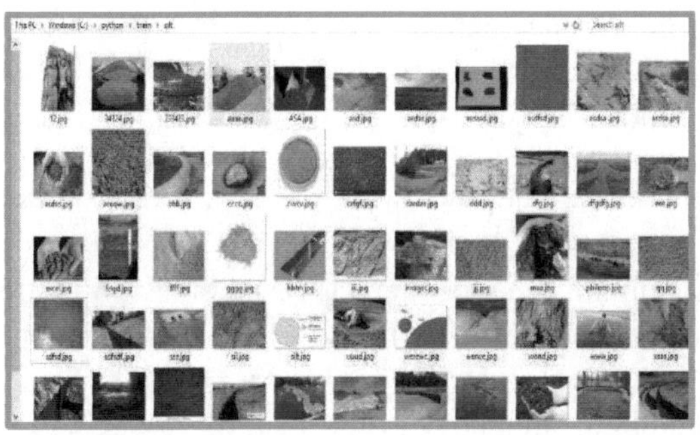

Figure 3. Silt images

3. Results and Discussions

3.1 Features of Ineception-v3

The Inception-v3 is a deep convolutional network intended for classification responsibilities on ImageNet. It has a dataset comprising of 1.2 million Red, Green, and Blue images from 1000 classes [7]. The Inception-v3 model comprises of two portions: a) feature mining layer. It is a portion where a convolutional neural network is located; and b) Classification layer. The said layer is portion where the fully-connected and softmax layers are located. The model extracts general features from input images in the first part and classifies them based on those features in the second part.

3.2 Classifying Soil Texture using Inception-v3

Firstly, to demonstrate that Inception-v3 is capable of classifying images, figure 4, 5, and 6 are hereby presented.

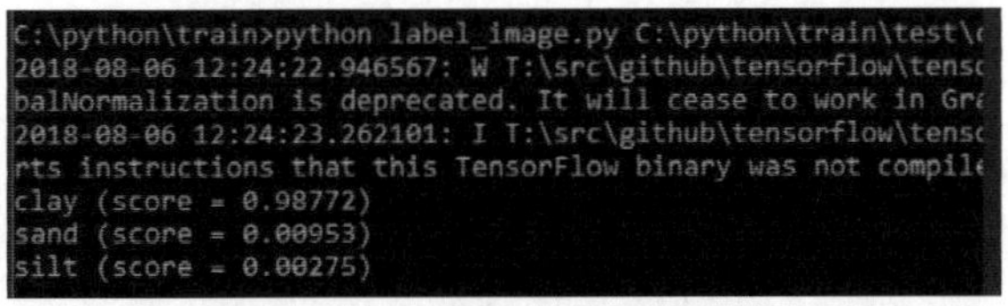

Figure 4. Clay prediction result

It can be gleaned on Fig 6 the prediction result of the classifier to a Clay image. Clay has a score of .98772, sand has a score of .00953, and silt has a score of .00275. If the values are to be converted in percentile, .98772 means 98 percent, .00953 means 0 percent, and lastly, .00275 is also 0 percent. Therefore, the classifier is 98 percent sure that the image is a clay, 0 percent sure that the image is sand and silt. This means that 98 being the highest in terms of percentile rating in the classifier prediction, the image fed is classified to be Clay.

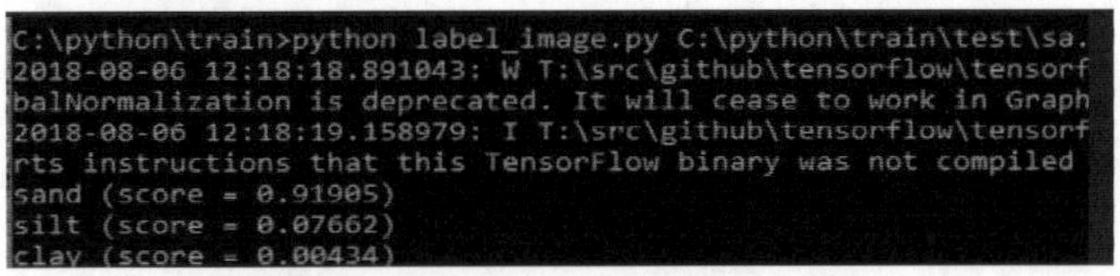

Figure 5. Sand prediction result

It can be seen on Fig 7 the prediction result of the classifier to a Sand image. Sand has a score of .91905, Silt has a score of .07662, and clay has a score of .00434. If the values are to be converted in percentile, .91905 means 92 percent, .07662 means 8 percent, and lastly, .00434 is 0 percent. Therefore, the classifier is 92 percent sure that the image is a sand, 8 percent sure that the image is silt, and 0 percent sure that the image is clay. This means that 92 percent being the highest in terms of percentile rating in the classifier, the image fed is classified to be Sand.

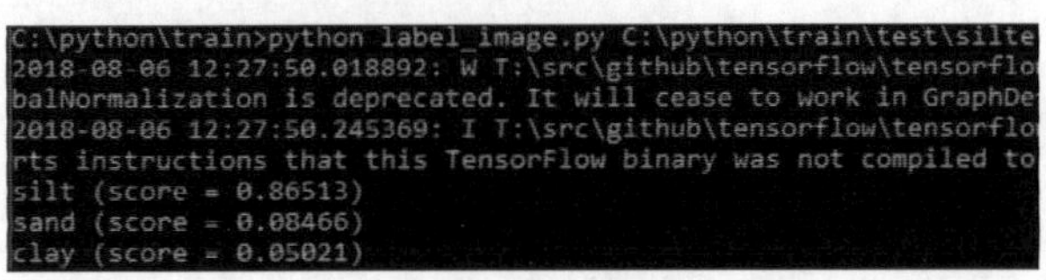

Figure 6. Silt prediction result

Fig 8 shows the prediction result to a Silt image. Silt has a score of .86513, Sand has a score of .08466, and clay has a score of .05021. If the values are to be converted in percentile, .86513 means 87 percent, .08466 means 8 percent, and lastly, .05021 is 5 percent. Therefore, the classifier is 87 percent sure that the image is Silt, 8 percent sure that the image is sand, and 5 percent sure that the image is clay. This means that 87 percent being the highest in terms of percentile rating in the classifier, the image fed is classified to be Silt.

To demonstrate the performance of Inception-v3 to images which were blurred, figure 7, 8 and 9 are hereby presented.

IOP Conf. Series: Materials Science and Engineering **482** (2019) 012042 doi:10.1088/1757-899X/482/1/012042

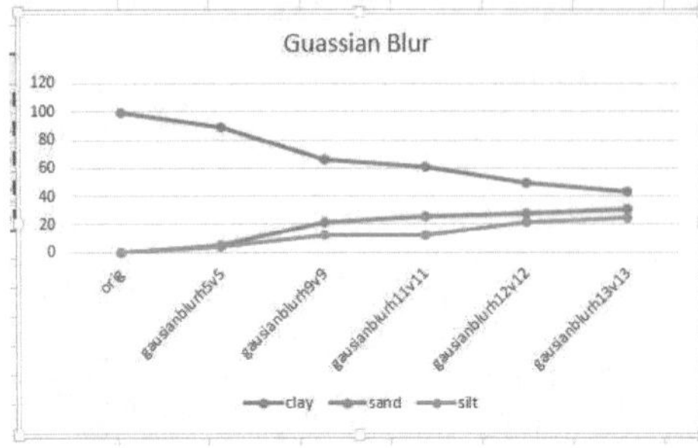

Figure 7. Gaussian Blur

Figure 7 shows the prediction pattern of model when tested with images applied with Gaussian blur. It can be seen that as the horizontal and vertical value of Gaussian blur increases, the percentile prediction rating of the Inception-v3 model of the image to be a clay decreases while its percentile prediction rating on sand and silt increases. Of the five tests, the average prediction rating of the model of the image to be clay is 61.6 percent.

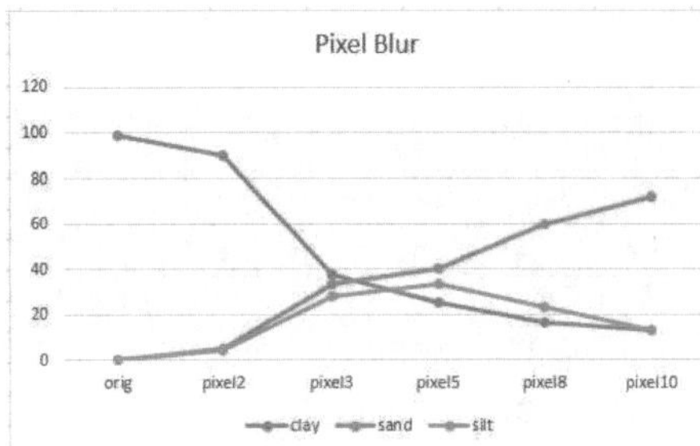

Figure 8. Pixel Blur

Figure 8 displays the prediction pattern of model when tested with images applied with Pixel blur. It can be seen that as Pixel value increases, the percentile prediction rating of the Inception-v3 model of the image to be a clay decreases while its percentile prediction rating on sand increases. Of the five tests, the average prediction rating of the model of the image to be clay is 36.2 percent.

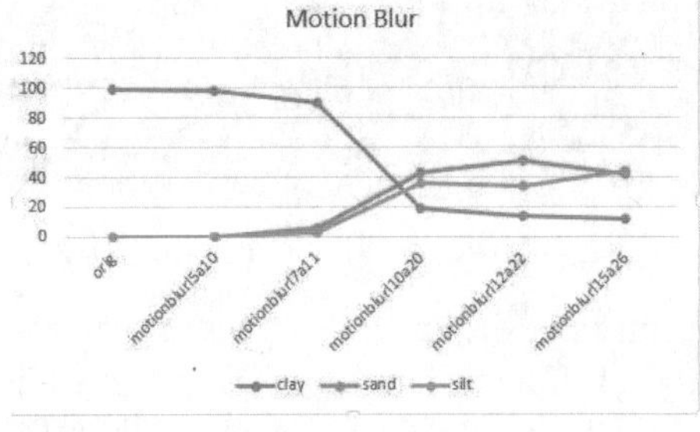

Figure 9. Motion Blur

Figure 9 shows the prediction pattern of model when tested with images applied with Motion blur. It can be seen that as the length and angle values increases, the percentile prediction rating of the Inception-v3 model of the image to be a clay decreases while its percentile prediction rating on sand and silt increases. Of the five tests, the average prediction rating of the model of the image to be clay is 46.6 percent.

The predictions of Inception-v3 is impressive when images are clear and recognizable. In this evaluation, it has an average percentile rating of 98% prediction accuracy to images that were not altered. However, when the image is blurred with Gaussian, Pixel and Motion blurs, Inception-v3 percentile prediction decreases as shown in figure 7, 8, 9. It can hardly recognizes objects that are altered.

4. Conclusions

Transfer Learning comprises of two portions: a) feature mining and b) classification. Transfer learning using Inception-v3 is effective in classifying soil texture images. Soil Texture images are just but an example of what could be impressively done with Transfer Learning. Definitely it could be used to classify other interesting objects.

The predictions of Inception-v3 is impressive when images are clear and recognizable. However, when the image is blurred with Gaussian, Pixel and Motion blurs, Inception-v3 percentile prediction decreases. It can hardly recognizes objects that are altered.

5. References

[1] TensorFlow 2018 *How to Retrain an Image Classifier for New Categories* retrieved from https://www.tensorflow.org

[2] Olivas E S, Guerrero J D M, Sober M M, Benedito J R M and Lopez A J S 2009 *Handbook Of Research On Machine Learning Applications and Trends: Algorithms, Methods and Techniques*-2 Volumes

[3] Brownlee J 2017 *Deep Learning* retrieved from https://machinelearningmastery.com

[4] Torrey L and Shavlik J 2010 Transfer learning *Handbook of Research on Machine Learning Applications and Trends: Algorithms, Methods, and Techniques* (IGI Global) pp 242-64

[5] Russakovsky O, Deng J, Su H, Krause J, Satheesh S, Ma S, Huang Z, Karpathy A, Khosla A, Bernstein M, Berg A C and Fei-fei L 2015 Imagenet large scale visual recognition challenge *Int. J. of Comp. Vision* **115(3)** pp 211-52

[6] Szegedy C, Vanhoucke V, Ioffe S, Shlens J and Wojna Z 2016 Rethinking the inception architecture for computer vision *Proc. of the IEEE Conf. on Computer Vision and Pattern Recognition* pp 2818-26

[7] Sharma R 2018 A note on the inception score *Preprint* arXiv:1801.01973

An analysis on the insights of the anti-vaccine movement from social media posts using k-means clustering algorithm and VADER sentiment analyzer

J Garay, R Yap and M J Sabellano

Department of Computer and Information Sciences, University of San Carlos, Talamban, Cebu City, Cebu 6000, Philippines

lawrencetecho@gmail.com

Abstract. This study analyzes the insights and sentiments of the anti-vaccine movements in the social media. The data, comprising of tweets and excerpts, are pre-processed to omit noise and irrelevant data. They are clustered using the k-means clustering algorithm. Each word belonging to a cluster is processed by VADER sentiment analyzer. Prevalent sentiments per cluster label the mood associated in the cluster. The results suggest insights about vaccines such as: side effects, post-shot injuries, ineffectiveness, damage from ingredients, unvaccinated elite, reinforcement of the right to not vaccinate, toxic ingredients, big pharmaceuticals' profit maximization, links to autism, and health issues after getting vaccine shots. To evaluate k-means results, the silhouette score is determined to indicate how far a point is to other nearby clusters. The resulting average silhouette score of all points is 0.013540022 which indicates that the points are close to the decision boundaries.

1. Introduction

Curbing preventable diseases is a political, social, economic, and health-related endeavor. Many countries and institutions mandate vaccination so people most especially in public places are guarded against preventable diseases. [1] However, a significant number of parents in the United States show little confidence for vaccines. [2] In 1998, the rise in the number of people this movement has skyrocketed due to a faux scientific paper that linked vaccination to autism [3]. Together with the advancement of social media and the rise of technologies that make it easier to spread misinformation for its cause, the movement has garnered more followers and had been proliferating their causes in an even faster rate [4].

While in many respects, machine learning techniques could not extend as much as humans can in interpreting behavior and insights especially those involving heavy qualitative data, however, combining natural language processing and machine learning to help tackle social issues can be used as an advantage.

The purpose of this paper is: a) to demonstrate the use of the k-means clustering method in discovering insights from a corpus; b) to understand the sentiment in every cluster by using VADER sentiment analyzer to further examine the mood incorporated for every insight (cluster); and c) to gain useful insight and sentiment of the anti-vaccine movement.

2. Review of Related Literature

2.1. K-Means Clustering

A popular algorithm for clustering would be K-means clustering. [5], [6], [7] show k-means to have a relatively low time complexity (O(nkd)) which indicates faster performance. It also works well with large data set. Though it is quite sensitive to noisy data. With the pros in mind, the researchers concluded that this would be a good approach.

A study conducted by [8] shows the difference of speed between K-means and Fuzzy K-means. In their experiment where they cluster 39,000 tweets in order to compare the speed, it is shown that K-means outperforms Fuzzy K-means overall.

The researchers need a way to thoroughly analyze tweets to be able to make sense of anti-vaccine sentiments. A study analyzing disaster risk reduction (DRR) responses from the Philippines uses K-means clustering which allowed them to find out some keywords produced by the clusters and with the use of open coding, were able to interpret them. [9]

2.2. Sentiment Analysis

Sentiment Analysis has been gaining traction lately with the appearance of a large source of opinionated data through the web and its wide range of applicability in many fields of society. [10]

A study was conducted to find a way to use sentiment analysis in analyzing student evaluations of teaching (SET) for universities. They compared the sentiment of comments together with quantitative scores. They found that the polarity of the comments generally follows the trend of the meanings of the scores, making sentiment analysis a viable method for dealing with SETs. They used VADER as their sentiment analysis tool [11].

2.3. The Anti-Vaccine Movement

The anti-vaccination movement's rapid growth can be attributed to some very persuasive websites and information online. A study conducted to find out why these sites are persuasive was conducted. These sites contain a large amount of misinformation, such as vaccines are dangerous and can cause autism and brain damage. They use scientific evidence and anecdotes to support these claims, and they also commonly promoted the use of alternative medicine, homeopathy, and an organic lifestyle. These combined with many effective persuasion techniques to convince people to adapt their beliefs. [12]

3. Methodology

3.1. Data Gathering and Pre-processing

The data in this study refer to tweets and post excerpts from popular public Facebook anti-vaccination groups.

3.1.1. Data Cleaning. The mined data were cleaned by removing "stopwords" – non-important words (e.g. "to"), non-alphanumeric characters, words with lengths less than 4, and non-English words were removed from the data

3.1.2. Data Pre-Processing. The data were converted into vectors using the term-TF-IDF vectorization method.

3.2. TF-IF Vectorization Method

Term frequency times inverse document frequency (TF-IDF) is a numerical statistic method used to determine how important a word is to a corpus. This method compares the frequency of the word in a document versus its frequency in the whole aggregate of documents (corpus). The study uses scikit's TF-IDF vectorizer to achieve vectorization.

3.3. K-Means Clustering Algorithm

The use of k-means clustering is to create k number of groups and to designate each word (vector) in the corpus a corresponding group. This method works by iteratively designating a data to one of the k groups based on a common mean vector. The clusters are formed as follows:

1. Set k
2. Designate random k locations; this will serve as the centroids.
3. For each data, do the following until there is no longer a significant shift after measuring the new centroids (shift is significant if it's above a threshold):
   i) Find a centroid whose mean has the least squared Euclidean distance with the current data
   ii) Find the new centroid within the space in this centroid by calculating the mean of all points in the space.

3.4. Validation Methods

3.4.1. Silhouette Score. The silhouette score describes how near a point is to its own cluster and how far the point is from other clusters. The study uses this to measure the correctness of the assignment of the points/vectors in their respective clusters. This is calculated using the mean intra-cluster distance i and the mean nearest-cluster distance j for each word (vector) in the data. In practice, a score nearing 1 indicates more distance from other clusters. A score nearing -1 indicates that the point is assigned in the wrong cluster. Nearing 0 means that the point is close to the decision boundary. The formula is as follows:

$$\frac{i - j}{max(i, j)} \tag{1}$$

3.4.2. Picking the Optimal Number of Clusters. The number of clusters that yields a silhouette score closest to 1 is the optimal number of clusters. Table 1 shows the number of clusters and their corresponding silhouette score. In the result, having 10 clusters has the silhouette score closest to 1. This means having 10 clusters yield clusters farther from each other with the least likely occurrence of overlap relative to other k number of clusters. The silhouette score started to decrease after the 10th cluster. Implied from this, any k cluster beyond 10 will yield a silhouette score father from 1 which indicates that weaker clusters are formed since points become nearer the decision boundary. The 10th cluster however, given 0.013540022 is the most optimal silhouette score relative to other k.

Table 1. Resulting silhouette score after K iterations of the corpus

Number of Clusters	Silhouette Score
2	0.005830464
3	0.007237809
4	0.006643489
5	0.009101259
6	0.010489809
7	0.012620094
8	0.012190529
9	0.012351144
10	0.013540022
11	0.012131123
12	0.011417081

3.5. VADER Sentiment Analyzer

Valence Aware Dictionary and sEntiment Reasoner (VADER) is a lexicon and rule-based sentiment analyzer specifically trained sentiments from social media. This API is used in the study by determining the corresponding sentiment for every word. The analyzer works by passing to it a word and it returns the sentiment associated with the word. This method is used in the study by determining the prevalent sentiment in every cluster. Every word in the a is processed by VADER sentiment analyzer and the cluster is then labelled a sentiment by identifying the most prevalent sentiment.

4. Results and Discussions

All results below are the results of k-means clustering algorithm with a set k of 10 to optimize resulting clusters after having tested various k and producing different silhouette scores.

Table 2. Clusters generated by K-means with the closest words it contains and the prevalent sentiment from all words in the cluster as generated by VADER analyzer.

Cluster	Nearest frequent keywords in the cluster	Sentiment
1 – Possible side effects of vaccines	kill, cripple, death, injection, disorder, neurological	negative
2 – Experience of vaccine injury	parents, children, kids, injured, sick, doctors	neutral
3 – A call to question whether vaccines really work	unnecessary, effectivity, expire, herd, immunity, gardasil	neutral
4 – Chronic damage caused by toxic vaccines	damage, brain, toxic, learn, cause, ones	neutral
5 – Emphasis on the elite who choose not to vaccinate	children, unvaccinated, elite, rich, doctors, unlikely	neutral
6 – Expression of their right to choose	child, choice, right, parent, decision, choose	neutral
7 – "toxic" Ingredients present, research	research, children, ingredients, wanting, kids, toxicity	neutral
8 – Capitalism of vaccine products	kids, sick, pediatricians, profit, pharmaceutical, big	negative
9 – Autism being linked to vaccines	autism, amish, cause, children, relation, link	neutral
10 – children started getting health issues after vaccine shots	health, cause, issues, daughter, shots, jab	neutral

4.1. Cluster 1: Possible side effects of vaccines (kill, cripple, death, injection, disorder, neurological; sentiment: negative)

The discouragement of people who fall into this cluster relies heavily on the prevalent fear-mongering in social media. The social media space has a subset of people who use fear to discourage people opting to vaccination. Although vaccines in fact may have side effects, side effects in this cluster are exaggerated to evoke fear within social media (kill, cripple, death). The negative sentiment supports such claim. It can be implied as though advocacies and campaigns should focus more on debunking the claimed side effects.

4.2. Cluster 2: Experience of vaccine injury (parents, children, kids, injured, sick, doctors; sentiment: neutral)

Campaigns and advocacies should start to raise awareness as to the improbability of vaccine injuries from happening.

4.3. Cluster 3: A question on the effectivity of vaccines (unnecessary, effectivity, expire, herd, immunity, Gardasil; sentiment: neutral)

People who fall into this cluster are skeptical of vaccine's effectivity; resulting to claims that vaccines are unnecessary. It becomes apparent that there is a need to create a discourse about the scientifically-tested benefits of vaccines. The neutral sentiment manifests that such skepticism does not come with other negative notions about vaccines.

4.4. Cluster 4: Chronic damage caused by toxic vaccines (damage, brain, toxic, learn, cause, ones; sentiment: neutral)

There is a prevalent claim in this cluster that vaccines could cause brain damage. Another way to interpret this result is to say that people in this cluster encourage to learn as to how toxicity in vaccines could cause brain damage. Either interpretation is necessitates debunking such claim.

4.5. Cluster 5: Emphasis on the rich, elite few whose choice is to opt out from vaccination (children, unvaccinated, rich, elite, doctors, unlikely; sentiment: neutral)

This cluster allows us to understand how the decisions of the rich and elite being unlikely to opt from vaccination has an impact on the middle class' decision whether or not they should vaccinate. Their influence does have a significant impact on the decisions of people who fall in this cluster.

4.6. Cluster 6: Expression of their right to choose (child, choice, right, parent, decision, choose; sentiment: neutral)

Vaccinations have become a requirement for different social and economic activities in many countries. It is vague to interpret the reason why parents might not vaccinate their child. It is apparent in this cluster however, that parents are reinforcement the value of choice and the ability to decide for themselves to opt to vaccination or not.

4.7. Cluster 7: Toxic ingredients present in vaccines (research, children, ingredients, wanting, kids, toxicity; sentiment: neutral)

In this cluster, research is introduced and is associated with other keywords like toxicity and ingredients. A way to interpret this cluster is that there may be "research" about the toxicity of the ingredients found in the vaccines that parents believe. This cluster allows us to understand that people in this cluster are not motivated by fear but are people who believe in findings in "research", although there has been no research so far that suggests with confidence about the significant harms of the ingredients present in vaccines.

4.8. Cluster 8: Capitalism of vaccine products (kids, sick, pharmaceutical, profit, big; sentiment: negative)

People who fall into this cluster are skeptical of the vaccines produced by pharmaceutical companies. This finding could be interpreted in a way that people have negative perception/sentiment as to how reliable vaccines are. The association of "profit" in this cluster may imply that people's perception of vaccine as a necessity only benefits pharmaceutical companies as it may be thought of as a marketing ploy.

4.9. Cluster 9: Autism linked to vaccines (autism, amish, cause, children, relation, link; sentiment: neutral)

This cluster reinforces the existence of the already prominent idea with in antivaxxers that vaccines cause autism in children.

4.10. Cluster 10: Vaccines caused other health-related issues in children (health, cause, issues, daughter, shots, jab; sentiment: neutral)

People in this cluster claim that vaccine shots caused health issues to their children.

5. Conclusion

The resulting average silhouette score computed for all points is 0.013540022. Any silhouette scores higher than zero indicates that the data were fit to the clusters they were assigned to.

Majority of the sentiments were neutral. Implied from this, the subjects were more cautious with the reasons provided per cluster than they are negative about it.

The study manifests that k-means clustering can be used as an approach to clustering a corpus of qualitative data such as tweets and posts from social media. Furthermore, the study shows that quantitative methods in machine learning, particularly k-means algorithm, can be used to analyze qualitative data such as opinions.

The methods in this study can be replicated in many other practical domains. The same method of clustering data and finding sentiment for each formed cluster can be applied to social, political, and business domains with data such as reviews and feedbacks. This opens possibilities of new types of case studies that require interpreting feedback by extracting features and understanding the sentiment associated with it.

We recommend the use of a stronger, context-sensitive machine learning algorithm in determining the associated emotion of a word. The used model — that VADER utilizes to determine the emotion — is trained with general, non-context-sensitive statements and topics from their own massive dataset. While massive datasets create a more neutral, less-biased, and less context-sensitive output, in a world political and social issues, context can be a strong factor for determining emotion within a text.

Acknowledgements
We wish to acknowledge the University of San Carlos and its Department of Computer and Information Sciences for the research opportunity.

References

[1] Walkinshaw E 2011 Mandatory vaccinations: The international landscape *Canadian Medical Assoc. J.* **183(16)** pp 1167-8
[2] Kennedy A, LaVail K, Nowak G, Basket M and Landry S 2011 Confidence about vaccines in the United States: understanding parents' perceptions *Health Affairs* **33(6)** pp 1151–9
[3] Olpiński M 2012 Anti-vaccination movement and parental refusals of immunization of children in USA *Pediatria Polska* **87(4)** pp 381–5
[4] Stien R 2017 The golden age of anti-vaccine conspiracies *Germs* **7**(4) pp168–70
[5] Xu D and Tian Y 2015 A comprehensive survey of clustering algorithms *Annals of Data Science* **2** pp165–93
[6] Fahad A, Alshatri N, Tari Z, Alamri A, Khalil I, Zomaya A, Foufou S and Bouras A 2014 A survey of clustering algorithms for big data: Taxonomy and empirical analysis *IEEE transactions on emerging topics in computing* **2** pp 267–79
[7] Jain E and Jain S 2014 Using Mahout for clustering similar Twitter users: Performance evaluation of k-means and its comparison with fuzzy k-means *Int. Conf. on Computer and Communication Technology (ICCCT)* pp 29-33
[8] Go Bui S M, Gorro K, Aquino G A and Sabellano M J 2017 An analysis of DRR suggestions using K-means clustering *Proc. of the 2017 Int. Conf. on Information Technology* pp 76-80
[9] Liu B 2012 Sentiment analysis and opinion mining *Synthesis Lectures on Human Language Technologies* vol **5** no **1** pp 1-67
[10] Newman H and Joyner D 2018 Sentiment analysis of student evaluations of teaching *Int. Conf. on Artificial Intelligence in Education* pp 246-50
[11] Moran M B, Lucas M, Everhart K, Morgan A and Prickett E 2016 What makes anti-vaccine websites persuasive? A content analysis of techniques used by anti-vaccine websites to engender anti-vaccine sentiment *J. of Communication in Healthcare* **9** pp 151–63
[12] Kata A 2012 Anti-vaccine activists, Web 2.0, and the postmodern paradigm – An overview of tactics and tropes used online by the anti-vaccination movement *Vaccine* **30** pp 3778–89

IOP Conf. Series: Materials Science and Engineering **482** (2019) 012044 doi:10.1088/1757-899X/482/1/012044

Poverty incidence identification of cities and municipalities using convolutional Neural Network as applied to satellite imagery

J Mesina[1], J Isanan[1] and C Maderazo[1]

[1]Department of Computer and Information Sciences, University of San Carlos, Talamban, Cebu City, Cebu 6000, Philippines

joshuaisanan@gmail.com

Abstract. Poverty remains to be a hindrance to national growth for many developing countries. This presents a problem in a country's resource management and urban planning of its people. Furthermore, there is a big gap when it comes to data collection and monitoring in developing countries such that the government fails to prioritize this area of responsibility. The authors aim to bridge this gap by classifying wealth through the use of satellite images and emerging technology, particularly Convolutional Neural Networks. The researchers' goal is to test whether AlexNet, a Convolutional Neural Network Architecture, can identify the poverty levels of municipalities in the Philippines, by estimating their poverty incidences, based on satellite images. With this research, it can measure a broader set of society growth indicators. Indicators such as the material of people's houses, the size of their houses, their living conditions, access to roads and also how highly urbanized an area is, are considered important factors to indicate poverty. The results shows an accuracy of 0.84 and average precision-recall of 0.86 and 0.84, respectively. This study will give deeper insight to the government as to which areas to improve on in certain sectors given that a poverty incidence is high in a certain municipality.

1. Introduction

An essential building block of a state are its people. The problem is that some countries lack the awareness of the state of their people in order to respond and act accordingly .Although a developing country, it still does not fully cater to the entirety of the poor. One reason for this is that the government is not aware of the state they are living in. This research was intended to test AlexNet, a Convolutional Neural Network Architecture, using satellite imagery in order to reduce manual information collection on poverty incidences with the help of a Neural Network. Data preparation started with gathering satellite images of municipalities in Cebu and other provinces. Convolutional Neural Network (CNN) is utilized to train the gathered images that will extract the vital features used to identify and predict poverty incidence of specific municipality. The data consists of RGB satellite images obtained from Google Earth. The model developed by the researchers did not take into account rich areas that were abundant in mountains since they were lacking during the gathering of data. An example for this would be privately owned lots and subdivisions located in mountainous areas that have few but wealthy residents. The time at which the images were made were not taken into account and these areas may contain obstructions, blur or noise.

2. Review of Related Literature

2.1. Basis for Poverty

For a study on predicting poverty by [1], a basis they had used for economic estimation was the Global Multidimensional Poverty Index also known as MPI. Their source of features and data were not conventional as opposed to other studies wherein they used satellite imagery alone. Instead, the researchers used environmental data which was connected to food security, economic activity, accessibility to facilities and call data records. Their empirical results prove the above average accuracy when using data that differ greatly, yielded a Pearson Correlation of 0.91. They then attempted to predict poverty in context of health, education, and standards of living. Surprisingly their Pearson Correlation showed that they are good indicators for poverty prediction with Pear-son Correlations through 0.84-0.86.

2.2. Poverty Prediction

A 2016 study conducted by [2] used artificial intelligence, particularly Convolutional Neural Networks and high-resolution satellite imagery to predict the economic well-being of 5 African countries. They found the idea to use satellite features instead of local data collected because collection of data manually is expensive and time-costly. Their study was able to predict 75 percent of the variation in local-level outcomes.This supports the current research greatly because of the high probability of its success and because according to [3], the Philippines suffers a higher poverty incidence than the general population which is estimated at 21.6 percent in the year 2015. [4] proposed a study to use Convolutional Neural Networks but had a different approach when it came to the features used in satellite imagery. They used both object and texture features extracted from satellite imagery. They took into account features such as the number and density of buildings using the existence of shadow area, which is used as a model for building height, the number of cars within those villages, the length and density of roads, paths and roof material to name a few. Another team of researchers [5], have also taken to using satellite imagery as basis for the prediction of poverty except they used publicly available, freely distributed satellite imagery. Their study differed from previous ones because they made use of multi-spectral satellite imagery.

3. Methodology

3.1. Data Gathering and Pre-processing

Data on poverty was gathered from the local office of the National Economic and Development Authority. The gathered data comprises the Poverty Incidences of the different municipalities of Cebu in the year 2012. Poverty is formally defined by [1] in Equation 1 where Q represents the number of families or individuals with per capita annual income or expenditure less than the per capita poverty threshold divided by total number of families or individuals. These incidences indicate the standard of living in a country. Satellite images of municipalities were retrieved from Google Earth for training. Municipalities were divided using the extracted coordinates of their boundaries using GADM database of Global Administrative Areas which provided a list of coordinates that form a polygon to map out each municipality for a given country.

$$P = (Q / n) * 100 \tag{1}$$

3.1.1. Data Treatment. The images were re sized to 227 by 227 pixels.The image data sets were cleaned by padding anything outside of the boundaries black.

3.1.2. Data Augmentation. After the images were cleaned, the images were mirrored or inverted which was a technique to gain more image data sets. 2 sets of data were obtained, and each image within the sets were rotated by 1 degree. They rotated each image until they reached 360 rotations to gain more data. In total, the researchers gained 72,000 images, enough for a substantial study to be conducted. They then split this data into 70 percent for training and 30 percent for testing.

3.2. Network Architecture

Similar to previous studies [2], [6], the researchers used Convolutional Neural Networks to classify objects using the municipalities extracted. AlexNet was used for the network architecture. The reason as to why AlexNet was the chosen network architecture is because AlexNet reduces over-fitting by using a dropout layer that is applied to every neuron. Another reason as to why AlexNet was the chosen architecture was because of how AlexNet uses Rectified Linear Unit over Sigmoid which lessens the possibility of the occurrence of the vanishing gradient problem. A large amount of data may contain overfitting which Rectified Linear Unit can prevent. Rectified Linear Unit provides a way for simplifying activation functions which is currently popular for deep neural network [7].

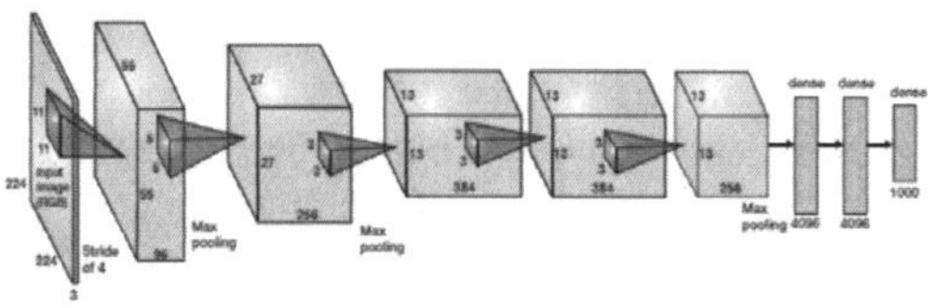

Figure 1. AlexNet CNN Diagram

Caffe, a deep learning framework that included support for AlexNet, was used. The epoch count was modified to 440 in the CNN to fit the nature of the data to avoid over training. 100 images were gathered which was augmented to 72,000 images. The 72,000 images were divided into 50,396 images for training and 21,604 images for testing. Each image would be labeled after its respective class which in our case is 0 for 0-10 poverty incidence, 1 for 11-20, 2 for 21-30, 3 for 31-40 and 4 for 41-beyond. The reason as to why the researchers grouped together municipalities with poverty incidences beyond 40 is because of the lack of municipalities beyond that poverty incidence. Per iteration during the training, 50 images were propagated through the network. The loss function that was used by the researchers was Softmax with Loss. The reason why softmax is useful is because it converts the output of the last layer in the neural network into what is essentially a probability distribution. The trained model was used to classify each image in the test set of the data set. To measure the performance of the trained model, the researchers calculated the Accuracy, Recall, Precision and the F1-Score of the classifier. In the words of [8], Recall in statistics, is the ability of a model to find all the relevant cases within a data set. Precision is defined as the number of true positives divided by the number of true positives plus the number of false positives. While recall expresses the ability to find all relevant instances in a data set, precision expresses the proportion of the data points our model says was relevant, actually were relevant [8]. The F1-Score takes into account both Precision and Recall. According to [9], F1 score is used if one needs to convey the balance between Precision and Recall. It is also important to use F1-Score if there is an uneven class distribution.

4. Results and Discussions

Initially, 50 images were extracted which were then augmented to 36,000. The initial dataset showed signs of under-fitting because of the lack of data. 50 more images of municipalities were extracted which totaled to a 100. This produced an augmented total of 72,000 images. The data was divided into 70 percent for training (50,396 images) and 30 percent for testing (21,604 images). The first idea was to make each unique poverty incidence a specific class and train the model but at an early stage of the training the model showed signs of underfitting because all of the classes only contained 720 or 1,440 images. This was not substantial enough data for features to be detected by the CNN. A new approach was attempted in creating the classes, which was to use two classes only. The classifier only classifies an image as above or below a certain poverty incidence threshold. The mean poverty incidence of 24 was taken from the data set. The model was trained with binary classification as its specification. It was discovered that the model didn't learn because there were too many features in both classes to conclude whether a certain municipality in an image was either above or below the poverty incidence threshold. Finally, the researchers tried grouping together poverty incidences by range of 10 such that the classes were 0-10, 11-20, 21- 30, 31-40, 41-beyond for poverty incidence classification and also adjusted the parameters by increasing the number of epochs for better results. This experimentation resulted in the loss value growing due to over-training but was continuously modified resulting in an epoch count of 440 with a loss value of 0.30. The trained model was used to classify the test set and was able to reach an accuracy of 84%, classifying 18,149 images correctly out of 21,604. In Table 3 there is a sample of experience results by the CNN. The statistical tests of the classifier were calculated with recall, precision and F1 Score and came up with these results which is shown on Table 1.

Table 1. Results of the Testing.

Classes	Precision	Recall	F1-Score	Support
Avg/total	0.86	0.84	0.84	21,604
0-10	0.91	0.95	0.93	6,265
11-20	0.85	0.73	0.78	3,457
21-30	0.63	0.92	0.75	3,241
31-40	0.88	0.85	0.86	5,184
41-beyond	0.98	0.67	0.79	3,457

4.1. Class 0-10

The statistical test for this class performed remarkably well in terms of all the statistical tests. In fact, it has one of the highest scores in each test compared to all the other classes. The kinds of images in this class consisted mostly of densely populated areas such as the Metro Manila area. There were more buildings and structures present in the images and contained very little barren land.

4.2. Class 11-20

Compared to 0-10 class, this class did not perform as high but its results are still considered above average. The pattern of the images that were identified in this class had buildings and structures but not as much as 0-10. The patches of green and empty land were still overwhelmed by the number of buildings.

Table 2. Samples of Images and Respective Actual Poverty Incidence.

Image	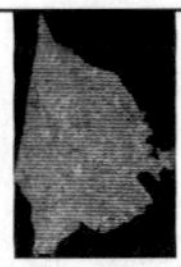			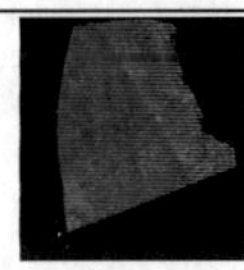	
Poverty Incidence Label	0-10	11-20	21-30	31-40	41-beyond

4.3. Class 21-30

The statistical tests for this class has substantial scores but something important to note is that, compared to the rest in this precision test, the classifier regarded many of the images as truly part of the 21-30 range but it also regarded many other images that are not in this class as part of this poverty incidence range. This could also be described as a high amount of true positives but an even higher amount of false positives. Images that were labelled with that class would usually consist of mountainous regions combined with very few amounts of buildings and sometimes no buildings at all. The cases where there were almost no buildings at all in the image allowed the model to classify it as 31-40 or 41-beyond. In cases where there were a few buildings present in the image, the model performed well and classified it correctly to the 21-30 class.

Table 3. Results Sample.

Image	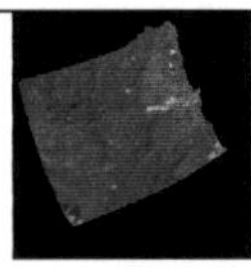	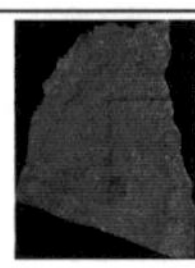	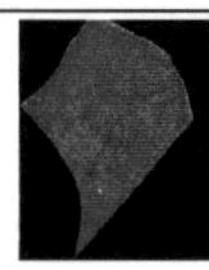	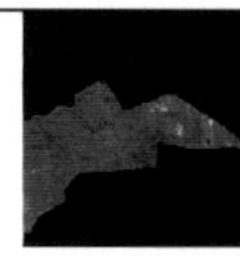	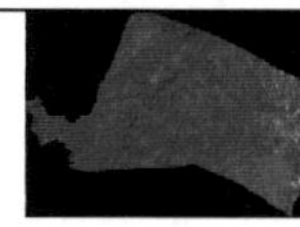
Poverty Incidence Label	20.7	27.5	27.9	9.7	16.6
Predicted Class/Range	41-beyond	31-40	41-beyond	41-beyond[a]	41-beyond[a]

[a] These are incorrectly classified images because mountainous regions take up majority of the area

4.4. Class 31-40

The model was able to classify precision and recall well for this class as it has picked up features such as municipalities that have really large mountainous regions while having small amounts of buildings visible within the satellite imagery. Compared to the Class 41-beyond which contained mostly no buildings present within the image.

4.5. Class 41-beyond

Almost all the actual poverty incidences within this class were what the classifier predicted correctly because of its high precision but many of the actual poverty incidences within this class were deemed as false or one could say this class contained high false negatives. The 41-beyond class has the highest precision among the classes, but this class suffers the lowest recall. The reason why the recall of this class is very low is because of the presence of mountainous regions in the image that were classified as 41-beyond. This meant that the model did not yet find features that could possibly classify wealthy areas that satisfied this criteria. This caused the amount of False Negatives to increase because most of the mountainous municipalities would be classified as 41-beyond which affected the model's recall.

5. Conclusion

A major problem of society today is the manual and costly labor of data gathering. The researchers' goal was to test if the use of a certain CNN architecture namely AlexNet could estimate the poverty levels within a certain country accurately by breaking it down into manageable images. From the results, the most significant reason why the system could not obtain a higher estimation rate was because the researchers lacked the data set for wealthy areas with low poverty incidences that contained mountainous regions within them. Advice for further studies would be to increase the variability of the data set since it lacked the accuracy for it to be reliable for government use and most probably use another method for coming up with results since the output comes in the form of brackets. It would be best if there was a specific predicted output for a given expected poverty level and then perform regression tests on it.

References

[1] Pokhriyal N and Jacques D C 2017 Combining disparate data sources for improved poverty

prediction and mapping *Proc. of the National Academy of Sciences* vol **114** no **46**

[2] Jean N, Burke M, Xie M, Davis W M, Lobell D B and Ermon S 2016 Combining satellite imagery and machine learning to predict poverty *Science* vol **353** pp 790-4

[3] Anonymous 2018 *Philippine Statistics Authority Glossary of Terms* https://psa.gov.ph/poverty-press-releases/glossary

[4] Engstrom R, Hersh J and Newhouse D 2016 Poverty from Space: Using High Resolution Satellite Imagery for Estimating Economic Well-Being and Geographic Targeting

[5] Perez A, Yeh C, Azzari G, Burke M Lobell D and Ermon S 2017 Poverty prediction with public landsat 7 satellite imagery and machine learning *Preprint* arXiv:1711.03654

[6] Wang W, Yang S, He Z, Wang M, Zhang J and Zhang W 2018 Urban perception of commercial activeness from satellite images and streetscapes *Companion of the The Web Conf. 2018 on The Web Conf. 2018* pp 647-54

[7] Ramachandran P, Zoph B and Le Q V 2018 Searching for activation functions *Under review as a Conf. paper at ICLR 2018*

[8] Koehrsen W 2018 *Beyond Accuracy: Precision and Recall* https://towardsdatascience.com/beyond-accuracy-precision-and-recall-3da06bea9f6c

[9] Shung K P 2018 *Accuracy, Precision, Recall or F1?* https://towardsdatascience.com/accuracy-precision-recall-or-f1-331fb37c5cb9

IOP Conf. Series: Materials Science and Engineering **482** (2019) 012045 doi:10.1088/1757-899X/482/1/012045

IoT based model for monitoring and controlling water distribution

J G Natividad[1] and T D Palaoag[2]

[1]Isabela State University – Ilagan Campus, City of Ilagan, Isabela, Philippines
[2]University of the Cordilleras, Baguio City, Philippines

[1]sirdyowi@gmail.com, [2]tpalaoag@gmail.com

Abstract. These days, because of increment in relocation from a provincial territory to urban ranges, the population in urban areas is obviously expanding quickly together with the requirement for comfortable living. With increase in population, urban areas have expanded, water becomes one of the major problems in a city particularly water distribution, interfered with water supply, water protection, water utilization and furthermore the water quality. To overcome water supply related problems proper monitoring and controlling system must be implemented. The developed system consist of different IoT devices like water pressure sensor, ultrasonic sensor, solid state relay switch, motorized electric water valve, Raspberry PI, GSM module and Arduino UNO micro-controller. This paper focused on the monitoring and controlling of water distribution using IOT based model. It aims to design and develop a low cost reliable and efficient technique to improve water distribution in the community. A prototype was developed to simulate the operation of a water distribution. Also, a web application was created as a front-end system for monitoring the status of the different pumping stations as well as controlling. Also, fuzzy logic algorithm was integrated into the developed prototype system to be more scientific in making decision. As a result, the experiment was successful and passed all the conditions set for monitoring and controlling water distribution using IoT based model.

1. Introduction

According to recent survey, growth in population caused cities to face water distribution issues. Many communities suffer from insufficient water supply for their day to day needs. Lack of monitoring and controlling water distribution becomes a serious problem. Some areas in a city will have enough supply of water while other areas do not have. This is due to some problems in the distribution line such as damage pipeline cause by over pressure or low water pressure where in water cannot reach consumers located on a high-ground areas or far away from the pumping stations or water tank. All of these issues concerning water distribution are because of lack in real-time monitoring and controlling mechanism and due to its manual or traditional operation.

Today, cities are now transforming and started to adapt smart technologies for sustainable communities. As they participate for economic advancement and the facilities that increase to their vibrancy, water has become a priority in their checklists [1]. Creating water sustainability requires a multidisciplinary approach. It also requires state of the art equipment to facilitate the operation and management especially in collecting and analyzing data to initiate an action for smart management, planning and decision making.

The City of Ilagan in the province of Isabela is under continues development to improve the quality of living for its constituents. It is the capital and the largest town in the province that has 91 barangays and targeting to become a smart city. The water distribution in the city particularly the barangays in the Poblacion area is under the management of the City of Ilagan Water District (CIWaD). The continues development in the city such as construction of commercial buildings, subdivisions, residential houses and business establishment, water supply becomes one of the priority. Currently, CIWaD continue in expanding their coverage to provide the needs of the community and providing the needed amount of water daily for the consumers is a serious matter. The poor mechanism in monitoring and maintaining water distribution has become the problem of the management due to its manual operation that requires human intervention. As to date, the geographical jurisdiction of CIWaD covers 24 barangays within and near the Poblacion area. Currently, CIWaD manage 7 water reservoirs where in 2 are located in a high ground and 5 elevated tank. There are also 22 pumping station deployed in the different areas within the jurisdiction connected in a series that supply waters to the community and to the water reservoirs

According to Mr. Moises P. Pascual, Water Utility Management and Development Officer, in an interview conducted, monitoring of facilities is still in a manual operation. It requires 2 operators to be deployed in every pumping station but due to lack of manpower, 1 operator maintains two pumping stations that cause maintenance and monitoring issues. Thus, this study is very timely because providing sustainable community is one of the city's priorities.

The focus of this paper is the use of IoT technologies for the development of intelligent water management system for CIWaD in monitoring the status of the different water tanks and pumping stations as well as water pressure. Controlling system is also included to the proposed system to avoid tanks from over filling, pipe leakage due to over pressure. A prototype was developed for simulation and testing. In particular, Arduino micro-controller and other IoT devices were used for monitoring and controlling. Also, a web application was developed as front-end for monitoring the status of the operation.

This paper is limited only to CIWaD water distribution and management. Result and evaluation of the study was based and conducted only on the developed prototype.

2. Review of Related Studies

The popularity of IoT technologies is growing due to its capability in developing various application and only little portions is currently available in the community. Its domain includes health-care, transportation, logistic, smart environment and many more [2].

Prachet Verma et al. [3] developed an IoT based for water distribution in a campus. They developed a system that will distribute enough water to each tank to satisfy local demands. In their study, ultrasonic sensor has been used to monitor the water level in a tank.

In a study conducted by Ejiofor Virginia Ebere and Oladipo Onaolapo Francisca [4], the automatic detection of water level in a tank is done by a comparator circuit. They used the electrical conductivity property of water as an advantage. The copper conductors serve as the water level sensor. When the copper sensor touched by water, voltage runs to the copper which in turn is transferred to the comparator circuit for further processing. Then a micro-controller uses this signal coming from the comparator to control the water pump.

Also another study for automatic water tank filling system was conducted suitable to be used in home activity to reduce energy consumption due to water spills [5]. This study also helps the community to analyze their water consumption. The prototype can be used to solve water pump problems due to the operations of the manual switch as well as the floating ball tap to stop the water tank filling.

3. System Architecture

In this work, the IoT-based model for monitoring and controlling of water distribution system consists of the following major units: ultrasonic sensors, pressure sensor, motorized electric water valve, GSM module, Arduino micro-controller, Raspberry Pi and the solid state relay switch. Figure 1 below describes the flow of operations in the system as well as their inter-operability.

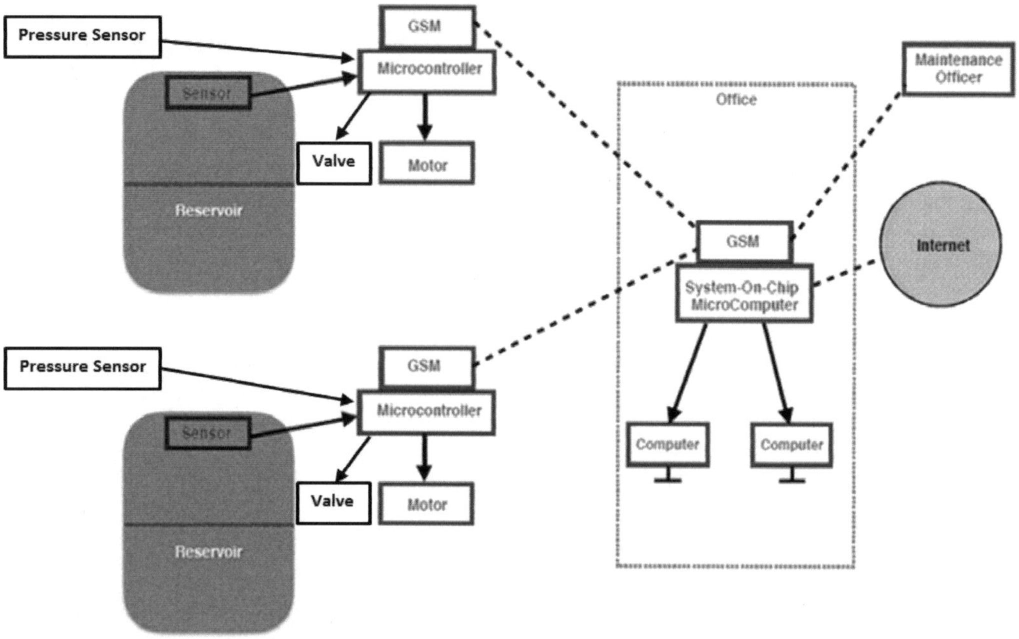

Figure 1. System architecture.

To monitor the level of water within the tank, a water level sensor was installed. As the level of water increases or decreases in a certain point, the micro-controller will activate the water pump relay switch to turn on or off to avoid water tank overflowing or unfilled. Same process also implemented in controlling water pressure to avoid damages in the pipeline. If the water pressure increases or decreases, the micro-controller also activate the motorized electric water valve to rotate at a certain point to control the flow of water.

Here, GSM is also used to send a message when there is an activity made by the micro-controller. An SMS will be sent to the office through the GSM module to report the status of activity in a particular water reservoir or pumping station. The System-on-chip microcomputer that will be located in the office will store the received data then sends an update to the maintenance officer regarding the received report through SMS. The System-on-Chip microcomputer will be configured also as a web server to install the developed web application for the management to view the status the different water reservoir and pumping stations. The web application can be access using a computer within the office connected to a network and can be access also through the Internet.

4. Methodology

The IoT based model for monitoring and controlling water distribution is consisting of two circuit designs, the Monitoring and Controlling System using micro-controller and Web Server for monitoring using system-on-chip micro-computer.

4.1 Monitoring and Controlling System

This module is composed of Ultrasonic sensor, 5volts Analog Water Pressure sensor, Arduino UNO micro-controller, SIM800 GSM Shield module, 220VAC 2-way Motorized Electric ball valve and 240VAC relay switch. For monitoring, ultrasonic sensor was used in the prototype because of its capability to probe inside water tank non-destructively. The ultrasonic sensor is connected to the

Arduino UNO micro-controller and the signal send by the sensor will be interpreted and processed by the micro-controller to evaluate the distance or level of water in the tank. Based on the uploaded code, if the level of water requires filling the tank then the Arduino micro-controller will send signal to turn on the solid state relay switch where the water pump is connected. As a control measure, the same process to avoid overflowing while fills up the water tank or before it gets empty. If the water is already in a full level, the Arduino micro-controller will send signal to the solid state relay switch to turn off and when the water level is in a minimum level it will turn on. Also, controlling water pressure is done using a water pressure sensor connected to the Arduino UNO micro-controller. This is to maintain the acceptable pressure to distribute water in the main line to avoid breakage. The micro-controller sends signal to the solid state relay switch where the motorized electric valve is connected to close and open if the pressure is too low or exceed in the acceptable level. The micro-controller will also activate the GSM module that is connected to send data to the server to report the status of the water reservoir and current activities of the pumping station. The data will be sent through SMS which is believe as the most practical and economical way of communication. The monitoring and controlling system is also allowed to receive SMS send by the server located in the CIWaD office to override the built-in automatic controlling system of this module. Using the designed web application installed in the server, it can manually control a particular pumping station or water reservoir by sending SMS with predefined keyword message.

4.2 Web Server

This module is composed of two different components, the hardware and the software. The hardware component includes the Raspberry Pi 3 Model B, 5V power adapter/supply, Class-10 Micro-SD card, LED monitor, SIM800 GSM module, USB keyboard and mouse. The software component includes Raspbian Operating System, Linux version of Apache, MySQL, PHP (LAMP) for the web services and Python for the script used in sending and receiving SMS from the GSM module.

In this project the Raspberry Pi will be configured as a web server for viewing status or activities and controlling the different water reservoirs and pumping stations of CIWaD. The received report messages will then be extracted by the developed web application to get the necessary data for viewing in the web interface of the application. The GSM module also used to send message report to the maintenance officer assigned to a particular water reservoir or pumping station to inform about the current activity. The web application is also design for controlling the deployed monitoring and controlling system of the different pumping stations and water reservoirs by sending an SMS using a predefine keyword to override the current activities such as switching on and off a particular pumping station or opening and closing the gate valve of a particular water reservoir.

5. Results and Discussions

Based on the data gathered, the researchers come up with a design to address the problems encountered by the CIWaD and develop a prototype for simulation, testing and evaluation.

The researchers purchased all the needed hardware or equipment in building the prototype. After assembling the prototype monitoring and controlling system, the researchers immediately writes the codes, compiled and uploaded into the Arduino micro-controller. Also, the researchers setup and configure the Raspberry Pi as a web server and developed the web application to be used as front-end system for monitoring and controlling.

5.1 Testing

The researchers tested the prototype monitoring and controlling system to evaluate the accuracy of the data and the code uploaded. The inputs are the water level of the tank and the pressure of water flows in the pipeline. In the designed setup, an additional tank was used as temporary source of ground water where in the electric motor pump used to get water to fill the temporary water reservoir. To imitate the CIWaD water system, the temporary water reservoir was also placed on a higher platform so that the gravitational force of the water increases the pressure in the pipeline to meet the required pressure for the testing. It was also labelled in inches as basis to measure the level of water. Figure 2 below shows the prototype layout design of a water distribution for the testing of the developed monitoring and controlling system.

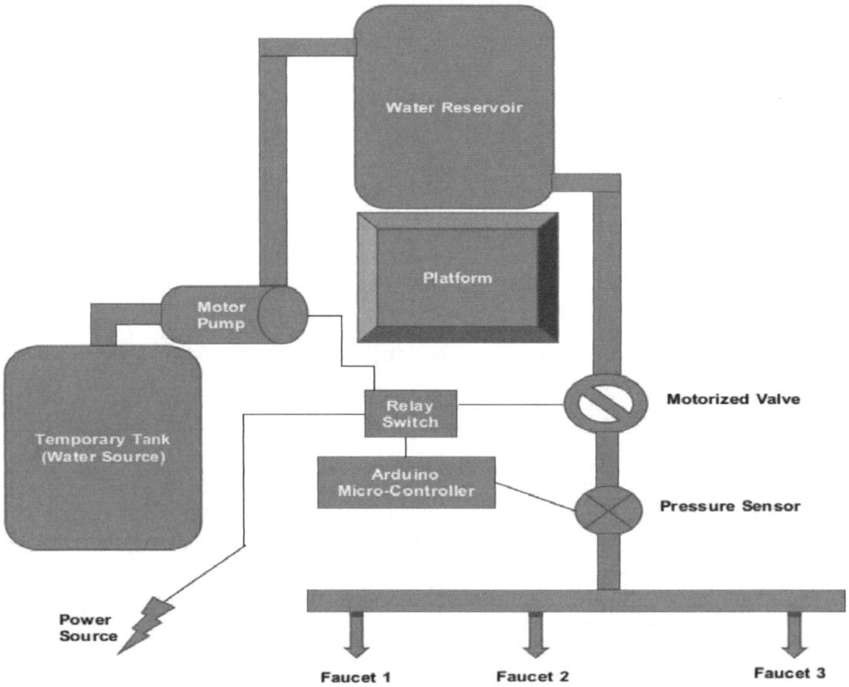

Figure 2. Prototype water distribution layout.

In the uploaded code, the researchers set a threshold values for monitoring water level and water pressure. This is for the system to decide what action to be taken.

To determine when to send report to the server or to turn on the motor pump to re-fill the water reservoir, several sub-parameters will serve as input. Also as a control measure while re-filling, the micro-controller will turn off the motor pump if it reaches the set maximum level of water. In sending SMS report to the server regarding the status of the water reservoir and what action to be taken to the motor pump, fuzzy logic algorithm was implemented. As shown in Table 1, the level of water measured in inches served as the input. Based on the uploaded rules into the micro-controller, it can provide SMS report or action to be taken as an output.

Table 1. Water level threshold.

Level	Status	Re-fill Action
20" and below	Low Level	Turn-On
21" to 35"	Normal	Turn-Off
36" to 40"	Full	Turn-Off

Likewise in monitoring and controlling for water pressure, the same process and algorithm was implemented. Here, pound per square inch (psi) is used as unit of measurement for water pressure. As shown in Table 2, the fuzzy logic algorithm used different input parameters measured in PSI as basis for the output.

Table 2. Water pressure threshold.

PSI	Status	Valve Action
3 and below	Low	Open
4 - 6	Normal	Open
7 and above	High	Close

On the server side, all the SMS reports received were posted in the web application. Also, a notification message was sent to the registered operator cell phone number to inform about the status of a particular pumping station or water tank. The web application also includes controlling mechanism to override the deployed monitoring and controlling system. Table 3 below shows the pre-define keywords used in SMS message sent by the developed web application to manually turn off or on the pumping station and open or close the motorized electric valve of the water tank. A toggle switch is provided in the web interface to easily switch on or off a pumping station motor or the motorized electric valve.

Table 3. Pre-define keywords.

Pumping Station Motor	Motorized Electric Valve
PSM Off	Valve Close
PSM On	Valve Open

6. Conclusions and Recommendations

Providing good quality services for a sustainable community is a big challenge to the leaders. Based from the problems that the CIWaD management are now facing in terms of water distribution, the developed system can be a big help for them to address these problems. It allows them to utilize low-cost smart technologies to easily manage the distribution of water to the community. The developed prototype functions properly according to the specification and purpose. It successfully passed all the tests and met the expected result based on the different input parameters.

Finally, the continues development in technologies particularly IoT devices in terms of providing smart communities will have a great chance for this study to further improve the ideas and concept for future use.

References

[1] Batten J J 2018 Realizing a more sustainable water future from a one water view *J. - American Water Works Assoc.*

[2] Kim S and Kim H 2011 Systematic optimal design of overlapped ultrasonic sensor ring for high performance obstacle detection *Proc. 15th Int. Conf. on Advanced Robotics* pp 600-5

[3] Verma P, Kumar A, Rathod N, Jain P, Mallikarjun S, Subramanian R, Amrutur B, Kumar M M and Sundaresan R 2015 Towards an IoT based water management system for a campus *Smart Cities Conf. (ISC2) 2015 IEEE First Int.* pp 1-6

[4] Ebere E V and Francisca OO 2013 Microcontroller based automatic water level control system *Int. J. of Innov. Res. in Comp. and Communication Eng.* vol **1** no **6**

[5] Prima E C, Munifaha SS, Salam R, Aziz M H and Suryani A T 2016 Automatic water tank filling system controlled using arduino based sensor for home application *Engineering Physics Int. Conf. 2016* vol **170** pp 373-7

IOP Conf. Series: Materials Science and Engineering **482** (2019) 012046 doi:10.1088/1757-899X/482/1/012046

Fittest Job First Dynamic Round Robin (FJFDRR) scheduling algorithm using dual queue and arrival time factor: a comparison

Jezreel Ian C. Manuel[1], Rey Benjamin M. Baquirin[2], Kier Sostenes Guevara[3], and Dionisio R. Tandingan Jr[4]
University of the Cordilleras, Baguio City, Philippines

[1]jezreelianmanuel@gmail.com
[2]reybenbaq@gmail.com
[3]kier.guevara @gmail.com
[4]dion_tand@yahoo.com

Abstract. The Fittest Job First Dynamic Round Robin (FJFDRR) was introduced as a CPU scheduling algorithm whose performance evaluation reduces the number of context switches (CS), average waiting time (AWT), and average turnaround time (ATAT) of processes in a single CPU environment. In this paper, we explored improvements on the FJFDRR by including the process arrival time as an algorithmic factor implemented using a dual queue. We then compared the performance of the proposed algorithm called enhanced Fittest Job First Dynamic Round Robin (eFJFDRR) as with the FJFDRR algorithm together with the other CPU scheduling algorithms. Trial results showed that eFJFDRR scheduling algorithm performed better in reducing average waiting time, average turnaround time, and average response time in some cases. It was also found to balance the number of context switches of the processor during execution.

1. Introduction

CPU scheduling determines the order in which processes are executed in the CPU. A computer system that implements an effective scheduling algorithm is one that optimizes CPU utilization, minimizes response time, and maximizes throughput [1]. As such, scheduling algorithms have remained to be the topic of interest in computer architecture and operating systems researches. With modern operating systems becoming more complex, there is a need to continually improve CPU allocation through scheduling algorithms to maximize system throughput, and CPU utilization [2].

This paper focuses on the improvement of the existing scheduling algorithms based on the Best Job First (which uses a fit factor 'f' where 'f' is given by the summation of priority, arrival time and burst time of a process) and the Fittest Job First Dynamic Round Robin (which combines the best job first (BJF) algorithm and other scheduling algorithms starting with the lowest Fit Factor 'f' where 'f' is given by the constant values of user priority weight and burst time priority weight.). The improvement of the two algorithms will be called as enhanced Fittest Job First Dynamic Round Robin (eFJFDRR). This algorithm implements using dual ready queues and arrival time factor to satisfy the scheduling criteria which decreases the average turnaround time, average waiting time, average response time and context switching. The proposed eFJFDRR scheduling algorithm is compared to the other CPU scheduling namely first come first serve (FCFS), shortest job first (SJF), round robin (RR), and best job first (BJF) algorithms.

2. Review of Related Literature

Several CPU scheduling algorithms have different properties and may favor one class of processes over another. Users must consider the characteristics of these algorithms to ensure a better performance metrics. Many criteria have been suggested for comparing CPU scheduling algorithms. The following criteria include:

CPU utilization: A user want to keep the CPU as busy as possible. It may range from 0 to 100%. In real time system, it suits range from 40% (for a lightly loaded system) to 90% (for heavily loaded system).

Throughput: If the CPU is busy executing processes, then work is being done. One measure of work is the number of processes completed per time unit or throughput. For long processes, this rate may be one process per hour; for short transactions, it may be ten processes per second.

Turnaround Time: From the submission time of a process to its completion time is turnaround time. It refers to the total time period spends waiting to get into memory, waiting in the ready queue, executing in the CPU and doing input/output operations.

Waiting Time: The CPU scheduling algorithm does not affect the amount of the time during which process execute or does input/output. It affects only the amount of time that a process spends waiting in the ready queue. Waiting time refers to the sum of time periods spends waiting in the ready queue.

Response Time: It refers to the time from the submission of a request until the first response is produced called response time. It is the time it takes to start responding, not the time it takes to output the response. The turnaround time is generally limited by the speed of the output device.

Context Switch: The process of storing the state of a process or of a thread, so that it can be restored and execution resumed from the same point later. This allows multiple processes to share a single CPU, and is an essential feature of a multitasking operating system [3].

Most of the newly proposed scheduling algorithms today are combinations or improvements of basic pre-emptive and non-pre-emptive solutions like First Come First Served (FCFS), Shortest Job First (SJF), Priority Scheduling, and Round Robin (RR). This is primarily done to overcome the limits posed by each individual algorithm. For example, FCFS scheduling solely relies on the execution of processes in the order they arrive in the ready queue, hence increasing the waiting time of all processes with short burst-times that follow processes with longer burst-times [4]. Conversely, SJF scheduling executes the shortest processes first but leaves longer processes starved [5]. Priority Scheduling executes processes with the highest priority first and starves processes with lower priority [6]. RR, on the other hand, executes each process equally on a constant time quantum, which may reduce system throughput if the time quantum is too small [7].

The limitations mentioned above motivate many programmers to come up with hybrid algorithms for process scheduling. Two such algorithms are imperative to this study. Foremost is the Best Job First (BJF) CPU Scheduling Algorithm [8] which was used as a benchmark for comparison against the proposed algorithm as shown in Section 3 of this paper. BJF uses a computed Factor 'f' to schedule processes where 'f' is given by:

f = Priority + Arrival Time + Burst Time ----- (equation1)

BJF schedules processes starting from the lowest 'f' score, then executes them in order, non-pre-emptively. Second is the Fittest Job First Dynamic Round Robin [8] (FJFDRR), an algorithm that combines BJF with other algorithms [4, 8] and schedules processes in order starting with the lowest Fit Factor 'f' score as given by:

f = UP * UW + SP * BW ----- (equation 2)

Where UP = User Priority, UW = User Priority weight, SP = Shorter Burst time Priority, and BW = Burst time Priority Weight. As user Priority weight(UW) was deemed to be more important than burst time priority weight(BW) [8], UW and BW have constant values of 60% and 40% respectively assuming that all processes have same arrival time. Once ordered, the algorithm then executes the processes on Dynamic Round Robin; computing variable time quantum based on the median of all processes' remaining burst time [9]. A performance evaluation done on the FJFDRR algorithm in 2011 yielded a decrease in the number of context switches, average waiting time, and average turnaround time [10] when compared against the Priority Based Static Round Robin (PBSRR) algorithm. However, the experimental test cases used did not factor in the arrival time for each process and the researchers recommended further studies be done on the algorithm to incorporate it.

Taking on the recommendation, the researchers present an improvement on the FJFDRR scheduling algorithm by incorporating the arrival time of processes into the algorithm and managing them through multiple queues. The researchers refer to our algorithm as eFJFDRR henceforth.

3. Methods

The methodology for eFJFDRR comprises the following:

(1) The original Fit Factor formula 'f' as given by

f = (UP * 0.6) + (SP * 0.4) ----- (equation 3)

(2) The inclusion of each process' Arrival Time as an algorithmic factor;

(3) The use of an additional queue to handle Newly Arrived processes; and

(4) A dynamic time quantum [10] implemented for 1 RR cycle per queue.

The algorithm can be described as follows:

(1) Before execution starts on the processes in the ready queue, each process' Fit Factor 'f' is computed using equation 3.

(2) They are rearranged ascendingly in the queue and the time quantum is computed by getting the median of all processes' remaining burst times.

(3) The queue starts process execution via RR using the computed time quantum.

(4) Meanwhile, all processes that enter the CPU while the first queue is executing are added to a separate queue.

(5) When the first queue completes 1 cycle of RR, it then switches to the other queue, and treats it as the new ready queue starting again with step 1 above with the difference that when it gets to step 4, the algorithm does not create another queue to add newly arrived processes, instead it uses the first queue (which is not executing at this point) and inserts the newly arrived processes there. The algorithm is repeated until no processes remain to be executed in either queue. Evidently, eFJFDRR relies heavily on the alternating execution of two queues via dynamic RR. Further illustration on how the algorithm works is shown below:

3.1. Pseudocode of the eFJFDRR algorithm

Let Q1 and Q2 be the queues to be executed alternately. Let TQ be the time quantum. Let NA be the newly arrived processes.

```
1. Let Q1 be the ready queue and Q2 the waiting queue.
2. Add the first process to arrive at the ready queue.
3. If there are processes with the same arrival time as the first process, add it to the ready queue.
4. Calculate the fit factor of each process in the ready queue using Equation (3).
5. Sort the ready queue in ascending order by the fit factor.
6. Set TQ = median of the remaining burst time of the ready queue.
7. Execute the ready queue for 1 cycle of RR.
8. While (ready queue is executing) {
   add NA to the waiting queue.
   remove process with no remaining burst time from queue.
   }
9. If (ready queue and waiting queue is empty)
        repeat steps 1-9.
```

```
Else if (waiting queue is empty)
      repeat steps 4-9.
Else if (waiting queue is not empty)
{
    If (Q1 = ready queue)
        {Let Q2 be the ready queue and Q1 be the waiting queue}
    Else
      {Let Q1 be the ready queue and Q2 be the waiting queue}
      repeat step 4-9
}
```

3.2. Illustration of the eFJFDRR algorithm

To further illustrate how the algorithm works, we present a simulation. Table 1 below shows 6 processes with arrival time, priority, and burst time all known before CPU execution.

TABLE 1. SAMPLE PROCESSES

Process	Arrival Time	Priority	Burst Time	Fit Factor
P0	1	1	20	8.6
P1	10	2	10	5.2
P2	15	1	15	6.6
P3	17	3	5	3.8
P4	22	1	5	2.6
P5	25	2	8	4.4

Process P0 arrived first with the arrival time of 1. Since no other processes have the same arrival time as P0, only P0 will be added to Q1. Since Q1 has only 1 process, TQ is 20. Q1 will be executed for 1 round. While Q1 is executing, processes P1, P2, and P3 arrive. They will be added to Q2. When Q1 is finished executing, P0 has no remaining burst time so it will be removed from the queue. The fit factor for P1, P2 and P3 are 5.2, 6.6, and 3.8 respectively. The TQ for Q2 is 10 which is the median of the remaining burst time of Q2. The process with the least fit factor will be scheduled first hence the order will be P3, P1, and P2. Q2 will be executed for 1 round. While Q2 is executing, processes P4 and P5 arrive. They will be added to Q1 since Q2 is executing. When Q2 is finished executing, only P2 remains with a remaining burst time of 5. The fit factor for P4 and P5 are 2.6 and 4.4 respectively. The time quantum for Q1 is 6. Q1 will be executed. Since P4 has a lower fit factor, it will be scheduled first. While Q1 is executing, there are no processes to add to Q2 since there are no processes that arrive. When Q1 finished, only P5 remains with a remaining burst time of 2. Q2 will begin executing with TQ as 5. When Q2 finished executing, it will be empty and only Q1 remains with 1 process which is P5. Q1 will begin executing with TQ as 2. When Q1 finished executing, there will be no remaining process to execute.

4. Test Cases

All experimental test cases were performed with the following assumptions: (1) Processes are executed in a single processor. (2) All the processes are CPU bound. (3) All parameters, namely total number of processes, arrival time, burst time, and priority for each process, are known before CPU execution starts. (4) BJF, FCFS, SJF, and RR are used as benchmark algorithms. RR will have a time quantum of 15 as per the performance evaluation of the original FJFDRR [9].

The metrics used for algorithm comparison are the average turnaround time (ATAT), average waiting time (AWT), average response time (ART) and number of context switches (CS). ATAT is the mean time from submission to completion of processes. AWT is the average amount of time processes are in a ready state waiting to be picked up for execution. ART is the average time between the submission of the processes and the first response to that request. CS is the number of times the CPU changes from one process to another. All of these metrics are evaluated per algorithm with the lower scores signifying better performance.

Test Case 1: We assumed 5 processes. Processes with longer burst times arrive earlier and have lower priority while processes with shorter burst times arrive later and have higher priority. Table 2 presents the test case parameters including the computed Fit Factor for each process.

TABLE 2. TEST CASE 1

Process	Arrival Time	Priority	Burst Time	Fit Factor
P0	1	10	100	46
P1	20	5	40	19
P2	31	4	17	9.2
P3	54	3	12	6.6
P4	70	1	6	3

Table 3 shows the comparative results of eFJFDRR against the algorithms assumed as benchmarks.

TABLE 3. E FJFDRR RESULT ON TEST CASE 1

Algorithm	TQ	ATAT	AWT	ART	CS
eFJFDRR	100, 14, 14, 12	95.4	60.4	57	7
BJF	N/A	95.8	60.8	60.8	5
RR	15	91	91.2	19.6	14
SJF	N/A	92.6	57.6	57.6	5
FCFS	N/A	114	79	79	5

eFJFDRR performed better than the remaining algorithms, except SJF, in ATAT and AWT in this test case as shown in Figure 1 below. It also scored better in ART, coming in second to RR. CS score was average in this case.

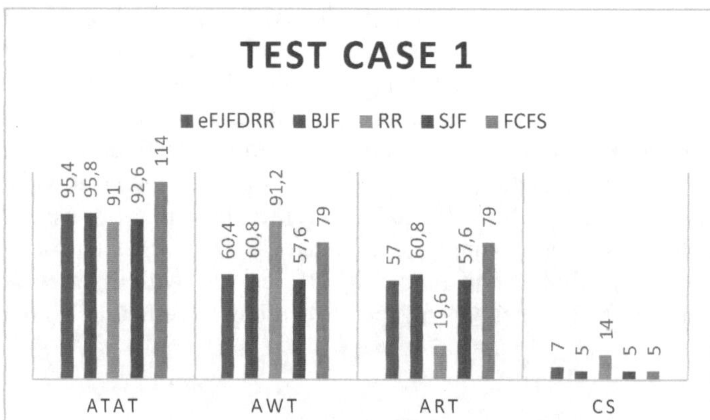

Figure 1. Comparative Results for Test Case 1

Test Case 2: The researchers assume 5 processes. Processes with shorter burst times arrive earlier and have lower priority while processes with longer burst times arrive later and have higher priority. Table 4 shows the case parameters including the computed Fit Factor for each process. Table 5 shows the comparative results of eFJFDRR algorithm against the other algorithms assumed as benchmarks.

IOP Conf. Series: Materials Science and Engineering **482** (2019) 012046 doi:10.1088/1757-899X/482/1/012046

TABLE 4. eFJFDRR RESULT ON TEST CASE 2

Process	Arrival Time	Priority	Burst Time	Fit Factor
P0	1	10	34	19.6
P1	11	8	58	28
P2	54	6	79	35.2
P3	88	3	98	41
P4	92	1	101	41

TABLE 5. eFJFDRR RESULT ON TEST CASE 2

Algorithm	TQ	ATAT	AWT	ART	CS
eFJFDRR	34, 58, 101	139.6	65.6	65.6	5
BJF	N/A	139	65	65	5
RR	15	223.4	198.6	22	27
SJF	N/A	139	65	65	5
FCFS	N/A	139	65	65	5

For all metrics, each algorithm performed roughly on equal ground in this test case except for RR which expectedly has excellent results in ART because of a constant time-quantum but has poor performance in the remaining metrics. Figure 2 below shows the comparative results of each algorithm.

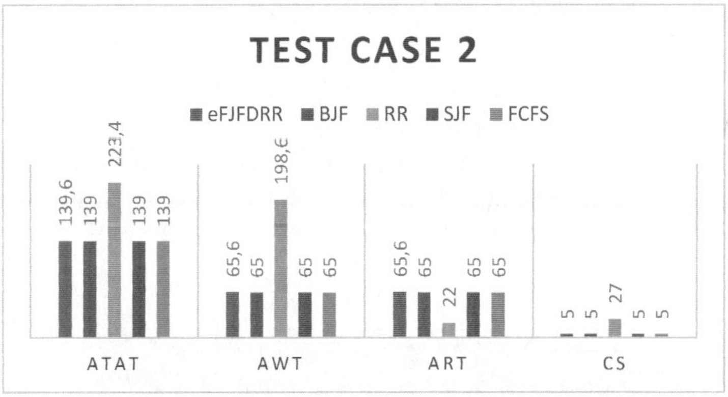

Figure 2: Comparative Results for Test Case 2

Test Case 3: The researchers assume 5 processes. Processes that have higher priorities arrive earlier with random burst times. Table 6 presents the test case parameters including the computed Fit Factor for each process. Table 7 shows the comparative results of eFJFDRR against the algorithms assumed as benchmarks.

TABLE 6. TEST CASE 3

Process	Arrival Time	Priority	Burst Time	Fit Factor
P0	1	1	80	32.6
P1	3	1	29	12.2
P2	20	1	40	16.6
P3	31	5	17	9.8

P4	70	10	12	10.8

TABLE 7. eFJFDRR RESULT ON TEST CASE 3

Algorithm	TQ	ATAT	AWT	ART	CS
eFJFDRR	80,23,11,6	101	65.4	59.6	7
BJF	N/A	107.8	72.2	72.2	5
RR	15	113.8	103.2	23.8	14
SJF	N/A	95.4	59.8	59.8	5
FCFS	N/A	112.4	76.8	76.8	5

Similar to Test Case 1, eFJFDRR has the best ART in this test case, second only to RR. ATAT and AWT scores were also better than others excluding SJF. The CS score was average. Figure 3 shows the comparative results of each algorithm.

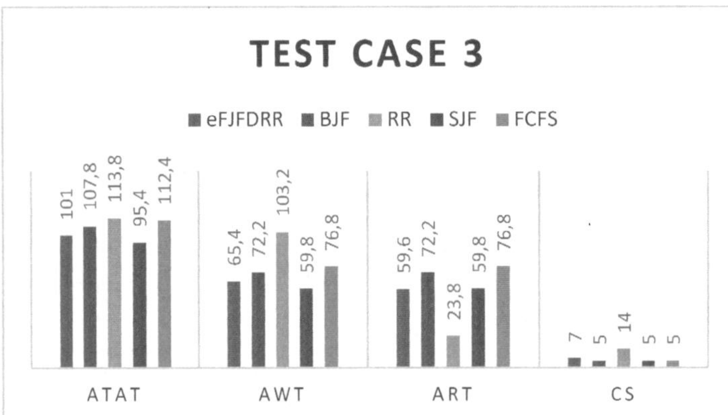

Figure 3. Comparative Results for Test Case 3

Test Case 4: The researchers assume 5 processes. Processes that have longer burst times arrive earlier with random priorities. Table 8 presents the test case parameters including the computed Fit Factor for each process. Table 9 shows the comparative results of eFJFDRR against the other algorithms assumed as benchmarks.

TABLE 8. TEST CASE 4

Process	Arrival Time	Priority	Burst Time	Fit Factor
P0	1	1	30	12.6
P1	3	10	40	22
P2	10	2	20	9.2
P3	30	1	10	4.6
P4	34	1	15	6.6

TABLE 9. eFJFDRR RESULT ON TEST CASE 4

Algorithm	TQ	ATAT	AWT	ART	CS
eFJFDRR	30, 20, 10, 20	55.4	32.4	30.4	6
BJF	N/A	51.4	28.4	28.4	5
RR	15	76.4	69	24.4	9
SJF	N/A	49.4	26.4	26.4	5
FCFS	N/A	67.4	44.4	44.4	5

This test case showed processes with higher fit factor scores arriving earlier. eFJFDRR performed closest to par with BJF and SJF in this test case. Figure 4 below shows the comparative results of each algorithm.

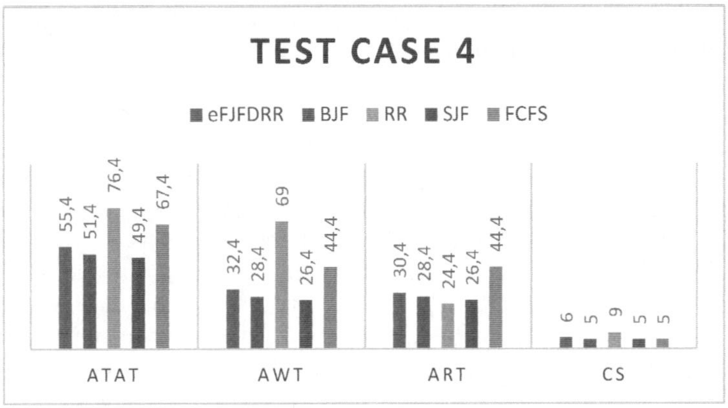

Figure 4. Comparative Results for Test Case 4

As the summary of the findings, we averaged the scores of each algorithm for all test cases. Results showed that eFJFDRR has reduced ATAT, AWT, and ART while maintaining a balanced CS. However, SJF yielded better results across all metrics as exemplified by figure 5.

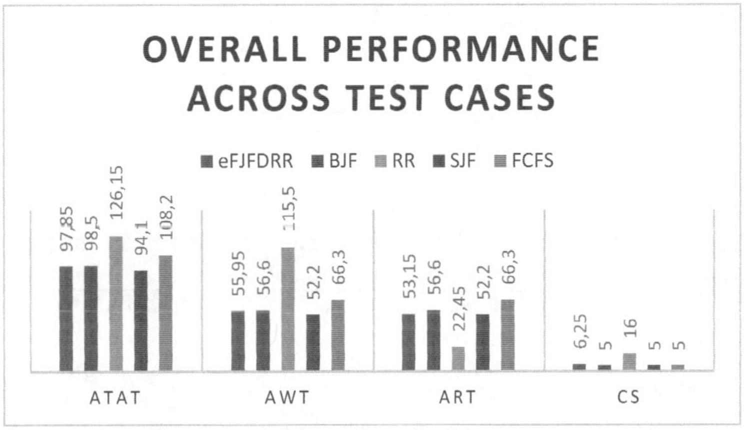

Figure 5. Comparative Results across all Test Cases

5. Discussions

Based on the Test Cases, the enhanced Fittest Job First Dynamic Round Robin (eFJFDRR) scheduling algorithm improves CPU scheduling on three metrics: the average turnaround time, average waiting time, and average response time. The average turnaround time and average waiting time improvements were evident in Test Cases 1 and 3, where the scores for both metrics were better than all algorithms except the shortest job first scheduling algorithm. This is despite the fact that processes with lower priorities arrived earlier in Test Case 1 and processes with higher priorities arrived earlier in Test Case 3. The researchers infer that the weight of priority in our algorithm is not as significant as it is in other algorithms like the best job first scheduling algorithm where due to it being non-preemptive and it schedules processes only using fit factor, priority is imperative. On the other hand, reduction on average response time was evident in Test Case 1, where our algorithm surpassed all others except round robin algorithm. However, we argue that round robin's excellent score in the metric can be offset by its poor performance in the remaining ones.

Results also show that eFJFDRR has the most balanced context switch score based on the number of processes submitted. The researchers attribute this to the dynamic computation of time quantum before a queue is executed; making our algorithm flexible enough to adjust throughput depending on a process' parameters. In contrast, the best job first, first come first serve, and shortest job first are all non-preemptive algorithms and round robin implements a static time quantum so flexibility is limited.

Furthermore, the eFJFDRR was observed to always be close to par with both best job first and shortest job first algorithms, sometimes surpassing them. In Cases 1 and 3, it was able to surpass best job first in three metrics: average turnaround time, average waiting time, and average response time as discussed above. It surpassed shortest job first only in one metric, the average response time, in Case 1.

While we succeeded in discovering individual test cases where our algorithm performed best, Figure 4 showed that it was only second to SJF in every metric when averaged across all test cases. The researchers surmise that this might be caused by the small difference in standard deviation of BJF and SJF on performance metrics [8]. As FJFDRR was originally based on BJF and our algorithm is a modification of FJFDRR, it is plausible that the experimental test cases we conducted were not sufficient in number and variety to distinguish why SJF performed better than eFJFDRR in this study.

6. Conclusions

Comparative results per test case showed that our algorithm can reduce average turnaround time and average waiting time. The researchers also discovered that the enhanced Fittest Job First Dynamic Round Robin scheduling algorithm is flexible enough to adjust throughput despite disparities in process' burst time and priority by maintaining a balanced context switch score. These results align with previous performance evaluations done with FJFDRR [10] where average turnaround time, average waiting time, and context switches were similarly reduced. In addition, our study revealed that the enhanced Fittest Job First Dynamic Round Robin scheduling algorithm can reduce average response time in some cases.

However, we also found that the number of our test cases and their varieties were not sufficient to properly distinguish why SJF performed better than eFJFDRR when averaged across all test cases. Thus, additional test cases are recommended.

References

[1] Zahedi M H and Naghibzadeh M 2008 Fuzzy round robin cpu scheduling (FRRCS) algorithm *Adv. in Comp. and Information Sci. and Eng.* pp 348-53

[2] Helmy T and Dekdouk A 2007 Burst round robin as a proportional-share scheduling algorithm *IEEEGCC* (King Fahed University)

[3] Silberchatz A, Galvini P B and Gagne G 2013 *Operating Systems Concept* (Courier Kendallville) pp 261-6

[4] El-Sayed E, Rodriguez J M, Sushimita M and Thampi S 2016 *Intelligent Systems Technologies and Applications* (Springer International Publishing) p 318

[5] Fatai A, Tarek H and Sallam E 2010 An efficient randomized algorithm for real-time process scheduling *PicOS Operating Syst. Adv. Techniques in Computing Sci. and Software Eng.* p 120

[6] Xu C 2005 *Scalable and Secure Internet Services and Architecture* (Chapman & Hall/CRC) p 62

[7] Reddy C M 2009 *Operating Systems Made Easy* (University Science Press) p 33

[8] Al-Husainy M 2007 Best-job-first CPU scheduling algorithm *Information Technol. J.* vol **6** no **2** pp 288-93

[9] Mohanty R, Das M, Lakshmi P and Sudhashree 2011 Design and performance evaluation of a new proposed fittest job first dynamic round robin (FJFDRR) scheduling algorithm *Int. J. of Comp. Information Sys.* vol **2** no **2**

[10] Matarneh R 2009 Self-Adjustment time quantum in round robin algorithm depending on burst time of the now running processes *American J. of Applied Sci.* vol **6** no **10** pp 1831-7

Bank Air Kami: Terban waterscape information system

B Syahputra[1], A Sujarwo[2] and I Maharika3

[1]Department of Informatics, Universitas Islam Indonesia
[2]Board of Information System, Universitas Islam Indonesia
[3]Department of Architecture, Universitas Islam Indonesia

[2]ari.sujarwo@uii.ac.id

Abstract. As an informal urban-rural mixture area in the middle of Yogyakarta, Terban holds a special living facility: water. Dividing the downtown with a popular river sourcing from Mt. Merapi volcano, Terban waterscape has nine springs watering the society in a traditional way. This paper offers a new approach for the Terban waterscape to show water quality data for researcher and tsociety through information system, and to increase the benefits of natural resources among people. The approach was conducted by designing and deploying a set of sensors to collect data which were then managed by information system for further processing by researchers and locals.

1. Introduction and overview

Terban stands as *desakota* (a mixture of urban and rural living in an area [1]) in Gondokusuman, a sub-district of Yogyakarta, Indonesia. There are nine springs as the part of its waterscape beside the famous Kali Code, a river that divides this cultural city. People around the springs have a long story regarding the usage of the springs. They have used it for their basic needs in traditional way: bathing, toileting, and laundering. Some researchers around the world have come and provided a number of solutions to increase the benefit of the springs, as published in [1], [2], [3], and [4].

The publication describes the job and the research that have been conducted for the installed water purification system for Terban Waterscape. This system is called *Bank Air Kami*, or *Our Water Bank*. Since then, people around the spring have been daily consuming the purified water. To ensure the water quality, the locals have periodically sent the water sample to the local health facility to be assessed. The results are then recorded and informed to the consumers manually.

Life has changed. Last decade marks the technology existence as the successor in almost of human daily life. Inevitably, technology has also intervened the way people in consuming the water produced by Bank Air Kami. This paper proposes a system to improve the water quality reporting system, and to digitally increase the awareness of water consumers in the framework of smart city development in Yogyakarta.

The important element of smart city development is smart technology such as smart sensor and wireless platform as stated by Frost and Sullivan as mentioned in [5]. By using technology, water quality such as in pH, turbidity, and temperature are sensed in real time by sensors. At the same time, it is reported and saved to the system.

Sensors placed in the spring in such a way depends on the functionality and controlled by Arduino. The data then are saved into a premise database system and synchronized to the central database in a remote datacenter. Based on this database, data are then used and visualized by the proposed information system. Water quality data. Beside to show the real-time water quality data, this system also facilitates researchers to download data in bulk for further need.

2. Terban Waterscape

Terban Bawah (Kampung Gondolayu) emerges as an informal residential area with a long story of neighborhood as this location at the beginning was a cemetery area. This area topology (**Figure 1**) shows the variety in the height of landscape, and slope surrounding the valley.

Figure 1. Riverside living

2.1. Terban Bawah as Informal City

Kampung, a Javanese terminology explains a living area in an informal residential location. With a basic lifestyle, people in Kampung are often living below the living standard [6]. The low monthly income, uncertain casual job, and many other factors have made people to struggle to fulfill their needs for food, health facility, and housing. In some areas, people still show a traditional living, as they still keep their close relationship with neighbors, high density housing, and limited access to clean water, sanitary, and rubbish management.

Above factors, and the lack of residential space have posed people to live insufficiently. People in Terban Bawah choose to live in an informal residential area rather than in proper residential area (**Figure 2**). Based on the regulation of Indonesia Government: PP No. 38/2011 it defines that in the area of river, 10-20 meters around the river needs to be kept clear of housing. The slope and the valley owned by government are not permitted for housing. This area is defined as illegal for housing. However, people does not have any options to live better, and face a high risk of disaster (**Figure 3**). They have built the houses and limited facilities. This area is then called as the informal city. It is easy to find the informal residential areas in Indonesia, even in Jakarta as the capital of Indonesia.

Java as the main island of Indonesia has the higher density of residents [7] compared to others caused by people movement. People from other cities or islands attempt to find job and seek a better living, and choose to migrate to Java. However, to find a good job often needs some time, but life never stops. This situation has forced people to live in informal residential area, as there is no legal procedure to be followed. In other words, people does not have any legal documents to proof the land and house ownership.

Yogyakarta has a special case. Most of lands are still owned by the traditional owner of the land: Karaton Ngayogyakarta Hadiningrat, the Sultan Palace. The King of Yogyakarta often gives people a formal letter called *Kekancingan* [8] to permit people to live in Sultan's land, even in so called informal city. The ownership of the Sultan's land is quite contradictive with the aforementioned PP No 38/2011.

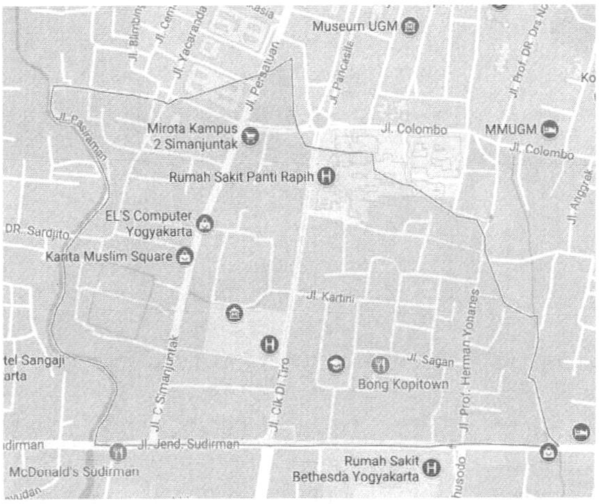

Figure 2. Terban Waterscape area around the river (Terban Bawah)

Figure 3. High-risk neighborhood

2.2. Water Springs Profile

There is no clear statement when people discovered springs around Terban neighborhood. What people know is that there is a water spring they can use to support their life for laundering, bathing, and drinking, without any awareness whether the water is clean and consumable. Maharika in 2014 [4] introduced a water filtering technology - installed in a water cleansing house as shown in **Figure 4**. Since then, people have managed to consume water from Bank Air Kami and sell it to surrounding people.

Figure 4. Bank Air Kami water filtration station

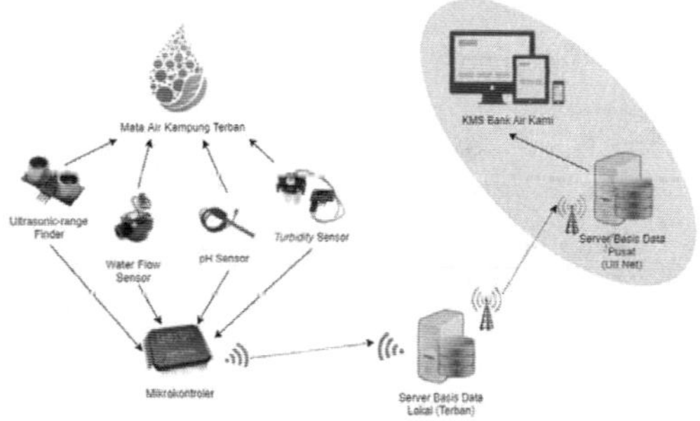

Figure 5. Waterbank Information System Architecture

There are about nine water springs on the slope of Kali Code, the Code River, one of which is called Belik Ayu. This spring is where Bank Air Kami is located. When this research was conducted, most of the water springs were found built with a small squared tank to collect water.

2.3. Challenges

Böhlen in [4] placed the first milestone of this potential research in Terban Bawah by installing a water filtering facility equipped with an e-coli assessment system situated in Puskesmas Gondokusuman, a local health care close to the springs. In turn, this research was conducted to observe more data from Terban Waterscape springs. We have placed various kinds of water sensors to get the picture of water, and built data management facility to be accessed remotely by water consumers and researchers.

As stated in Indonesia Ministry of Health Decree No 492/MENKES/PER/IV/2010 about consumable water quality, water production system should obey the water quality specifications below:

Table 1. Consumable water quality specification

No	Parameters	Unit	Max Threshold
1	pH		6.5 – 8.5
2	Turbidity	NTU	5
3	Temperature	°C	10 °C – 15 °C
4	Salinity	%	0.05
5	*Coliform*	per 100 ml sample	0
6	*E. Coli*	per 100 ml sample	0

The water source of Bank Air Kami would be monitored and reported to the system and focused on the first four parameters above.

3. Water Quality Data Management Sensors and Ubiquitous Computing

During the early step of the research, we conducted data and functionality analysis soon after the interview stage. This process resulted in the design of the system, which must have several parts: ranging from data collection units to datacenter and Bank Air Kami Knowledge Management System. **Figure 5** shows the architecture of the system.

The figure illustrates the process flow of data collection by a set of sensors to procedure of data saving in the datacenter. We deployed pH sensors, water flow sensors, turbidity and ultrasonic range finder sensors. Data from sensors were managed by a microcontroller, and processed to get the data in some standardized units and saved into the local database.

The local database kept the data from all sensors located in water springs temporarily, since the final data saving was in the datacenter. The local database enabled a controller to save real- time data nearby. The aim of placing this facility close by was to avoid any data loss during the communication between components. A mechanism was installed between the local database and datacenter to sync the data between those two parties. In the end, data saved in the datacenter would become the data supplier to Bank Air Kami's knowledge management system (KMS).

The KMS [9] manages and visualizes the data in various formats: graph and table. Data are also available to be downloaded in PDF, TXT, or CSV formats for further data processing.

4. KMS design and deployment

4.1. Actors and its role

This system has been designed for various actors: water consumers, researchers, and Bank Air Kami employees as the administrator. The system has been designed into four parts: (1) sensors and controller system to produce data, (2) robust synchronization system to ensure that the local saved data have been sent to datacenter successfully, (3) robust network between sensors and controller to central system, and central system to datacenter, (4) and last but not least, the web-based information system itself as an interface for the end user to consume data. **Figure 6** shows the users and their role to the system.

Here, the administrators are able to manage the user's login, add or remove sensors, manage the list of monitored water springs and its locations. The other users (researchers and water consumers), meanwhile, can monitor and download data.

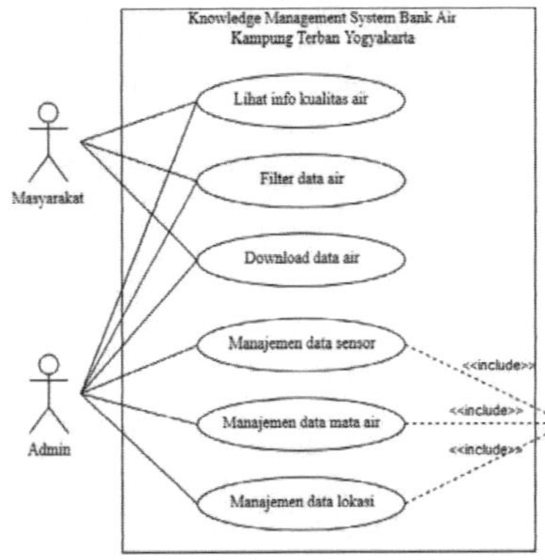

Figure 6. System users and its role

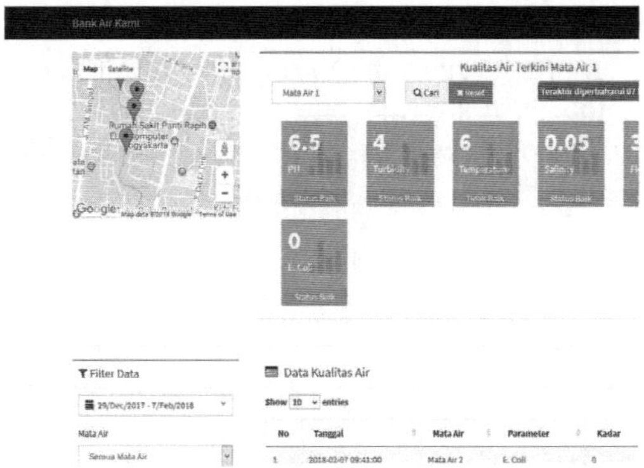

Figure 7. Main interface of real time water monitoring system

4.2. System features for researchers and water consumers

The main interface shows the water data quality based on sensors we have deployed. The users are possible to find data by applying various data filters (**Figure 7**) based on the water spring names, range of time, and water quality data parameters. IoT produces more data than other systems do [10]. In this case, an interface shows data in the form of an easy-to-see mini dashboard at the top of the page, and brief table in the middle of the page with an available download button. To see further data in details, another interface is provided as shown in **Figure 8**.

Figure 8. Water quality data in detail

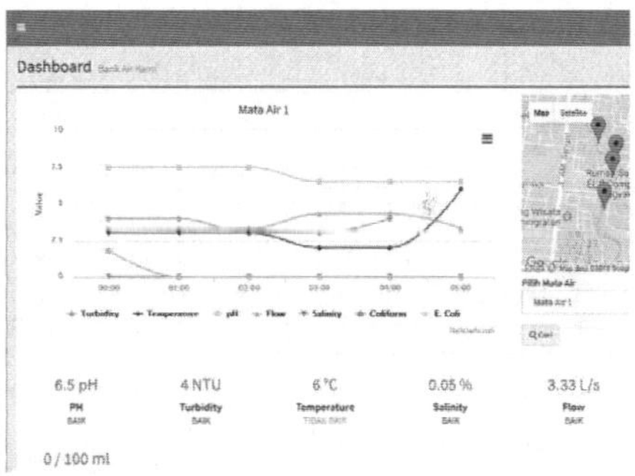

Figure 9. Dashboard to show specific water spring data

4.3. Dashboard for Bank Air Kami administrator

Administrators play an important role at the beginning of new added set of sensors in a water spring to the data management and output. This system was designed for multi-location and multi-devices support. At the beginning, we deployed two sets of sensors placed in Belik Ayu as the central system and Mata Air 4 chosen as the water spring to communicate with.

The details of water springs e.g. latitude, longitude, location description, and sensors specifications needed to be registered to the system. System was then ready to accept data sent by the controller. During its operations, administrator could add more data collection units to be placed in water spring area, and did a similar procedure to register it to the system. When data started to arrive, system would visualize it as shown in **Figure 9**.

4.4. Usability Test

Usability test began with a step to write the set of questions, and applied the validity and reliability test to ensure the questions used proper in this test. The groups of locals were involved to help us to test the prototype of information system. We did a set of usability tests to ensure that our system could work well and was beneficial. The result shown in **Table 2** depicts that system was able to fulfill stakeholder's needs, although there might be a need for further improvement.

Table 2. Usability test result

No	Question	Mark
1	*Learnability*	4.2
2	*Efficiency*	4.1
3	*Memorability*	4.3
4	*Errors*	4.1
5	*Satisfaction*	3.9

5. Conclusion

System is able to show the real-time water quality data in various formats: table and graph, and provided data downloadable in a bulk mode. Turbidity, temperature, pH, and salinity sensors worked well to send data for visualization in KMS Bank Air Kami. Learnability, efficiency, memorability, and errors within usability test showed a good result. In general, it can be stated that system is quite usable to facilitate the stakeholder's needs for water quality data.

Acknowledgement

This research is a part of PTUPT (Penelitian Terapan Unggulan Perguruan Tinggi) Indonesia Government funding plan No. 03/ST-DirDPPM/70/DPPM/Penelitian Terapan Unggulan Perguruan Tinggi Kemristekdikti/IV/2017 by Dr. Ilya Fadjar Maharika, IAI as the team leader, and Ari Sujarwo, S.Kom., MIT (Hons), Medila Kusriyanto, ST., M.Eng. and DR.-Ing. Widodo B.Msc as the team members.

References

[1] Böhlen M and Maharika I F 2018 *Learning from WaterBank* vol **10** no **2** pp 138–56
[2] Arum H R T, Sastrosasmita S, and Iskandar D A 2006 Clean water governance in a small community: a case study of community development approach in the water provision in Yogyakarta, Indonesia *Proc.: Regional Conf. on Urban Water and Sanitation in Southeast Asian Cities* p 61
[3] Seftyono C 2017 *Dilema Implementasi Kebijakan Pembangunan Bantaran Kali Code-Yogyakarta* Master Thesis (Yogyakarta: Gadjah Mada University)
[4] Böhlen M, Maharika I, Yin Z and Hakim L 2014 Biosensing in the Kampung *2014 Int. Conf. on Intelligent Environments (IE)* pp 23–30
[5] Letaifa S B 2015 How to strategize smart cities: Revealing the SMART model *J. Bus. Res.* vol **68** no **7** pp 1414–9
[6] McCallum D and Benjamin S 1985 Low-income urban housing in the third world: broadening the economic perspective *Urban Stud.* vol **22** no **4** pp 277–87
[7] Yulianto F, Tjahjono B and Anwar S 2014 Detection settlements and population distribution using GIS and remotely sensed data, in the surrounding area of Merapi Volcano, Central Java, Indonesia *Int. J. Emerg. Technol. Adv. Eng.* vol **4** no **3** pp 1–10
[8] Kurniadi B D 2009 Yogyakarta in decentralized Indonesia: integrating traditional institution in democratic transitions *J. Ilmu Sos. dan Ilmu Polit.* vol **13** no **2** pp 190–203
[9] Alavi M and Leidner D E 2001 Knowledge management and knowledge management systems: conceptual foundations and research issues *MIS Q.* pp 107–36
[10] Jernigan S, Ransbotham S and Kiron D 2016 Data sharing and analytics drive success with IOT *MIT Sloan Manag. Rev.* September

A decision support system with 3D visualization for box space optimization using the locust-based particle swarm optimization algorithm

C Janea[1], B L Alcisto[2] and C Maderazo[3]

Department of Computer and Information Sciences, School of Arts and Sciences, University of San Carlos, Cebu City, Philippines

[1] janeacarlo@gmail.com
[2] bojoluis21798@gmail.com
[3] cvmaderazo@usc.edu.ph

Abstract. The non-stop increase in world population has resulted to an increase of the demand of space for human habitability. Space allocation is a problem mostly dealt with those with limited to scarce resources, and those who cannot afford or are not willing to spend for an alternative or an additional space as a solution. The objective, then, is to aid people in maximizing their current available space without having to sacrifice cost, time, and effort. This study has developed a decision support system that provides the optimum arrangement between a set of objects, selected by the user from sample models from the ShapeNet dataset presented by the system through 3D Visualization, to maximize the space of the given box. The developed system obtained the optimal arrangement, which maximized 60% to 91% space of the given box, of the set of objects by utilizing the locust-based Particle Swarm Optimization algorithm.

1. Introduction

Space is a resource that few disregard as important but is actually critical with the limited resources made available by the environment. Space allocation is a problem dealt with those with limited to scare resources, and those who cannot afford or are not willing to spend for an alternative or an additional space as a solution. This problem comes into play when objects within a given space is misarranged and then take more space than they are intended to, which then cause the person to sacrifice cost, time, and effort in the solution of finding more space. This study proposes a decision support system, aided by 3D Visualization, which provides the optimum arrangement between a set of objects from sample models from the ShapeNet dataset, to maximize the space within a given box by utilizing the locust-based Particle Swarm Optimization algorithm..

2. Related works

These are the past literatures that were utilized by the researchers in the creation of this study.

2.1. Space optimization

The problem of space optimization is not new in the computing field, as there have been many studies made to solve different variations and applications of such problem such as the research of [1] regarding the use of a greedy algorithm paired up with a tabu search algorithm in solving the office space allocation problem which is to optimally arrange entities in an office. The problem of space optimization can also be linked up with the well-known knapsack problem due to its spatial scope. The knapsack problem was solved by [2] through a two-step ants algorithm, which is a swarm-based algorithm, where the approach started by making use of an anti-pheromone technique that was capable of warning ants that this specific food is not a good choice to include in the knapsack, and then, applying the ant colony optimization algorithm which does the decision making on what objects to put in the knapsack after such filtering by the anti-pheromone technique.

2.2. Locust-based Particle Swarm Optimization architecture

The locust-based particle swarm optimization architecture was made by [3] with the goal of solving combinatorial optimization problems with a general architecture. According to their model, the algorithm is divided into four different stages which mirrors on how locust swarms are formed during an attack – Identification, Verification, Analysis, and Solution.

Identification dwells on the swarm's initialization and the corresponding parameters such as dimensions, number of iterations, the size of search space, and etc. Verification is checking the variation of the data compared to the last iteration and/or the current best solution of the swarm. This also includes updating the overall best for the entire group whenever possible. Analysis is saving the best solutions to a separate location for comparison for future iterations. Solution is the stage where the distribution of the best solution towards the problem happens.

This general architecture has been adapted to several combinatorial optimization problems such as the knapsack problem, traveling salesman, cellular robotic systems, and routing telecommunication networks due to its flexibility and usability.

2.3. Improving the Particle Swarm Optimization algorithm

The hyper parameters – such as the number of particles, the cognitive and social constants, the inertia constant, and etc. – of most optimization algorithms like the Particle Swarm Optimization (PSO) affect the solutions and result of the optimization significantly as these factors dictate the behavior and the interaction between particles within the swarm, and even the swarm itself.

Automatic tuning of parameters was used by [4] to obtain the ideal preliminary parameters for the Particle Swarm Optimization. They discovered that having a relatively larger population, contrast to standard PSO parameters, does improve the quality of the solutions given by the algorithm as tested by the researchers of the said study. Table 1 shows the obtained parameter changes from the study made by [4].

Table 1. Tuned parameters for PSO by [4].

Parameter	Standard PSO	Meta-optimized PSO
Population Size	40	246
Inertia Factor (w)	$\frac{1}{2 \cdot \ln(2)}$	0.687378
Cognitive constant (c_1)	$0.5 + \ln(2)$	0.246448
Social constant (c_2)	$0.5 + \ln(2)$	0.701503

The performance of the standard PSO was improved by [5] through determining a stop condition for the swarm iterations on the basis of the swarm behavior during convergence. This allows the algorithm to put a dynamic limit on the swarm iterations and not exhaust the system for the same quality of solutions.

The quality of the solutions found by the swarm was enhanced by [6] through a stagnation detection method wherein the swarm's search process is restarted when the fitness of the solutions stagnate. This allows the swarm to potentially avoid local optima within the search space.

3. Proposed method

In this section, the conceptual framework of the proposed decision support system is discussed. The summary of the conceptual framework is shown in figure 1.

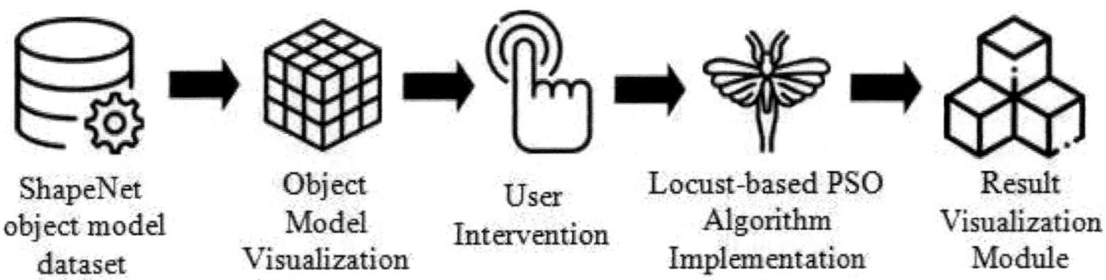

Figure 1. Development approach for the proposed decision support system.

3.1. ShapeNet object model dataset

The system uses object model data from ShapeNet, a collaborative effort by Princeton, Stanford, and TTIC that provides large-scale and defined datasets for researches in computer graphics, computer vision, robotics, and other fields related. The researchers then filtered out object model data, from the said dataset, that had incomplete fields in their metadata, and those that were relatively larger than the large-sized moving box (18 x 18 x 24 inches).

3.2. Object model visualization

The system visualizes a sample handpicked by the researchers from the filtered dataset through Panda3D so the user can proceed in selecting from the visualized sample.

3.3. User intervention

The user can now then select a set of objects from this visualized sample and also set the dimensions of the box in inches whose space will be optimized. These dimensions are used in constructing the box, represented as a three dimensional array, for the algorithm to work on. With array's index having to only handle discrete numbers, the system converts the given dimensions to millimetre and disregards the remaining decimal points to minimize inaccuracy of solutions.

3.4. Locust-based Particle Swarm Optimization

The system receives the set of objects, with their nearest bounding box dimensions, selected by the user and the box-container dimensions and then proceeds with the optimization with the locust-based particle swarm optimization algorithm. This optimization generates an array of inserted objects and an array of coordinates – a pair between object and best location within the box.

Table 2. Stages in solving the box space optimization based on the architecture by [3].

Locust-based Particle Swarm Optimization			
Identification	**Verification**	**Analysis**	**Solution**
Initialization of parameters	Check number of free space around object	Analysis of locations (coordinates) with smaller free spaces around it	Insert object to the best location with minimum free space around the object
Determine problem dimensions	Check box's available space	Store location (coordinates) with minimum free space	
Object combination			

3.4.1. Finding the best combination of objects. Before the system determines the optimal arrangement of objects within the box, it first finds the best combination of objects that will occupy the most space within the box through a simple greedy algorithm.

3.4.2. Finding the best arrangement. After determining what combination of objects to insert, the locust-based Particle Swarm Optimization algorithm, with the same parameters as [4], then comes into play. The basis of the system to determine which location and what position of the object is optimal is through calculating the free space around the object. The system considers 6 rotations of the object which is shown in figure 2 and calculates the free space around the object per rotation for each location, and since all objects are treated as rectangles (nearest bounding box) the system counts the free space within the 6 sides per object.

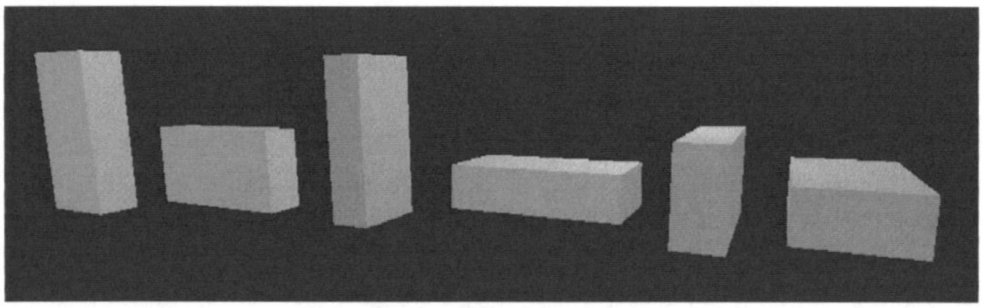

Figure 2. Six rotations of the object covered by the system.

The lesser the free space around the object of the given location and the specific rotation, the more likely the swarm will choose it as the solution. The objective function whose goal is to find the optimal location with the least amount of free space around the object can then be stated as

$$\min f(x) = \sum_{i=1}^{6} \sum x \in p_i * q_i * r, \tag{1}$$

where i refers to the sides of the object's bounding box, p_i and q_i are the length of the two edges of the ith side being operated on, r is the distance either between the ith side and the corresponding border of the box-container opposite to the object's i^{th} side, or between the ith side and the nearest opposite bounding box side of another inserted object, and x is the free cell found within the volume of the rectangular prism calculated by $p_i * q_i * r$. After the swarm finishes the analysis between the solutions, the best solution is then selected, thus, inserting the object within the box through updating the three-dimensional array representation of the box-container and the list of inserted objects with their corresponding coordinate locations. The four stages of the locust-based Particle Swarm Optimization algorithm used by the system are shown in table 2.

3.4.3. Finding the percentage of optimized box space. After an object is inserted, the system checks the box-container's optimized space through

$$box_{percentage} = \frac{\sum_{i=0}^{n} v_i}{t} * 100, \tag{2}$$

where n is the number of objects inserted in the box, v is the solid volume of the object's bounding box, and t is the total volume of the box-container. If the calculated percentage is already at its maximum (100%) then the system stops any further insertion since the box-container's space are already fully occupied, but if the space of the given box isn't fully optimized, the process resumes with section 3.4.2 until the box space is fully maximized or until no object in the given combination can be further inserted.

This shows that the system supports the possibility of having to insert only one object, given that it occupies the entire box space or somewhere near 100%, since the proposed system focuses on maximizing box space, not the number of inserted objects. This also shows that if the given combination's total volume doesn't sum up to the total box-container's volume, there is no way the system can maximize the box-container's space to 100%.

3.5. Visualization of result

The system then passes the array of inserted objects and the array of coordinates to Panda3D for the visualization of the optimal arrangement as a result from the optimization. The system can also notify the user when there is not enough space to insert an object/s or when the box is too small to fit anything inside it.

4. Results

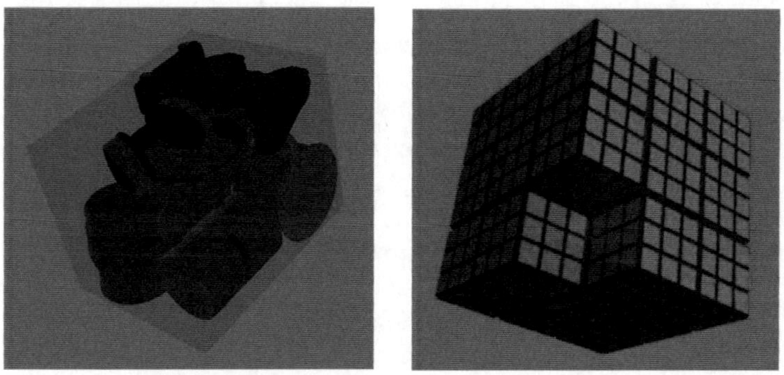

(a) (b)

Figure 3. Visualization of results (best) using Panda3D for (a) objects with varying dimensions and (b) objects with uniform dimensions.

To test the effectiveness of the proposed methodology, the algorithm was tested on two kinds of object combinations – the first five combinations were composed of objects with varying dimensions, and the second five combinations were composed of objects with uniform dimensions. Both combinations totalled or exceeded the given box volume and was ran 5 times per combination with the same initial parameters for the swarm by [4]. For each run, the percentage of the box space optimization, calculated by equation 2, was saved and was averaged when the 5 runs were completed. The results of the tests are shown in table 3 and figure 3.

Table 3. Results obtained by the proposed methodology.

Combination #	Average Space Optimized (%)	Best Space Optimized (%)
1[a]	68.8220	72.4241
2[a]	65.8629	68.8544
3[a]	69.8358	72.4614
4[a]	69.3207	77.7381
5[a]	60.8461	72.1862
6[b]	78.7142	81.0294
7[b]	84.8720	88.9788
8[b]	72.3733	79.1583
9[b]	75.7440	77.7728
10[b]	90.2290	91.4008

[a]Combination composed of objects with varying dimensions
[b]Combination composed of objects with uniform dimensions

The results show that the first five combinations that were comprised of objects with varying dimensions were optimized only by 77% at best. This is mainly caused by the locusts converging to local optima since the objective function used only takes into account the free space around a current object. It should take into account the effect of inserting such object in that position to the rest of the objects to be inserted in the future in order for it to peak the solution towards the global optima. The second half of the test, which was comprised of objects with uniform dimensions, show promising results as it peaked to 91% space optimization. This is possible because having to deal with objects that have uniform dimensions, the swarm would not have to worry about how the position of the current object would affect the future objects to be inserted.

5. Conclusion and recommendations

The use of the simple greedy algorithm combined with the locust-based PSO algorithm showed a 60% to 77% space optimization for object combinations with varying dimensions and 72% to 91% space optimization for object combinations with uniform dimensions as seen from the results from table 3. Although having objects with varying dimensions resulted to a low space optimization, the cause was determined and was due to the used objective function in the locust-based algorithm. This was proven by testing the algorithm on objects with uniform dimensions, as the objective function would have little to no effect on the solutions, granted 91% space optimization, at best.

This research can be further improved by utilizing a different data structure to represent the box where the swarm can operate on since a 3D array showed inaccuracy and inefficiency for locating the optimal solution per object. Another improvement is by using a more exclusive objective function that considers not only the optimal space for an object, but also what object is the most optimal to insert at that specific location and iteration. This will allow the algorithm to peak its solution towards the global maxima. Another possible improvement towards this study is considering more test cases during testing as the researchers showed only the minimal combinations that prove that the algorithm can perform its objective. Lastly, to improve the realism of the research, the researchers recommend to operate on object's real dimensions, not just its bounding box, and to handle containers which allow insertion of one object inside another object.

References

[1] Castillo F, Riff M-C and Montero E 2016 New bounds for office space allocation using tabu search *Proc. of the Genetic and Evolutionary Computation Conf. 2016* (Denver, Colorado)
[2] Rojas N, Montero E and Riff M-C 2015 Using anti-pheromone to identify core objects for multidimensional knapsack problems: a two-step ants based approach *Proc. of the Companion Publication of the 2015 Annual Conf. on Genetic and Evolutionary Computation* (Madrid, Spain)
[3] Ribiero T and Paucar V P 2017 Problem solving in combinatorial optimization based on the algorithmic model of attack behavior of locust swarm *Proc. of the 2nd Int. Conf. on Internet of Things, Data and Cloud Computing* (Cambridge, United Kingdom)
[4] Ugolotti R and Cagnoni S 2015 Automatic tuning of standard pso versions *Proc. of the Companion Publication of the 2015 Annual Conf. on Genetic and Evolutionary Computation* (Madrid, Spain)
[5] Perroni P F, Weingaertner D and Delgado M R 2017 Estimating stop conditions of swarm based stochastic metaheuristic algorithms *Proc. of the Genetic and Evolutionary Computation Conf.* (Berlin, Germany)
[6] Ahmed H R 2014 An efficient fitness-based stagnation detection method for particle swarm optimization *Proc. of the Companion Publication of the 2014 Annual Conf. on Genetic and Evolutionary Computation* (Vancouver, BC)

Acknowledgements

The researchers would like to thank the Department of Computer and Information Sciences of the University of San Carlos for the research opportunity and for the knowledge and guidance during the said study.

Digital conversion model for hand-filled forms using optical character recognition (OCR)

J E M Adriano[1], K A S Calma[2], N T Lopez[3], J A Parado[4], L W Rabago[5], and J M Cabardo[6]

Asia Pacific College, 3 Humabon Place, Magallanes, Makati City, Philippines 1232

[1]jmadriano2@student.apc.edu.ph
[2]kscalma@student.apc.edu.ph
[3]ntlopez@student.apc.edu.ph
[4]japarado@student.apc.edu.ph
[5]lorenar@apc.edu.ph
[6]jayveec@apc.edu.ph

Abstract. The process of manual data entry used by several industries garners a high error rate. This is because the manual process relies too heavily on a human's capability to interpret handwritten forms. To reduce the high error rate of data entry, the researchers explored the different processes that comprise optical character recognition (OCR) and used it on a novel digital conversion model for hand-filled forms. The OCR process is made up of 4 major phases. The techniques for each stage are as follows: Sauvola binarization for image pre-processing; blob analysis for character segmentation; pre-trained Convolutional Neural Networks: GoogLeNet, AlexNet, and VGG16 for feature extraction, and Support Vector Machines (SVM), K-Nearest Neighbor (KNN), and Naïve Bayes for classification. The novel combination of Convolutional Neural Networks for feature extraction coupled with SVM for character classification showed promising results, going up to 98.62% in accuracy and 65.31% in F-Score.

1. Introduction

An educated human can look at a paper and read its contents as easy as it comes, but having a computer do the same is much more complicated than most people would think. One would first need to have a digital image of a document, process it to remove unnecessary information, get the computer to locate the characters and segment them so each individual letter can be identified. Only then can it get an output of a series of machine readable characters. This process is called Optical Character Recognition (OCR) which may be used to copy, search or edit digital text so the data can be utilized in various ways from something as simple as data storage, to more complicated applications like generation of reports or listing data-driven decisions.

As it stands in the modern age, the process of manual data entry is still widely used among industries such as healthcare, banking & finance, and real estate. These are industries whose integrity and profit rely on the accuracy of the data encoded to their databases from physical records. A study by Barchard [1] cracked down on the consequences of human data entry errors in the study they conducted in 2011. It's quite surprising how data entry errors, no matter the amount, largely affect the statistical results and conclusions of the higher management of companies. Furthermore, there is a concept known as 1-10-

100 developed by Labovitz et al. in 1992, quantifying the cost of errors in terms of units such as labor, money, time, etc. [2]. That is, a company spends 1 unit for preventing an error (Prevention Cost), 10 units for correcting an error (Correction Cost), and 100 units for working with data that were never corrected (Failure Cost). This means that the costs an organization must pay for data entry errors increase exponentially the later it is dealt with.

The average benchmark for data entry error rate is generally acknowledged to be 1% [3]. It may seem insignificant, but if 1 out of 100 documents from a total of thousands get an error, the cost they will accrue overtime will be heavy, especially for larger companies. The survey results on data entry errors in clinical databases showed that double-entry method ranged from 2.3% to 26.9% [4]. These errors were caused by mistakes in data entry and misinterpretation of the information from the original documents. Unfortunately, in the end, the ones who undertake these jobs are still human, who are naturally prone to error. This is especially pronounced when the data needed to be stored come from documents with handwritten elements because of the different ways letters can be written. Not to mention the different markings a writer can create, like a smudge or a correction, for example. To solve this, the multiple facets of optical character recognition is explored to be used in a Digital Conversion Model that is tailored for functioning on forms that contain handwritten content. This can potentially remove the human element which makes operations susceptible to errors, or at the very least, decrease data entry errors and increase efficiency. Additionally, the researchers noted that this is the only OCR methodology that utilizes pre-trained image CNNs for feature extraction instead of the usual global and local feature extraction techniques such as Local Binary Pattern (LBP), Features from Accelerated Segment Test (FAST), and HOG (Histogram of Gradients). The reason for this will be discussed later in the results and discussion portion of the paper.

Though the research deals with handwritten input, the type of characters to be recognized is limited to block (sample) characters only, as cursive (*sample*) writings require a different methodology from what is being proposed. Additionally, special characters (ñ, ~m, %) and punctuation marks such as periods (.) commas (,) exclamation points (!) etc. are not included. Finally, some character classes have been merged. These are characters that look similar regardless of case. All in all, there is a total of 47 classes of characters.

2. Related Works

ImageNet Large-Scale Visual Recognition Challenge (ILSVRC) is an annual competition that evaluates different algorithms for large-scale object detection and image classification. Some notable entries and winners of the ILSVRC are pre-trained convolutional neural networks (CNN) such as AlexNet, VGG 16, and GoogLeNet. During the ILSVRC 2012 AlexNet was the winning entry. It contains 5 convolutional layers and 3 fully connected layers and makes use of Relu to add non-linearity which accelerates speed by 6 times while maintaining constant accuracy. It also makes use of dropout to deal with overfitting and reduces the size of network through overlap pooling [5]. During the ILSVRC 2014, there were two promising entries. One of these entries was the VGG-16, a pre-trained convolutional neural network that contains 16 weight layers. Its simple architecture uses a 3x3 convolutional layer stacked on top of each other and the depth increases by each layer. It reduces volume size by max pooling [6]. On the other hand, GoogLeNet is a pre-trained CNN developed by Google. It was the entry on ILSVRC 2014 that won that year's competition. It makes use of a new variant of CNN called Inception, which is used for classification; while R-CNN was used for detection. Its network was designed to increase the depth and width resulting to a 22-layer deep network while at the same time, keeping the computational budget constant [7]. Both VGG-16 andGoogLeNet gives a matrix output of [1000 x n] where n stands for the number of data entries.

All the pre-trained convolutional neural networks mentioned (AlexNet, VGG 16, and GoogLeNet) were trained using a subset of the ImageNet database. They were trained using more than a million images from ImageNet and can classify images into 1000 object categories. Furthermore, these pre-trained CNNs can be applied to the current study's (digital conversion model) feature extraction phase where the researchers intend to use convolutional neural networks.

The study done by Malon [8] successfully applied SVMs as their character classifier for mathematical symbols. Their objective was to address was the low success rate of conventional OCRs at classifying mathematical characters. A multiclass SVM classifier along with five different kernels were compared against a Naïve Bayes classifier reducing the overall error rate of InftyReader by 41%. Another study constructed an OCR methodology to be applied on Ottoman characters with SVM as the image classifier [9]. Three SVM kernels were used: linear, quadratic, and radial basis function (RBF). Of all the kernels the quadratic one was the most accurate giving a recognition rate of 87.32%. Lastly, a survey done by Thome [10] discusses the application of SVMs in the field of OCR. He focuses on Multiclass SVM which can be done in two ways: one-vs-one and one-vs-all. Additionally, Thome explains the advantage of SVMs over neural networks by stating the former's requirement of a smaller training data set requirement, greater ability to generalize, and the higher probability of generating good classifiers.

3. Proposed Method

Optical Character Recognition (OCR) is a complex combination of many processes that work subsequently to produce a machine-readable string of characters. Each major phase of OCR, consisting of Pre-processing, Character Segmentation, Feature Extraction, and Classification may utilize different techniques to fit different types of documents depending on their identifying characteristics. In fact, most studies today usually focus on improving only one stage without mentioning the others. **Figure 1** below shows the methods that the proponents have deemed feasible for the process.

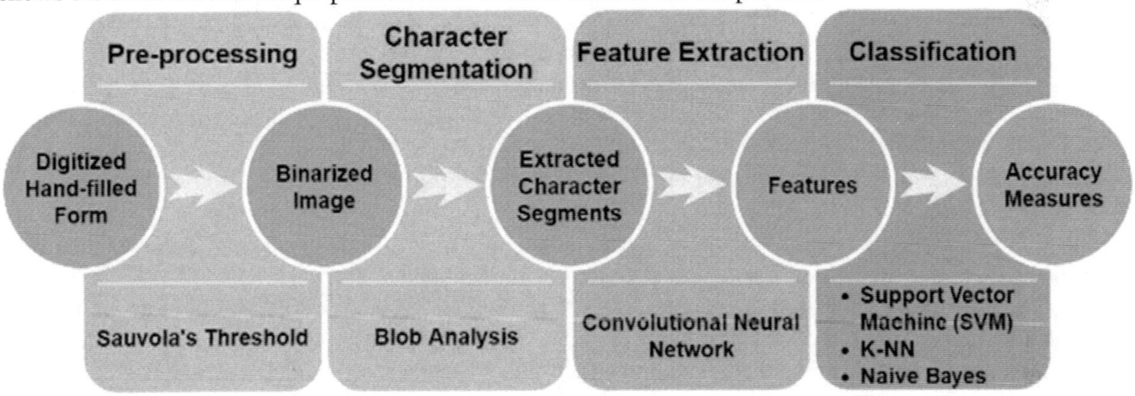

Figure 1. The Digital Conversion Model

In pre-processing, the image obtained through digitization is grayscaled. It changes the value of each pixel to represent only the intensity information such that its value ranges from 0-255. It is then binarized; that is, each pixel is classified into either 0 (black) or 1 (white), depending on the value of each pixel as compared to the threshold value. Without image binarization, it would be very hard for any device to properly identify characters because of the various noise that usually surround the relevant text area. In this case, Sauvola's Threshold, with its nature as a local thresholding technique, has been chosen as the most suitable for binarizing handwritten forms. Since it is exceptional in separating the foreground and the background even in noise cluttered areas of the image, which is apparent in some forms possessing color shading, background patterns, and unintended markings from the writer. The formula for Sauvola's threshold is,

$$T_S = m * \left(1 - k * \left(1 - \frac{s}{R}\right)\right) \tag{1}$$

Where m is the median and s is the standard deviation of the area [11].

The resulting binarized image is then ready for character segmentation phase where it will be segmented into areas. The objective of this phase is to determine the segments of the image that contains individual characters from a line of text. Blob Analysis, the technique to be used, computes the statistics for connected regions within the binary image to identify blobs (region or area of connected pixels in an image) based on the values 0 or 1 [12]. The output of this phase is a collection of images containing a single character which will be used in the next phase.

During the feature extraction phase, text segments are analyzed for differentiating features. These differentiating features are extracted from the matrices of segmented characters. They are selected to uniquely identify a given text segment [13]. These features can then be used for machine learning training and predicting purposes. Both training data and test data are transformed into feature vectors prior to training the machine and having it make predictions. For this phase, Convolutional Neural Networks (CNN) were used; CNN is a deep learning algorithm or a type of machine learning algorithm that learns to classify images, video, text, or sound. Its architecture is trainable and is composed of multiple layers or stages. Each stage's input and output are a feature map of a set of arrays [14]. Furthermore, CNN incorporates constraints and deformation invariance through local receptive fields, shared weights and spatial subsampling [15]. An attractive quality of CNN is that its capacity can be controlled through depth and breadth variations. It contains fewer connections and parameters as compared to standard feedforward neural networks [16]. These characteristics make CNN less complicating to train than other neural networks thus it is opted to be the algorithm used for the digital conversion model's feature extraction phase.

4. Experiment and Results

In this section, results will be discussed according to the data and tools used and the optical character recognition pipeline. The data used for OCR applications for this research came from the National Institute of Standards and Technology Special Database 19 [17]. This dataset already provides a form to apply the character extraction techniques as well as individual character images to train the classifiers. As for the training and testing of classifiers, the researchers used MATLAB, a proprietary, numerical, computing environment and language that has a variety of "apps" that provide a graphical user interface for different functions like machine learning algorithms, algorithm optimization, and image processing, among others. GNU Octave, an open-source implementation of MATLAB whose syntax and formatting are largely compatible to MATLAB, was also used to some extent.

First, images are binarized using Sauvola's thresholding technique. The picture is reduced to grayscale to eliminate colors, background noise, and significantly lessen the impact of uneven lighting or shading. Hand-filled forms from the National Institute of Standards and Technology Special Database 19 were used.

After pre-processing is segmentation where character segments are determined from the image and line of texts using blob analysis. From the binary image having (0) as black and (1) as white, blob analysis detects objects by bounding (0) black connected pixels. The MATLAB function *vision.BlobAnalysis()* is used for segmentation. One of the outputs is the region of interest (x, y, width, height) of each blob detected where x and y are the coordinates of where the blob is in the image. To find the characters in an image, the program or function for detecting is modified. Blob analysis is not created specifically for character segmentation, it bounds any (0) black pixels that it could recognize no matter how big, long, or small it is. The proponents limited the aspect ratio, and area of the blob to suit the size of a handwritten character. Originally, there are no constraints indicated which lead to the blob analysis bounding anything that it detects. This leads to redundant miscalculations of extracted segments. From observations, a handwritten character in the form is not lower than an area of 20 pixels. The area constraint is needed so most small characters like the period, and comma is excluded in the extracts. Constraining the aspect ratio between 0.2 and 2.0 of the blobs, which is the size of an individual character further increased the segmentation accuracy and shows a segmented image without constraints.

The researchers also used segmentation to collect testing data for classification to ensure that it truly tests the extracted segments from blob analysis and not from the pre-segmented individual characters NIST data set. From 11 randomly selected forms from the NIST 19 data set, there were 10,319 extracted segments. 1,742 of the extracted images were excluded due to being outside of the research's scope or being unidentified segments. This leaves 8,577 segments for the classification phase.

The feature extraction process for both test and training sets follow the exact same procedure. The data set images are stored in a folder which contain subfolders that correspond to their character class. These images are then turned into a MATLAB image datastore class object. After labelling the data within the image datastore, it is resized (only within the datastore) into whatever dimensions the CNN requires it to be. The images are then transformed into feature vectors of variable size depending on the network being utilized. **Table 1** below shows the feature vector dimensions for the CNNs.

Table 1. Feature vector dimensions by CNN

CNN	Training Data	Test Data
AlexNet	65800 x 4096	8057 x 4096
GoogLeNet	65800 x 1000	8057 x 1000
VGG-16	65800 x 1000	8057 x 1000

Lastly, classification utilizes the output of the feature extraction as well as the data set labels as inputs for training and testing the classification models. The testing covers the determination of the success metrics: precision, recall, accuracy, and F-score. **Table 2** below shows the final tabulated results of the research.

Table 2. Consolidated test results

Convolutional Neural Network	Algorithm		Metrics (%)				Training Time (sec)	
			Precision	Recall	Accuracy	F-Score	Machine Training	Feature Extraction
AlexNet	SVM	1v1 Linear	71.17	68.39	98.62	65.31	1286	221
		1vAll Linear	66.25	63.07	98.35	59.99	5860	
	K-NN		69.19	54.60	98.08	56.83	9	
	Naïve Bayes		51.92	35.86	97.43	37.86	82	
GoogLeNet	SVM	1v1 Linear	59.20	54.24	98.07	51.79	237	686
		1vAll Linear	59.97	46.08	97.66	48.32	12500	
	K-NN		51.72	43.86	97.56	43.15	3	
	Naïve Bayes		40.03	24.47	96.87	25.02	18	
VGG - 16	SVM	1v1 Linear	64.24	58.17	98.22	58.17	356	1980
		1vAll Linear	61.87	51.63	97.94	53.80	11637	
	K-NN		44.18	36.19	97.42	35.56	2	
	Naïve Bayes		29.72	21.44	96.84	20.77	18	

5. Conclusion

To remove the high error rate produced by manual entry process of hand-filled forms, the proponents of the study explored the use of a digital conversion model using Optical Character Recognition technology which is tailored for functioning on forms containing handwritten content. This digital conversion model for hand-filled forms was designed using a unique combination of major OCR phase techniques – Sauvola's threshold (pre-processing), Blob Analysis (character segmentation), Pre-trained Convolutional Neural Networks (feature extraction), and Support Vector Machine (classification). AlexNet, GoogLeNet, and VGG-16 were the pre-trained CNNs used for feature extraction. The results of characters classified using the 1v1 Linear and 1vAll Linear Approach of SVM were also compared with the results of characters classified using K-Nearest Neighbor (K-NN) and Naïve Bayes Classifier. Furthermore, the precision, recall, accuracy and f-score of each combination of pre-trained CNN and classifier were determined. The best performing pipeline is feature extraction via AlexNet and a one-vs-one coding design, linear kernel support vector machine with an F score of 66.31%. Meanwhile, feature extraction via VGG-16 and classification via Naïve Bayes garnered the lowest accuracy, precision, recall and F-score. Overall, using convolutional neural network as a means of feature extraction turned out to give very positive results which could be a solid foothold for eliminating manual data entry.

6. Recommendations

The researchers encountered technological limitations during the implementation of the methodology. One of the problems encountered was the limitation on RAM. Some MATLAB variables are so large, they fully occupy the entire RAM space, even before any script was ran. As such, the team has determined a minimum of 16GB and recommends 32GB of RAM should future researchers attempt to emulate the procedure. Stronger hardware equates to increased capability for more training data as shown by direct relation of the training data with accuracy. There are many more CNNs that can be used other than the three mentioned in this paper. For example, Resnet50, Resnet101 and Squeezenet. Importing ready-made ones or creating one's own is also possible and should be explored. In addition, one can also use other SVM kernels aside from linear, such as Gaussian and RBF. If these limitations are surpassed, a wide avenue of possibilities may be explored by future researchers.

References

[1] Barchard K A and Pace L A 2011 Preventing human error: The impact of data entry methods on data accuracy and statistical results *Computers in Human Behavior* no **27** pp 1834-9

[2] Labovitz G, Chang Y S and Rosansky V 1992 Making quality work : a leadership guide for the results-driven manager *Chichester: Essex Junction* (VT : Omneo)

[3] Rongala A 2015 Top 6 Manual Data Entry Challenges Companies Face *Invensis Technologies 27 April 2015* Online Available: https://www.invensis.net/blog/data-processing/top-6-manual-data-entry-challenges-companies-face/ Accessed 13 July 2018

[4] Goldberg S I, Niemierko A and Turchin A 2008 Analysis of data errors in clinical research databases *AMIA Annu Symp Proc.* vol **v.15** no **3** pp 242–6

[5] Gao H 2017 A walk-through of AlexNet *Madium 7 August 2017* Online Available: https://medium.com/@smallfishbigsea/a-walk-through-of-alexnet-6cbd137a5637 Accessed 16 August 2018

[6] Simonyan K and Zisserman A 2015 Very deep convolutional networks for large-scale image recognition *ICLR*

[7] Szegedy C, Liu W, Jia Y, Sermanet P, Reed S, Anguelov D, Erhan D, Vanchoucke V and Rabinovich A 2014 *Going Deeper with Convolutions* Online Available: https://arxiv.org/pdf/1409.4842.pdf Accessed 16 August 2018

[8] Malon C, Uchida S and Suzuki M 2008 Mathematical symbol recognition using support vector machines *Pattern Recognition Letters* no **29** pp 1326-32

[9] Kilic N, Gorgel P, Ucan O and Kala A 2008 Multifront Ottoman character recognition using support vector machine *IEEE Xplore* pp 328-33

[10] Thome A C G 2012 Advances in character recognition *SVM Classifiers - Concepts and Applications to Character Recognition* (IntechOpen) pp 25-50

[11] Kalaiselvi T, Nagaraja P and Indhu V 2017 A comparative study on thresholding techniques for gray image binarization *Int. J. of Advanced Research in Computer Science* vol **8** no **7**

[12] Sookman S 2006 Blob analysis and edge detection in the real world *Evaluation Engineering 1 August 2006* Online Available: https://www.evaluationengineering.com/blob-analysis-and-edge-detection-in-the-real-world Accessed 16 August 2018

[13] Verma R and Ali D J 2012 A-survey of feature extraction and classification techniques in OCR systems *Int. J. of Computer Applications & Information Technology* vol **1** no **3**

[14] LeCun Y, Kavukcuoglu K and Farabet C 2010 Convolutional networks and applications in vision *IEEE Int. Symp. on Circuits and Systems: Nano-Bio Circuit Fabrics and Systems* pp 253-6

[15] Lawrence S, Giles L, Ah C T and Back A D 1997 Face recognition: a convolutional neural-network approach *IEEE Transactions on Neural Networks* vol **8** no **1**

[16] Krizhevsky A, Sutskever I and Hinton G E ImageNet classification with deep convolutional

neural networks *Proc. of the 25th Int. Conf. on Neural Information Processing Systems* vol 1 pp 1097-05
[17] Grother P 2016 *NIST Special Database 19 September 2016* Online Available: http://doi.org/10.18434/T4H01C Accessed 30 July 2018

Modern file transfer protocol using lossless compression, lattice-based encryption, and a data integrity hashing function

Vince Jerald A. Villamora[1], Marvin S. Lim[2], and Archival J. Sebial[3]

University of San Carlos, Nasipit, Talamban, Cebu City, Cebu, Philippines

[1]vincevillamora@gmail.com
[2]mrvn_lm@yahoo.com
[3]ajsebial@usc.edu.ph

Abstract. With the recent advances in technology, the creation and access to voluminous amounts of data has become more and more common. From consumer computers and smartphones to big data, the transfer of data entailed a longer download and upload time as the size gets bigger, thus taking a significant amount of time for data to reach its destination. Moreover, data can be corrupted as it is being transferred over the network and most file transfer protocols do not have a file integrity check. Lastly, most file transfer protocols employ the RSA cryptosystem which can be broken through the use of a quantum algorithm known as Shor's algorithm. The proponents have implemented a file transfer protocol that integrates file hashing, compression, and a quantum-resilient cryptosystem. The results show that the proposed file transfer protocol is faster than plain FTP and even faster against FTP over TLS.

1. Introduction

With the advent of the digital age, there has been an exponential growth in the number of internet users in the world. Along with this is the emergence of cloud technologies such as cloud storage, Software as a Service (SAAS), hosting, cloud-based databases, content delivery networks, and big data analytics. Data has become increasingly more common. Anyone with a smartphone or a computer can create data or content such as a photo, a video, or a message which can be transferred or shared over the internet. This transfer of data utilizes a file transfer protocol which are anything but fast, secure, and reliable.

Lossless compression is a type of data compression where a data is compressed without any loss or degradation from the original uncompressed data [1]. Compression is a possible solution for a faster file transfer protocol. One such compression algorithm is called the Zstandard Compression, also known as "Zstd". It is a lossless compression algorithm that employs asymmetric numeral systems, an entropy coding method 30 times faster than other methods. Compressing the file prior to sending reduces transfer time by sending less bytes over the network. An appropriate lossless compression algorithm applied to an internet protocol must not saturate CPU time so it would not bottleneck the throughput.

A data's integrity may be compromised as it is being sent over the network. One high-speed hashing algorithm being used is called xxHash. It is a hashing algorithm that works at speeds close to RAM limits. This hashing algorithm verifies the file as it is successfully transmitted over the network.

TCP/IP (Transmission Control Protocol / Internet Protocol) is a group of protocols that allows a collection of computers to communicate and exchange data within a network in general, and the Internet in particular [2]. A protocol is a set of rules that allows a computer to connect with one another, usually in the form of time, sequence during data transmission [3]. Computers are able to connect to the internet and communicate because of this protocol.

Encryption is the process of encoding a message or data in such a way that only authorized parties can access it [4]. Most file transfer protocols rely on three core cryptographic functionalities to protect data: public key encryption, digital signatures, and key exchange [5]. Peter Shor of Bell Laboratories showed that quantum computers can efficiently solve these problems, rendering most public key cryptosystems ineffective [5]. Quantum computers pose a great risk to modern cryptosystems: being able to break current cryptosystems in use, thus compromising data security and integrity. One quantum-resilient cryptosystem is a lattice-based encryption algorithm that is an alternative to RSA (Rivest-Shamir-Adleman) and ECC (Elliptic-curve cryptography) which relies on lattice-based constructions that cannot be solved efficiently even with the use of quantum computers. In time, quantum computers will supersede classical computers and as long as old protocols are not keeping up, data confidentiality is in great peril.

The proponents have confidence that the methods stated above are crucial features for modern file transfer protocols. Reliability, performance, and security have never been more important. The proponents hope to pave the way for the widespread adoption of lattice-based encryption, real-time compression, and hashing.

2. Related Works

2.1. Network Protocol
Packet loss and performance degradation are effects of packet corruption in data center networks. To try to mitigate packet corruption, a research observed hundreds of thousands of links across data centers [6]. The study concluded that there is a significant loss of corruption in packets in data center networks and that packet corruption can be mitigated through the simple act of cleaning the connectors or replacing the cables.

2.2. Compression
The study of File Management System for P2P Environment using Lossless Compression Algorithm aimed to reduce file transfer time by applying a lossless compression on the fly as a file is being sent through the network [7]. The research showed that even a zip compression is sufficient enough for an improved file transfer speed and that the newly-developed file management system provided better file accountability and management.

2.3. Encryption
Regardless of the exact time of the arrival of the quantum computing era, information security systems as early as now should be able to resist quantum computing [5]. A study was conducted on lattice-based cryptography using a four-degree polynomial over z [8]. The research concluded that without sacrificing security, more speed can still be achieved in the encryption and decryption process.

2.4. Hashing
Storage, server, networking, database, and more are the services delivered by cloud computing over the internet and provides virtualization scalability, and fine-grained data access control and performance benefit. Despite the benefits of cloud computing, there are various issues of security and data integrity for the cloud [9]. Adding a generic hash function to determine the integrity of the data being transferred over the network ensures the data transmitted is corruption-free because TCP alone already exhibits a high amount of error rate [10].

3. Methodology

The flow of implementation in Figure 1 shows the key exchange and dictionary initialization.

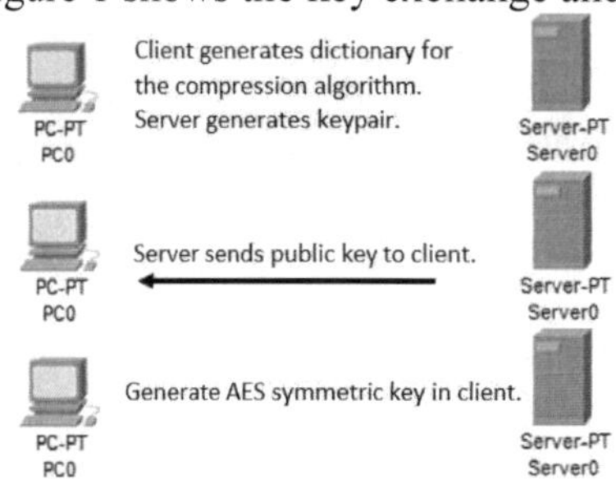

Figure 1. Initial handshake

A detailed description of Figure 1 from top to bottom:

1. The sending end generates a dictionary based on the data type which tunes the algorithm for the data being sent. This results to a faster compression and decompression, also increasing the compression ratio.
2. The server generates the lattice-based encryption public key and sends it to the client.
3. The client generates the AES (Advanced Encryption Standard) key and encrypts it with the server's public key and sends that encrypted key back to the server.

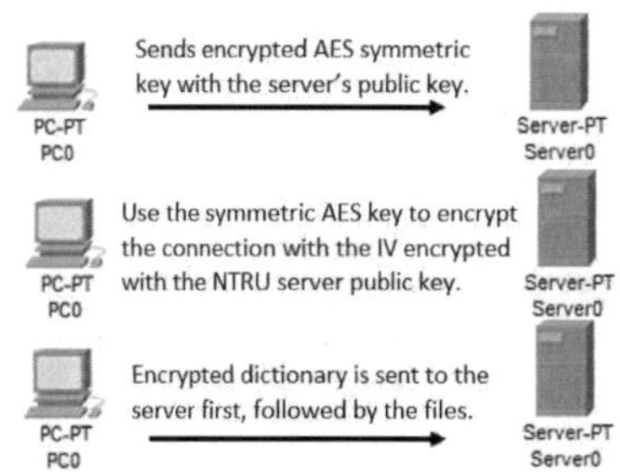

Figure 2. Key exchange and connection establishment.

A detailed description of Figure 2 from top to bottom:

1. The connection tunnel is being encrypted by the AES key.
2. The AES encrypted payload is appended with an IV (initialization vector) encrypted with the server's public key. This approach is a hybrid encryption of lattice-based encryption and AES since the payload needs the IV along with the AES key in order for it to be decrypted.
3. The generated dictionary and other files being sent are encrypted on the methods described above.

3.1. Encryption

Lattice-based cryptography is an encryption and decryption algorithm that involves lattices. This cryptography arose from the idea that factoring large numbers or finding the discrete logarithm is mathematically proven to be easily deciphered by quantum-based computers with the help of Shor's algorithm [11].

1. Generate a random arbitrary matrix A with size $m \times n$ satisfying the definition of a lattice.
2. Generate an arbitrary binary $\vec{x}$ with size m to be used as public key.
3. To get the private key, matrix A is multiplied to $\vec{x}$ to obtain vector $\vec{u}$.
4. Encrypting the data requires the sending end to encapsulate the data in $\overrightarrow{b_1}$ and b_2 wherein $\overrightarrow{b_1} = A \times \vec{s} + \overrightarrow{e_1}$ where $\vec{s}$ is an arbitrary with size m and $\overrightarrow{e_1}$ with the same size but the magnitude of its numbers must be smaller than the numbers found in $\vec{s}$ and $b_2 = \vec{s} \times \vec{u} +$

$e_2 + bit \times \frac{9}{2}$, where bit is either 0 or 1, the data that's going to be encrypted and e_2 is any arbitrary value less than $\frac{9}{4}$.

5. To decrypt the data, the following equation is used $b_2 - \overrightarrow{b_1} \times \vec{x}$, the resulting number would either be close to 0 or 1.

3.2. Hashing

xxHash is general purpose hashing function. It has high Memory bandwidth usage and works closely to the speed of RAM [12]. The following is the 32-bit implementation algorithm of the hashing function:

1. Initialize internal $accumulator$ based on optional seed input.
2. Process stripes, in which a stripe is a contiguous segment of 16 bytes evenly divided into 4 lanes of 4 bytes each. Each lane read is associated to its 32-bit value using little-endian convention.
3. For each $\{lane, accumulator\}$, the bits are shuffled so that any bit from input $lane$ impacts several bits in output $accumulator$. Input is consumed one full stripe at a time. Step 2 is looped as many times as necessary to consume the whole input, except the last remaining bytes which cannot form a stripe.
4. All 4 lane $accumulators$ from the previous steps are merged to produce a single remaining accumulator of same width (32-bit).
5. Adding an input length so that it participates to the final mixing.
6. Remaining inputs are consumed, which ensures that all input bytes are present in the final mix.
7. Final mix. The final mix ensures that all input bits have a chance to impact any bit in the output digest, resulting in an unbiased distribution.
8. The hashing function produces an unsigned 32-bit value as output.

3.3. Compression

The Zstandard compression algorithm is a modified LZ77 lossless compression algorithm that is designed for fast compression and decompression speed while maintaining a high compression ratio by tuning the compression algorithm to a specific data set or type [13]. A LZ77 algorithm implements compression through storing references of repetitive bits of data in a dictionary [14]. The reference to that data is stored as two integers in memory occupying 4 bytes each, one for the number of jumps in the file to reach the referred data and the other as the length of the data that is being referenced to. The resulting output is a series of literals (original bytes) and length/distance pairs (references). Figure 3 demonstrates the algorithm explained above and its output.

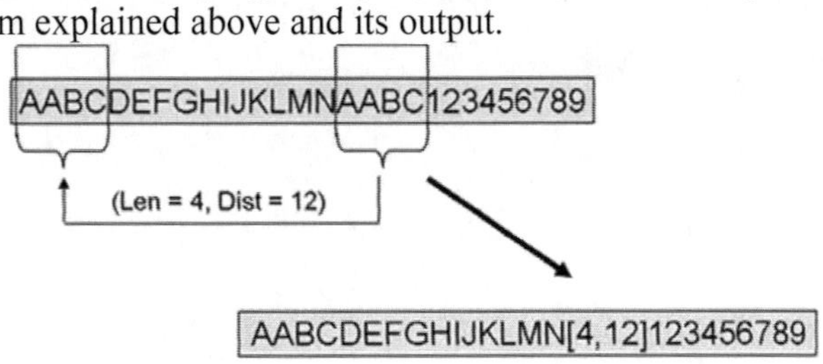

Figure 3. LZ77 lossless compression technique

To test the performance of the proposed protocol, the researchers opted to test using the Windows operating system and write the protocol in the Java programming language and with the critical parts of the protocol written in Java Native Interface. There are ten scenarios with different file types and sizes.

Compression ratio is calculated by the formula $compression\ ratio = \frac{size\ after\ compression}{size\ before\ compression}$. The compression ratio in percent is the inverse of the compression ratio formula. The lower the percentage, the better.

Space savings is calculated by the formula $space\ savings = 1 - \frac{compressed\ size}{uncompressed\ size}$ [15].

4. Results and Discussion

The following tables show the results of the conducted tests. Table 1 shows the difference in bytes using the file transfer protocol.

Table 1. Tabulated results

Scenarios	File Type	No. of Files	Actual Bytes Read (MB)	Bytes Transferred Over the Network (MB)	Difference (Bytes)	Difference (MB)	Space Savings (Percent)	Compression Ratio	Compression Ratio (Percent)
			Proposed Protocol Results						
Video	AVI	85	7,252.25	7,101.85	150396538	150.40	2.07	1.02:1	97.92
Audio	WAV	800	13,828.015	10,382.88	3445132269	3,445.13	24.91	1.33:1	75.09
Photos	PNG, JPEG	3,500	1,455.31	1,438.73	16581109	16.58	1.14	1.01:1	98.86
Text	TXT	2,688	1,217.33	268.84	948487489	948.49	77.92	4.53:1	22.08
Executables	EXE	700	2,811.26	2,709.06	102206929	102.21	3.64	1.04:1	96.36
Mixed	AVI, WAV, PNG	500	6,612.54	5,738.08	874457679	874.46	13.22	1.15:1	86.78
Mixed	MP3, TXT	500	1,466.89	1,394.46	72433322	72.43	4.94	1.05:1	95.06
Mixed	AVI, WAV, EXE, TXT	4,101	20,833.00	16,760.92	4072078124	4,072.08	19.55	1.24:1	80.45
Photo	JPEG	2,415	2219.60	2184.13	35469794	35.469794	1.60	1.02:1	98.40
Image	PNG	1,812	1.80	1200295	596008	0.596008	33.18	1.50:1	66.82

The data in Table 2 shows the difference in time with other FTP protocols in 100 MBPS and 1 GBPS connections respectively and figure 4 presents a graphical representation of this data.

Table 2. Comparison in speed between TCP protocols and the proposed protocol

Scenarios	File Type	No. of Files	Plain FTP		FTP over TLS		Windows SMB v3		Proposed Protocol	
			100MBPS	1GBPS	100MBPS	1GBPS	100MBPS	1GBPS	100MBPS	1GBPS
Video	AVI	85	10:04.000	01:01.000	10:10.000	01:02.000	10:17.439	01:07.105	10:03.448	01:32.262
Audio	WAV	800	19:25.000	01:58.000	19:27.000	01:58.000	21:35.315	02:37.140	14:39.892	01:32.711
Photos	PNG, JPEG	3,500	03:11.000	00:28.000	02:23.000	00:47.000	04:41.779	01:30.728	02:02.449	00:17.566
Text	TXT	2,688	01:46.000	01:04.000	02:20.000	01:06.000	03:43.299	02:11.556	00:36.124	00:25.471
Executables	EXE	700	01:38.000	00:55.000	01:55.000	01:07.000	03:04.951	01:12.451	01:02.208	00:22.864
Mixed	AVI, WAV, PNG	500	09:16.000	00:56.000	09:17.000	00:58.00	09:42.661	01:08.580	08:05.428	01:28.243
Mixed	MP3, TXT	500	02:00.000	00:13.000	02:02.000	00:21.000	03:04.951	00:27.797	01:02.208	00:19.991
Mixed	AVI, WAV, EXE, TXT	4,101	29:58.000	03:50.000	29:24.000	03:21.000	32:18.320	05:58.632	23:38.694	03:39.951
Photo	JPEG	2,415	03:07.000	00:23.000	03:08.000	00:36.000	04:54.202	01:38.555	03:05.847	00:21.692
Image	PNG	1,812	00:09.000	00:07.000	00:23.000	00:20.000	01:51.100	00:56.919	00:04.281	00:04.625

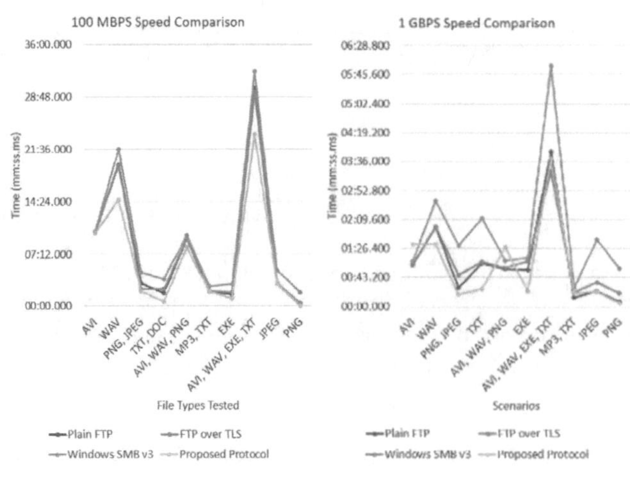

Figure 4. Graphical representation derived from the time data

5. Conclusion

In both 100MBPS and 1GBPS uplinks, results show significant space savings in terms of bytes transferred over the network but compressing uncompressible data can possibly lead to higher transfer time due to compression overhead but this overhead is reduced due to the usage of a real-time compression algorithm Zstd and its dictionary training function.

Lattice-based encryption did not add any noticeable slowdown in encryption and even performs faster. By combining AES and the lattice-based encryption, the data being transferred over the network is securely transmitted and is safe from quantum-based attacks using Shor's algorithm.

The exchange of files through the network is a crucial role in the field of computing. Extensive researches and studies have been done regarding the uses and real-world applications of a quantum computer and there are serious long-term threats to modern cryptosystems. As hardware continues to evolve, the proponents believe that the software and protocols that work together with it must evolve too. The results shown in this research demonstrates significant improvements over mainstream protocols and paves the way for the adoption of such algorithms in a modern file transfer protocol.

References

[1] McAnlis C and Haecky A 2016 *Understanding Compression: Data Compression for Modern Developers 1st Edition* (Sebastopol, CA: O'Reilly Media)

[2] Rahim R et al 2018 *Int. J. Eng. Technol.* 7 pp 44–7

[3] Nurdiyanto H, Rahim R, Ahmar A S, Syahril M, Dahria M, and Ahmad H 2018 *J. Phys. Conf. Ser.* **1028** pp 1-11

[4] Barenghi A, Beretta M, Di Federico A and Pelosi G 2014 *IEEE Int. Conf. on High Performance Computing and Communications (Paris)* pp 763-70

[5] Chen L, Jordan S, Liu Y K, Moody D, Peralta R, Perlner R, Smith-Tone D 2016 *Report on Post-Quantum Cryptography* (Gaithersburg: NIST Pubs) pp 1-10

[6] Zhuo D, Ghobadi M, Mahajan R, Forster K, Krishnamurthy A and Anderson T 2017 *SIGCOMM '17 Proc. of the Conf of the ACM Special Interest Group on Data Communication (Los Angeles)* (New York: ACM) pp 362-75

[7] Maderazo C, Ceniza A, Abangan B and Apor J 2016 *Proc. of the 14th National Conference on IT Education (NCITE 2016) (Dipolog)* (Philippines: NCITE) pp 295-300

[8] Baligasa C 2018 *Cryptography Using a Four Degree Polynomial Over Z* (Philippines : University of San Carlos)

[9] Singh S, Sharma P and Arora D 2017 *Int. J. of Advanced Research in Computer Science* **8** pp 1974-8

[10] Duarte R P 2008 Transport protocols for large bandwidth-delay product networks *Preprint* arXiv:1705.08929v1

[11] Bahaa-Eldin A, Mohsen A and Sobh M 2017 *12th Int. Conf. on Computer Engineering Systems (Cairo)* (New Jersey: IEEE) pp 462-7

[12] Thejaswi S and Westrup C 2016 *Approximate Computation of Frequency Moments* (Helsinki: Aalto University) pp 1-17

[13] Najmabadi S M, Tungal H S, Tran T and Simon S 2017 *Design and Architectures for Signal and Image Processing (DASIP) (Dresden)* (New Jersey: IEEE) pp 1-6

[14] Dong Q, Li B, Zhang L and Shang, Z 2014 *Electronics Letters (IET)* **50** pp 1522-4

[15] Sparsh M and Vetter J 2016 *IEEE Transactions on Parallel and Distributed Systems* **27** pp 1524-36

Q-DAR: quick disaster aid and response model using Naïve Bayes and Bag-of-Words algorithm

R A Camama[1], J M Baldelomar[2], J M Claveria[3], M C Espinas[4], J M Cabardo[5], and L W Rabago[6]

Asia Pacific College, Magallanes, Makati City, Philippines

[1]racamama@student.apc.edu.ph
[2]jmbaldelomar@student.apc.edu.ph
[3]jmclaveria@student.apc.edu.ph
[4]mcespinas@student.apc.edu.ph
[5]jayveec@apc.edu.ph
[6]lorenar@apc.edu.ph

Abstract. A real-time approach in decision-making for disaster management here in the Philippines is important, given that the country is vulnerable to disasters. Many studies show that data from social media can be of good use especially in times of disaster. This research aims to develop a model that will serve as a decision support tool for the government to respond during disasters with the use of Twitter hashtags that are based from the DRRM disaster phases; Disaster Response, Disaster Preparedness, Disaster Rehabilitation and Recovery, and Disaster Mitigation. The result of the study provides an overview of the critical level of the disaster phases with the use of Naïve Bayes and Bag-of-Words algorithm. The process of this study has four phases; the Training phase includes the labeling of the hashtags and list of Bag-of-Words, the next phase is Data Collection where Twitter data automatically updates and can be extracted every hour. The third phase is Data Pre-processing; where the tweets go through tokenization and normalization. Lastly, the dataset goes through Q-DAR with two sub-processes: The Disaster Phase Analysis using Bag-of-Words and Critical Level Analysis using Naïve Bayes in WEKA.

1. Introduction

Disasters are evident here in the Philippines. Many natural and man-made disasters have been experienced throughout the years. This is because of the current location of the country given that it is along the Ring of Fire [1], where frequent earthquakes and volcanic eruptions occur. Also, being beside the largest body of water, the Pacific Ocean, the Philippines encounters many disastrous typhoons with an average number of 20 typhoons per year [2].

The National Risk Reduction and Management Council handles various disasters with a similar framework focusing on the following disaster phases; Disaster Prevention and Mitigation, Disaster Preparedness, Disaster Response, and Disaster Rehabilitation and Recovery. It has the objective of having *safer, adaptive, and disaster-resilient Filipino communities toward sustainable development* [3].

Disaster Management in a community or area is aided by initial forecasts by the National Disaster Risk Reduction and Management Council that involves full cooperation of other government agencies (e.g. Department of Health, Department of Science and Technology and Department of Social Welfare and Development), private agencies, as well as the community or the public. Each disaster phase is handled by a different government agency that is responsible for the overall disaster phase, along with other various implementing partners or government and local agencies. However, these local agencies are often understaffed, and a significant gap exists as the NDRRMC cannot supervise all the local councils [4]. In order to decide for the actual disaster response, it must go through all these sub-departments. Thus, this results to delayed decision-making in terms of responding and giving assistance to the people in need.

On the other hand, many studies have taken consideration of social media as an avenue for involvement of the community in times of disasters. Twitter is often noted as a source of big data that can be transformed to multiple uses such as disaster management [5] [6]. Thus, social media, on the other hand, can be an avenue for being an aid to disaster management through manipulation of data from social media sites and making use of them for decision-making. The government uses Social Networking Sites when it comes to disasters through official hashtags (e.g. #ReliefPH, #RescuePH, #PreparednessPH). People resort to these hashtags to make the government and other people aware that they are in need of help. Although the government has emergency hotlines available for the community, in the actual time of disaster, these hotlines are often not reliable.

Objectives of this research is QDAR model as a decision support tool for the government through classifying Twitter hashtags and providing an overview of the critical level of tweets per disaster phase with the use of Naïve Bayes and Bag of Words algorithm.

2. Related Works

2.1. Social Media Use during Emergency Response

Twitter, as the center of interest of the study, is a venue to find out about what is happening in the world in real-time. This type of information is important because the faster the information is available, the faster the decisions can be made. In times of extreme events such as natural disasters, real-time updates have the potential to prevent the loss of lives [7]. Based on the study conducted by C. David, J. Ong and E. F. Legara, the functions of Twitter evolved overtime showing a disaster awareness cycle on a social media platform. They examined these throughout the 20-day period before the arrival of typhoon Haiyan- the 20-foot biggest and deadliest storm surge that hit the island of Leyte in the Philippines, using Twitter users' tweets and retweets about the typhoon [2] [3]. They also found out that people are not only relying on twitter for the news, but they also engage themselves by tweeting and retweeting stories that interest them. Basically, Twitter is a site where people share information regarding the disaster relief, locally organized fundraising and finding potential volunteers and donors for the disaster [2].

2.2. S-Sense: A Sentiment Analysis Framework for Social Media Sensing

Feedback from the public cannot be utilized if not thoroughly gathered, filtered, and analyzed. This is where social media comes in. Specifically, Twitter alone has millions of users, which that there are means millions of potential sources of feedback or data. What this related literature and the current study conducted by the researchers have in common is that they both gather data and have to go through the process of filtering the data based on their relevance to the topic. To be specific, Naive Bayes is used in both studies to determine whether, for instance, a tweet, has something to do with the product or service being offered by a certain business or in the case of Q-DAR, something that pertains to the current disaster being experienced in a locality. If executed correctly, it can yield more accurate results which can then be used for wiser and faster strategic planning or decision-making.

In accordance to the purpose of this study which is to incorporate Natural Language Processing on Twitter feeds, several studies support and have taken the idea into further research. Alongside these methods, techniques, and algorithms gathered, this study also focused on the idea of establishing the disaster model of being an aid to the decision-making of the government in terms of responding to various disaster phases. This would be with regard to how a government is disaster-prepared and how disasters are responded unto.

3. Proposed Method

The following is the methodology of the Q-DAR model as shown in Figure 1 below. It consists of the following phases: (1) Training, (2) Data Collection, (3) Data Pre-processing and lastly, and (4) Q-DAR: Quick Disaster Aid and Response.

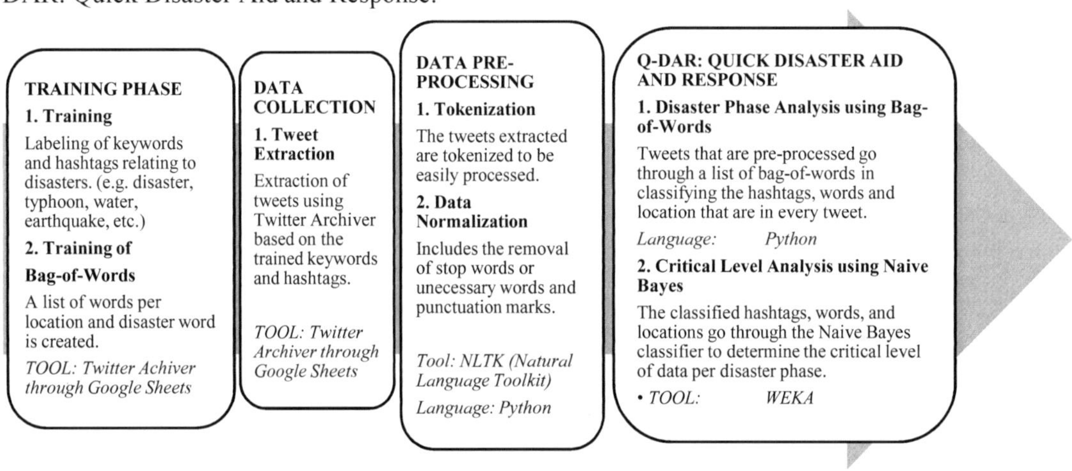

Figure 1. Q-DAR model

3.1. Training Phase

The initial step in the methodology is the training phase where the labeling of the necessary keywords, hashtags, etc. are supplied to the Twitter Archiver. These labels and hashtags are manually supplied to Twitter Archiver. The team supplied the appropriate hashtags and labels that are used to collect all the tweets.

3.2. Data Collection

Data Collection phase includes Tweet Extraction. It is where all the data that match the supplied labels and hashtags from the training phase are gathered. The tweets collected are automatically updated every hour.

This phase includes the retrieval of the tweets from Twitter using the Twitter Archiver add-on from Google Sheets. It will be achieving its real-time approach of gathering data from Twitter because Twitter Archiver is capable of being updated automatically every hour. Gathered data are collected based from the labels that are supplied in the training phase. The data is saved into a .csv file.

3.3. Data Pre-processing

Under this phase are the following processes: (1) Tokenization and (2) Data Normalization. The tool that will be used in this phase are the packages in the Natural Language Tool Kit (NLTK).

3.3.1. Tokenization. Tokenization is used in this phase to split up the texts that are gathered. For this process, the nltk.tokenize package, a package under NLTK, is used. The saved .csv file contains the tokenized tweets from the 520 tweets that are gathered, including the 100 dummy data that are produced for this research study.

3.3.2. Data Normalization. Data Normalization includes the removal of the stop words from the tweets. In the removal of stop words, punctuation marks are removed in the tweets, leaving out necessary words.

3.4. Q-DAR: Quick Disaster Aid and Response

3.4.1. Disaster Phase Analysis using Bag-of-Words. After data pre-processing, the tweets go through two sets of Bag-of-Words; location and disaster words. Tweets are classified by location and by words that appear in the tweet. This is also the process where the tweets are categorized into their respective disaster phases.

There are four attributes in this analysis phase, namely month, word, location, hashtag, and disaster phase. The month attribute is from the date the tweet is created. Word attribute is from the words that are used in the tweet. Location attribute is from the tweet that mentions a location that is inside the Bag-of-Words, and hashtag is also from the tweet when it mentions the specific hashtag. The tweets are then categorized into their distinctive disaster phases. The output data is then converted to an .arff file that will be processed by WEKA.

3.4.2. Critical Level Analysis using Naïve Bayes. After pre-processing, the .arff file is opened in WEKA and the following results are immediately shown in this process:

Figure 2. Critical level per disaster phase

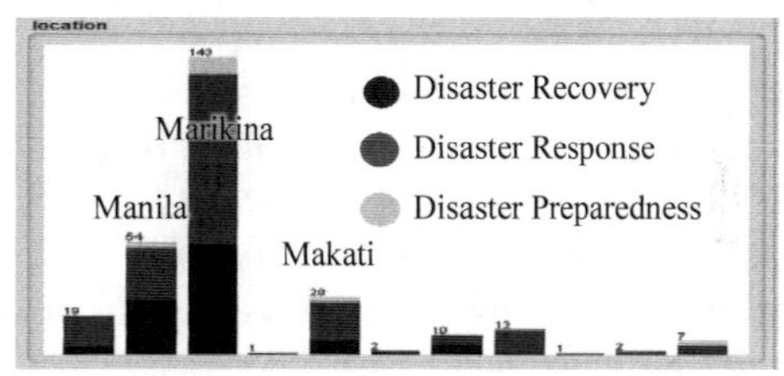

Figure 3. Critical level of disaster phases by location

In Figure 2 above, the four disaster phases (Mitigation, Response, Preparedness, and Recovery) are shown with the number of tweets per phase. Figure 3 shows the critical level of the phases per location, tested with a total of 520 tweets, including the 100 dummy tweets that are created for this study. The values show that Marikina has the highest bar in the location graph. Thus, it has the greatest number of tweets with the Disaster Response phase as its most critical phase.

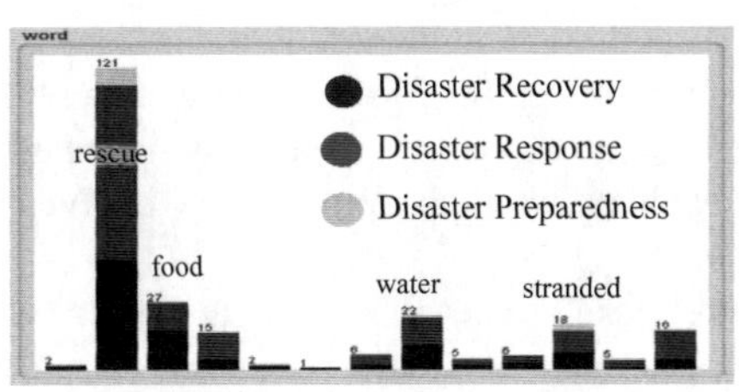

Figure 4. Critical level of words used per disaster phase

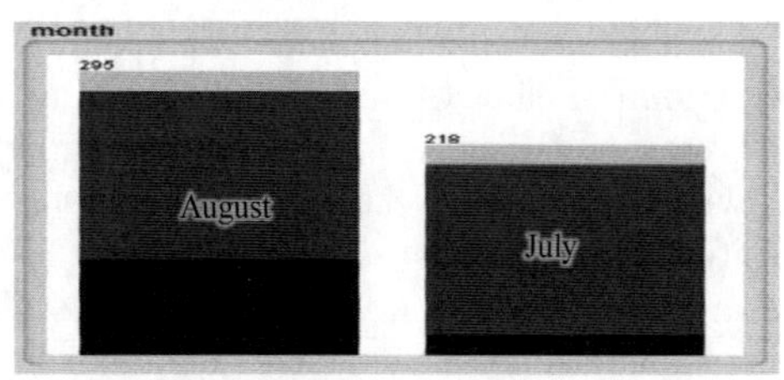

Figure 5. Critical level of disaster phases per month of tweet

Figure 4 above shows the most used word from the tweets gathered. Based on the result, the word mostly used is *'rescue'*, followed by *'food'* and *'water'*. Figure 5 shows the critical level of the disaster phases monthly. Given the date of the tweets that are processed, the bar graph shows the results in classifying the tweets in each of the disaster phases.

Figure 6 shows the hashtag attribute in determining the critical level of disaster phases. The first bar is #RescuePH and falls under the Disaster Respond phase. The second bar is #ReliefPH and is under Disaster Recovery phase. The third bar is #PreparednessPH that falls under the Disaster Preparedness phase.

Figure 6. Critical level of hashtags used per disaster phase

In order to get the best testing option for Naïve Bayes algorithm, the dataset goes through Percentage Splits and K-Cross Validation. Both testing options are evaluated based on their accuracy and kappa statistics. After getting the best testing option, the dataset is then classified using Naïve Bayes with the testing option that attained the highest accuracy for Naïve Bayes.

Results from the Naïve Bayes classification process determined the critical level of the disaster phases (Disaster Response, Disaster Preparedness, and Disaster Recovery) based on the following attributes: *disaster word used, month, location, and hashtag.*

4. Experiment and Results

The hashtags (#RescuePH, #ReliefPH, and #PreparednessPH) managed to collect disaster-related tweets that are classified by different disaster phases *(Disaster Preparedness, Disaster Response, and Disaster Recovery and Rehabilitation).* The data collection phase extracted 520 tweets from Twitter. Data Pre-processing phase managed to tokenize and remove stop words from all the tweets accordingly.

In order to analyze the best testing option for the Naïve Bayes algorithm in determining the critical level of each disaster phase, Percentage Split and K-Fold Cross Validation are compared using Naïve Bayes. The following are the results of the evaluation of two testing options.

Table 1 shows the accuracy and kappa statistic of the Percentage Splits 50-50 percentage of training data and test data. Table 2 shows Cross Validation Folds with 10 folds. Result shows that having 10 folds is more accurate between the two testing options.

Table 1. Disaster Phase percentage split

Percentage Split	Kappa Statistic	Accuracy
50-50	0.88	93.33%

Table 2. Disaster Phase Cross Validation Folds

Cross-Validation Folds	Kappa Statistic	Accuracy
10	0.90	94.44%

With a total number of 520 extracted tweets, it can be seen in Figures 3-6 that the overview of the critical level of disaster phases is given four attributes; location, word, month and hashtag. The following tables show the percentage of attributes to each disaster phase using the Naïve Bayes classifier with 10 cross validation folds.

Table 3. Tweet Location with the Highest Critical Level per Disaster Phase Using Naive Bayes

Location	Disaster Response	Disaster Preparedness	Disaster Recovery	Total
Marikina	83.0	9.0	54.0	146.0
Manila	26.0	4.0	27.0	57.0
Makati	20.0	3.0	8.0	31.0

The locations that are listed on Table 3 are the top 3 locations with the greatest number of tweets and are classified based on the disaster phase of each tweet. The results show that Marikina has the greatest number of tweets with a total of 146.0. Disaster response in that area has the highest level with the result of 83.0, followed by Manila and Makati.

Table 4 shows that the word 'rescue' is the most frequently used disaster word, with Disaster Response as its most critical disaster phase. This shows that tweets gathered and processed in that hour needs rescue.

Table 4. Words with the Highest Critical Level per Disaster Phase Using Naive Bayes

Words	Disaster Response	Disaster Preparedness	Disaster Recovery	Total
rescue	71.0	8.0	45.0	124.0
food	12.0	17.0	1.0	30.0
stranded	10.0	8.0	3.0	21.0
flood	12.0	5.0	1.0	18.0

Table 5. Hashtag with the Highest Critical Level per Disaster Phase Using Naive Bayes

Hashtag	Disaster Response	Disaster Preparedness	Disaster Recovery	Total
#RescuePH	353.0	1.0	1.0	355.0
#ReliefPH	1.0	1.0	121.0	123.0
#PreparednessPH	1.0	42.0	1.0	44.0

Table 5 shows that #RescuePH is the most used hashtag and Disaster Response as the most critical phase. This also shows that most tweets are related to rescue and the need for help.

Table 6. Result of the Highest Attributes with the Most Critical Disaster Phase

Attribute	Has the Highest Result	Disaster Response	Disaster Preparedness	Disaster Recovery
Location	Marikina	**83.0**	9.0	54.0
Word	rescue	**71.0**	8.0	45.0
Hashtag	#RescuePH	**353.0**	1.0	1.0

Table 6 shows all attributes with the highest results. It shows that Marikina is the location that needs rescue the most, given that Disaster Response is its highest disaster phase along with the most used hashtag #RescuePH. The most used word was rescue. All attributes have Disaster Response as its highest disaster phase as well.

Table 7. Month with the highest level of Disaster Phases using Naive Bayes

Month	Disaster Response	Disaster Preparedness	Disaster Recovery	Total
July	**178.0**	21.0	22.0	221.0
August	176.0	22.0	100.0	298.0

Given that the most critical disaster phase is Disaster Response as seen in Table 6, it can be concluded that it is in the month of July, as seen in Table 7, that the Philippines has encountered numerous typhoons.

5. Conclusion and Recommendations

With the Quick Disaster Aid and Response Model, the government can have a visual overview of the critical level of each of the disaster phases (Disaster Mitigation, Disaster Response, Disaster Preparedness, and Disaster Recovery) through processing and classifying the unified hashtags. It can be an aid to their decision-making, as to where and what actions are needed to prioritize, given the critical level of the disaster phases that are classified by location, word, month, and hashtags.

With the results of Q-DAR, the most critical disaster phase per location can be easily detected. The local government unit would have an idea of what location needs immediate rescue and assistance in that hour with its real-time approach of gathering and processing tweets from Twitter. The Naïve Bayes algorithm successfully classified the tweets and showed the critical level of each disaster phase as shown in Tables 3-7.

Q-DAR, on the other hand, is a study that can be further improved and capable of determining location. The concept of Geographical Information Systems can be applied to Q-DAR. The overview of the result generated can be processed and evaluated thoroughly. It can be created as a system that the government can use for a more efficient decision-making. Image processing can also be a study that can be applied to Q-DAR. In the process of collecting tweets, only texts are gathered and processed by the Q-DAR study. On the other hand, it can be further enhanced by including image processing in collecting and processing these data. Another would be the use of other data sources aside from Twitter. Lastly, Artificial Intelligence can be supported in the study through the Bag-of-Words algorithm. Whereas, in this process, the list of bag-of-words can be trained to automatically learn new words based on the tweets that will be gathered.

References

[1] UNICEF 2018 *Disaster Risk Reduction* Online Available: https://www.unicef.org/philippines/risk.html#.Wy2qwu6FPIU Accessed 23 June 2018

[2] Wingard J and Brandlin A S 2013 *Philippines: A Country Prone to Natural Disasters* 11 October 2013 Online Available: http://www.dw.com/en/philippines-a-country-prone-to-natural-disasters/a-17217404 Accessed 23 June 2018

[3] NDRRMC 2011 *National Risk Reduction and Management Council* Online Available: http://www.ndrrmc.gov.ph/attachments/article/41/NDRRM_Plan_2011-2028.pdf Accessed 25 June 2018

[4] Alcayna T, Bollettino V, Dy P and Vinck P 2016 *Resilience and Disaster Trends in the Philippines: Opportunities for National and Local Capacity Building* Online Available: http://currents.plos.org/disasters/article/resilience-and-disaster-trends-in-the-philippines-opportunities-for-national-and-local-capacity-building/ Accessed 22 August 2018

[5] David C C, Ong J C and Legara E F T 2016 *Tweeting Supertyphoon Haiyan: Evolving Functions of Twitter during and after a Disaster Event* Online Available: http://journals.plos.org/plosone/article/file?id=10.1371/journal.pone.0150190&type=printable Accessed 23 June 2018

[6] Lindsay B 2011 *Social Media and Disasters: Current Uses, Future Options, and Policy Considerations* Online Available: https://fas.org/sgp/crs/homesec/R41987.pdf Accessed 21 August 2018

[7] Avci H 2016 *10 Ways Twitter Is Revolutionizing Real-Time Information* Online Available: https://digitalculturist.com/10-ways-twitter-is-revolutionizing-real-time-information-704a894d1d0f Accessed 21 August 2018